Heavy Metal Detoxification from the Environment: Uncovering Molecular Approaches

Edited by

Kanika Khanna

Department of Microbiology, D.A.V University
Jalandhar, Punjab, India
Department of Botanical and Environmental Sciences
Guru Nanak Dev University, Amritsar
Punjab, India

Sukhmeen Kaur Kohli

School of Life Sciences, Central University Hyderabad
Hyderabad, Telangana, India
Deptartment of Botanical and Environmental Sciences
Guru Nanak Dev University, Amritsar
Punjab, India

&

Renu Bhardwaj

Department of Botanical and Environmental
Sciences, Guru Nanak Dev University, Amritsar
Punjab, India

Heavy Metal Detoxification from the Environment: Uncovering Molecular Approaches

Editors: Kanika Khanna, Sukhmeen Kaur Kohli & Renu Bhardwaj

ISBN (Online): 979-8-89881-117-4

ISBN (Print): 979-8-89881-118-1

ISBN (Paperback): 979-8-89881-119-8

Published by Bentham Science Publishers Pte. Ltd. Singapore, in collaboration with Eureka Conferences, USA. All Rights Reserved.

First published in 2025.

need for a court order if at any point you breach any terms of this License Agreement. In no event will any delay or failure by Bentham Science Publishers in enforcing your compliance with this License Agreement constitute a waiver of any of its rights.

3. You acknowledge that you have read this License Agreement, and agree to be bound by its terms and conditions. To the extent that any other terms and conditions presented on any website of Bentham Science Publishers conflict with, or are inconsistent with, the terms and conditions set out in this License Agreement, you acknowledge that the terms and conditions set out in this License Agreement shall prevail.

Bentham Science Publishers Pte. Ltd.
No. 9 Raffles Place
Office No. 26-01
Singapore 048619
Singapore
Email: subscriptions@benthamscience.net

**BENTHAM
SCIENCE**

CONTENTS

PREFACE

Heavy metal remediation has gained much more attention in the past few years with the exploration of varied methods and detoxification approaches. They have acquired immense importance due to their hazardous effects on the environment. Heavy metals occur naturally in the environment, but their concentrations have been exponentially increased due to anthropogenic activities. They are continuously persisting in the environment in enormous quantities, thereby, entering the food chain through crops and consequent accumulation in the living cells via biomagnification. Novel remediation strategies have attained a specific position in the present era and also attracted interests of researchers through the implementation of bioremediation processes as well as various assistants to remediate the polluted soils. This in turn has become the talk of the town owing to its sustainable properties. The molecular insights provide excellent and new perspectives to unravel the heavy metal accumulation and detoxification. The metagenomic studies as well as screening strategies are used to extract the genomic information for detoxification mechanisms. Moreover, the development of next-generation sequencing studies has also provided us the unparalleled perspectives through key remediation mechanisms. This book mainly focuses on the impact of heavy metals in the environment, and their detoxification potential with special emphasis on the sustainable approaches, molecular patterns, omics approaches and metagenomics analysis. This book discusses the novel and new generation strategies for heavy metal detoxification and remediation by the use of eco-friendly methods such as microbes, plant hormones, earthworms, mycorrhizae, microalgae, probiotics etc. It further focuses on the function and characteristics of the remediating agents used for metal detoxification at molecular or genetic levels.

Kanika Khanna
Department of Microbiology, D.A.V University
Jalandhar, Punjab, India
Department of Botanical and Environmental Sciences
Guru Nanak Dev University, Amritsar
Punjab, India

Sukhmeen Kaur Kohli
School of Life Sciences, Central University Hyderabad
Hyderabad, Telangana, India
Deptartment of Botanical and Environmental Sciences
Guru Nanak Dev University, Amritsar
Punjab, India

&

Renu Bhardwaj
Department of Botanical and Environmental
Sciences, Guru Nanak Dev University, Amritsar
Punjab, India

INTRODUCTION

For decades, the environment has been found disparaging due to heavy metal toxicity and sustainability has become a crucial aspect in the present scenario. In other words, environmental engineering ensures societal development and the use of environmental resources in a crucial manner. This requires the collaborative role of biologists, researchers, chemical engineers, industrialists, biotechnologists, microbiologists, botanists, molecular biologists, biochemists, and genetic engineers. All these environmental engineers are directly linked with problems associated with air, water, and soil contamination and they provide the technical solutions in order to resolve or attenuate such situations in a sustainable way that complies with legislative, social, economic, and political concerns. Moreover, the researchers should find a reasonable way to check the heavy metal contamination of groundwater, surface water, and soil along with the remediation of the contaminated sites, bioremediation, and waste management. Heavy metal toxicity is the most crucial concern that damages the entire environment and affects the plants followed by the complete food chain. It is therefore the need of the hour to find a suitable solution for the detoxification of heavy metals from the environment. Over the decades, biologists and researchers have contributed a lot in this realm and generated a vast information. The life sciences have provided descriptive approaches and this phenomenon is now better understood for the benefit of society. We have now focussed on the approaches that provide spectacular reductions in cost with minimal risks and more effectiveness. Our aim has been now more oriented towards realistic solutions with advanced biotechnological and molecular approaches in order to find the solutions.

This book is written by multiple authors concerned with heavy metal detoxification after the advent of molecular techniques. It intends to serve the professionals as well as encourage researchers to conduct various research in this field. The book is organised into various areas that are important for the environmentalists, biologists, and researchers that are working in this field for sustainability and removal of heavy metals and other toxic compounds from the environment. This book has covered all the main aspects and methods for heavy metal toxicity as this field is changing and moving very rapidly. The authors have included all the fundamental topics and aspects of heavy metal toxicity, and detoxification strategies through novel approaches. In addition, the chapters on bottlenecks in sustainable technology of heavy metal remediation along with future prospects are also included. Further, the topics associated with sustainable amelioration of heavy metal contaminated areas through the use of probiotics, microbes, genetic engineering, nanotechnology, biotechnology, phytoremediation, etc. along with the hurdles and futuristic strategies are also included. The book deals with some hitherto neglected areas for sustainable treatment technologies for heavy metal detoxification. It is most likely expected that these contributors will target many other researchers to contribute their research and facilitate in disseminating of knowledge in this discipline.

Dr. Kanika Khanna

Dept. of Microbiology, D.A.V University, Jalandhar, Punjab, India, 144001 Dept. of Botanical and Environmental Sciences,Guru Nanak Dev University, Amritsar, Punjab, India, 143005

List of Contributors

Arun Dev Singh	Department of Botanical and Environmental Sciences, Guru Nanak Dev University, Amritsar, Punjab, India
Ashutosh Sharma	Department of Agricultural Sciences, DAV University, Sarmastpur, Jalandhar, Punjab, India
Anil Kumar	Serum Institute of India Ltd., Hadapsar, Pune, Maharashtra, India
Deepak Sharma	Department of Zoology, Guru Nanak Dev University, Amritsar, Punjab, India
Diksha Kalia	Department of Biotechnology, CSIR-Institute of Himalayan Bioresource Technology, Palampur, Himachal Pradesh, India Academy of Scientific and Innovative Research (AcSIR), Ghaziabad, Ghaziabad, Uttar Pradesh, India
Dev Singh	Department of Botanical and Environmental Sciences, Guru Nanak Dev University, Amritsar, Punjab, India
Deepak Kumar	Department of Zoology, Guru Nanak Dev University, Amritsar, Punjab, India
Gaurav Zinta	Department of Biotechnology, CSIR-Institute of Himalayan Bioresource Technology, Palampur, Himachal Pradesh, India Academy of Scientific and Innovative Research (AcSIR), Ghaziabad, Ghaziabad, Uttar Pradesh, India
Gurvarinder Kaur	Department of Botany, Punjabi University, Patiala, Punjab, India
Geetika Sirhindi	Department of Botany, Punjabi University, Patiala, Punjab, India
Harpreet Kaur	P.G. Department of Botany, Khalsa College Amritsar, Amritsar, Punjab, India
Isha Madaan	Department of Botany, Punjabi University, Patiala, Punjab, India Government College of Education, Jalandhar, Jalandhar, Punjab, India
Indu Sharma	Department of Life Sciences & Allied Health Sciences, Sant Baba Bhag Singh University, Jalandhar, Punjab, India
Jaspreet Kour	Department of Botanical and Environmental Sciences, Guru Nanak Dev University, Amritsar, Punjab, India
Kulwinder Singh	Sanmati Govt. College of Science Education and Research, Jagraon, Jagraon, Punjab, India
Komal Goel	Department of Biotechnology, CSIR-Institute of Himalayan Bioresource Technology, Palampur, Himachal Pradesh, India Academy of Scientific and Innovative Research (AcSIR), Ghaziabad, Ghaziabad, Uttar Pradesh, India
Kanika Khanna	Department of Botanical and Environmental Sciences, Guru Nanak Dev University, Amritsar, Punjab, India
Manish Kumar	Department of Botany, SD College Barnala, Barnala, Punjab, India
Mohd Ali	Department of Zoology, Guru Nanak Dev University, Amritsar, Punjab, India
Madhu Chandel	P.G. Department of Botany, Khalsa College Amritsar, Amritsar, Punjab, India
Neha Guleri	Department of Plant Sciences, Central University of Himachal Pradesh, Shahpur, Himachal Pradesh, India

Nandni Sharma	Department of Zoology, DAV University, Jalandhar, Punjab, India
Nitika Kapoor	Department of Botanical and Environmental Sciences, Guru Nanak Dev University, Amritsar, Punjab, India
Parkirti	Department of Zoology, Guru Nanak Dev University, Amritsar, Punjab, India
Puja Ohri	Department of Zoology, Guru Nanak Dev University, Amritsar, Punjab, India
Parul Preet Gill	Postgraduate Department of Botany, Dev Samaj College for Women, Ferozepur, Punjab, India
Pooja Sharma	Department of Microbiology, DAV University, Sarmastpur, Jalandhar, Punjab, India Department of Botanical and Environmental Sciences, Guru Nanak Dev University, Amritsar, Punjab, India
Priyanka Sharma	School of Bioengineering Sciences & Research, MIT Art Design and Technology University, Pune, Maharashtra, India
Rajesh Kumar Singh	Department of Biotechnology, CSIR-Institute of Himalayan Bioresource Technology, Palampur, Himachal Pradesh, India Academy of Scientific and Innovative Research (AcSIR), Ghaziabad, Ghaziabad, Uttar Pradesh, India
Ravinderjit Kaur	Department of Zoology, S.R. Government College (Women), Amritsar, Punjab, India
Roohi Sharma	Department of Zoology, Guru Nanak Dev University, Amritsar, Punjab, India
Raman Tikoria	Department of Zoology, Guru Nanak Dev University, Amritsar, Punjab, India
Raj Bala	Department of Botany and Environment Studies, DAV University, Jalandhar, Punjab, India
Renu Bhardwaj	Department of Botanical and Environmental Sciences, Guru Nanak Dev University, Amritsar, Punjab, India
Sonia Sharma	P.G. Department of Botany, Khalsa College Amritsar, Amritsar, Punjab, India
Sandeep Kour	Department of Zoology, Guru Nanak Dev University, Amritsar, Punjab, India
Sahil Dhiman	Department of Agricultural Sciences, DAV University, Sarmastpur, Jalandhar, Punjab, India
Sandeep	Department of Agricultural Sciences, DAV University, Sarmastpur, Jalandhar, Punjab, India Department of Botany and Environment Studies, DAV University, Jalandhar, Punjab, India
Sanjay Kumar	Department of Life Sciences & Allied Health Sciences, Sant Baba Bhag Singh University, Jalandhar, Punjab, India
Sukhwinder Kaur	Department of Life Sciences & Allied Health Sciences, Sant Baba Bhag Singh University, Jalandhar, Punjab, India
Sakshi Verma	Department of Zoology, Hans Raj Mahila Maha Vidyalaya, Jalandhar, Punjab, India
Shalini Dhiman	Department of Botanical and Environmental Sciences, Guru Nanak Dev University, Amritsar, Punjab, India
Sandeep Kumar	Department of Physics, DAV University, Jalandhar, Punjab, India

Tanuja Verma   Government College of Education, Jalandhar, Jalandhar, Punjab, India

Tamanna Bhardwaj   Department of Botanical and Environmental Sciences, Guru Nanak Dev University, Amritsar, Punjab, India

CHAPTER 1

Hazardous Effects of Heavy Metals towards Plants: Recent Trends and Challenges

Madhu Chandel[1,*], Sonia Sharma[1], Harpreet Kaur[1] and Manish Kumar[2]

[1] *P.G. Department of Botany, Khalsa College Amritsar, Amritsar, Punjab, India*

[2] *Department of Botany, SD College Barnala, Barnala, Punjab, India*

Abstract: The prevalence of heavy metal soil pollution has increased as a result of increased geology and human activities. Metal pollution issues are becoming more prevalent in India and other nations. Numerous reports of metal toxicity in coal-burning power plants, foundries, smelters, mining, and agriculture have been made. As a result of industrialization and urbanisation processes, pesticides, petroleum products, acids, and heavy metals have been integrated into natural resources. Due to this, the quality of the environment has declined, affecting both biotic and abiotic components and consequently has an impact on the ecosystem. Some metals are required in trace levels for plant metabolism. However, they can be dangerous to plants when present in larger amounts. Lead, nickel, cadmium, copper, cobalt, chromium, and mercury are heavy metals that are significant environmental contaminants and have hazardous effects on plants. Plant growth, performance, and output are all reduced in heavy metal-contaminated soil-grown plants. Plants that are exposed to heavy metals experience oxidative stress, which damages their cellular structure. Metal ions build up in plants and upset the balance of cells. Plants have developed detoxifying systems to decrease the harmful effects of exposure to the accumulation of heavy metals (HMs). To treat heavy metals-contaminated soils, several in-situ and ex-situ remediation methods have been used, but they also have a number of drawbacks, such as high capital costs, toxicity, and environmental health hazards. The risks that heavy metals pose to plants are the main topics of the current chapter.

Keywords: Environmental contamination, Heavy metals, Plant metabolism, Toxicity.

INTRODUCTION

Different concentrations of heavy metals in soil, water, and air cause pollution after a certain threshold is achieved. Different factors contribute to this type of pollution. The natural weathering of rocks, volcanic eruptions, industrial proces-

* **Corresponding author Madhu Chandel:**P.G. Department of Botany, Khalsa College Amritsar, Amritsar, Punjab, India; E-mail: madhuchandel11.bot@gmail.com

Kanika Khanna, Sukhmeen Kaur Kohli & Renu Bhardwaj (Eds.)

ses including mining, the burning of fossil fuels, and the discharge of sewage release metals into the atmosphere on a constant basis [1 - 3]. Copper, lead, zinc, and cadmium are particularly prevalent in agricultural soils [4]. Because of changes in biochemical processes, such as the suppression of enzyme activity, protein penetration, and decreased nutrition, *etc.*, these metals are frequently readily taken up by plants and prove hazardous to them, which can be seen as growth failure [5]. The persistence of hazardous compounds in the environment, as well as their bioaccumulation and biomagnification in the trophic chain—where plants are the primary producers at the base of terrestrial ecosystems—pose a serious threat to human health [5]. Patients who consumed food contaminated with Cd, Pb, As, or Hg manifested neurological problems, stomach discomfort, and various ailments related to the heart, kidney, liver, digestive system, and several forms of cancer [6 - 9].

In plants, the toxic symptoms may include DNA deterioration and cell death, as well as the suppression of root development, photosynthesis, and mitochondrial respiration impairment. Under different conditions (Cd, Hg, As, or Cu), different plant species display different stress signals [10 - 13].

To cope with the stress, the primary mechanism for maintaining cellular redox homeostasis in plants is the SOD-mediated conversion of superoxide radicals into hydrogen peroxide, which in turn metabolized into non-toxic products by catalase and ascorbate peroxidize [14]. Glutathione reductase (GR) is another important antioxidant enzyme that converts GSSG into Glutathione (GSH) by employing NADPH as an electron source and preserving the GSH cellular pool [15].

TOXIC EFFECTS OF CADMIUM

A very toxic metal pollutant of soils, Cadmium (Cd) impairs homeostasis and nutrient absorption, impedes crop production and root and shoot growth, enters the food chain, and poses a major risk to human and animal health [16 - 18]. The development of seedlings and seed germination are both influenced by cadmium.

Effects of Cadmium on Germination and Plant Growth

Plant tissues that have accumulated Cd exhibit various toxicological signs as well as decreased development. Cadmium stress reduced the development of *Nicotiana tobacum*; adverse results included plant dwarfism, loss of green leaves, leaf detachment, and even plant mortality [19]. Rao *et al.* [20] investigated the effects of various cadmium concentrations (10 mg/L, to 250 mg/L) on *Brassica napus* and discovered a decline in different growth parameters. Similar results were observed in plants like *Cucumis sativus*, *Hordeum distichum*, *Lactuca sativa, etc.* [21 - 23]. When *Peganum harmala* seedlings were exposed to 100, 200, or 300 µM concentrations of Cd for 15 days, Nedjimi [24] observed a decrease in germination rate, length of hypocotyl, and index of tolerance. When *Pisum sativum* was grown and treated with Cd concentrations of 500, 750, 1000, and 1250 mg kg-1 in order to study the rate of germination, length of hypocotyls and root along with dry biomass, and tolerance index. After 15 days of treatment with various Cd doses, the researchers observed that all metrics decreased as Cd concentrations increased [25]. After 6 days of treatment, Zayneb *et al.* [26] found that *Trigonella foenum graecum's* germination rate, root length, and hypocotyl elongation decreased as Cd concentrations increased from 0.1, 0.5, 1 to 10 µM concentrations. El Rasafi *et al.* [27] discovered that after 7 days of Cd treatment, (10, 50, 100, 250, 500, 750, 1000 mg/l) *Triticum aestivum* showed a decrease in morphological parameters. Goel *et al.* [28] investigated the harmful effects of cadmium on the development and metabolic responses of rice and maize. Heidari and Sarani [29] evaluated that in mustard plant, seed germination and root development steadily decreased as Cd concentration increased.

Effects of Cadmium on Lipid Peroxidation

Dey *et al.* [30] identified a link between metal toxicity and oxidative stress. Malondialdehyde (MDA) and hydrogen peroxide concentration are quantified to determine the level of oxidative stress using the Cd application. Wu *et al.* [31] found that treating barley with 1 or 5M Cd increased the MDA content. Zayneb *et al.* [26] reported an increase in the content of MDA in plants (fenugreek) treated with Cd for 30 days. In a hydroponic experiment, Meng *et al.* [32], investigated the impact of cadmium and discovered a rise in the amount of MDA in *Lactuca sativa* and *Chrysanthemum coronarium*. According to the investigations, Cd-induced lipid peroxidation damages membranes, as shown by an increase in electrolyte leakage. Numerous authors have discussed the effects of cadmium at various concentrations on a variety of plant families, including Poaceae [33], Brassicaceae [34 - 37], Fabaceae [45, 38, 39], Asteraceae [32, 40] and Lamiaceae [41].

Effects of Cadmium on Photosynthesis

It has been observed that heavy metals can prevent the production of chlorophyll, especially by preventing the activity of the enzymes protochlorophyllide reductase and δ-aminolevulinic acid dehydrogenase [42]. Ekmekci *et al.* [43] studied the effect of Cd in two different varieties of maize *i.e.* 3223 and 32D99 and observed that eight days of treated seedlings of both cultivars showed signs of membrane damage and chlorophyll and carotenoid loss with an increase in Cd concentration. Chlorophyll fluorescence measurements showed that 3223 was significantly more impacted by Cd levels than 32D99 in terms of photochemical efficiency.

Chloroplast size and quantity were both drastically decreased, and their ultrastructure was altered. Thylakoid form is also altered by Cd stress, which causes thylakoid enlargement. The results of the investigation demonstrated that cadmium damages the structure of chloroplasts and reduces their capacity to perform. When the effects of cadmium exposure on two radish cultivars were examined, it was discovered that the rate of photosynthesis and electron transport were reduced [44]. Barley seedlings on exposure to the Cd concentrations of 20 and 30 mM showed a reduction in chlorophyll a, b, and carotenoids [45]. Many researchers claimed that Cd exposure decreased the amount of chlorophyll in wheat [46], tobacco [47], lettuce [48], tomatoes [49], soybeans [50], and peas [51].

Effects of Cadmium on Plant Antioxidant Defence System

The cellular antioxidant machinery changes in response to ROS brought on by heavy metal contamination [52, 53]. In fact, changes in the activity of antioxidant defence enzymes have been linked to some of the stress responses induced by those contaminants [54]. These enzymes' activity changed depending on the kind of plant, its age, the organs that were tested, the amount of metal, and the length of exposure [17, 52]. Numerous studies have documented changes in the antioxidant levels in various plants [53]. Anjum *et al.* [38] observed that exposure to cadmium concentrations for 15 days increased the enzyme activity of GST, GPX, and GR in seedlings of *Brassica compestris*. After 30 days of Cd exposure, Mobin and Khan [35] studied the impact of various Cd doses *i.e.* 25 mg kg^{-1} to 100 mg kg^{-1} on the enzyme activity in *B. juncea*. The activities of ascorbate peroxidise (APX), glutathione reductase (GR), catalase (CAT), and superoxide dismutase (SOD) increased in the exposed plants. Gill *et al.* [36] and Sheetal *et al.* [55] also reported the same findings. Whereas, Meng *et al.* [32] found no significant differences in the SOD, POD, and CAT activities in *Chrysanthemum coronarium*, *Lactuca sativa*, and *Spinacia oleracea* treated with 4 mg/L of cadmium in a hydroponics experiment. Additional research on *Brassica compestris* and *Sorghum bicolor* by Wu and group [56] and Soudek *et al.* [57] revealed a high potential for a few antioxidant enzymes while a decrease in others, respectively.

LEAD TOXICITY

Among hazardous substances in the environment, lead is a widely used and potentially harmful environmental chemical on a global scale [58]. Due to its non-biodegradable nature and continuous usage, its concentration increases at an alarming rate in the environment. Lead poisoning causes chlorosis, necrosis, root development suppression, and stunted growth. Lead affects various physiological functions in plants [59 - 61].

Toxic Effects of Pb in Growth and Germination

Many researchers observed that even micromolar concentrations of Pb may negatively affect the growth and germination of plants [62, 63]. Kabir *et al.* [64] studied the effects of Pb on the growth parameters of *Thespesia populnea* L. and observed that different concentrations of Pb (10 µmolL^{-1} to 70 µmolL^{-1}) reduced germination and growth rate as compared to controls. When compared to control, seed germination decreased from 88% to 14% at the maximum tested concentration of Pb. Other metrics, including root, shoot, and seedling lengths, were decreased from 2.49 cm to 0.08 cm, 3.90 cm to 3.08 cm, and 6.39 cm to 3.16 cm, respectively. Negative effects of Pb on growth parameters have been found in *Vigna ambacensis* L [65], *Triticum aestivum* L [66], *Zea mays* [67, 68], *Brassica rapa* [69], *Thespesia populnea* [70], and *Leucaena leucocephala* [71]. Similar findings of other researchers are given in Table **1**.

Effects of Pb on Lipid Peroxidation

Kumar *et al.* [81] studied the effect of Pb in lipid peroxidation assay and observed an increase in unsaturated fatty acids and a decrease in saturated fatty acids in the plasma membrane of various plant species, which resulted in anomalies in the cytoskeleton and organelle ultrastructure. At various Pb concentrations, Bhardwaj *et al.* [82] detected changes in lipid peroxidation in *Phaseolus. vulgaris* leaves. Kaur *et al.* studied an increase in the content of MDA and hydrogen peroxide in wheat roots treated with 500 µM to 2500 µM Pb concentrations [83]. Similar findings were also reported in other plants like *Zea mays* L [84], *Triticum aestivum* L [85], and *Oryza sativa* L [86]. under Pb stress. Another research found that *Arachis hypogaea* L. roots and leaves produced more free radicals and accumulated more MDA when exposed to Pb stress [87]. Wheat roots exposed to Pb (40 mg/l) showed an increase in MDA level, which implies ROS generation and oxidative stress [88]. Relative Electrolyte Leakage (REL) has risen in cellular membranes, which is evidence of Pb-induced lipid peroxidation. REL is a sign of membrane deterioration and develops as a result of membrane peroxidation brought on by an oxidative burst [89, 90].

Toxic Effects of Pb on Photosynthesis and Photosynthetic Pigments

According to reports, Pb in plants decreases photosynthetic rate by affecting the production of carotenoids, plastoquinone, and chlorophyll, as well as a decreases the activity of numerous enzymes of the TCA cycle and electron transport [91]. *Ceratophyllum demersum* cultivated in aqueous media containing $Pb(NO_3)_2$

revealed structural alterations in the chloroplast. In these Pb-exposed plants, Rebechini *et al.* [92] found that there were fewer grana stacks, stroma amounts were lower, and there were no starch grains. Changing the lipid content of thylakoid membranes and inhibiting chlorophyll production owing to the decreased absorption of vital nutrients like magnesium and iron are further impacts observed in plants exposed to Pb [93, 94]. Many other workers reported the toxic effects of Pb on photosynthesis and photosynthetic pigments as given in Table **2**.

Table 1. Effect of lead on growth and germination in plants.

Plant	Family	Concentration	Treatment duration	Effects	References
Leucaena leucocephala (Lam.)	Fabaceae	25, 50, 75 and 100 ppm	10 days	At the highest concentration (100 ppm) many growth parameters like germination rate, hypocotyl and root length decreased as compared to control in *L. leucocephala*	[72]
Albizia lebbeck	Fabaceae	10, 30, 50, 70 and 90 μmol/L	12 days	At the highest tested concentration of Pb, the seed germination decreased from 96% to 54%, root, shoot and seedling length from 5.1 to 1.4, 8.8 to 6.4, 13 to 7.9 cm respectively. root shoot ratio decreased from 0.63 to 0.22 and dry biomass decreased from 39.58 to 29.62 respectively	[73]
Triticum aestivum L. cv. Achtar	Poaceae	0.15, 0.3, 1.5 and 3 mM	6 days	Lead concentration (3 mM) reduced the radical length by 2.3 cm on the 6th day of treatment, as compared to control (7.7 cm). The variation in root length was highly significant (P <0.001).	[74]
Jatropha curcas L	Euphorbiaceae	0.5t o 4 mM /kg soil	60 days	Leaf area of seedlings, root growth and number of root hairs decrease in dose dependent manner	[75]

(Table 1) cont.....

Plant	Family	Concentration	Treatment duration	Effects	References
Lolium perenne	Poaceae	0.5, and 3.2 mM	7 days	Reduction in root and shoot length of perennial ryegrass with increasing concentrations of Pb	[76]
Abutilon indicum		0, 25, 50, 75 and 100 μM	15-30 days	In 15 and 30 days of treatment, shoot length, leaf size and number of leaves per plant decrease at 100 μM concentration.	[77]
Vigna unguiculata L. (Walp.)	Fabaceae	0.5 mM EDTA-Pb	7 days	Reduction in the growth of roots and leaves was observed in 0.5 mM EDTA-Pb treated plants.	[78]
Capsicum annuum L. cv. Semerkand	Solanaceae	0.1mM PbCl$_2$	21 days	Total dry mass of root and shoot of pepper plants decreased under stress.	[79]
Cynodon dactylon	Poaceae	50, 100, 300, 500 and 1000 mg/kg	30 days	With increase in concentration of Pb(NO$_3$)$_2$ from 0 to 1000 mg/kg, the rate of germination of seeds reduced from 91% to 27%.	[80]

Table 2. Effects of Pb on Photosynthesis and Photosynthetic pigments.

Plant	Family	Concentration	Treatment duration	Effects	References
Thuidium delicatulum T. Sparsifolium (mosses) leafy liverwort *Ptychanthus striatus*	Bryophytes	10–10 to 10–2 M	1 to 72 hours	Decrease in chlorophyll content was observed in investigated liverwort. At highest tested concentration (10^{-4}M), Chlorophyll content decreased by 54%, 22%, 42% in *T.delicatulum*, *T. sparsifolium*, and *P. striatus* respectively.	[95]
Brassica rapa	Brassicaceae	0.5–5 mM	20 days	Due to inhibition of δ-aminolevulinic acid dehydratase activity a decrease in photosynthetic pigment was observed	[96]
Myrica rubra	Myricaceae	2, 4, and 6 mM	10 and 30 days	After the 30 days treatment of pb (6 mM), content of Chl *a* and Chl *b* reduced by 24.9 and 12.3%, respectively.	[97]

(Table 2) cont.....

Plant	Family	Concentration	Treatment duration	Effects	References
Morus alba L.	Moraceae	100 and 200 µmol L-1	5 days	Decreasing trend in chlorophyll content was observed in treated mulberry leaves	[98]
Triticum durum, T. aestivum, Hordeum vulgare and Avena sativa	Poaceae	0.15, 0.30 and 0.60 g/L.	8 weeks	At the highest concentration (0.60 g/L) the amount of chlorophyll *a*, b and c decreased by 23.3% 55.7% and 36.3% respectively.	[99]

Toxic Effects of Pb on Antioxidant Enzyme Activity

Under metal stress, plants generate defence systems (superoxide dismutase, glutathione reductase, guaiacol peroxidise, *etc.*) to scavenge ROS buildup [100, 101]. The main defensive enzyme known to transform superoxide radicals into nontoxic products is SOD, a metalloenzyme [102]. Due to its unpaired electrons, H_2O_2 is a strong oxidant that is eliminated by CAT activities, which convert H_2O_2 into water and molecule oxygen [103]. Additionally, peroxidases can potentially remove H_2O_2 molecules by converting them into H_2O and O_2 [104, 105]. Monodehydroascorbate radical (MDHA) catalyses the enzymatic conversion of MDHA to AsA and DHA [106]. DHAR produces glutathione disulfide (GSSG) by converting DHA to ascorbate while utilising glutathione (electron donor). Finally, employing NADPH-dependent reduction, the GR enzyme catalyses the conversion of GSSG to glutathione [105]. Many other authors also observed similar findings in different plants [107 - 110].

Hg Toxicity

Higher aquatic plants may easily uptake Hg^{2+}, which at high concentrations affects the functioning of mitochondria and causes oxidative stress by producing ROS. In plants, this causes degradation of biomembrane lipids and cell metabolism [110 - 114]. It is one of the most dangerous metals that affect both living beings and the environment worldwide. Mercury is a strong neurotoxic and also has a number of negative impacts on all of the major bodily systems.

The Agency for Toxic Substances and Disease registered Hg at third place in the "priority list of hazardous compounds" [115]. Hg is a metal that may readily be converted into a variety of oxidation states and can permeate a wide range of ecosystems [116, 117]. Hg is known to interfere with the regulation of enzymatic and non-enzymatic antioxidants [118]. Hg in several forms has been linked to loss of seed viability and growth [119].

Toxic Effects of Hg on Growth

Mondal and group [120] evaluated the role of various concentrations of $HgCl_2$ (0.1 to .5 ppm) on *Vigna radiate* and reported that the higher mercury salt concentrations severely hindered the root development of *Vigna radiate* seedlings. An increase in the mercury content results in a decrease in leaf area. The size and quantity of nodules/plants have gradually decreased with an increase in mercury concentrations. Cargnelutti *et al*. [114] investigated the effect of 0.5, 50, 250, and 500 μM concentrations of Hg on *Cucumis sativus* after 10 and 15 days. It was noted that the length of the root and shoot was significantly decreased by Hg exposure. Hamim and group [121] studied the impact of 0.5, 1, 2, and 3mM $Hg(NO_3)_2$ concentration on *Ricinus communis, Jatropha curcas, Reutealis trisperma* and *Melia azedarach,* in 3-week pot experiments and observed a decrease in root dry weight in *M. azedarach* (38%), *J. curcas* (26%) and *R. communis* (17%), whereas *Reutealis trisperma* showed no change. Many other researchers *viz*. Elbaz *et al*. [122] in *Chlamydomonas reinhardtii,* Devi *et al* [123] in *Vigna unguiculata,* Mei *et al* [124] in *Gossypium hirsutum* and Israr *et al* [113] in *Sesbania drummondii* reported a significant reduction in different growth parameters when treated with 1-8 μM, 40 mg g^{-1}; 0.05-2.0 mM; 1, 10, 50, 100 μM and 10-100 mg/l concentrations, respectively.

Toxic Effects of Hg on Lipid Peroxidation

Hg is known to cause lipid peroxidation in plant tissues [125]. *Gossypium hirsutum* (Malvaceae) seedlings were grown for 7 days under various concentrations of 1, 10, 50, and 100 μM Hg and observed the accumulation of MDA [124]. Hamim *et al*. [121] studied the effect of 0.5, 1, 2, 3 mM $Hg(NO_3)_2$ on *Ricinus communis, Jatropha curcas, Melia azedarach,* and *Reutealis trisperma* and reported that MDA content in leaves of all the four plant species increased with concentrations. *Cucumis sativus* seedlings on treatment with 0.5, 50, 250, and 500 μM Hg for 10 and 15 days showed an increase in the amount of lipid peroxides, as determined by TBARS level, at the maximum concentration of Hg (500 μM $HgCl_2$) [114].

Toxic Effects of Hg on Photosynthesis

The rate of transpiration, water intake, and chlorophyll synthesis can all be slowed down by Hg exposure. It has been demonstrated that Hg, both organic and inorganic, results in potassium, magnesium, and manganese loss as well as iron buildup [116]. Deng *et al*. [126] cultured *Microsorium pteropus* and treated them with 0–330 μg L^{-1} for 3 days. Results revealed that as compared to the control, mercury concentration (3.3 g L^{-1}) decreased the content of Chlorophyll and carotenoids in *M. pteropus*. Hg^{2+} severely reduced PSI's ability to function. Winter wheat was grown in pots containing 100-500 mg Hg/kg for 14, 21, 28, and 34 days by Liu *et al.* [127], and observed that on 28 and 34 days, the amount of total chlorophyll reduced as the Hg level increased. Total chlorophyll content was reduced to 1.36 and 1.05 on days 28 and 34 of Hg treatment, respectively at the maximum measured dose of Hg (control values: 1.72 and 1.65 mg Hg/kg).

Toxic Effects of Hg on Antioxidant Enzymes

The enzymatic antioxidants SOD, APX, and GR as well as the nonenzymatic GSH and NPSH are known to be modulated by mercury [125, 128 - 130]. Cargnelutti and group [114] studied the impact of Hg on *Cucumis sativus* after 10 and 15 days, in comparison to control, treated seedlings showed decreased catalase activity and ascorbate peroxidase activity. Priya *et al.* [131] studied the effect of 1 µg/ml Hg on *Clitoria ternatea* seedlings for 10 days. On Hg treatment, SOD, CAT, and POD content increased as compared to control. *Sesbania drummondii* (Fabaceae) were exposed to 10 to 100 mg/l for 10 days [113]. It was noted that all the antioxidant enzyme activities increased with Hg treatment except GR, which showed no significant change on treatment with Hg.

Challenges or New Insights

Utilizing plants' innate capacity to absorb mineral nutrients from the soil, phytoremediation is envisioned as a low-cost, environmentally beneficial process. In extremely polluted soils, phytostabilization is especially helpful and the most practical phytoremediation approach. Prior to using phytostabilization techniques, which rely on the activation of various defence mechanisms, it is necessary to choose plants that are resistant to harmful metal(loid)s. The main challenges in this field are the lack of interactions (synergistic, suppressive) among different heavy metals in plants and the lack of research on genetic variability among crops, which can affect the susceptibility and tolerance capacity against different heavy metals toxicity. By resolving these challenges, we can conserve our environment and promote sustainable agricultural practices.

CONCLUSION

The development of technologies to purify our environment is crucial since heavy metal contamination is a big global issue. Approaches to remediation that are effective are required. Phytoremediation is a solar-powered, environmentally friendly technology that is well-liked by the locals. Transgenic plants have been shown to have phytoremediation potential, but further research is required to realise the potential of this technique. This green remediation technology's future depends on coordinated research in the domains of environmental engineering, soil microbiology, plant physiology/biochemistry, and genetics. It is important to perform more research to better understand the interactions between the primary components of the rhizosphere, which include metals, soil, microorganisms, and plant roots. Exploring new spectroscopic and chromatographic technologies will help us understand how metal ions behave in plant tissues, which will help us understand how metal hyperaccumulation and plant tolerance work.

REFERENCES

[1]   De Abreu CA, De Abreu MF, De Andrade JC. Distribution of lead in the soil profile evaluated by DTPA and Mehlich-3 solutions. Bragantia 1998; 57: 185-92.

[2]   Forstner U. Land contamination by metals: global scope and magnitude of problem. In: Allen HE, Huang CP, Bailey GW, Bowers AR, Eds. Metal Speciation and Contamination of Soil. Boca Raton, FL: Lewis Publishers 1995; pp. 1-33.

[3]   Shanker A, Cervantes C, Lozatavera H, Avudainayagam S. Chromium toxicity in plants. Environ Int 2005; 31(5): 739-53.
[http://dx.doi.org/10.1016/j.envint.2005.02.003] [PMID: 15878200]

[4]   Alloway BJ. EdsHeavy metals in soils: trace metals and metalloids in soils and their bioavailability. Springer Science & Business Media 2012; p. 22.

[5]   Järup L. Hazards of heavy metal contamination. Br Med Bull 2003; 68(1): 167-82.
[http://dx.doi.org/10.1093/bmb/ldg032] [PMID: 14757716]

[6]   Järup L, Åkesson A. Current status of cadmium as an environmental health problem. Toxicol Appl Pharmacol 2009; 238(3): 201-8.
[http://dx.doi.org/10.1016/j.taap.2009.04.020] [PMID: 19409405]

[7]   Needleman H. Lead Poisoning. Annu Rev Med 2004; 55(1): 209-22.
[http://dx.doi.org/10.1146/annurev.med.55.091902.103653] [PMID: 14746518]

[8]   Kapaj S, Peterson H, Liber K, Bhattacharya P. Human health effects from chronic arsenic poisoning--a review. J Environ Sci Health Part A Tox Hazard Subst Environ Eng 2006; 41(10): 2399-428.
[http://dx.doi.org/10.1080/10934520600873571] [PMID: 17018421]

[9]   Ekino S, Susa M, Ninomiya T, Imamura K, Kitamura T. Minamata disease revisited: an update on the acute and chronic manifestations of methyl mercury poisoning. J Neurol Sci 2007; 262(1-2): 131-44.

[10] Sobrino-Plata J, Ortega-Villasante C, Laura Flores-Cáceres M, Escobar C, Del Campo FF, Hernández LE. Differential alterations of antioxidant defenses as bioindicators of mercury and cadmium toxicity in alfalfa. Chemosphere 2009; 77(7): 946-54.
[http://dx.doi.org/10.1016/j.chemosphere.2009.08.007] [PMID: 19732935]

[11] Cuypers A, Karen S, Jos R, *et al.* The cellular redox state as a modulator in cadmium and copper responses in *Arabidopsis thaliana* seedlings. J Plant Physiol 2011; 168(4): 309-16.
[http://dx.doi.org/10.1016/j.jplph.2010.07.010] [PMID: 20828869]

[12] Sobrino-Plata J, Herrero J, Carrasco-Gil S, *et al.* Specific stress responses to cadmium, arsenic and mercury appear in the metallophyte *Silene vulgaris* when grown hydroponically. RSC Advances 2013; 3(14): 4736-44.
[http://dx.doi.org/10.1039/c3ra40357b]

[13] Mészáros P, Rybanský Ľ, Spieß N, *et al.* Plant chitinase responses to different metal-type stresses reveal specificity. Plant Cell Rep 2014; 33(11): 1789-99.
[http://dx.doi.org/10.1007/s00299-014-1657-9] [PMID: 25023875]

[14] Nakano Y, Asada K. Purification of ascorbate peroxidase in spinach chloroplasts: Its inactivation in ascorbate-depleted medium and reactivation by monodehydroascorbate radical. Plant Cell Physiol 1987; 28: 131-40.

[15]   Gill SS, Anjum NA, Hasanuzzaman M, *et al.* Glutathione and glutathione reductase: A boon in disguise for plant abiotic stress defense operations. Plant Physiol Biochem 2013; 70: 204-12.
       [http://dx.doi.org/10.1016/j.plaphy.2013.05.032] [PMID: 23792825]

[16]   Xiao H, Yin L, Xu X, Li T, Han Z. The iron-regulated transporter, MbNRAMP1, isolated from *Malus baccata* is involved in Fe, Mn and Cd trafficking. Ann Bot (Lond) 2008; 102(6): 881-9.
       [http://dx.doi.org/10.1093/aob/mcn178] [PMID: 18819951]

[17]   Sanità di Toppi L, Gabbrielli R. Response to cadmium in higher plants. Environ Exp Bot 1999; 41(2): 105-30.
       [http://dx.doi.org/10.1016/S0098-8472(98)00058-6]

[18]   Mengel K, Kirkby EA, Kosegarten H, Appel T. Elements with more toxic effects. Princip Plant Nutr. 2001; pp. 657-73.
       [http://dx.doi.org/10.1007/978-94-010-1009-2_20]

[19]   Keller C, Marchetti M, Rossi L, Lugon-Moulin N. Reduction of cadmium availability to tobacco (*Nicotianatabacum*) plants using soil amendments in low cadmium-contaminated agricultural soils: a pot experiment. Plant Soil 2005; 276(1-2): 69-84.
       [http://dx.doi.org/10.1007/s11104-005-3101-y]

[20]   Rao G, Huang S, Ashraf U, *et al.* Ultrasonic seed treatment improved cadmium (Cd) tolerance in *Brassica napus* L. Ecotoxicol Environ Saf 2019; 185: 109659.
       [http://dx.doi.org/10.1016/j.ecoenv.2019.109659] [PMID: 31541946]

[21]   Wang B, Li Y, Zhang WH. Brassinosteroids are involved in response of cucumber (*Cucumis sativus*) to iron deficiency. Ann Bot (Lond) 2012; 110(3): 681-8.
       [http://dx.doi.org/10.1093/aob/mcs126] [PMID: 22684685]

[22]   Nouri M, El Rasafi T, Haddioui A. Responses of two barley subspecies to induced heavy metal stress: Seeds germination, seedlings growth and cytotoxicity assay. Agric (Pol'nohosp) 2019; 65(3): 107-18.
       [http://dx.doi.org/10.2478/agri-2019-0011]

[23]   Wang M, Chen L, Chen S, Ma Y. Alleviation of cadmium-induced root growth inhibition in crop seedlings by nanoparticles. Ecotoxicol Environ Saf 2012; 79: 48-54.
       [http://dx.doi.org/10.1016/j.ecoenv.2011.11.044] [PMID: 22189214]

[24]   Nedjimi B. Germination characteristics of *Peganum harmala* L. (Nitrariaceae) subjected to heavy metals: implications for the use in polluted dryland restoration. Int J Environ Sci Technol 2020; 17(4): 2113-22.
       [http://dx.doi.org/10.1007/s13762-019-02600-3]

[25]   Majeed A, Muhammad Z, Siyar S. Assessment of heavy metal induced stress responses in pea (Pisum sativum L.). Acta Ecol Sin 2019; 39(4): 284-8.
       [http://dx.doi.org/10.1016/j.chnaes.2018.12.002]

[26]   Zayneb C, Bassem K, Zeineb K, *et al.* Physiological responses of fenugreek seedlings and plants treated with cadmium. Environ Sci Pollut Res Int 2015; 22(14): 10679-89.
       [http://dx.doi.org/10.1007/s11356-015-4270-8] [PMID: 25752634]

[27]   El Rasafi T, Nouri M, Bouda S, Haddioui A. The effect of Cd, Zn and Fe on seed germination and early seedling growth of wheat and bean. Ekologia (Bratisl) 2016; 35(3): 213-23.
       [http://dx.doi.org/10.1515/eko-2016-0017]

[28]   Goel S, Malik J, Awasthi R, Sandhir R, Nayyar H. Growth and metabolic responses of maize (C_4 species) and rice (C_3 species) genotypes to cadmium toxicity. Cereal Res Commun 2012; 40(2): 225-34.
       [http://dx.doi.org/10.1556/CRC.40.2012.2.7]

[29]   Heidari M, Sarani S. Effects of lead and cadmium on seed germination, seedling growth and antioxidant enzymes activities of mustard (Sinapisarvensis L.). ARPN J AgricBiol Sci 2011; 6(1): 44-7.

[30]   Dey SK, Dey J, Patra S, Pothal D. Changes in the antioxidative enzyme activities and lipid peroxidation in wheat seedlings exposed to cadmium and lead stress. Braz J Plant Physiol 2007; 19(1): 53-60.
[http://dx.doi.org/10.1590/S1677-04202007000100006]

[31]   Wu F, Zhang G, Dominy P. Four barley genotypes respond differently to cadmium: lipid peroxidation and activities of antioxidant capacity. Environ Exp Bot 2003; 50(1): 67-78.
[http://dx.doi.org/10.1016/S0098-8472(02)00113-2]

[32]   Meng Y, Zhang L, Wang L, Zhou C, Shangguan Y, Yang Y. Antioxidative enzymes activity and thiol metabolism in three leafy vegetables under Cd stress. Ecotoxicol Environ Saf 2019; 173: 214-24.
[http://dx.doi.org/10.1016/j.ecoenv.2019.02.026] [PMID: 30772711]

[33]   Chiao WT, Syu CH, Chen BC, Juang KW. Cadmium in rice grains from a field trial in relation to model parameters of Cd-toxicity and -absorption in rice seedlings. Ecotoxicol Environ Saf 2019; 169: 837-47.
[http://dx.doi.org/10.1016/j.ecoenv.2018.11.061] [PMID: 30597783]

[34]   Anjum NA, Ahmad I, Pereira ME, Eds. The plant family Brassicaceae: contribution towards phytoremediation. Dordrecht: Springer Netherlands 2012.
[http://dx.doi.org/10.1007/978-94-007-3913-0]

[35]   Mobin M, Khan NA. Photosynthetic activity, pigment composition and antioxidative response of two mustard (*Brassica juncea*) cultivars differing in photosynthetic capacity subjected to cadmium stress. J Plant Physiol 2007; 164(5): 601-10.
[http://dx.doi.org/10.1016/j.jplph.2006.03.003] [PMID: 16621132]

[36]   Gill SS, Khan NA, Tuteja N. Differential cadmium stress tolerance in five Indian mustard (*Brassica juncea* L.) cultivars. Plant Signal Behav 2011; 6(2): 293-300.
[http://dx.doi.org/10.4161/psb.6.2.15049] [PMID: 21744661]

[37]   Bauddh K, Singh RP. Growth, tolerance efficiency and phytoremediation potential of *Ricinus communis* (L.) and *Brassica juncea* (L.) in salinity and drought affected cadmium contaminated soil. Ecotoxicol Environ Saf 2012; 85: 13-22.
[http://dx.doi.org/10.1016/j.ecoenv.2012.08.019] [PMID: 22959315]

[38]   Anjum NA, Umar S, Iqbal M. Assessment of cadmium accumulation, toxicity, and tolerance in Brassicaceae and Fabaceae plants—implications for phytoremediation. Environ Sci Pollut Res Int 2014; 21(17): 10286-93.
[http://dx.doi.org/10.1007/s11356-014-2889-5] [PMID: 24756685]

[39]   Metwally A, Safronova VI, Belimov AA, Dietz KJ. Genotypic variation of the response to cadmium toxicity in *Pisum sativum* L. J Exp Bot 2005; 56(409): 167-78.
[PMID: 15533881]

[40]   Mani D, Sharma B, Kumar C. Phytoaccumulation, interaction, toxicity and remediation of cadmium from *Helianthus annuus* L. (sunflower). Bull Environ Contam Toxicol 2007; 79(1): 71-9.
[http://dx.doi.org/10.1007/s00128-007-9153-3] [PMID: 17549427]

[41]   Azimychetabi Z, Sabokdast Nodehi M, Karami Moghadam T, Motesharezadeh B. Cadmium stress alters the essential oil composition and the expression of genes involved in their synthesis in peppermint (*Mantha piperita* L.). Ind Crops Prod 2021; 168: 113602.
[http://dx.doi.org/10.1016/j.indcrop.2021.113602]

[42]   Ouzounidou G, Moustakas M, Eleftheriou EP. Physiological and ultrastructural effects of cadmium on wheat (*Triticum aestivum* L.) leaves. Arch Environ Contam Toxicol 1997; 32(2): 154-60.
[http://dx.doi.org/10.1007/s002449900168] [PMID: 9069190]

[43]   Ekmekçi Y, Tanyolaç D, Ayhan B. Effects of cadmium on antioxidant enzyme and photosynthetic activities in leaves of two maize cultivars. J Plant Physiol 2008; 165(6): 600-11.
[http://dx.doi.org/10.1016/j.jplph.2007.01.017] [PMID: 17728009]

[44] Xin J, Zhao XH, Tan QL, Sun XC, Zhao YY, Hu CX. Effects of cadmium exposure on the growth, photosynthesis, and antioxidant defense system in two radish (*Raphanus sativus* L.) cultivars. Photosynthetica 2019; 57(4): 967-73.
[http://dx.doi.org/10.32615/ps.2019.076]

[45] Gubrelay U, Agnihotri RK, Singh G, Kaur R, Sharma R. Effect of heavy metal Cd on some physiological and biochemical parameters of Barley (*Hordeumvulgare* L.). Int J Agric Sci 2013; 5(22): 27-43.

[46] Li S, Yang W, Yang T, Chen Y, Ni W. Effects of cadmium stress on leaf chlorophyll fluorescence and photosynthesis of *Elsholtzia argyi*—a cadmium accumulating plant. Int J Phytoremediation 2015; 17(1): 85-92.
[http://dx.doi.org/10.1080/15226514.2013.828020] [PMID: 25174428]

[47] Zhang H, Xu Z, Guo K, *et al.* Toxic effects of heavy metal Cd and Zn on chlorophyll, carotenoid metabolism and photosynthetic function in tobacco leaves revealed by physiological and proteomics analysis. Ecotoxicol Environ Saf 2020; 202: 110856.
[http://dx.doi.org/10.1016/j.ecoenv.2020.110856] [PMID: 32629202]

[48] Monteiro MS, Santos C, Soares AMVM, Mann RM. Assessment of biomarkers of cadmium stress in lettuce. Ecotoxicol Environ Saf 2009; 72(3): 811-8.
[http://dx.doi.org/10.1016/j.ecoenv.2008.08.002] [PMID: 18952284]

[49] Hédiji H, Djebali W, Cabasson C, *et al.* Effects of long-term cadmium exposure on growth and metabolomic profile of tomato plants. Ecotoxicol Environ Saf 2010; 73(8): 1965-74.
[http://dx.doi.org/10.1016/j.ecoenv.2010.08.014] [PMID: 20846723]

[50] Xue Z, Gao H, Zhao S. Effects of cadmium on the photosynthetic activity in mature and young leaves of soybean plants. Environ Sci Pollut Res Int 2014; 21(6): 4656-64.
[http://dx.doi.org/10.1007/s11356-013-2433-z] [PMID: 24352551]

[51] Agrawal SB, Mishra S. Effects of supplemental ultraviolet-B and cadmium on growth, antioxidants and yield of *Pisum sativum* L. Ecotoxicol Environ Saf 2009; 72(2): 610-8.
[http://dx.doi.org/10.1016/j.ecoenv.2007.10.007] [PMID: 18061671]

[52] Schützendübel A, Nikolova P, Rudolf C, Polle A. Cadmium and H_2O_2-induced oxidative stress in *Populus* × *canescens* roots. Plant Physiol Biochem 2002; 40(6-8): 577-84.
[http://dx.doi.org/10.1016/S0981-9428(02)01411-0]

[53] Gratão PL, Polle A, Lea PJ, Azevedo RA. Making the life of heavy metal-stressed plants a little easier. Funct Plant Biol 2005; 32(6): 481-94.
[http://dx.doi.org/10.1071/FP05016] [PMID: 32689149]

[54] Sharma SS, Dietz KJ. The relationship between metal toxicity and cellular redox imbalance. Trends Plant Sci 2009; 14(1): 43-50.
[http://dx.doi.org/10.1016/j.tplants.2008.10.007] [PMID: 19070530]

[55] Sheetal KR, Singh SD, Anand A, Prasad S. Heavy metal accumulation and effects on growth, biomass and physiological processes in mustard. Indian J Plant Physiol 2016; 21(2): 219-23.
[http://dx.doi.org/10.1007/s40502-016-0221-8]

[56] Wu Z, Liu S, Zhao J, *et al.* Comparative responses to silicon and selenium in relation to antioxidant enzyme system and the glutathione-ascorbate cycle in flowering Chinese cabbage (*Brassica campestris* L. ssp. chinensis var. *utilis*) under cadmium stress. Environ Exp Bot 2017; 133: 1-11.
[http://dx.doi.org/10.1016/j.envexpbot.2016.09.005]

[57] Soudek P, Petrová Š, Vaňková R, Song J, Vaněk T. Accumulation of heavy metals using Sorghum sp. Chemosphere 2014; 104: 15-24.
[http://dx.doi.org/10.1016/j.chemosphere.2013.09.079] [PMID: 24268752]

[58] Mahaffey KR. Environmental lead toxicity: nutrition as a component of intervention. Environ Health Perspect 1990; 89: 75-8.

[http://dx.doi.org/10.1289/ehp.908975] [PMID: 2088758]

[59]   Sharma P, Dubey RS. Lead toxicity in plants. Braz J Plant Physiol 2005; 17(1): 35-52.
        [http://dx.doi.org/10.1590/S1677-04202005000100004]

[60]   Seregin IV, Ivanov VB. Physiological aspects of cadmium and lead toxic effects on higher plants.
        Russ J Plant Physiol 2001; 48(4): 523-44.
        [http://dx.doi.org/10.1023/A:1016719901147]

[61]   Mishra S, Srivastava S, Tripathi RD, Kumar R, Seth CS, Gupta DK. Lead detoxification by coontail
        (*Ceratophyllum demersum* L.) involves induction of phytochelatins and antioxidant system in response
        to its accumulation. Chemosphere 2006; 65(6): 1027-39.
        [http://dx.doi.org/10.1016/j.chemosphere.2006.03.033] [PMID: 16682069]

[62]   Kopittke PM, Asher CJ, Menzies NW. Prediction of Pb speciation in concentrated and dilute nutrient
        solutions. Environ Pollut 2008; 153(3): 548-54.
        [http://dx.doi.org/10.1016/j.envpol.2007.09.012] [PMID: 17959287]

[63]   Islam E, Liu D, Li T, *et al.* Effect of Pb toxicity on leaf growth, physiology and ultrastructure in the
        two ecotypes of *Elsholtzia argyi.*. J Hazard Mater 2008; 154(1-3): 914-26.
        [http://dx.doi.org/10.1016/j.jhazmat.2007.10.121] [PMID: 18162296]

[64]   Kabir M, Iqbal MZ, Shafiq M, Farooqi ZR. Reduction in germination and seedling growth of
        Thespesiapopulnea L., caused by lead and cadmium treatments. Pak J Bot 2008; 40(6): 2419-26.

[65]   Al-Yemeni MN. Effect of cadmium, mercury and lead on seed germination and early seedling growth
        of *Vignaambacensis* L. Indian J Plant Physiol 2001; 6(2): 147-51.

[66]   Yang Y, Wei X, Lu J, You J, Wang W, Shi R. Lead-induced phytotoxicity mechanism involved in
        seed germination and seedling growth of wheat (*Triticum aestivum* L.). Ecotoxicol Environ Saf 2010;
        73(8): 1982-7.
        [http://dx.doi.org/10.1016/j.ecoenv.2010.08.041] [PMID: 20833428]

[67]   Hussain A, Abbas N, Arshad F, *et al.* Effects of diverse doses of Lead (Pb) on different growth
        attributes of *Zea-Mays* L. Agric Sci 2013; 4: 262-5.

[68]   Vasilachi IC, Minut M, Betianu C, Gavrilescu M. Investigation of the toxic effects of lead on maize
        germination and growth (Zea mays). In 2021 International Conference on e-Health and Bioengineering
        EHB. 1-4.

[69]   Siddiqui MM, Abbasi BH, Ahmad N, Ali M, Mahmood T. Toxic effects of heavy metals (Cd, Cr and
        Pb) on seed germination and growth and DPPH-scavenging activity in *Brassica rapa* var. *turnip*.
        Toxicol Ind Health 2014; 30(3): 238-49.
        [http://dx.doi.org/10.1177/0748233712452605] [PMID: 22872632]

[70]   Kabir M, Iqbal MZ, Shafiq M, Farooqi ZR. Effects of lead on seedling growth of *spesia populnea.*.
        Plant Soil Environ 2010; 56(4): 194-9.
        [http://dx.doi.org/10.17221/147/2009-PSE]

[71]   Kabir M, Iqbal MZ, Shafiq M. The effects of lead and cadmium individually and in combinations on
        germination and seedling growth of *Leucaena leucocephala* (Lam.) de Wit. ASRJETS 2018; 43(1):
        33-43.

[72]   Shafiq M, Iqbal MZ, Mohammad A. Effect of lead and cadmium on germination and seedling growth
        of *Leucaena leucocephala.*. J Appl Sci Environ Manag 2008; 12(3): 61-6.

[73]   Farooqi ZR, Iqbal MZ, Kabir M, Shafiq M. Toxic effects of lead and cadmium on germination and
        seedling growth of *Albizialebbeck* (L.) Benth. Pak J Bot 2009; 41(1): 27-33.

[74]   Lamhamdi M, Bakrim A, Aarab A, Lafont R, Sayah F. Lead phytotoxicity on wheat (*Triticum
        aestivum* L.) seed germination and seedlings growth. C R Biol 2011; 334(2): 118-26.
        [http://dx.doi.org/10.1016/j.crvi.2010.12.006] [PMID: 21333942]

[75]   Shu X, Yin L, Zhang Q, Wang W. Effect of Pb toxicity on leaf growth, antioxidant enzyme activities,

and photosynthesis in cuttings and seedlings of *Jatropha curcas* L. Environ Sci Pollut Res Int 2012; 19(3): 893-902.
[http://dx.doi.org/10.1007/s11356-011-0625-y] [PMID: 21964550]

[76]   Li H, Luo H, Li D, Hu T, Fu J. Antioxidant enzyme activity and gene expression in response to lead stress in perennial ryegrass. J Am Soc Hortic Sci 2012; 137(2): 80-5.
[http://dx.doi.org/10.21273/JASHS.137.2.80]

[77]   Sahoo SL, Mohanty S, Rout S, Kanungo S. The effect of lead toxicity on growth and antioxidant enzyme expression of *Abutilon indicum* L. Int J Pharma Sci 2015; 7(2): 357-64.

[78]   Bezerril Fontenele NM, Otoch MLO, Gomes-Rochette NF, *et al.* Effect of lead on physiological and antioxidant responses in two *Vigna unguiculata* cultivars differing in Pb-accumulation. Chemosphere 2017; 176: 397-404.
[http://dx.doi.org/10.1016/j.chemosphere.2017.02.072] [PMID: 28278428]

[79]   Kaya C, Akram NA, Sürücü A, Ashraf M. Alleviating effect of nitric oxide on oxidative stress and antioxidant defence system in pepper (*Capsicum annuum* L.) plants exposed to cadmium and lead toxicity applied separately or in combination. Sci Hortic (Amsterdam) 2019; 255: 52-60.
[http://dx.doi.org/10.1016/j.scienta.2019.05.029]

[80]   Xiong Z, Yang J, Zhang K. Effects of lead pollution on germination and seedling growth of turfgrass, *Cynodon dactylon.*. Pak J Bot 2021; 53(6): 2003-9.
[http://dx.doi.org/10.30848/PJB2021-6(6)]

[81]   Kumar A, Prasad MNV. Plant-lead interactions: Transport, toxicity, tolerance, and detoxification mechanisms. Ecotoxicol Environ Saf 2018; 166: 401-18.
[http://dx.doi.org/10.1016/j.ecoenv.2018.09.113] [PMID: 30290327]

[82]   Bhardwaj P, Chaturvedi AK, Prasad P. Effect of enhanced lead and cadmium in soil on physiological and biochemical attributes of *Phaseolus vulgaris* L. Nat Sci 2009; 7(8): 63-75.

[83]   Kaur G, Singh HP, Batish DR, Kohli RK. Growth, photosynthetic activity and oxidative stress in wheat (*Triticum aestivum*) after exposure of lead to soil. J Environ Biol 2012; 33(2): 265-9.
[PMID: 23033692]

[84]   Gupta DK, Nicoloso FT, Schetinger MRC, *et al.* Antioxidant defense mechanism in hydroponically grown *Zea mays* seedlings under moderate lead stress. J Hazard Mater 2009; 172(1): 479-84.
[http://dx.doi.org/10.1016/j.jhazmat.2009.06.141] [PMID: 19625122]

[85]   Yang Y, Zhang Y, Wei X, *et al.* Comparative antioxidative responses and proline metabolism in two wheat cultivars under short term lead stress. Ecotoxicol Environ Saf 2011; 74(4): 733-40.
[http://dx.doi.org/10.1016/j.ecoenv.2010.10.035] [PMID: 21074856]

[86]   Thakur S, Singh L, Zularisam AW, Sakinah M, Din MFM. Lead induced oxidative stress and alteration in the activities of antioxidative enzymes in rice shoots. Biol Plant 2017; 61(3): 595-8.
[http://dx.doi.org/10.1007/s10535-016-0680-9]

[87]   Nareshkumar A, Veeranagamallaiah G, Pandurangaiah M, *et al.* Pb-stress induced oxidative stress caused alterations in antioxidant efficacy in two groundnut (*Arachishypogaea* L.) cultivars. Agric Sci 2015; 6(10): 1283.

[88]   Kaur G, Singh HP, Batish DR, Kohli RK. Lead (Pb)-induced biochemical and ultrastructural changes in wheat (Triticum aestivum) roots. Protoplasma 2013; 250(1): 53-62.
[http://dx.doi.org/10.1007/s00709-011-0372-4] [PMID: 22231903]

[89]   Bajji M, Kinet JM, Lutts S. The use of the electrolyte leakage method for assessing cell membrane stability as a water stress tolerance test in durum wheat. Plant Growth Regul 2002; 36(1): 61-70.
[http://dx.doi.org/10.1023/A:1014732714549]

[90]   Pirzadah TB, Malik B, Tahir I, Hakeem KR, Alharby HF, Rehman RU. Lead toxicity alters the antioxidant defense machinery and modulate the biomarkers in Tartary buckwheat plants. Int Biodeterior Biodegradation 2020; 151: 104992.

[http://dx.doi.org/10.1016/j.ibiod.2020.104992]

[91]   Stefanov K, Seizova K, Popova I, Petkov V, Kimenov G, Popov S. Effects of lead ions on the phospholipid composition in leaves of *Zea mays* and *Phaseolus vulgaris*.. J Plant Physiol 1995; 147(2): 243-6.
[http://dx.doi.org/10.1016/S0176-1617(11)81511-7]

[92]   Rebechini HM, Hanzely L. Lead-induced ultrastructural changes in chloroplasts of the hydrophyte, *Ceratophyllum demersum*.. Z Pflanzenphysiol 1974; 73(5): 377-86.
[http://dx.doi.org/10.1016/S0044-328X(74)80106-6]

[93]   Ahmed A, Tajmir-Riahi HA. Interaction of toxic metal ions Cd^{2+}, Hg^{2+}, and Pb^{2+} with light-harvesting proteins of chloroplast thylakoid membranes. An FTIR spectroscopic study. J Inorg Biochem 1993; 50(4): 235-43.
[http://dx.doi.org/10.1016/0162-0134(93)80050-J] [PMID: 8120530]

[94]   Drazkiewicz M. Chlorophyll–occurrence, functions, mechanism of action, effects of internal and external factors. Photosynthetica 1994; 30: 321-31.

[95]   Shakya K, Chettri MK, Sawidis T. Experimental investigations of five different mosses on accumulation capacities of Cu, Pb and Zn. Toxicol Environ Chem 2008; 90(3): 585-601.
[http://dx.doi.org/10.1080/02772240701609780]

[96]   Cenkci S, Ciğerci İH, Yıldız M, Özay C, Bozdağ A, Terzi H. Lead contamination reduces chlorophyll biosynthesis and genomic template stability in *Brassica rapa* L. Environ Exp Bot 2010; 67(3): 467-73.
[http://dx.doi.org/10.1016/j.envexpbot.2009.10.001]

[97]   He B, Gu M, Wang X, He X. The effects of lead on photosynthetic performance of waxberry seedlings (Myrica rubra). Photosynthetica 2018; 56(4): 1147-53.
[http://dx.doi.org/10.1007/s11099-018-0800-1]

[98]   Huihui Z, Xin L, Zisong X, *et al*. Toxic effects of heavy metals Pb and Cd on mulberry (*Morus alba* L.) seedling leaves: Photosynthetic function and reactive oxygen species (ROS) metabolism responses. Ecotoxicol Environ Saf 2020; 195: 110469.
[http://dx.doi.org/10.1016/j.ecoenv.2020.110469] [PMID: 32179235]

[99]   Souahi H, Chebout A, Akrout K, Massaoud N, Gacem R. Physiological responses to lead exposure in wheat, barley and oat. Environ Chall (Amst) 2021; 4: 100079.
[http://dx.doi.org/10.1016/j.envc.2021.100079]

[100]   Hattab S, Hattab S, Flores-Casseres ML, *et al*. Characterisation of lead-induced stress molecular biomarkers in *Medicago sativa* plants. Environ Exp Bot 2016; 123: 1-12.
[http://dx.doi.org/10.1016/j.envexpbot.2015.10.005]

[101]   Alamri SA, Siddiqui MH, Al-Khaishany MYY, *et al*. Ascorbic acid improves the tolerance of wheat plants to lead toxicity. J Plant Interact 2018; 13(1): 409-19.
[http://dx.doi.org/10.1080/17429145.2018.1491067]

[102]   Hasanuzzaman M, Hossain MA, da Silva JAT, Fujita M. Plant response and tolerance to abiotic oxidative stress: Antioxidant defense is a key factor. Crop Stress Manag 2012; 261-315.
[http://dx.doi.org/10.1007/978-94-007-2220-0_8]

[103]   Gill SS, Tuteja N. Reactive oxygen species and antioxidant machinery in abiotic stress tolerance in crop plants. Plant Physiol Biochem 2010; 48(12): 909-30.
[http://dx.doi.org/10.1016/j.plaphy.2010.08.016] [PMID: 20870416]

[104]   Singh R, Tripathi RD, Dwivedi S, Kumar A, Trivedi PK, Chakrabarty D. Lead bioaccumulation potential of an aquatic macrophyte *Najas indica* are related to antioxidant system. Bioresour Technol 2010; 101(9): 3025-32.
[http://dx.doi.org/10.1016/j.biortech.2009.12.031] [PMID: 20053550]

[105]   Potters G, Horemans N, Jansen MAK. The cellular redox state in plant stress biology – A charging concept. Plant Physiol Biochem 2010; 48(5): 292-300.

[http://dx.doi.org/10.1016/j.plaphy.2009.12.007] [PMID: 20137959]

[106] Malecka A, Piechalak A, Tomaszewska B. Reactive oxygen species production and antioxidative defense system in pea root tissues treated with lead ions: the whole roots level. Acta Physiol Plant 2009; 31(5): 1053-63.
[http://dx.doi.org/10.1007/s11738-009-0326-z]

[107] Rucińska R, Waplak S, Gwóźdź EA. Free radical formation and activity of antioxidant enzymes in lupin roots exposed to lead. Plant Physiol Biochem 1999; 37(3): 187-94.
[http://dx.doi.org/10.1016/S0981-9428(99)80033-3]

[108] Mohamed AA. Changes in non protein thiols, some antioxidant enzymes activity and ultrastructural alteration in radish plant (*Raphanus sativus* L.) grown under lead toxicity. Not Bot Horti Agrobot Cluj-Napoca 2010; 38(3): 76-85.

[109] Dias MC, Mariz-Ponte N, Santos C. Lead induces oxidative stress in *Pisum sativum* plants and changes the levels of phytohormones with antioxidant role. Plant Physiol Biochem 2019; 137: 121-9.
[http://dx.doi.org/10.1016/j.plaphy.2019.02.005] [PMID: 30772622]

[110] Han FX, Su Y, Monts DL, Waggoner CA, Plodinec MJ. Binding, distribution, and plant uptake of mercury in a soil from Oak Ridge, Tennessee, USA. Sci Total Environ 2006; 368(2-3): 753-68.
[http://dx.doi.org/10.1016/j.scitotenv.2006.02.026] [PMID: 16569422]

[111] Kamal M, Ghaly AE, Mahmoud N, Côté R. Phytoaccumulation of heavy metals by aquatic plants. Environ Int 2004; 29(8): 1029-39.
[http://dx.doi.org/10.1016/S0160-4120(03)00091-6] [PMID: 14680885]

[112] Greger M, Wang Y, Neuschütz C. Absence of Hg transpiration by shoot after Hg uptake by roots of six terrestrial plant species. Environ Pollut 2005; 134(2): 201-8.
[http://dx.doi.org/10.1016/j.envpol.2004.08.007] [PMID: 15589647]

[113] Israr M, Sahi S, Datta R, Sarkar D. Bioaccumulation and physiological effects of mercury in *Sesbania drummondii.*. Chemosphere 2006; 65(4): 591-8.
[http://dx.doi.org/10.1016/j.chemosphere.2006.02.016] [PMID: 16564071]

[114] Cargnelutti D, Tabaldi LA, Spanevello RM, *et al.* Mercury toxicity induces oxidative stress in growing cucumber seedlings. Chemosphere 2006; 65(6): 999-1006.
[http://dx.doi.org/10.1016/j.chemosphere.2006.03.037] [PMID: 16674986]

[115] Rahman Z, Singh VP. The relative impact of toxic heavy metals (THMs) (arsenic (As), cadmium (Cd), chromium (Cr)(VI), mercury (Hg), and lead (Pb)) on the total environment: an overview. Environ Monit Assess 2019; 191(7): 419.
[http://dx.doi.org/10.1007/s10661-019-7528-7] [PMID: 31177337]

[116] Boening DW. Ecological effects, transport, and fate of mercury: a general review. Chemosphere 2000; 40(12): 1335-51.
[http://dx.doi.org/10.1016/S0045-6535(99)00283-0] [PMID: 10789973]

[117] Clarkson TW, Magos L. The toxicology of mercury and its chemical compounds. Crit Rev Toxicol 2006; 36(8): 609-62.
[http://dx.doi.org/10.1080/10408440600845619] [PMID: 16973445]

[118] Sparks DL. Toxic metals in the environment: the role of surfaces. Elements (Quebec) 2005; 1(4): 193-7.
[http://dx.doi.org/10.2113/gselements.1.4.193]

[119] Patra M, Bhowmik N, Bandopadhyay B, Sharma A. Comparison of mercury, lead and arsenic with respect to genotoxic effects on plant systems and the development of genetic tolerance. Environ Exp Bot 2004; 52(3): 199-223.
[http://dx.doi.org/10.1016/j.envexpbot.2004.02.009]

[120] Mondal NK, Das C, Datta JK. Effect of mercury on seedling growth, nodulation and ultrastructural deformation of *Vigna radiata* (L) Wilczek. Environ Monit Assess 2015; 187(5): 241.

[http://dx.doi.org/10.1007/s10661-015-4484-8] [PMID: 25861903]

[121] Hamim H, Mutyandini A, Sulistyani YC, Putra HF, Saprudin D, Setyanings L. Effect of mercury on growth, anatomy and physiology of four non-edible oil-producing species. Asian J Plant Sci 2019; 18(4): 164-74.
[http://dx.doi.org/10.3923/ajps.2019.164.174]

[122] Elbaz A, Wei YY, Meng Q, Zheng Q, Yang ZM. Mercury-induced oxidative stress and impact on antioxidant enzymes in *Chlamydomonas reinhardtii.*. Ecotoxicology 2010; 19(7): 1285-93.
[http://dx.doi.org/10.1007/s10646-010-0514-z] [PMID: 20571879]

[123] Devi PU, Murugan S, Akilapriyadharasini S, Suja S, Chinnaswamy P. Effect of mercury and effluents on seed germination, root-shoot length, amylase activity and phenolic compounds in *Vignaunguiculata.*. Nat Environ PollutTechnol 2007; 6(3): 457.

[124] Mei L, Zhu Y, Zhang X, *et al.* Mercury-induced phytotoxicity and responses in upland cotton (*Gossypiumhirsutum* L.) seedlings. Plants 2021; 10(8): 1494.
[http://dx.doi.org/10.3390/plants10081494] [PMID: 34451539]

[125] Ahmad P, Ahanger MA, Egamberdieva D, Alam P, Alyemeni MN, Ashraf M. Modification of osmolytes and antioxidant enzymes by 24-epibrassinolide in chickpea seedlings under mercury (Hg) toxicity. J Plant Growth Regul 2018; 37(1): 309-22.
[http://dx.doi.org/10.1007/s00344-017-9730-6]

[126] Deng C, Zhang D, Pan X, Chang F, Wang S. Toxic effects of mercury on PSI and PSII activities, membrane potential and transthylakoid proton gradient in *Microsoriumpteropus*. J PhotochemPhotobiol B. Biology (Basel) 2013; 127: 1-7.
[PMID: 23920143]

[127] Liu D, Wang X, Chen Z, Xu H, Wang Y. Influence of mercury on chlorophyll content in winter wheat and mercury bioaccumulation. Plant Soil Environ 2010; 56(3): 139-43.
[http://dx.doi.org/10.17221/210/2009-PSE]

[128] Shaw BP. Effects of mercury and cadmium on the activities of antioxidative enzymes in the seedlings of *Phaseolusaureus.*. BiolPlantarum 1995; 37: 587-96.

[129] Ansari MKA, Ahmad A, Umar S, Iqbal M. Mercury-induced changes in growth variables and antioxidative enzyme activities in Indian mustard. J Plant Interact 2009; 4(2): 131-6.
[http://dx.doi.org/10.1080/17429140802716713]

[130] Zhou ZS, Wang SJ, Yang ZM. Biological detection and analysis of mercury toxicity to alfalfa (*Medicago sativa*) plants. Chemosphere 2008; 70(8): 1500-9.
[http://dx.doi.org/10.1016/j.chemosphere.2007.08.028] [PMID: 17905409]

[131] Priya M, Balakrishnan V, Kiruthika Lakshmi A, Aruna R, Ravindran KC. Mercury induced oxidative stress of antioxidants in Clitoriaternatea L. Int Lett Nat Sci 2014; (18):

CHAPTER 2

Bioaccumulation and Detoxification of Heavy Metals in Plants

Gurvarinder Kaur[1], Isha Madaan[1,2], Harpreet Kaur[3], Kulwinder Singh[4], Diksha Kalia[5], Komal Goel[5,6], Neha Guleri[7], Manish Kumar[8], Rajesh Kumar Singh[5], Gaurav Zinta[5,6], Renu Bhardwaj[9] and Geetika Sirhindi[1,*]

[1] *Department of Botany, Punjabi University, Patiala, Punjab, India*

[2] *Government College of Education, Jalandhar, Punjab, India*

[3] *P.G. Department of Botany, Khalsa College Amritsar, Amritsar, Punjab, India*

[4] *Sanmati Govt. College of Science Education and Research, Jagraon, Punjab, India*

[5] *Department of Biotechnology, CSIR-Institute of Himalayan Bioresource Technology, Palampur, Himachal Pradesh, India*

[6] *Academy of Scientific and Innovative Research (AcSIR), Ghaziabad, Uttar Pradesh, India*

[7] *Department of Plant Sciences, Central University of Himachal Pradesh, Shahpur, Himachal Pradesh, India*

[8] *Post Graduate Department of Botany, Dev Samaj College for Women, Ferozepur, Punjab, India*

[9] *Department of Botanical and Environmental Sciences, Guru Nanak Dev University, Amritsar, Punjab, India*

Abstract: Rapid industrialization and urbanization have exacerbated the accumulation of HMs (HMs) in the soil. Their greater mobility enables them to readily accumulate in the body of living organisms, which is referred to as bioaccumulation, thereby leading to their magnification in the biological systems. As a result, they have become an inevitable part of our food chain affecting the lives of organisms at all trophic levels; however, plants are at greater risk due to their sessile habitat. The bioaccumulation of HMs is affected by several factors like the type of plant species, its development stage and plant parts as well the environmental factors like temperature, pH, and composition of soil as well as the form of HM present in water and soil. Furthermore, the entry of noxious HMs into the plant system triggers several detrimental effects on the plants ROS burst, inhibition of enzymatic activities, reduced mineral nutrition, and many more resulting in reduced growth and development of plants. Therefore, in order to tolerate HMs in their system, plants incorporate several detoxification mechanisms like the efflux of HMs from the plant body, sequestration and compartmentalization through GSH (GSH), phytochelatins (PCs), and metallothionein (MTs), chelation of HMs with some organic compounds and activating antioxidant defense system to

* **Corresponding author Geetika Sirhindi:** Department of Botany, Punjabi University, Patiala, Punjab, India; E-mail: geetika@pbi.ac.in

Kanika Khanna, Sukhmeen Kaur Kohli & Renu Bhardwaj (Eds.)

mitigate the deleterious effects produced by ROS generated due to HM stress. The role of miRNAs in helping the plants battle against the HMs and their toxicity has also been acknowledged. The bioaccumulation potential and detoxification mechanisms advocated by the plants enable them to become potential candidates for phytoremediation. This chapter will unveil the bioaccumulation potential of plants along with presenting detailed mechanisms underlying the detoxification of HMs that protects plants from their toxic effects.

Keywords: Bioremediation, HMs, Hyperaccumulation, ROS, Phytochelatins.

INTRODUCTION

The inclination of the human race towards rapid industrialization and urbanization has led to the contamination of natural resources thereby making it a global concern threatening the survival of biological creatures [1]. Also, the abrupt rise in the global population has exerted significant pressure on agriculture to enhance food production which further deteriorates the condition of soil and water contamination [2]. Among pollutants, Heavy Metals (HMs) have become an inevitable part of the biological food chain thereby posing detrimental health effects [3]. Plants rooted in a single spot are most prone to HM contamination in different natural sources. The uptake of HMs by plants is affected by several factors like soil composition, pH, microbiotics, type of ions present in the soil and water, and several others [4, 5]. Such factors impact the amount of uptake of HMs in plants and influence the type of plant parts accumulating HMs. Based on the bioaccumulation capacity of plants, they can be classified into excluders, metal indicators, and metal accumulators or hyperaccumulators [6]. This becomes the basis for the exploitation of plants for the process of bioremediation of HM-contaminated soils.

Upon the entry of HMs into the plant body, they pose several hazardous impacts including oxidative stress, enzyme activity inhibition, reduced photosynthetic pigments and photosynthetic efficiency, and affecting overall metabolic machinery [7 - 9]. Therefore, to curb the deleterious effects of HMs upon their accumulation into the plant body, plants regulate a wide array of responses against HMs. Firstly, the plants tend to incorporate efflux mechanisms that tend to use metal transporters for the same. Furthermore, upon entry into the plant body, HMs disturb cellular homeostasis. The biochemical responses of plants to maintain cellular integrity are through the process of osmoregulation, which involves the synthesis of osmoprotectants like proline, glycine betaine, and sugars [10, 11]. Further, plants tend to immobilize the HMs and render them inactive through the process of sequestration and chelation. They also synthesize several organic phytochelatins that sequester HMs and restrict their deleterious impacts by rendering them inactive [12]. Metallothioneins (MTs) play a major role in chelation of HMs along with GSH (GSH) [13, 14]. Furthermore, the HMs disturb metabolic machinery and lead to oxidative stress in plants. In order to scavenge oxidative species, plants enhance their antioxidant machinery which comprises both enzymatic and non-enzymatic antioxidants that neutralize the reactive oxygen species (ROS) [1, 7, 15]. Lastly, the role of miRNAs (microRNAs) has greatly been acknowledged in altering the gene expression of plants facing HM stress. Therefore, the chapter will summarize the HM bioaccumulation tendency of plants and the various factors impacting the uptake of HMs along with discussing the mechanisms initiated as a response to HM accumulation in plants in order to shield their metabolic machinery.

Factors Affecting the uptake of HMs in Plants

With the advent of industrialisation and population explosion, a large number of anthropogenic activities are being carried out on a large scale, which is responsible for HMs contamination of the basic components of the environment may it be air, water, or soil. These HMs make their way into plant systems through roots or foliar adsorption thereby posing a serious threat to the normal metabolic machinery of plants. The entry of HMs into the plant body by either means is controlled by various factors grouped as physical, chemical, biological, and meteorological factors (Fig. **1**). Each of these factors impacts the amount and type of HM that gets accumulated within the root system and aerial parts of plants.

The soil-root transfer of HMs is regulated by a variety of factors including the ionic state of metals, soil particle size, soil pH, soil microbiota composition, organic matter content, and cation exchange capacity of plant roots [16]. Mostly, HMs are uptaken by plants at low pH levels of soil (4.5-5). Also, the diverse microbial composition of soil lessens the movement of HMs to plant systems as microbes adsorb their large amounts within themselves [17]. The foliar uptake of HMs depends on a variety of aspects *i.e.*, leaf size, surface area, cuticular and epidermal composition, physical form, trichome length and number, speciation of metal ion, plant habitat, duration of HM exposure, stomatal density and rate of gaseous exchange [18]. Plants with smaller leaves of rough surfaces are found to be more potent accumulators of atmospheric HMs than those with larger and smooth-surfaced leaves [19]. Mostly small-sized and lipophilic HMs enter through cuticular penetration and large-sized and hydrophilic ones make their entry *via* stomatal pores or other aqueous pores in the cuticle [20].

The permeability of the cuticle and the ionic state of HMs are held accountable for the rate of foliar adsorption of HMs by the plants. Usually, metals forming ionic compounds are easily absorbed as compared to those forming covalent precipitates [21]. Plant age is another factor that affects the uptake of HMs *via* leaves, young plants tend to adsorb more HM particles in comparison to the old and mature plants [22, 23]. A cuticle is an external coating consisting of wax embedded within it in the form of a polymer of cutin cross-linked through ester bonds. The reason behind this might be the presence of thin cuticles and epicuticular wax in younger leaves. The accumulation of HMs in plants also varies with the difference in the distance of the plants from the source of HMs (Fig. **1**).

The external factors *i.e.*, climatic conditions also exert an impact on the status of HM uptake by plants as they affect the physico-chemical properties of the soil as well as the leaf's surface. Under the conditions of high relative humidity, the rate of HMs uptake increases due to enhanced wetness and decreased hydrophobic potential of the cuticle that results in the infiltration of HM particles into the leaf surface whereas the low levels of relative humidity cause cuticle drying and shrinkage and closure of stomata, which impedes the entry of HMs into plants [24]. Similarly, the duration of rainfall is also considered to directly affect the penetration of HMs into plant systems through roots as well as leaves due to obvious reasons [25]. However, the accumulation of HMs above their permissible limits within the plant parts under the control of any of the above-discussed may prove toxic to the plant as well as human health, which in turn leads to imbalances in ecosystem stability.

Assessment of HM Bioaccumulation Tendency in Plants

Plants' survival, growth, and reproduction depend upon the physical and chemical characteristics and changes in the soil. To survive in changing environmental conditions, plants have to modify themselves otherwise they will die out. The area that stresses the plant is the pressure of increased levels of HMs. HMs like Iron (Fe), Manganese (Mn), Copper (Cu), Nickel (Ni), Cobalt (Co), Cadmium (Cd), Zinc (Zn), Mercury (Hg), and Arsenic (As) are concentrated in soils for long by anthropogenic activities such as metal mining, casting, smelting and other metal-related industries, leaching of waste from the sources like waste site, dumping, sewage sludge, and automobile runoffs [26] (Fig. **2**). The main contact site is the plant root for heavy metals but can be absorbed directly into the plant leaves due to their deposition on its surfaces [27]. Increased levels of metals can lead to changes in physiological and biochemical processes in plants like chlorosis, inhibition of growth, root browning, and ultimately death of plants [28, 29].

Interestingly, some plants act as bioaccumulators. They can accumulate metals from the water and soil which may or may not be necessary for growth and survival. The accumulation rate and tolerance of plants to heavy metals are different in different species, with some becoming toxic at a higher rate [30]. Plants have been categorized into three main categories based on their growth on soil contaminated with HMs; excluders, metal indicators, and metal accumulators or hyperaccumulators. Excluder plant species contain a high amount of metal in the root area. There is a very small amount of metal translocation in them maintaining effectively low concentration in the shoot over a large range of soil levels [31]. Metal indicator plant species can extract and concentrate HMs in the above-ground tissues. These metal levels in the tissue generally reflect soil metal levels. These plants offer biological and ecological functions in that they are potential indicators of pollution and are useful in pollution absorption [32]. The plant species growing in metal-containing highly toxic environments are regarded as hyper tolerance plants. Accumulator (Hyperaccululator) plants can accumulate HMs in the above-ground tissues and shoot to levels far exceeding those present in soil or the non-accumulating plants growing close to them. They are extensively used in the phytoremediation technology. Studies show that concentrated HMs in their tissues play physiological and ecological functions like protection against bacterial and fungal diseases. Hyperaccumulator plants have special characteristic features like the presence of aerial tissue, which can accumulate some metal at levels higher than normal. They have a variety of ways to minimize the stress of metal toxicity like changes in the expression of the gene, particularly in membrane transporter, responsible for absorption, translocation, and sequestrating of nutrients and minerals. This can be grouped into four categories:

(1) transfer from root to shoot

(2) transportation of heavy metal to root

(3) chelating of metal

(4) remediation in vacuole.

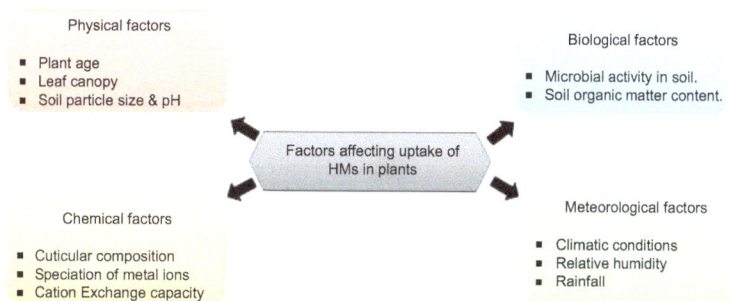

Fig. (1). Factors affecting uptake of heavy metals in plants.

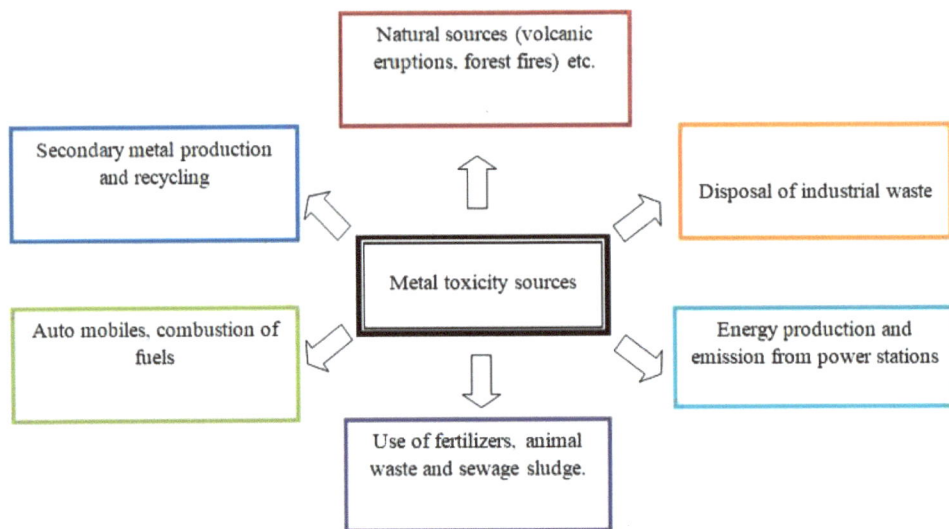

Fig. (2). Heavy metal accumulation sources in soil and water.

Throughout the world, it has been reported that more than 45 families act as hyperaccumulators in both tropical and temperate zones. Different hyperaccumulating plants have been listed in Table **1**. They can extract and concentrate Chromium, Nickel, or lead >1 mg/g (0.1%), Selenium and Titanium > 0.1 mg/g (0.01%), Zinc (0.3%) >10 mg/g (1%), Cobalt or Copper >0.3 mg/g (0.05%) in shoot and leaves [33]. Plants hyperaccumulate HMs by their redistribution and reallocation in fruit and seeds with the help of phloem transport. The transfer of HMs can also be performed by the xylem to the phloem [34].

Table 1. Metal accumulation from contamination sites by the hyperaccumulator plants.

Plants	HMs	Reference
Alyssum beetolonii	Ni	[35]
Alyssum murale	Ni	[36]
Arabidopsis halleri	Cd, Zn	[37]
Axonopus compressus	Hg	[38]
Brassica napus, Glycine max	Zn, Cd, and Pb	[39]
Eichhorina crassipes	Fe, Cd, B, and Cr	[40]
Euphorbia cheiradenia	Cu, Fe, Pb, Zn	[41]
Helianthus annuus	Cd, Pb, Zn	[42]
Helianthus annuus	Pb	[43]

(Table 1) cont.....

Plants	HMs	Reference
Minuartia verna	Zn, Cd, Pb	[44]
Phragmites	Zn, Cu, Mg	[45]
Pteris vittata	As	[46]
Pteris vittata	Hg	[47]
Sedum alfredii	Pb	[48]
Spartina maritime	As, Cu, Pb, Zn	[49]
Viola boashanensis	Pb, Zn, Cd	[50]

Insights into HM Detoxification Mechanisms in Plants

Efflux Mechanisms

One of the most important environmental issues on a global basis is heavy metal pollution of agricultural soils [51]. Contaminant metals can frequently accumulate in large proportions in plant tissues that are harmful to humans and animals [52]. A strategy is needed to lessen heavy metal buildup in plants that may involve the efflux of HMs from the plant system. Several studies showed that efflux transporters have a function in metal detoxification by excreting heavy metal ions outside of cells [53]. As(III) efflux has been demonstrated to be mediated by the aquaporin *Lsi1* [*Oryza sativa* nodulin 26-like intrinsic protein 2;1 (OsNIP2:1)], which is involved in As(III) uptake. *O. Sativa's* nodulin 26-like intrinsic protein 2;1, 3;2 and other plant aquaporins like *Arabidopsis thaliana's* nodulin 26-like intrinsic protein 3;1, 5;1, 6;1, and 7;1 as well as *Pteris vitata* aquaporin gene 4;1 (*PvTIP4*;1) are also involved in the bidirectional transport of As(III) [54 - 56]. Yeast's As(III) efflux permeases (*ACR3*) are responsible for releasing As(III) into the environment [57]. ACR3 proteins are also present in the As hyperaccumulator plant *Pteris vitata* but are absent from rice and other flowering plants. HMAs, or P1B-type heavy metal ATPases are transmembrane proteins that transport metal and are crucial for maintaining metal homeostasis. Zn, Pb, and Cu efflux have been linked to *O. sativa* heavy metal ATPase 9 (*OsHMA9*), a member of the P1B-type ATPase family [62]. By infusing hydrogen ion (H^+), cucumber gene *CsMTP9*, a member of the plant metal tolerance protein (MTP) family, effuses Mn^{2+} and Cd^{2+} from the endodermis into the vascular tissue [58].

Osmoregulation in Heavy Metal Stress

The term osmolytes refers to various low molecular weight compounds or metabolites namely, sugars, polyamines, secondary metabolites, amino acids, and polyol [59, 60]. These molecules are also known as cytoprotectants due to their ability to protect cell contents against abiotic stresses such as cold, heat, UV, submergence, wounding, extremes of temperature, drought, salinity, high metal concentrations, water logging, nutrient deficiency stress, and biotic stresses such as phytopathogenic viruses, bacteria, fungi, algae, nematodes, and insects [61 - 63]. To cope with different stresses and protect themselves, plants have evolved into complex and well-organized mechanisms, where the SOS pathway, ABA signaling, calcium signaling, and the MAP kinase network respond rapidly and synergistically in various abiotic stresses for establishing osmotic homeostasis by regulating the production and accumulation of osmolyte [64, 65]. The genes involved in the SOS pathway are referred to as SOS genes. The maintenance of ion homeostasis, which may also involve the manufacture of suitable solutes, is achieved through the overexpression of SOS genes in plants, increasing their capacity to withstand salt. Proline has been reported to accumulate in tissues/organs of plants subjected to drought, salt or temperature stress, heavy metal stress, and/or infected by some pathogens. Proline is known to be an essential osmoticum found in cells that are underwater, salt, and heavy metal stress. In recent years, proline's function as a ROS scavenger during stressful conditions has also come to light. Pyroline-5-carboxylate reductase (P5CR) and pyrroline-5-carboxylate synthase (P5CS), which are both enzymes, catalyze two successive reductions of glutamate to produce proline. Another potential source of proline is ornithine (Orn), which can be converted into proline by the enzyme orn-d-aminotransferase (OAT), which is found in the mitochondrion [66].

Cellular Homeostasis

Proline can be accumulated in the cytosol under numerous biotic and abiotic stresses. It is observed that the exogenous application of proline may enhance the endogenous proline level under heavy metal stress conditions, which helps maintain the potential of intracellular redox homeostasis, which protects enzymes, the 3-D structure of proteins and vital organelles including the cell membranes and also reducing the risk of peroxidation of lipids and proteins [67 - 70] (Table 2).

Table 2. Implications of Proline in inculcating heavy metal stress protection.

Plant Name	Heavy metal	Action
Sunflower	Arsenic	Protection by scavenging hydroxyl radical and reducing metal uptake of As that increased endogenous proline [71].
Rice	Mercury	Proline is implicated in the Hg2+ tolerance and has a reduced water potential [72].
Indian senna	Lead	Implicated in systems for stress resistance [73].

(Table 2) cont.....

Plant Name	Heavy metal	Action
Microalgae, Tobacco, Chickpea, Brinjal	Cadmium	Protects membranes, enhances growth, reduces oxidative stress through antioxidant mechanism, provides molecular mechanisms of proline-mediated tolerance [67, 69, 74, 75].
Bean	Selenium	Redox balance, a signalling molecule, minimises oxidative stress to lessen the effects of phytotoxicity [76].
Wheat	Nickel	Plays an osmoprotective role and protects plants from Ni stress [77].

Sequestration and Chelation

Sequestration is the suppression of a metal's characteristic or response without the metal being eliminated from the system or phase. The word "chelation" refers to a particular type of chemical bonding.

Plant roots secrete exudates into the soil matrix as their first line of defence against HMs. One of the major functions of root exudates is to chelate metals and to prevent their uptake inside the cells. By binding to HMs and facilitating their transport into the vacuoles, specific low molecular weight chelators used in this approach include GSH, phytochelatins (PCs), and metallothionein (MT) [78, 79].

Glutathione (GSH)

Many cellular detoxification pathways of xenobiotics and HMs depend critically on GSH. GSH does this by first activating and conjugating with these substances [80]. The enzyme Glutathione-S-transferase controls the conjugation of GSH with these compounds. The conjugates are then sent to the vacuole, where they shield the plant cell from their negative effects [81]. One protective role of GSH in plants during heavy metal stress exposure is the quenching of ROS.

Phytochelations (PCs)

PCs have the general structure (-Glu-Cys) n-Gly and are derived from GSH (GSH) (where n=2 to 11)14. When the PC synthase enzyme was activated in the presence of HMs, PCs were generated by trans-peptiding -glutamyl-cysteinyl dipeptides from GSH [82]. PCs' cysteine component coordinates metals, and their glutamic acid and thiol groups make them water-soluble [83]. Among the various HMs, Cd is the strongest inducer of PCs. Not all of the metals that start the synthesis of PCs can form complexes with it. In general, PCs are linked to heavy metal detoxification, essential metal trafficking, and equilibrium. Once the PC-metal complex had been created, the hazardous metals were isolated from metal-sensitive enzymes by metal/H+ antiporters or ABC transporters before being transferred to vacuoles [84]. As inorganic sulphide and sulfite ions are incorporated into vacuoles, PC metal complexes become more resistant to proteolytic breakdown [83]. When conditions are right, metals from PC-metal sulfide complexes are released, degrading PC.

Metallothioneins

Metallothioneins are low molecular weight polypeptides with Cys-rich residues and bind to metals using their thiol functional group. Based on how the Cys residues are arranged, this diverse set of molecules has been categorized into three classes: There are three types of Cys-Cys motifs:

(1) Cys-Cys

(2) Cys-X-Cys

(3) Cys-X-X-Cys

There are numerous isoforms of metallothionein that can form compounds with different metal ions. The exact function of these metal-binding ligands is still unknown; however, their probable functions of metallothionein are thought to take part in maintaining homeostasis of essential ions transported through the plant. In addition, they are hypothesized to have a role in the sequestration of HMs and defense against reactive oxygen species-induced intracellular damage. The synthesis of metallothionein is induced during heavy metal toxicity and is controlled at the transcriptional level [85].

Chelation of HMs with Some Organic Compounds

Organic acid ligands are linked to enhanced mental absorption, translocation, resistance, and high accumulation in accumulator plants [86]. A plant leaf vacuole could contain ligands like citrate or malate with the highest metal concentration. The ability of organic acid inverters to control how much metal they can store. The largest malate concentration in the *T. caerulescens* was followed by succinate, oxalate, and citrate. Both oxalate and malate were associated with the plant shoot soluble Zn [87]. While Zn exists as Zn phosphate in the non-hyperaccumulating, non-tolerant *Arabidopsis lyrata,* it forms compounds with malate in the aerial organs of *Alternanthera helleri* [88]. In the vacuoles of *T. goingense*, Ni was primarily discovered as an organic compound, likely with citrate [89]. Cu is significantly bonded to oxalate in the relatively Cu-tolerant *N. caerulescens*, compared to Cu-sensitive people [90]. The *Sedum alfredii* hyperaccumulator eco-type showed that organic acids were the cause of the increased mobility of Cd.

Reactive Oxygen Species (ROS), and Heavy Metal Tolerance

Stress brought on by HMs in plants triggers the immediate or abnormal production of ROS (*e.g.* superoxide radicals, hydrogen peroxide along with hydroxyl radicals). ROS build-up results in oxidative damage to proteins, lipid bilayers, and nucleic acids. The detoxification of ROS, which is facilitated by antioxidant chemicals and antioxidant enzyme systems, is one of the HM resilience systems [91]. Antioxidants form sophisticated intracellular as well as extracellular networks ensuring protection against oxidation and thus shape stress signalling [92]. Ascorbate, GSH, carotenoids, xanthophylls, tocopherol, and other non-enzymatic foraging mechanisms are just a few examples of the non-enzymatic foraging mechanisms used by plants to combat oxidative stress (Fig. **3**) [93].

Antioxidants

Heavy metal toxicity has become a major consideration in today's world because of increased environmental pollution. HMs are non-biodegradable and bio-accumulative, which frequently leads to deleterious biological effects. Plants require certain HMs for their growth and development but their excessive amounts can become toxic to plants by triggering the reactive oxygen species (ROS) generation such as $(O_2^{.-})$, $(OH^.)$, (H_2O_2), *etc.* that causes the oxidative stress that leads to various disorders like protein and lipid oxidation, DNA damage and denaturation of cell structure and membrane that finally results in the programmed cell death (PCD). To combat heavy metal stress, plants have evolved different defence modifications *viz.*, less uptake of metals, phytochelatins binding, and the activation of numerous antioxidants. Plants naturally develop a number of defense mechanisms against metal-induced stresses inside the plant body. Plants can reduce ROS production and may protect themselves from the negative effects of ROS through an antioxidant defence system [94]. The scavenging system to control ROS comprises enzymatic (SOD, POD, APX, CAT, GR, *etc*) and non-enzymatic (ASA, GSH, *etc*) components, and some of the studies (as shown in Table **3**) indicated an antioxidant defence system in higher plants exposed to metals (Fig. **4**).

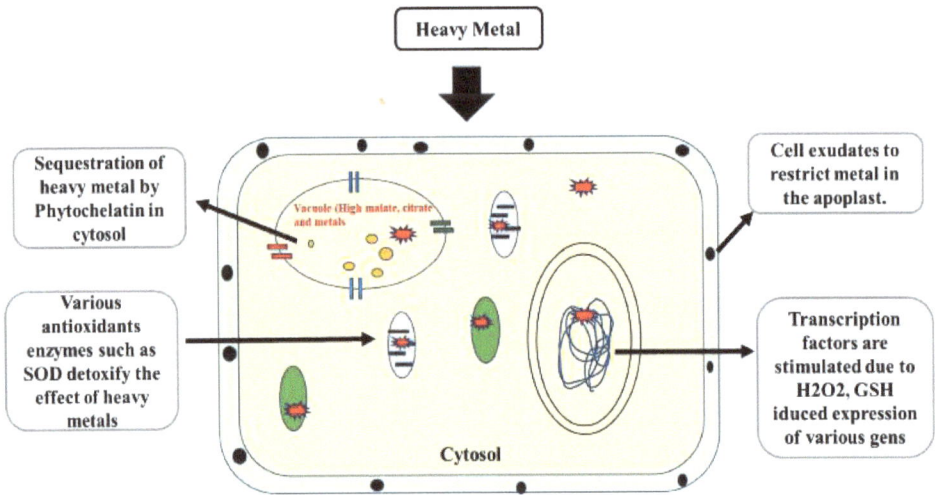

Fig. (3). Various mechanisms involved in heavy metal-induced tolerance.

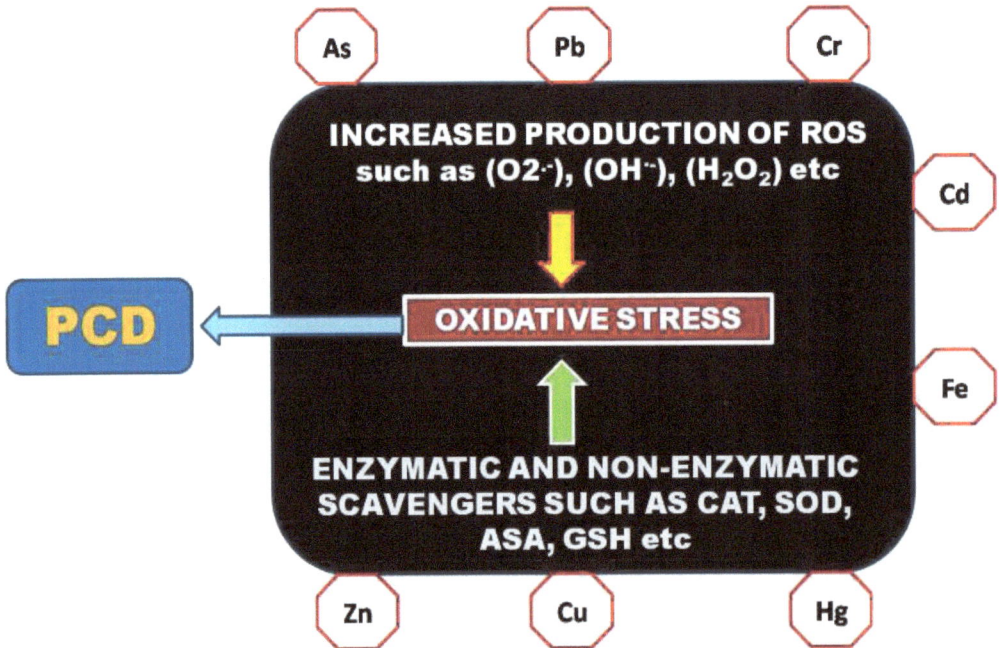

Fig. (4). Role of antioxidants under heavy metal stress in plant cells.

Table 3. Role of antioxidants in alleviating HMs stress in different plants.

Metal	Plant species	Plant Organ	Induced Antioxidants	References
Arsenic	*Brassica napus*	Leaf	CAT, POD, SOD, GST	[95]
	Spinacia oleracea	Leaf	SOD, GST	[96]
	Oryza sativa	Leaf	APX, GST, POD	[97]
	Brassica napus	Leaf	GR, SOD, CAT	[98]
Lead	*Tetraena qataranse*	Seedling	CAT, SOD, APX, GPX, GR,	[99]
	Zea mays	Root	DHAR, MDHAR, APX	[100]
	Triticum aestivum	Seedlings	SOD, CAT, POD, DHAR, MDHAR, GR	[101]
	Coronopus didymus	Seedlings	SOD, CAT, APX, GPX, GR	[102]
Chromium	*Lemna minor L.*	Seedlings	SOD, CAT, APX, POD	[103]
	Triticum aestivum L.	Seedlings	CAT, APX	[104]
	Plantago ovata	Shoot	DPPH assay, Ascorbic Acid	[105]
Cadmium	*Sorghum cultivars*	Shoot	SOD, POD, and CAT	[106]
	Vigna aconitifolia	Seedlings	Ascorbic acid, GSH	[107]
	Zea mays	Root	SOD, POD, CAT, APX, GPX	[108]
Iron	*Oryza sativa*	Shoot	SOD, POD, CAT, GR, AsA, GSH	[109]
	Glycine max	Roots, leaves	SOD, POD, and CAT	[110]
	Lycopersicon esculentum	Shoot	CAT, and APOX	[111]
Mercury	*Spirodela polyrhiza*	Seedlings	SOD, CAT, GPOD	[112]
	Raphanus sativus	Seedling	SOD, POD, APOX, MDHAR, DHAR, GR, GST, GPOX	[113]
	Fagopyrum tataricum	Seedlings	SOD, CAT, POD, APOX, GR, GST	[114]
	Triticum aestivum	Seedlings	SOD, CAT, POD, and APOX	[115]
Copper	*Colobanthus quitensis*	Explants	CAT, APX, POD, GSH, Ascorbic acid	[116]
	Citrus aurantium	Leaf	Antioxidants	[117]
	Olea europaea	Leaf	SOD, POD, CAT	[118]
Zinc	*Brassica juncea*	Seedlings	APOX, SOD, GR, CAT, MDHAR, DHAR, GPX, GST	[119]
	Solanum lycopersicon	Seedlings	SOD, APOX, CAT, GPX	[120]
	Coffea Arabica	Seedlings	Ascorbic acid	[121]

(Table 3) cont....

	Ipomoea batatas	Seedlings	CAT, POD, SOD, and APX, GSH	[122]
Nickel	*Solanum lycopersicum*	Leaf	SOD, CAT, APOX, GR	[123]
	Solanum lycopersicum	Seedlings	SOD, CAT, APOX, GR	[124]
	Zea mays	Seedlings	SOD, CAT, APOX, GR	[125]
	Soybean	Leaf, seed	SOD, POD, CAT	[126]

Mental stress is a major abiotic issue that reduces the crop productivity of agricultural crops by increasing environmental pollution. Excessive amounts of these metals may become toxic to plants by causing an oxidative burst. At the cellular level, plants might have developed a potential antioxidant defence mechanism of detoxification and thus give tolerance to heavy metal stress.

Role of miRNA

HMs Heavy Metals are needed by the plants in a trace amount to carry out various cellular activities, however, at higher concentrations, they may induce a toxic effect on plant growth and development. Therefore, to maintain this homeostasis, plants respond to heavy metal stress in an extremely coordinated way at all levels (physiological, biochemical, and molecular) through regulation transcriptionally and post-transcriptionally. It is a prerequisite to identify and understand the key players involved in the HM responses along with the regulatory network [127]. Several studies have reported the role of small non-protein-coding RNAs in the stress signaling cascade and among them, miRNAs are known for post-transcriptional gene regulation during stressful conditions [128].

miRNAs are an evolutionary conserved short set of 20-24 nucleotide non-coding protein RNAs [129]. They sit at the heart of regulatory pathways and trigger gene silencing by either cleavage or repression of target mRNAs [130]. miRNAs are processed from the hairpin precursors structure which is further used to silence target mRNAs *via* RNA-induced silencing complex (RISC) [131]. There is a negative correlation between the miRNA and the target mRNAs. Owing to their regulation of target genes, miRNAs are involved in all developmental and physiological processes and have been used for developing stress-resilient crops [132]. Several studies were conducted on plants like *Brassica juncea, Brassica napus, Medicago truncatula, Oryza sativa* and *Phaseolus vulgaris,* which have identified a few miRNAs families (miR$_{159}$, miR$_{156}$, miR$_{162}$, miR$_{166}$, miR$_{171}$, miR$_{390}$, miR$_{393}$, miR$_{395}$, and miR$_{396}$,) responsive to HM accumulation [133]. miRNAs usually target transcriptional factors or phytohormones, for instance, miR$_{171}$—Scarecrow-Like (SCL) TF, miR$_{396}$-Growth-regulating TFs, and Osa-miR$_{604}$ under Cd stress [134]. miRNAs have a key role in uptake, translocation, chelation, homeostasis, and oxidative and antioxidant responses during HM stress.

On exposure to individual metal stress, miRNAs are known to have altered gene expression. For example, during Al and Mn stress, ARF transcription factors targeted by miR_{172} have altered expression [135]. Similarly, miR_{171} has altered expression in Hg stress while miR_{319} expression is altered in Cd toxicity [136]. A lot has been studied in the field of miRNAs during individual stress, but it will be interesting to understand the expression pattern of these miRNAs in combined HM stress. A scenario of combined stress where they are up-regulated by one stress and down-regulated by another has also been observed. Such a pattern was observed in miR156, which is up-regulated by Cd and Hg while down-regulated by As and Al. Studying this combined stress will help us to understand gene regulation in the natural environment.

mi-RNAs are emerging as the key regulators of stress tolerance during HM accumulation, and hence, mi-RNAs-based genetic modification can be used as a strategy for resilient crops (Fig. **5**). Several studies have already been done in this field where the transgenic plant overexpressing miR166 in rice shows Cd tolerance [137]. Similarly, miR156 overexpression lines show improved antioxidant enzyme activity hinting and conferring tolerance to Cd-mediated oxidative stress.

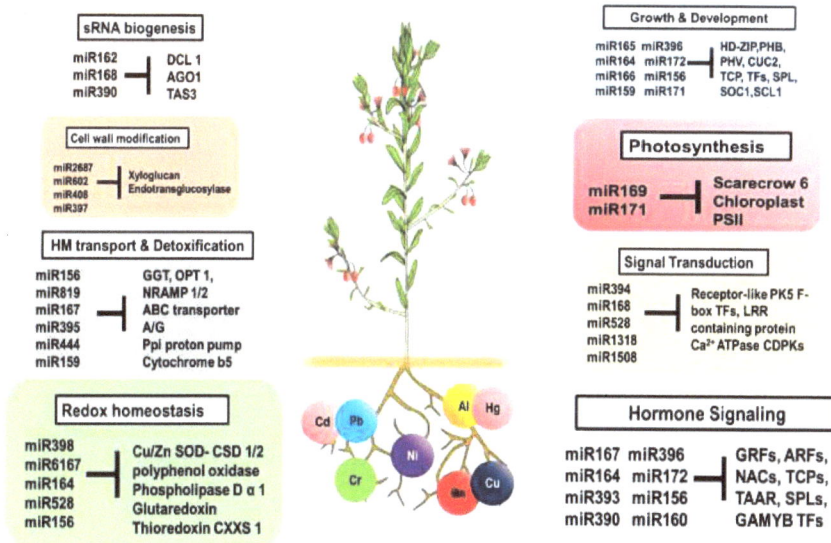

Fig. (5). Schematic illustration of expressed miRNA and their target genes involved in various metabolic processes during heavy metal stress.

CONCLUSION

Plants have always been prone to environmental stress conditions specifically because of their sessile nature. HMs toxicity is one such stress that has deleterious impact on the overall health of plants. HMs from different sources like soil and water enter into the plant body. However, there are several factors that affect their uptake into the plant system thereby impacting their bioaccumulation potential. There are wide array of biochemical responses that are governed by the plant body to curb the deleterious impacts produced by HMs thereby making them effective candidates for phytoremediation, which can be employed for the phytoextraction of HMs from the contaminated soils as the biochemical machinery of plants help them to shield themselves against HMs. However, the use of several biostimulants can be enhanced in order to enhance the protective potential of plants against HM stress.

REFERENCES

[1]   Ali H, Khan E, Ilahi I. Environmental chemistry and ecotoxicology of hazardous heavy metals: environmental persistence, toxicity, and bioaccumulation. J Chem 2019; 2019: 1-14.
[http://dx.doi.org/10.1155/2019/6730305]

[2]   Maximillian J, Brusseau ML, Glenn EP, Matthias AA. Pollution and environmental perturbations in the global system. Environ Pollut Sci 2019; 457-76.
[http://dx.doi.org/10.1016/B978-0-12-814719-1.00025-2]

[3]   Afonne OJ, Ifediba EC. Heavy metals risks in plant foods – need to step up precautionary measures. Curr Opin Toxicol 2020; 22: 1-6.
[http://dx.doi.org/10.1016/j.cotox.2019.12.006]

[4]   Zwolak A, Sarzyńska M, Szpyrka E, Stawarczyk K. Sources of soil pollution by heavy metals and their accumulation in vegetables: A review. Water Air Soil Pollut 2019; 230(7): 164.
[http://dx.doi.org/10.1007/s11270-019-4221-y]

[5]   Zhang Q, Chen H, Xu C, Zhu H, Zhu Q. Heavy metal uptake in rice is regulated by pH-dependent iron plaque formation and the expression of the metal transporter genes. Environ Exp Bot 2019; 162: 392-8.
[http://dx.doi.org/10.1016/j.envexpbot.2019.03.004]

[6]   Awa SH, Hadibarata T. Removal of heavy metals in contaminated soil by phytoremediation mechanism: a review. Water Air Soil Pollut 2020; 231(2): 47.
[http://dx.doi.org/10.1007/s11270-020-4426-0]

[7]   Kaur G, Dogra N, Kaushik S, Madaan I, Sidhu A, Sirhindi G. Biochemical responses of plants towards heavy metals in soil. Hazard Trace Mater Soil Plants 2022; 179-95.
[http://dx.doi.org/10.1016/B978-0-323-91632-5.00026-4]

[8]   Yaashikaa PR, Kumar PS, Jeevanantham S, Saravanan R. A review on bioremediation approach for heavy metal detoxification and accumulation in plants. Environ Pollut 2022; 301: 119035.
[http://dx.doi.org/10.1016/j.envpol.2022.119035] [PMID: 35196562]

[9]   Goyal D, Yadav A, Prasad M, *et al.* Effect of heavy metals on plant growth: An overview. Contam Agric 2020; 79-101.
[http://dx.doi.org/10.1007/978-3-030-41552-5_4]

[10]   Xin J, Ma S, Zhao C, Li Y, Tian R. Cadmium phytotoxicity, related physiological changes in Pontederia cordata: antioxidative, osmoregulatory substances, phytochelatins, photosynthesis, and

chlorophyll fluorescence. Environ Sci Pollut Res Int 2020; 27(33): 41596-608.
[http://dx.doi.org/10.1007/s11356-020-10002-z] [PMID: 32691317]

[11] Shakeri E, Mozafari AA, Sohrabi F, Saed-Moucheshi A. Role of proline and other osmoregulatory compounds in plant responses to abiotic stresses. Handb Plant Crop Stress 2019; 165-73.
[http://dx.doi.org/10.1201/9781351104609-9]

[12] Ahmad J, Ali AA, Baig MA, Iqbal M, Haq I, Qureshi MI. Role of phytochelatins in cadmium stress tolerance in plants. Cadmium Toxicol Toler 2019; 185-212.
[http://dx.doi.org/10.1016/B978-0-12-814864-8.00008-5]

[13] Das U, Rahman MA, Ela EJ, Lee KW, Kabir AH. Sulfur triggers glutathione and phytochelatin accumulation causing excess Cd bound to the cell wall of roots in alleviating Cd-toxicity in alfalfa. Chemosphere 2021; 262: 128361.
[http://dx.doi.org/10.1016/j.chemosphere.2020.128361] [PMID: 33182109]

[14] Raychaudhuri SS, Pramanick P, Talukder P, Basak A. Polyamines, metallothioneins, and phytochelatins—natural defense of plants to mitigate heavy metals. Stud Nat Prod Chem 2021; 69: 227-61.

[15] AbdElgawad H, Zinta G, Hamed BA, *et al.* Maize roots and shoots show distinct profiles of oxidative stress and antioxidant defense under heavy metal toxicity. Environ Pollut 2020; 258: 113705.
[http://dx.doi.org/10.1016/j.envpol.2019.113705]

[16] Shahid M, Dumat C, Pourrut B, Abbas G, Shahid N, Pinelli E. Role of metal speciation in lead-induced oxidative stress to Vicia faba roots. Russ J Plant Physiol 2015; 62(4): 448-54.
[http://dx.doi.org/10.1134/S1021443715040159]

[17] Rai PK, Chutia BM. Biomonitoring of atmospheric particulate matter (PM) using magnetic properties of Ficus bengalensis tree leaves in Aizawl, Mizoram, North-East India. Int J Environ Sci 2015; 5(4): 856-69.

[18] Kumar R, Sharma S, Kaundal M, Sharma S, Thakur M. Response of damask rose (Rosa damascena Mill.) to foliar application of magnesium (Mg), copper (Cu) and zinc (Zn) sulphate under western Himalayas. Ind Crops Prod 2016; 83: 596-602.
[http://dx.doi.org/10.1016/j.indcrop.2015.12.065]

[19] Pourrut B, Shahid M, Douay F, Dumat C, Pinelli E. Molecular mechanisms involved in lead uptake, toxicity and detoxification in higher plants. Heavy Metal Stress Plants 2013; 121-47.
[http://dx.doi.org/10.1007/978-3-642-38469-1_7]

[20] Uzu G, Sauvain JJ, Baeza-Squiban A, *et al. in vitro* assessment of the pulmonary toxicity and gastric availability of lead-rich particles from a lead recycling plant. Environ Sci Technol 2011; 45(18): 7888-95.
[http://dx.doi.org/10.1021/es200374c] [PMID: 21800914]

[21] Ahmad I, Akhtar MJ, Asghar HN, Ghafoor U, Shahid M. Differential effects of plant growth-promoting rhizobacteria on maize growth and cadmium uptake. J Plant Growth Regul 2016; 35(2): 303-15.
[http://dx.doi.org/10.1007/s00344-015-9534-5]

[22] Bondada BR, Tu S, Ma LQ. Absorption of foliar-applied arsenic by the arsenic hyperaccumulating fern (Pteris vittata L.). Sci Total Environ 2004; 332(1-3): 61-70.
[http://dx.doi.org/10.1016/j.scitotenv.2004.05.001] [PMID: 15336891]

[23] Schreck E, Laplanche C, Le Guédard M, *et al.* Influence of fine process particles enriched with metals and metalloids on Lactuca sativa L. leaf fatty acid composition following air and/or soil-plant field exposure. Environ Pollut 2013; 179: 242-9.
[http://dx.doi.org/10.1016/j.envpol.2013.04.024] [PMID: 23694728]

[24] Luo C, Liu C, Wang Y, *et al.* Heavy metal contamination in soils and vegetables near an e-waste processing site, south China. J Hazard Mater 2011; 186(1): 481-90.

[http://dx.doi.org/10.1016/j.jhazmat.2010.11.024] [PMID: 21144651]

[25]   Briffa J, Sinagra E, Blundell R. Heavy metal pollution in the environment and their toxicological effects on humans. Heliyon 2020; 6(9): e04691.
[http://dx.doi.org/10.1016/j.heliyon.2020.e04691] [PMID: 32964150]

[26]   Mganga N, Manoko ML, Rulangaranga ZK. Classification of plants according to their heavy metal content around North Mara gold mine, Tanzania: implication for phytoremediation. Tanzan J Sci 2011; 37.

[27]   Ozturk M, Yucel E, Gucel S, Sakçali S, Aksoy A. 28 Plants as Biomonitors of Trace Elements Pollution in Soil.

[28]   Öztürk M, Ashraf M, Aksoy A, Ahmad MS, Hakeem KR, Eds. Plants, pollutants and remediation. Berlin/Heidelberg, Germany: Springer 2015.
[http://dx.doi.org/10.1007/978-94-017-7194-8]

[29]   Ozturk M, Altay V, Karahan F. Studies on trace elements distributed in taxa in Hatay-Turkey Glycyrrhiza. Stud 2017; 3(2): 1-7.
[http://dx.doi.org/10.18811/ijpen.v3i02.10431]

[30]   Sun Z, Chen J, Wang X, Lv C. Heavy metal accumulation in native plants at a metallurgy waste site in rural areas of Northern China. Ecol Eng 2016; 86: 60-8.
[http://dx.doi.org/10.1016/j.ecoleng.2015.10.023]

[31]   Kvesitadze G, Khatisashvili G, Sadunishvili T, Ramsden JJ. Biochemical mechanisms of detoxification in higher plants: basis of phytoremediation. Springer Sci Bus Media 2006.

[32]   Reeves RD, Baker AJM, Jaffré T, Erskine PD, Echevarria G, van der Ent A. A global database for plants that hyperaccumulate metal and metalloid trace elements. New Phytol 2018; 218(2): 407-11.
[http://dx.doi.org/10.1111/nph.14907] [PMID: 29139134]

[33]   Turgeon R, Wolf S. Phloem transport: cellular pathways and molecular trafficking. Annu Rev Plant Biol 2009; 60(1): 207-21.
[http://dx.doi.org/10.1146/annurev.arplant.043008.092045] [PMID: 19025382]

[34]   Mengoni A, Cecchi L, Gonnelli C. Nickel hyperaccumulating plants and Alyssum bertolonii: model systems for studying biogeochemical interactions in serpentine soils. Bio-Geo Interact Metal-Contam Soils 2011; 279-96.

[35]   Broadhurst CL, Chaney RL. Growth and metal accumulation of an Alyssum murale nickel hyperaccumulator ecotype co-cropped with Alyssum montanum and perennial ryegrass in serpentine soil. Front Plant Sci 2016; 7: 451.
[http://dx.doi.org/10.3389/fpls.2016.00451] [PMID: 27092164]

[36]   Zhang Z, Wen X, Huang Y, Inoue C, Liang Y. Higher accumulation capacity of cadmium than zinc by Arabidopsis halleri ssp. germmifera in the field using different sowing strategies. Plant Soil 2017; 418(1-2): 165-76.
[http://dx.doi.org/10.1007/s11104-017-3285-y]

[37]   Liu Z, Chen B, Wang L, *et al.* A review on phytoremediation of mercury contaminated soils. J Hazard Mater 2020; 400: 123138.
[http://dx.doi.org/10.1016/j.jhazmat.2020.123138] [PMID: 32947735]

[38]   Delil AD, Köleli N, Dağhan H, Bahçeci G. Recovery of heavy metals from canola (Brassica napus) and soybean (Glycine max) biomasses using electrochemical process. Environ Technol Innov 2020; 17: 100559.
[http://dx.doi.org/10.1016/j.eti.2019.100559]

[39]   Elias SH, Mohamed MA, Ankur AN, *et al.* Water hyacinth bioremediation for ceramic industry wastewater treatment-application of rhizofiltration system. Sains Malays 2014; 43(9): 1397-403.

[40]   Nematian MA, Kazemeini F. Accumulation of Pb, Zn, Cu and Fe in plants and hyperaccumulator

choice in Galali iron mine area, Iran. Int J Agric Crop Sci 2013; 5(4): 426-32. [IJACS].

[41] Angelova VR, Perifanova-Nemska MN, Uzunova GP, Ivanov KI, Lee HQ. Potential of sunflower (Helianthus annuus L.) for phytoremediation of soils contaminated with heavy metals. World J Sci Eng Technol 2016; 10(9): 1-8.

[42] Chauhan P, Rajguru AB, Dudhe MY, Mathur J. Efficacy of lead (Pb) phytoextraction of five varieties of Helianthus annuus L. from contaminated soil. Environ Technol Innov 2020; 18: 100718.
[http://dx.doi.org/10.1016/j.eti.2020.100718]

[43] Bothe H. Plants in heavy metal soils. Detox Heavy Metals 2011; 35-57.
[http://dx.doi.org/10.1007/978-3-642-21408-0_2]

[44] Sharma P, Tripathi S, Chaturvedi P, Chaurasia D, Chandra R. Newly isolated Bacillus sp. PS-6 assisted phytoremediation of heavy metals using Phragmites communis: Potential application in wastewater treatment. Bioresour Technol 2021; 320(Pt B): 124353.
[http://dx.doi.org/10.1016/j.biortech.2020.124353] [PMID: 33202343]

[45] Nalla S, Hardaway CJ, Sneddon J. Phytoextraction of selected metals by the first and second growth seasons of Spartina alterniflora. Instrum Sci Technol 2012; 40(1): 17-28.
[http://dx.doi.org/10.1080/10739149.2011.633143]

[46] Su Y, Han FX, Chen J, Sridhar BBM, Monts DL. Phytoextraction and accumulation of mercury in three plant species: Indian mustard (Brassica juncea), beard grass (Polypogon monospeliensis), and Chinese brake fern (Pteris vittata). Int J Phytoremediation 2008; 10(6): 547-60.
[http://dx.doi.org/10.1080/15226510802115091] [PMID: 19260232]

[47] Chen L, Wang T, Zhao M, Tian Q, Zhang WH. Identification of aluminum-responsive microRNAs in Medicago truncatula by genome-wide high-throughput sequencing. Planta 2012; 235(2): 375-86.
[http://dx.doi.org/10.1007/s00425-011-1514-9] [PMID: 21909758]

[48] Mesa J, Rodríguez-Llorente ID, Pajuelo E, *et al.* Moving closer towards restoration of contaminated estuaries: Bioaugmentation with autochthonous rhizobacteria improves metal rhizoaccumulation in native Spartina maritima. J Hazard Mater 2015; 300: 263-71.
[http://dx.doi.org/10.1016/j.jhazmat.2015.07.006] [PMID: 26188869]

[49] Zhuang P, Ye ZH, Lan CY, Xie ZW, Shu WS. Chemically assisted phytoextraction of heavy metal contaminated soils using three plant species. Plant Soil 2005; 276(1-2): 153-62.
[http://dx.doi.org/10.1007/s11104-005-3901-0]

[50] Wuana RA, Okieimen FE. Heavy metals in contaminated soils: a review of sources, chemistry, risks and best available strategies for remediation. Int Scholarly Res Not 2011.
[http://dx.doi.org/10.5402/2011/402647]

[51] D'Amore JJ, Al-Abed SR, Scheckel KG, Ryan JA. Methods for speciation of metals in soils: a review. J Environ Qual 2005; 34(5): 1707-45.
[http://dx.doi.org/10.2134/jeq2004.0014] [PMID: 16151225]

[52] Singh AP, Dixit G, Mishra S, *et al.* Salicylic acid modulates arsenic toxicity by reducing its root to shoot translocation in rice (*Oryza sativa* L.). Front Plant Sci 2015; 6: 340.
[http://dx.doi.org/10.3389/fpls.2015.00340] [PMID: 26042132]

[53] Bienert GP, Thorsen M, Schüssler MD, *et al.* A subgroup of plant aquaporins facilitate the bi-directional diffusion of As(OH)3 and Sb(OH)3across membranes. BMC Biol 2008; 6(1): 26.
[http://dx.doi.org/10.1186/1741-7007-6-26] [PMID: 18544156]

[54] Xu W, Dai W, Yan H, *et al.* Arabidopsis NIP3; 1 plays an important role in arsenic uptake and root-t--shoot translocation under arsenite stress conditions. Mol Plant 2015; 8(5): 722-33.
[http://dx.doi.org/10.1016/j.molp.2015.01.005] [PMID: 25732589]

[55] He Z, Yan H, Chen Y, *et al.* An aquaporin Pv TIP 4;1 from *Pteris vittata* may mediate arsenite uptake. New Phytol 2016; 209(2): 746-61.
[http://dx.doi.org/10.1111/nph.13637] [PMID: 26372374]

[56] Wysocki R, Bobrowicz P, Ułaszewski S. The *Saccharomyces cerevisiae* ACR3 gene encodes a putative membrane protein involved in arsenite transport. J Biol Chem 1997; 272(48): 30061-6.
[http://dx.doi.org/10.1074/jbc.272.48.30061] [PMID: 9374482]

[57] Migocka M, Papierniak A, Kosieradzka A, *et al.* Retracted: Cucumber metal tolerance protein Cs MTP 9 is a plasma membrane H $^+$-coupled antiporter involved in the Mn^{2+} and Cd^{2+} efflux from root cells. Plant J 2015; 84(6): 1045-58.
[http://dx.doi.org/10.1111/tpj.13056] [PMID: 26485215]

[58] Roychoudhury A, Banerjee A. Endogenous glycine betaine accumulation mediates abiotic stress tolerance in plants. Trop Plant Res 2016; 3(1): 105-11.

[59] Hussain TM, Ch T, Hazara M, Sultan Z, Saleh BK, Gopal GR. Recent advances in salt stress biology a review. Biotechnol Mol Biol Rev 2008; 3(1): 8-13.

[60] Yancey PH. Organic osmolytes as compatible, metabolic and counteracting cytoprotectants in high osmolarity and other stresses. J Exp Biol 2005; 208(15): 2819-30.
[http://dx.doi.org/10.1242/jeb.01730] [PMID: 16043587]

[61] Khan SH, Ahmad N, Ahmad F, Kumar R. Naturally occurring organic osmolytes: From cell physiology to disease prevention. IUBMB Life 2010; 62(12): 891-5.
[http://dx.doi.org/10.1002/iub.406] [PMID: 21190292]

[62] Groppa MD, Benavides MP. Polyamines and abiotic stress: recent advances. Amino Acids 2008; 34(1): 35-45.
[http://dx.doi.org/10.1007/s00726-007-0501-8] [PMID: 17356805]

[63] Roychoudhury A, Chakraborty M. Biochemical and molecular basis of varietal difference in plant salt tolerance. Annu Res Rev Biol 2013; 3(4): 422-54.

[64] Roychoudhury A, Banerjee A. Abscisic acid signaling and involvement of mitogen activated protein kinases and calcium-dependent protein kinases during plant abiotic stress. Mech Plant Hormone Signal Stress 2017; 1: 197-241.
[http://dx.doi.org/10.1002/9781118889022.ch9]

[65] Aslam M, Saeed MS, Sattar S, *et al.* Specific role of proline against heavy metals toxicity in plants. International Journal of Pure & Applied Bioscience 2017; 5(6): 27-34.
[http://dx.doi.org/10.18782/2320-7051.6032]

[66] Hayat S, Hayat Q, Alyemeni MN, Ahmad A. Proline enhances antioxidative enzyme activity, photosynthesis and yield of Cicer arietinum L. exposed to cadmium stress. Acta Bot Croat 2013; 72(2): 323-35.
[http://dx.doi.org/10.2478/v10184-012-0019-3]

[67] Hoque MA, Banu MNA, Nakamura Y, Shimoishi Y, Murata Y. Proline and glycinebetaine enhance antioxidant defense and methylglyoxal detoxification systems and reduce NaCl-induced damage in cultured tobacco cells. J Plant Physiol 2008; 165(8): 813-24.
[http://dx.doi.org/10.1016/j.jplph.2007.07.013] [PMID: 17920727]

[68] Islam MM, Hoque MA, Okuma E, *et al.* Exogenous proline and glycinebetaine increase antioxidant enzyme activities and confer tolerance to cadmium stress in cultured tobacco cells. J Plant Physiol 2009; 166(15): 1587-97.
[http://dx.doi.org/10.1016/j.jplph.2009.04.002] [PMID: 19423184]

[69] Okuma E, Murakami Y, Shimoishi Y, Tada M, Murata Y. Effects of exogenous application of proline and betaine on the growth of tobacco cultured cells under saline conditions. Soil Sci Plant Nutr 2004; 50(8): 1301-5.
[http://dx.doi.org/10.1080/00380768.2004.10408608]

[70] Yadav G, Srivastava PK, Singh VP, Prasad SM. Light intensity alters the extent of arsenic toxicity in Helianthus annuus L. seedlings. Biol Trace Elem Res 2014; 158(3): 410-21.
[http://dx.doi.org/10.1007/s12011-014-9950-6] [PMID: 24699829]

[71] Wang F, Zeng B, Sun Z, Zhu C. Relationship between proline and Hg^{2+}-induced oxidative stress in a tolerant rice mutant. Arch Environ Contam Toxicol 2009; 56(4): 723-31.
[http://dx.doi.org/10.1007/s00244-008-9226-2] [PMID: 18787889]

[72] Qureshi MI, Abdin MZ, Qadir S, Iqbal M. Lead-induced oxidative stress and metabolic alterations in Cassia angustifolia Vahl. Biol Plant 2007; 51(1): 121-8.
[http://dx.doi.org/10.1007/s10535-007-0024-x]

[73] Singh S, Prasad SM. Growth, photosynthesis and oxidative responses of Solanum melongena L. seedlings to cadmium stress: Mechanism of toxicity amelioration by kinetin. Sci Hortic (Amsterdam) 2014; 176: 1-10.
[http://dx.doi.org/10.1016/j.scienta.2014.06.022]

[74] Siripornadulsil S, Traina S, Verma DPS, Sayre RT. Molecular mechanisms of proline-mediated tolerance to toxic heavy metals in transgenic microalgae. Plant Cell 2002; 14(11): 2837-47.
[http://dx.doi.org/10.1105/tpc.004853] [PMID: 12417705]

[75] Aggarwal M, Sharma S, Kaur N, *et al.* Exogenous proline application reduces phytotoxic effects of selenium by minimising oxidative stress and improves growth in bean (*Phaseolus vulgaris* L.) seedlings. Biol Trace Elem Res 2011; 140(3): 354-67.
[http://dx.doi.org/10.1007/s12011-010-8699-9] [PMID: 20455031]

[76] Gajewska E, Skłodowska M. Differential biochemical responses of wheat shoots and roots to nickel stress: antioxidative reactions and proline accumulation. Plant Growth Regul 2008; 54(2): 179-88.
[http://dx.doi.org/10.1007/s10725-007-9240-9]

[77] Navarrete A, González A, Gómez M, *et al.* Copper excess detoxification is mediated by a coordinated and complementary induction of glutathione, phytochelatins and metallothioneins in the green seaweed Ulva compressa. Plant Physiol Biochem 2019; 135: 423-31.
[http://dx.doi.org/10.1016/j.plaphy.2018.11.019] [PMID: 30501930]

[78] Chaudhary K, Agarwal S, Khan S. Role of phytochelatins (PCs), metallothioneins (MTs), and heavy metal ATPase (HMA) genes in heavy metal tolerance. Fungal Biol 2018; 2: 39-60.
[http://dx.doi.org/10.1007/978-3-319-77386-5_2]

[79] Chia JC. Phytochelatin synthase in heavy metal detoxification and xenobiotic metabolism. Biodegrad Technol Org Inorg Pollut 2021; 10

[80] Pradedova EV, Nimaeva OD, Karpova AB, *et al.* Glutathione in intact vacuoles: Comparison of glutathione pools in isolated vacuoles, plastids, and mitochondria from roots of red beet. Russ J Plant Physiol 2018; 65(2): 168-76.
[http://dx.doi.org/10.1134/S1021443718020048]

[81] Chen J, Zhou J, Goldsbrough PB. Characterization of phytochelatin synthase from tomato. Physiol Plant 1997; 101(1): 165-72.
[http://dx.doi.org/10.1111/j.1399-3054.1997.tb01833.x]

[82] Bertrand M, Guary JC, Schoefs B. How plants adapt their physiology to an excess of metals. Handb Plant Crop Physiol 2001; 773-84.

[83] Cobbett C. Phytochelatin biosynthesis and function in heavy-metal detoxification. Curr Opin Plant Biol 2000; 3(3): 211-6.
[http://dx.doi.org/10.1016/S1369-5266(00)00066-2] [PMID: 10837262]

[84] Hossain MA, Piyatida P, da Silva JA, Fujita M. Molecular mechanism of heavy metal toxicity and tolerance in plants: central role of glutathione in detoxification of reactive oxygen species and methylglyoxal and in heavy metal chelation. J Bot 2012.
[http://dx.doi.org/10.1155/2012/872875]

[85] Boominathan R, Doran PM. Cadmium tolerance and antioxidative defenses in hairy roots of the cadmium hyperaccumulator, *Thlaspi caerulescens*. Biotechnol Bioeng 2003; 83(2): 158-67.
[http://dx.doi.org/10.1002/bit.10656] [PMID: 12768621]

[86]   Sarret G, Saumitou-Laprade P, Bert V, *et al.* Forms of zinc accumulated in the hyperaccumulator Arabidopsis halleri. Plant Physiol 2002; 130(4): 1815-26.
[http://dx.doi.org/10.1104/pp.007799] [PMID: 12481065]

[87]   McNear DH Jr, Chaney RL, Sparks DL. The hyperaccumulator Alyssum murale uses complexation with nitrogen and oxygen donor ligands for Ni transport and storage. Phytochemistry 2010; 71(2-3): 188-200.
[http://dx.doi.org/10.1016/j.phytochem.2009.10.023] [PMID: 19954803]

[88]   Krämer U, Pickering IJ, Prince RC, Raskin I, Salt DE. Subcellular localization and speciation of nickel in hyperaccumulator and non-accumulator Thlaspi species. Plant Physiol 2000; 122(4): 1343-54.
[http://dx.doi.org/10.1104/pp.122.4.1343] [PMID: 10759531]

[89]   Mijovilovich A, Leitenmaier B, Meyer-Klaucke W, Kroneck PMH, Götz B, Küpper H. Complexation and toxicity of copper in higher plants. II. Different mechanisms for copper versus cadmium detoxification in the copper-sensitive cadmium/zinc hyperaccumulator Thlaspi caerulescens (Ganges Ecotype). Plant Physiol 2009; 151(2): 715-31.
[http://dx.doi.org/10.1104/pp.109.144675] [PMID: 19692532]

[90]   Variyar PS, Banerjee A, Akkarakaran JJ, Suprasanna P. Role of glucosinolates in plant stress tolerance. Emerg Technol Manag Crop Stress Toler 2014; 271-91.
[http://dx.doi.org/10.1016/B978-0-12-800876-8.00012-6]

[91]   Demidchik V. Mechanisms of oxidative stress in plants: From classical chemistry to cell biology. Environ Exp Bot 2015; 109: 212-28.
[http://dx.doi.org/10.1016/j.envexpbot.2014.06.021]

[92]   Kruszka K, Pieczynski M, Windels D, *et al.* Role of microRNAs and other sRNAs of plants in their changing environments. J Plant Physiol 2012; 169(16): 1664-72.
[http://dx.doi.org/10.1016/j.jplph.2012.03.009] [PMID: 22647959]

[93]   Hasanuzzaman M, Bhuyan MHMB, Parvin K, *et al.* Regulation of ROS metabolism in plants under environmental stress: A review of recent experimental evidence. Int J Mol Sci 2020; 21(22): 8695.
[http://dx.doi.org/10.3390/ijms21228695] [PMID: 33218014]

[94]   Farooq MA, Hong Z, Islam F, *et al.* Comprehensive proteomic analysis of arsenic induced toxicity reveals the mechanism of multilevel coordination of efficient defense and energy metabolism in two Brassica napus cultivars. Ecotoxicol Environ Saf 2021; 208: 111744.
[http://dx.doi.org/10.1016/j.ecoenv.2020.111744] [PMID: 33396070]

[95]   Amna S, Qamar S, Turab Naqvi AA, Al-Huqail AA, Qureshi MI. Role of sulfur in combating arsenic stress through upregulation of important proteins, and *in-silico* analysis to study the interaction between phosphate transporter (PHO1), arsenic and phosphate in spinach. Plant Physiol Biochem 2020; 157: 348-58.
[http://dx.doi.org/10.1016/j.plaphy.2020.11.002] [PMID: 33189055]

[96]   Chauhan R, Awasthi S, Indoliya Y, *et al.* Transcriptome and proteome analyses reveal selenium mediated amelioration of arsenic toxicity in rice (*Oryza sativa* L.). J Hazard Mater 2020; 390: 122122.
[http://dx.doi.org/10.1016/j.jhazmat.2020.122122] [PMID: 32006842]

[97]   Farooq MA, Islam F, Yang C, *et al.* Methyl jasmonate alleviates arsenic-induced oxidative damage and modulates the ascorbate–glutathione cycle in oilseed rape roots. Plant Growth Regul 2018; 84(1): 135-48.
[http://dx.doi.org/10.1007/s10725-017-0327-7]

[98]   Usman K, Abu-Dieyeh MH, Zouari N, Al-Ghouti MA. Lead (Pb) bioaccumulation and antioxidative responses in Tetraena qataranse. Sci Rep 2020; 10(1): 17070.
[http://dx.doi.org/10.1038/s41598-020-73621-z] [PMID: 33051495]

[99]   Kaur G, Kaur S, Singh HP, Batish DR, Kohli RK, Rishi V. Biochemical adaptations in Zea mays roots to short-term Pb²⁺ exposure: ROS generation and metabolism. Bull Environ Contam Toxicol 2015;

95(2): 246-53.
[http://dx.doi.org/10.1007/s00128-015-1564-y] [PMID: 26048438]

[100] Hasanuzzaman M, Nahar K, Rahman A, Mahmud JA, Alharby HF, Fujita M. Exogenous glutathione attenuates lead-induced oxidative stress in wheat by improving antioxidant defense and physiological mechanisms. J Plant Interact 2018; 13(1): 203-12.
[http://dx.doi.org/10.1080/17429145.2018.1458913]

[101] Sidhu GPS, Singh HP, Batish DR, Kohli RK. Effect of lead on oxidative status, antioxidative response and metal accumulation in Coronopus didymus. Plant Physiol Biochem 2016; 105: 290-6.
[http://dx.doi.org/10.1016/j.plaphy.2016.05.019] [PMID: 27214085]

[102] Sallah-Ud-Din R, Farid M, Saeed R, *et al.* Citric acid enhanced the antioxidant defense system and chromium uptake by *Lemna minor* L. grown in hydroponics under Cr stress. Environ Sci Pollut Res Int 2017; 24(21): 17669-78.
[http://dx.doi.org/10.1007/s11356-017-9290-0] [PMID: 28600794]

[103] Adrees M, Ali S, Iqbal M, *et al.* Mannitol alleviates chromium toxicity in wheat plants in relation to growth, yield, stimulation of anti-oxidative enzymes, oxidative stress and Cr uptake in sand and soil media. Ecotoxicol Environ Saf 2015; 122: 1-8.
[http://dx.doi.org/10.1016/j.ecoenv.2015.07.003] [PMID: 26164268]

[104] Kundu D, Dey S, Raychaudhuri SS. Chromium (VI) – induced stress response in the plant *Plantago ovata* Forsk *in vitro*. Genes Environ 2018; 40(1): 21.
[http://dx.doi.org/10.1186/s41021-018-0109-0] [PMID: 30349616]

[105] Jawad Hassan M, Ali Raza M, Ur Rehman S, *et al.* Effect of cadmium toxicity on growth, oxidative damage, antioxidant defense system and cadmium accumulation in two sorghum cultivars. Plants 2020; 9(11): 1575.
[http://dx.doi.org/10.3390/plants9111575] [PMID: 33203059]

[106] Vijendra PD, Huchappa KM, Lingappa R, Basappa G, Jayanna SG, Kumar V. Physiological and biochemical changes in moth bean (*Vigna aconitifolia* L.) under cadmium stress. J Bot 2016; 2: 1-13.

[107] Anjum SA, Tanveer M, Hussain S, *et al.* Cadmium toxicity in Maize (Zea mays L.): consequences on antioxidative systems, reactive oxygen species and cadmium accumulation. Environ Sci Pollut Res Int 2015; 22(21): 17022-30.
[http://dx.doi.org/10.1007/s11356-015-4882-z] [PMID: 26122572]

[108] Kabir AH, Begum MC, Haque A, *et al.* Genetic variation in Fe toxicity tolerance is associated with the regulation of translocation and chelation of iron along with antioxidant defence in shoots of rice. Funct Plant Biol 2016; 43(11): 1070-81.
[http://dx.doi.org/10.1071/FP16068] [PMID: 32480527]

[109] Delias DS, Da-Silva CJ, Martins AC, de Oliveira DSC, do Amarante L. Iron toxicity increases oxidative stress and impairs mineral accumulation and leaf gas exchange in soybean plants during hypoxia. Environ Sci Pollut Res Int 2022; 29(15): 22427-38.
[http://dx.doi.org/10.1007/s11356-021-17397-3] [PMID: 34791629]

[110] Ghasemi S, Khoshgoftarmanesh AH, Afyuni M, Hadadzadeh H. Iron(II)–amino acid chelates alleviate salt-stress induced oxidative damages on tomato grown in nutrient solution culture. Sci Hortic (Amsterdam) 2014; 165: 91-8.
[http://dx.doi.org/10.1016/j.scienta.2013.10.037]

[111] Singh H, Kumar D, Soni V. Copper and mercury induced oxidative stresses and antioxidant responses of *Spirodela polyrhiza* (L.) Schleid. Biochem Biophys Rep 2020; 23: 100781.
[http://dx.doi.org/10.1016/j.bbrep.2020.100781] [PMID: 32715102]

[112] Kapoor D, Rattan A, Gautam V, Bhardwaj R. Alleviation of cadmium and mercury stress by supplementation of steroid hormone to *Raphanus sativus* seedlings. Proc Natl Acad Sci, India, Sect B Biol Sci 2016; 86(3): 661-6.
[http://dx.doi.org/10.1007/s40011-015-0501-5]

[113] Pirzadah TB, Malik B, Tahir I, Irfan QM, Rehman RU. Characterization of mercury-induced stress biomarkers in *Fagopyrum tataricum* plants. Int J Phytoremediation 2018; 20(3): 225-36.
[http://dx.doi.org/10.1080/15226514.2017.1374332] [PMID: 29172663]

[114] Ibrahim AS, Ali GAM, Hassanein A, Attia AM, Marzouk ER. Toxicity and uptake of CuO nanoparticles: evaluation of an emerging nanofertilizer on wheat (*Triticum aestivum* L.) plant. Sustainability (Basel) 2022; 14(9): 4914.
[http://dx.doi.org/10.3390/su14094914]

[115] Contreras RA, Pizarro M, Köhler H, Sáez CA, Zúñiga GE. Copper stress induces antioxidant responses and accumulation of sugars and phytochelatins in Antarctic Colobanthus quitensis (Kunth) Bartl. Biol Res 2018; 51(1): 48.
[http://dx.doi.org/10.1186/s40659-018-0197-0] [PMID: 30428921]

[116] Giannakoula A, Therios I, Chatzissavvidis C. Effect of lead and copper on photosynthetic apparatus in citrus (*Citrus aurantium* L.) plants. The role of antioxidants in oxidative damage as a response to heavy metal stress. Plants 2021; 10(1): 155.
[http://dx.doi.org/10.3390/plants10010155] [PMID: 33466929]

[117] Liang J, Yang W. Effects of zinc and copper stress on antioxidant system of olive leaves. IOP Conf Ser: Earth Environ Sci. 300(5): 052058.
[http://dx.doi.org/10.1088/1755-1315/300/5/052058]

[118] Ahmad P, Alyemeni MN, Abass Ahanger M, *et al.* Upregulation of antioxidant and glyoxalase systems mitigates NaCl stress in *Brassica juncea* by supplementation of zinc and calcium. J Plant Interact 2018; 13(1): 151-62.
[http://dx.doi.org/10.1080/17429145.2018.1441452]

[119] Aazami MA, Rasouli F, Ebrahimzadeh A. Oxidative damage, antioxidant mechanism and gene expression in tomato responding to salinity stress under *in vitro* conditions and application of iron and zinc oxide nanoparticles on callus induction and plant regeneration. BMC Plant Biol 2021; 21(1): 597.
[http://dx.doi.org/10.1186/s12870-021-03379-7] [PMID: 34915853]

[120] dos Santos JO, Andrade CA, Dázio de Souza KR, *et al.* Impact of zinc stress on biochemical and biophysical parameters in *Coffea arabica* seedlings. J Crop Sci Biotechnol 2019; 22(3): 253-64.
[http://dx.doi.org/10.1007/s12892-019-0097-0]

[121] Kumar S, Wang M, Liu Y, *et al.* Vanadium stress alters sweet potato (*Ipomoea batatas* L.) growth, ROS accumulation, antioxidant defense system, stomatal traits, and vanadium uptake. Antioxidants 2022; 11(12): 2407.
[http://dx.doi.org/10.3390/antiox11122407] [PMID: 36552615]

[122] Altaf MA, Shahid R, Ren MX, Altaf MM, Jahan MS, Khan LU. Melatonin mitigates nickel toxicity by improving nutrient uptake fluxes, root architecture system, photosynthesis, and antioxidant potential in tomato seedling. J Soil Sci Plant Nutr 2021; 21(3): 1842-55.
[http://dx.doi.org/10.1007/s42729-021-00484-2]

[123] Amjad M, Raza H, Murtaza B, *et al.* Nickel toxicity induced changes in nutrient dynamics and antioxidant profiling in two maize (*Zea mays* L.) hybrids. Plants 2019; 9(1): 5.
[http://dx.doi.org/10.3390/plants9010005] [PMID: 31861411]

[124] Amjad M, Ameen N, Murtaza B, *et al.* Comparative physiological and biochemical evaluation of salt and nickel tolerance mechanisms in two contrasting tomato genotypes. Physiol Plant 2020; 168(1): 27-37.
[http://dx.doi.org/10.1111/ppl.12930] [PMID: 30684269]

[125] Barcelos JPQ, Reis HPG, Godoy CV, *et al.* Impact of foliar nickel application on urease activity, antioxidant metabolism and control of powdery mildew (*Microsphaera diffusa*) in soybean plants. Plant Pathol 2018; 67(7): 1502-13.
[http://dx.doi.org/10.1111/ppa.12871]

[126] Chen B, Ai W, Gong H, Gao X, Qiu B. Cleaning up of heavy metals-polluted water by a terrestrial hyperaccumulator *Sedum alfredii* Hance. Front Biol (Beijing) 2013; 8(6): 599-605.
[http://dx.doi.org/10.1007/s11515-013-1274-y]

[127] Li RY, Ago Y, Liu WJ, *et al.* The rice aquaporin Lsi1 mediates uptake of methylated arsenic species. Plant Physiol 2009; 150(4): 2071-80.
[http://dx.doi.org/10.1104/pp.109.140350] [PMID: 19542298]

[128] Zhang B. MicroRNA: a new target for improving plant tolerance to abiotic stress. J Exp Bot 2015; 66(7): 1749-61.
[http://dx.doi.org/10.1093/jxb/erv013] [PMID: 25697792]

[129] Zeng H, Wang G, Hu X, Wang H, Du L, Zhu Y. Role of microRNAs in plant responses to nutrient stress. Plant Soil 2014; 374(1-2): 1005-21.
[http://dx.doi.org/10.1007/s11104-013-1907-6]

[130] Jones-Rhoades MW, Bartel DP, Bartel B. MicroRNAs and their regulatory roles in plants. Annu Rev Plant Biol 2006; 57(1): 19-53.
[http://dx.doi.org/10.1146/annurev.arplant.57.032905.105218] [PMID: 16669754]

[131] Wani SH, Kumar V, Khare T, *et al.* miRNA applications for engineering abiotic stress tolerance in plants. Biologia (Bratisl) 2020; 75(7): 1063-81.
[http://dx.doi.org/10.2478/s11756-019-00397-7]

[132] Min Yang Z, Chen J. A potential role of microRNAs in plant response to metal toxicity. Metallomics 2013; 5(9): 1184-90.
[http://dx.doi.org/10.1039/c3mt00022b] [PMID: 23579282]

[133] Mendoza-Soto AB, Sánchez F, Hernández G. MicroRNAs as regulators in plant metal toxicity response. Front Plant Sci 2012; 3: 105.
[http://dx.doi.org/10.3389/fpls.2012.00105] [PMID: 22661980]

[134] Valdés-López O, Yang SS, Aparicio-Fabre R, *et al.* MicroRNA expression profile in common bean (*Phaseolus vulgaris*) under nutrient deficiency stresses and manganese toxicity. New Phytol 2010; 187(3): 805-18.
[http://dx.doi.org/10.1111/j.1469-8137.2010.03320.x] [PMID: 20553393]

[135] Zhou ZS, Song JB, Yang ZM. Genome-wide identification of *Brassica napus* microRNAs and their targets in response to cadmium. J Exp Bot 2012; 63(12): 4597-613.
[http://dx.doi.org/10.1093/jxb/ers136] [PMID: 22760473]

[136] Ding Y, Gong S, Wang Y, *et al.* MicroRNA166 modulates cadmium tolerance and accumulation in rice. Plant Physiol 2018; 177(4): 1691-703.
[http://dx.doi.org/10.1104/pp.18.00485] [PMID: 29925586]

CHAPTER 3

Molecular Interactions During Heavy Metal Detoxification in Plants

Jaspreet Kour[1], Arun Dev Singh[1], Shalini Dhiman[1], Isha Madaan[2,3], Deepak Sharma[4], Roohi Sharma[4], Kanika Khanna[1,*], Puja Ohri[4] and Renu Bhardwaj[1,*]

[1] *Department of Botanical and Environmental Sciences, Guru Nanak Dev University, Amritsar, Punjab, India*

[2] *Department of Botany, Punjabi University, Patiala, Punjab, India*

[3] *Government College of Education, Jalandhar, Punjab, India*

[4] *Department of Zoology, Guru Nanak Dev University, Amritsar, Punjab, India*

Abstract: Heavy metal stress is one of the abiotic stresses that are damaging plants. Heavy metals cause toxicity in plant cells by targeting key molecules and essential processes. The toxicity of these heavy metals is caused by their discharge from both natural and anthropogenic sources. When these heavy metals are absorbed by plants, they activate multiple components, including reactive oxygen species (ROS), phytohormones, and nitric oxide (NO), thereby triggering the Mitogen-Activated Protein Kinases (MAPK) cascade. Signals detected by the MAPK cascade through the cell membrane receptor cells are then transmitted to various transcriptional factors and proteins that aid in stress management. This chapter discusses the activation of the MAPK cascade by heavy metals. In addition, the involvement of transcriptional genes and factors regulated by the MAPK cascade as well as their interconnections have been examined in depth.

Keywords: Heavy metals, MAPKs, ROS, Sources, Toxicity.

INTRODUCTION

In addition to other soil pollutants, heavy metal (HM) pollution is a global environmental concern. Since these HMs cannot be decomposed, they are always present in the soil, and their concentration continues to increase dramatically [1]. Several organisms, including plants, are adversely affected by these HMs emitted by natural and anthropogenic sources. When plants are subjected to persistent HM

* **Corresponding authors Kanika Khanna & Renu Bhardwaj:** Department of Botanical and Environmental Sciences, Guru Nanak Dev University, Amritsar, Punjab, India;
E-mails: kanika.27590@gmail.com; renubhardwaj82@gmail.com

stress, their biological systems are irreparably damaged. Low yield and productivity are the results [2]. To overcome this HM stress, plants endure several biochemical and physiological changes, including a complex series of signalling pathways [3]. Through phosphorylation and dephosphorylation, surface receptors are amplified, and these signals are transmitted to cells when the cell membrane detects the signals [4]. When signals reach the cell, specific proteins such as enzymes, kinases, and several transcriptional factors are activated in the nucleus and cytoplasm to regulate gene expression.

Due to HM stress, various transcriptional factors are activated in plants. This causes the expression of several antioxidant enzymes. Protein phosphorylation is a crucial method of signal transduction in plants. These reactions are catalysed by mitogen-activated protein kinases (MAPKs). Some authors have reported that these MAPKs act downstream to coordinate numerous cellular responses required for proper plant growth and development [5]. Multiple MAPK pathways are present in plant cells. Each signal transmission pathway is independent and intertwined for normal transmission. In order to form a molecular network, these pathways are interconnected with other transduction pathways in the cell [6]. During times of pressure, MAPKs regulate the transcriptional factors bZIP, MYC, MYB, and WRKY [7]. These MAPKs are predominantly found in the cytoplasm and/or nucleus, although in some instances, they may be transferred from the cytoplasm to the nucleus for signalling purposes. This chapter discusses the sources and effects of HMs on plants. In addition, background information on plants under HM stress and new molecular approaches to transcriptional factors for accumulation and tolerance of HMs have been presented.

SOURCE SOF HEAVY METALS

Soils are often susceptible to environmental changes and HM contamination has become a global issue due to the related ecological dangers [8]. Due to the bioavailability, bioaccumulation, toxicity, and incapacity to break down HMs, soil pollution with HMs is a major concern [9, 10]. It is important to note that the physical, chemical, and biological characteristics of the soil affect the bioavailability of HMs there [11, 12]. Studies have demonstrated that prolonged contact with contaminated soils can have hazardous effects on living things like plants and animals, especially in areas where environmental protection laws are not strictly enforced [13 - 15]. In rural or urbanized areas, most soils could accumulate HMs over the permitted average levels, significantly enough to pose health risks to people, flora and fauna, ecosystems, or other media. The distribution of HMs in soils is associated with both natural and anthropogenic sources [16, 17] (Fig. **1**).

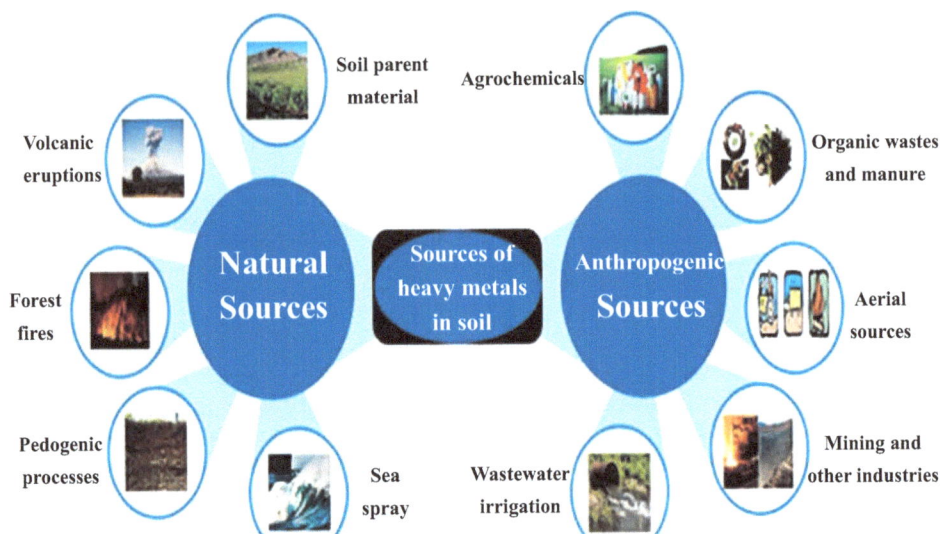

Fig. (1). Different sources of heavy metals in soil.

Natural Sources

The main natural sources of HM pollution in soil have been identified as soil parent materials, pedogenic processes, mineral degradation, volcanic eruptions, forest fires, sea spray, and evaporation from soil and water surfaces [18, 19]. The parent substance from which soils are formed is the soil's main source of HMs. The release of HMs into the soil is also caused by biological causes such as destructive forest fires, earthy weathering of metal-bearing rocks and other geological substances, volcanic occurrences, and natural processes including erosion and surface winds blowing dust particles [20 - 22]. Another source for the movement of HMs in nature is the gaseous exchange and bubbles that create bursts in the water [23]. However, anthropogenic sources are more substantial than natural ones.

Anthropogenic Sources

Metals are natural elements but increased levels of HMs have been found in soils as a result of anthropogenic inputs such as extensive use of agrochemicals (fertilizers and pesticides), wastewater irrigation/sewage sludge supplementation, use of organic wastes and manures, higher atmospheric depositions by industrial units, combustion of fossil fuels and other aerial sources [24, 25]. Mining and other industries also add up to HMs pollution in soil [26, 27]. The leading sub-sections provide explanations for some of these sources.

Agrochemicals

In agriculture, the use of conventional agrochemicals (fertilizers and pesticides) which have high metal concentrations, has increased over the past few years. Fertilizers provide various essential nutrients to improve plant productivity and growth as well as increase soil organic matter, thereby increasing soil fertility [28]. However, the excessive use of fertilizers both organic and inorganic is responsible for HMs accumulation in the soil. Phosphorus is frequently utilized in fertilizer production, but it also contributes significantly to the buildup of HMs in the soil [29]. It has been demonstrated that water-insoluble phosphorus fertilizers create phosphate rocks, which are crucial for the soil's ability to immobilize metals through the precipitation of metal phosphates. HM deposition in agricultural soils caused by prolonged, excessive fertilizer application diminishes soil fertility, which in turn reduces plant growth and output [30]. The main inorganic fertilizer types that contribute to the release of HMs in agricultural soil and are then absorbed by plants include phosphate fertilizers, liming materials, and bio-fertilizers [31]. Further, due to the widespread use of copper-based antifungal treatments in viniculture in nations like France, Brazil, Croatia and Spain, a significant Cu accumulation in the soil of grapevine orchards was observed in the 19th century [32]. Because of the prolonged usage of agrochemicals, agricultural soil has a larger potential for accumulating Cu, Zn, and Cd, and after HMs have contaminated the soil, it is very difficult to restore the soil ecosystem [33].

Mining and other Industries

The mining and processing of metal ores in tandem with industry have left many nations with the inheritance of extensive distribution of metal contaminants in soil. HMs enter the environment through two different processes: dry deposits (wind) and wet deposits (rainfall). Higher pollutant concentrations were caused by the direct release of mining and tailings, which are larger and heavier particles and settle at the bottom of the flotation process during mining, into nearby wetlands [10]. For instance, to investigate the level of soil-HM contamination and the risk to human health in mining sites [34], measured the concentrations of soil As, Cd, Cu, Cr, Hg, Ni, Pb, and Cr(VI) in an abandoned gold mining region. The results revealed that

1) The average concentrations of Cd, Cr, As, Cu, Hg, and Ni in the mining region were higher than the background levels of the soil elements.

2) The mining region has varying levels of HM pollution and significant potential ecological risks.

3) The overall carcinogenic risk of HMs was higher than the threshold for health risk.

It was concluded that Cd, Cr, As and Hg were the primary pollutants in the mining area [34]. Similarly, soil in the vicinity of the Barapukuria coal mining area of Bangladesh was found to be contaminated by Zn, Ni, and Cu [35]. Another study analyzed the degree of HM pollution in the soil surrounding the tungsten-molybdenum mining area in southwest Luoyang, Henan Province, identified the presence of HMs there, and determined the hazards these pollutants pose to human health. To evaluate these potential threat levels, the pollution index, potential ecological hazards, and health risk assessment techniques suggested by The United States Environmental Protection Agency (USEPA) were used. The findings showed that the average soil level of Zn, Cr, Cd and As was higher than the soil pollution risk screening threshold [36]. Thus, mining serves as the main root of contaminating the nearby land with heavy metals, whereas vapor forms of such HMs mixed with water make aerosols on condensation.

A considerable number of pollutants are released into the environment as a result of industrial activities, including HMs like lead, zinc, and copper. Further, increased industrial activity is a major source of soil contamination with HMs, such as Pb, Zn, and Cu along with significant amounts of other pollutants in the surrounding environment [37]. Extra materials are created by a variety of industries, such as the tanning, textile, petrochemical, pesticide and pharmaceutical industries, as well as petrochemicals from unintended oil spills or consumption of petroleum-based products. Although some items are left on the ground after use, others have benefits for forestry and agriculture. Additionally, some are occasionally, if ever, given to land may be hazardous because of the presence of HMs like Cr, Pb, and Zn or other toxic organic compounds in them. Some of them lack plant nutrients or have no influence on soil conditions [38, 27]. A study reported that industrial activities contributed more than 60% of Pb, Hg, and Cd levels in soils Jing-Jin-Ji Metropolitan Region, Pearl River Delta, and Yangtze River Delta in China [39]. Another study determined the effects of various industrial operations on the level of HM pollution, possible ecological risk, and risk to human health in Panzhihua, soil HM concentrations (As, Cr, Cu, Cd, Pb, V and Zn) in various locations were investigated. Findings demonstrated that soil enrichment in Cu and V was a result of ore smelting. Coal ore mining was connected to soil Cr build-up. The concentration of Cd, Zn, As, and Pb in the soil was linked to high-temperature coal combustion. Similarly, a study focused on a suburban industrial area adjacent to Bangladesh's Dhaka Export Processing Zone (DEPZ) [40]. They investigated the levels of HMs (Cd, Co, Cr, Fe, Mn Ni, Pb and Zn) in agricultural soils and plants. Results revealed that Fe, Cu, Co, Ni and Zn contamination levels were determined to be moderate, while Mn, Cr, Pb and Cd contamination levels were lower. In addition, the plant samples revealed higher amounts of Cr and comparable levels of Cd, Co, Cu, Ni, Pb and Zn compared to Bangladeshi industrial polluted areas [40].

Wastewater Irrigation/Sewage Sludge Supplementation

Discharging municipal, industrial, and related wastewater onto land has long been a tradition in many parts of the world [41]. Studies have demonstrated the advantages of applying sludge and composted sludge as fertilizers to various crops [42]. However, wastewater irrigation is one of the main sources of soil contamination with HMs, posing a risk to the environment and human health [43]. Continuous irrigation with sewage effluent could result in a progressive buildup of HMs in the soil, which could then result in a greater uptake of the HMs into the various portions of developing vegetables. A study examined the levels of Cu, Cd, Pb, and Zn in sewage effluent (Embu sewage treatment plant, Kenya) as well as in the vegetables and soil [44]. They reported that the concentration of HMs was above the WHO permissible levels in the sludge leading to accumulation of HMs in soil irrigated with this sewage sludge. Further, it was reported that sewage sludge supplementation elevated the build-up of Cr and Cd in the soil along with increased levels of Zn and Cd beyond the safe level in the leaves of *Spinacia oleracea* grown in the soil supplemented with sewage sludge [45]. Similar, results were obtained in another study which reported that irrigation with waste water obtained from the wastewater treatment plant in Marrakech, Morocco resulted in the contamination of soil with Cu (17.70 mg/kg), Cd (11.22 mg/kg), Pb (57.36 mg/kg) and Zn (112.71 mg/kg) [46]. Results also revealed high-risk indices, food crops tainted with HMs in the study area, and as a result, a serious health danger to the local human and animal populations [46]. Therefore, in the agricultural sector, wastewater irrigation generates both opportunities and issues.

Organic Wastes and Manure

In contrast to conventional farming, which heavily relies on the use of synthetic fertilizers and pesticides, organic farming has been shown to improve soil health through the appropriate application of organic amendments, with the goal of increasing crop productivity while preserving the health of agroecosystems [47, 48]. The restoration and maintenance of soil organic matter content, which significantly contributes to long-term soil fertility and functioning, is one of the key advantages of applying exogenous organic waste to agricultural soil. Given that soil organic matter supports the physical, chemical and biological aspects of soil fertility and health, it may be the most significant soil feature [49]. In terms of the potential advantages of organic waste amendments for the biological characteristics of agricultural soils, it is well known that organic amendments may either directly stimulate microbial proliferation by providing energy and key nutrients, or indirectly by plant growth promotion and, as a result, the number of root exudates increases in the rhizosphere [50].

Despite all the advantages listed above, adding organic amendments to agricultural soil may potentially have some negative consequences on the health of the soil ecosystem. The continuous use of organic amendments may cause metals to build up in the soil, posing dangers of metal bioaccumulation and biomagnification throughout the food chain [51]. A range of organic wastes and manures, including compost, municipal garbage, sewage waste and livestock manure, contribute to the deposition of HMs in soil, including Cd, Cr, Cu, Hg, Mo, Ni, Pb, Sb, Se, and Zn. For instance, a study reported the presence of Zn, Cu, Cd and As in pig manure [52]. Similarly, to measure specific HM concentrations (Cu, Cr, Cd, Ni, Pb and Zn), samples of dairy feed, manure and soil were taken from four intensive dairy farms in China [53]. All feed samples tested positive for the specified HMs, proving the widespread usage of chemicals during intensive dairy production, according to the results. The average values of Cr and Pb were 6.1 to 17.1 times higher than the recommended limits. Further, HM concentration was also reported in the manure samples. Last but not the least, the results of principal component analysis showed that soil HM levels were mostly derived from the parent soil components and then increased by anthropogenic causes [53]. Furthermore, the application of sheep manure and organic residue increased the level of As, Cr, Cd, Hg, and Pb in the soil and maize grains compared to the control treatment leading to greater health risks [54].

Aerial Sources

The primary aerial sources of HMs are the effuse from industrial sources, the burning of waste and traffic, and vehicular emissions, such as the wear on automobile brake and clutch liners, the state of the roads, other road amenities, and construction materials [55]. Along with pile or tunnel emissions of air, gas, or condensate streams, metals can also be released into the atmosphere through flyer emissions like dust from storage areas or trash heaps. Metals discharged into the atmosphere cause the metal-shaped particles in the gas stream. Automobile exhaust gases are the greatest contributors to lead contamination in soil, vegetation, and air. Pb had been added to gasoline for a long time as lead tetraethyl, which was emitted during the operation of mechanical engines [56]. Another study suggests that tyre dust, lubricants containing Cd and Zn, diesel engine exhaust, car washes containing Cd, Zn and Cu, Zn and vapor fumes including Zn, Cr, and Zn are other sources of HM contamination from automobiles [57, 58].

MECHANISM OF HEAVY METAL UPTAKE BY PLANTS

With the advent of industrialisation and population explosion, a large number of anthropogenic activities are being carried out on a large scale, which are responsible for HM contamination of the basic components of the environment may it be air, water, or soil. These HMs make their way into plant systems through roots or foliar adsorption (Fig. **2**) thereby posing a serious threat to the normal metabolic machinery of plants.

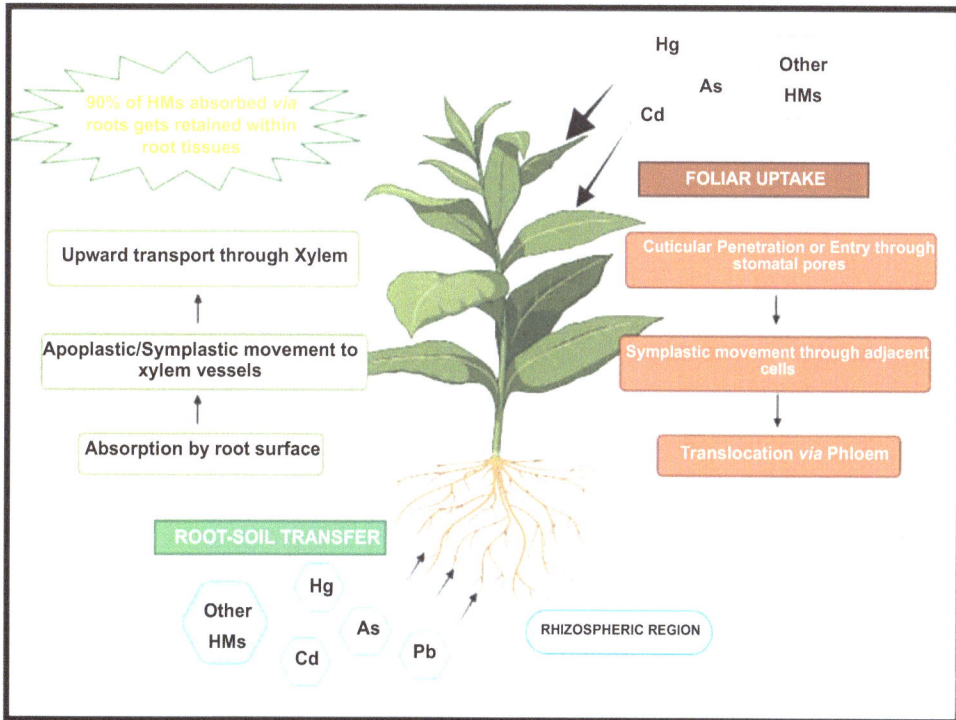

Fig. (2). Mechanism of uptake of HMs by plants.

The majority of HMs are concentrated in soil and water bodies, therefore, soil-root transport forms the basic pathway of entry of HMs into various parts of plants [59]. The HMs found in soil solution are adsorbed on the surface of the root hair followed by their linkage with polysaccharide compounds or CO groups of mucilage uronic acid present in the rhizodermal membrane [60]. Here, the HM ions are in contact with the cell walls of root cells, which provides a flexible medium for ion exchange. The further movement of HMs is mediated passively across the plasma membrane or may involve the employment of active cationic carriers or channels to mediate the uphill movement of the same [61, 62]. Then, translocation occurs through the xylem *via* passive transport by crossing the intercellular spaces (apoplastic pathway) or *via* the plasma membrane (symplast pathway). More than 90% of HMs adsorbed by roots get retained within the tissues of the root system in the form of precipitates of carbonates, sulphates, or phosphates as a result of their failure to cross the suberised walls of vascular tissues [63]. Therefore, the metal ions that get sequestered within the vacuoles are transported into the xylem vessels *via* the symplast pathway, and from xylem vessels, they are further translocated to aerial parts of plants, *i.e.*, shoots and leaves [64].

There are a number of factors that affect the uptake of HMs from the soil through roots including the ionic state of metals, soil particle size, soil pH, soil microbiota composition, organic matter content, and cation exchange capacity of plant roots [65]. The HM transporters that aid the uptake and translocation of HMs from soils into roots and then to upper parts of plants are Zn-regulated transporters (ZRT), cation diffusion facilitators (CDFs), natural resistance-associated macrophage proteins (NRAMPs), ATPases, cation antiporters, and the ZIP family [66, 67].

Foliar uptake of HMs has not been extensively studied yet. It occurs due to the deposition of HM compounds on the aerial parts of plants. It has been studied that the foliar adsorption of HMs occurs in plants through two methods a) adsorption and internalisation through leaf cuticle and b) entry through stoma [68]. The first method of entry *via* cuticle includes four sequential steps:

 i. Adherence of HM-containing particles to the leaf cuticle.
 ii. Entrance into cuticular space *via* endocytosis.
 iii. Desorption of HMs in the apoplastic regions
 iv. Transport to the adjacent cells

The foliar uptake of HMs depends on a variety of aspects *i.e.*, leaf size, cuticular and epidermal composition, phyllotaxy, and speciation of metal ions. Mostly small-sized and lipophilic HMs enter through cuticular penetration and large-sized and hydrophilic ones make their entry *via* stomatal pores or other aqueous pores in the cuticle [69]. These HMs are transported actively through the symplast pathway (through the plasma membrane) after their one-time penetration [70]. Accumulation of HMs above the optimum levels within the plant parts by any means of uptake may prove toxic to plant health as well as overall ecosystem sustainability.

TOXICITY CAUSED BY HEAVY METALS TO PLANTS

The toxicity of agriculture and plants has become a crucial and significant concern as a result of adverse ecological repercussions [71]. It has been noted that the land used for crop farming in many parts of the globe is damaged by an accumulation of HMs including Cd, Cr, As, Hg, Al, Co, Cu, Fe, Pb and Zn [72]. Although some HMs are essential for plant growth and sustenance, excessive amounts of various metals can be unsafe for plants [73]. Additionally, it is known that farmland with high concentrations of HMs has negative consequences for flora and fauna such as plant growth and its metabolic activity, soil physical and chemical properties and its fertility, bioactivity, biodiversification, animal and human health as well as plant's overall parameters [74]. Just because the plants are capable of accumulating crucial metals, they can also acquire other non-necessary metals [75]. Since metals are difficult to break down, therefore when levels within a plant exceed optimal levels, they have a detrimental direct and indirect effect on the plant's overall health [76, 77]. Globally, exhaustive research has been carried out to determine how these hazardous HMs influence plants [78, 79].

The adverse effect of HM toxicity is directly or may be influenced indirectly. Sometimes, the decline of crop development brought on by these detrimental consequences (direct and indirect) might lead to plant fatality [80]. Regulation of cytoplasmic enzymes and cellular structure breakdown carried on by oxidative stress are two examples of the direct detrimental effects of HM poisoning [81, 82]. An example of an indirect negative effect is the restoration of vital nutrients at cation exchange sites in plants. It is anticipated that HMs' unfavourable impacts on soil microbiota formation and function might indirectly influence plant growth [83, 84]. Exemplary, elevated metal content may lead to a decrease in the number of helpful soil bacteria, which may also impede the degradation of organic waste and lower soil elements [85]. HMs may hamper plant metabolism-related activity of the enzyme by interfering with the functions of microbial species [86, 87]. Furthermore, other biological and physiologic consequences of HM exposure on plants include decreased photosynthetic productivity and membrane instability, nutritive and hormone disequilibrium, lowered leaf chlorophyll content formation, and suppression of cellular division, cell cycle, and gene expression levels [88, 89].

The induction of different stress in floral cells depends on the developmental stage of plants, and the amount as well as the nature of HMs ingested [90]. When expressed in abundance, HMs harm and alter different biological structures, especially cell membranes, nucleic acids, and proteins [91]. Moreover, HMs cause pro-mutagenic degradation in the nucleus, which encompasses DNA strand breakage, DNA base modifications, intra- and intermolecular protein and DNA cross-linking, conformational changes, and depurination [92, 88]. This triggers the activation of genotoxic and oxidative stress responses by induced abiotic stress in plants [93]. For illustration, enhanced oxidative stress was noted in tobacco (*Nicotiana benthamiana*) plants under Manganese stress in terms of hydrogen peroxide (H_2O_2) concentration and malondialdehyde (MDA) whereby antioxidant enzyme activity, such as peroxidase (POD) and ascorbate peroxidase (APOX) was increased through exogenous treatment with melatonin [94]. A similar study was reported in *Oryza sativa* by HM Zn and Cd toxicity where it induced the accumulation of H_2O_2, MDA, and superoxide radicals [95].

Cr toxicity resulted in decreased root and shoot length and reduced seed germination rate and survival of *Triticum aestivum* [96]. It has also been reported in *Capsicum annuum* that 200μm Cd inhibited the overall germination of the plant [97]. Similar results were obtained in a study where increased electrolyte leakage led to stunted growth of rice plants by treating it with Cr concentrations of 50 and 100mg/kg [98]. Increased electrolyte leakage, MDA, and enzymatic antioxidant activity of superoxide dismutase (SOD), catalase (CAT), glutathione reductase (GR), and APOX were also observed after the Cr application. Delias and his co-workers reported reduced content of leaf gas exchange, enhanced antioxidants (SOD, APOX, and CAT) and their activity, and impaired mineral treatment leading to decreased *Glycine max* by Fe treatment. A decline in biomass and photosynthetic pigment content and upregulated oxidative stress in *Triticum aestivum* in soil contaminated with As was also reported [99]. In a concentration-dependent assay of foliar application, the biomass of *Coriandrum sativum* growing in Pb-contaminated soil was reported to decrease as flavonoids, MDA and antioxidant activity increased, with the greatest decline shown on treatment with 1500 mg/kg soil [100]. Further, an increase in the activity of antioxidant enzymes to counterbalance the heightened oxidative stress markers was observed in *Cajanus cajan* grown in soil having 1 and 5mM Cu [101]. Cu-mediated reduced growth, enhanced bioaccumulation of flavonoids, alkaloids, lignin, and total phenolics, and altered regulation of homeostasis were also observed in *Citrus grandis* [102]. In a study, lowered stomatal conductance, net photosynthetic rate, transpiration rate, and total soluble proteins and sugar content were reported on the application of 100 and 200mM Ni on wheat [103]. Ni accumulation also reduced plant dry weight by decreasing shoot length, leaf number, and leaf area, shoot weight in *Sorghum bicolor* [104]. In a similar study on Ni, increased protein, proline, and H_2O_2 content and increased activity of antioxidant enzymes like SOD, POD, CAT, and APOX were observed in *Alyssum inflatum* by varying concentrations (100, 200, and 400μm) of the HM [105].

Another study demonstrated that treatment of *Saccharum officinarum* seedlings with Mn resulted in decreased expression of *GluTR* (encoding glutamyl-tRNA reductase), an enzyme linked with chlorophyll biosynthesis leading to leaf chlorosis [106]. In the photosynthetic apparatus, increased lipid peroxidation and disrupted membrane integrity were observed in *Citrus aurantium* by treatment of Pb and Cu [107]. In leaves, the concomitant effects of such HMs (Pb and Cu) on growth, MDA concentration, hydrogen peroxide (H_2O_2), chlorophylls, flavonoids, carotenoids, phenolics, chlorophyll fluorescence, and photosynthetic parameters were also studied.

A summary of the harmful effects of several metals on the growth, biochemistry, and physiology of many plants has been modelled in Fig. (3).

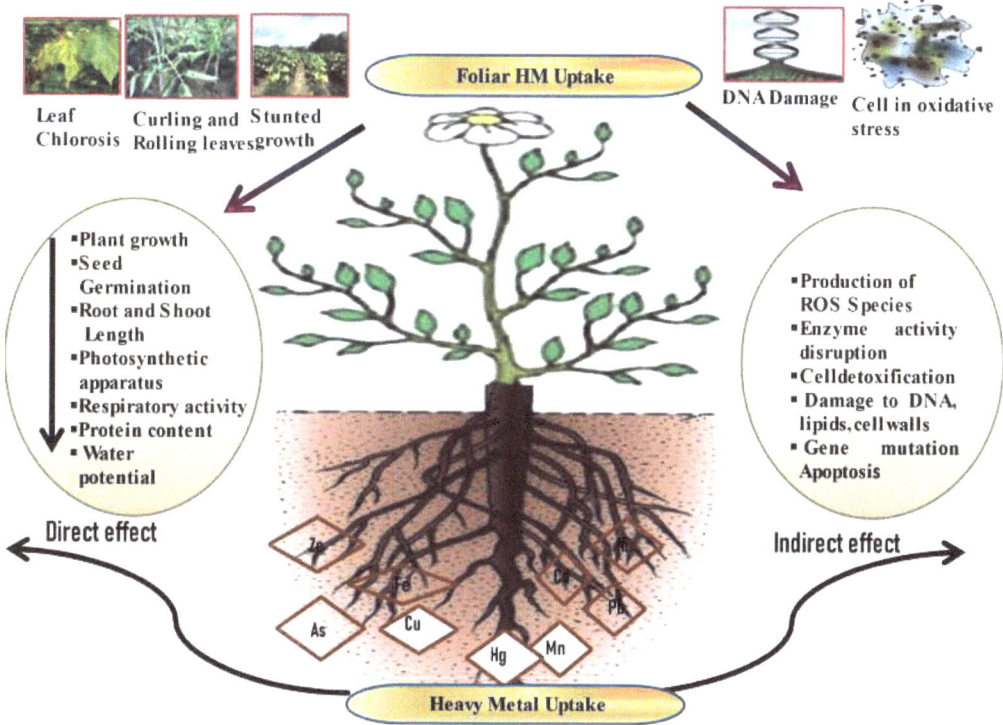

Fig. (3). Impacts of Heavy Metals on Growth, Biochemistry and Physiological parameters.

ACTIVATION OF MAPKS BY HEAVY METALS THROUGH ROS, NO, AND PHYTOHORMONES

MAPK is a critically important signalling component of plant immunity when it comes to responding to challenging stresses like HMs stress and promoting healthy growth and development of plants. A detailed illustration of MAPK's location and activation through various stress-associated stimuli is given in Fig. (4). It has been reported that under HM stress, many signalling molecules like ROS, NO, and phytohormones are directly or indirectly responsible for the activation of MAPK signalling pathways [108].

HMs trigger the production of reactive oxygen species (ROS), nitric oxide (NO), and plant growth regulators inside plants' bodies, which further act as a signalling molecule that in turn is responsible for the activation of MAPK kinases pathway inside the plant bodies, which finally ensure plant survival under severe stress [109]. Two important MAPK cascades (MEKK1-MKK4/5-MPK3/6 and MEKK1-MKK2-MPK4/6) work after ROS, which has been found to be involved in HM stress [110, 108]. Exogenous application of signalling molecules like NO and phytohormones individually or in combination or either through their cross-talks was reported to activate the MAPK pathways in order to safeguard the survival of the plant under toxic metal stress [108]. Additionally, a variety of hormones, including auxin (IAA), ethylene (ET), salicylic acid (SA), melatonin, brassinosteroids, and abscisic acid (ABA), are also involved in critical HM stress amelioration-related MAPK activation process in plants [111 - 114]. All these experimental findings (as shown in Table **1**) point to the possibility that the MAPK cascade can be triggered by various signalling molecules like ROS, NO, or various phytohormones that might play a significant functional role under diverse metal stresses.

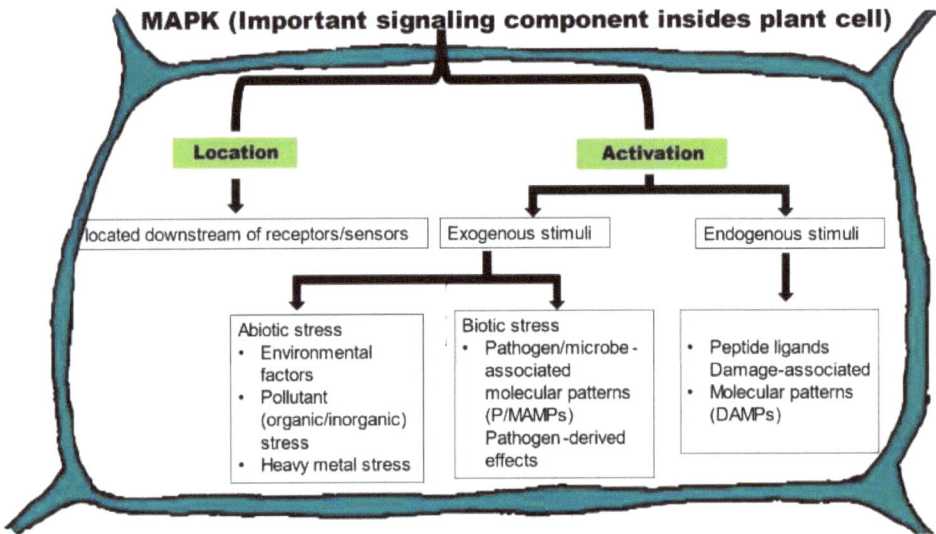

Fig. (4). Illustration of MAPK (an important signalling component) activation and location sites inside plant cells.

Table 1. Role of different signaling molecules either individually or through crosstalk in the activation of MAPK cascade under heavy metal stress.

S.No.	Plant Species	Signalling Molecules Involved	Heavy Metal	MAPK	References
1	*Glycine max*	Nitric oxide (NO) and hormone (Melatonin)	Cd, Pb	MAPKs cascades	[113]
2	*Capsicum annuum*	Hormone (24-Epibrassinolide)	Cr (VI)	MAPK cascade	[114]
3	*Zea mays* (Roots)	Reactive oxygen species (ROS)	Cd	ZmMPK3-1 and ZmMPK6-1	[115]
4	*Oryza sativa* L. cv. Zhonghua No. 11 (Roots)	Hormone (Abscisic acid (ABA), auxin)	Cd	MAPK	[111]
5	*Arabidopsis thaliana*	Nitric oxide (NO)	Cd	MAPK	[116]
6	*Arabidopsis thaliana*	Nitric oxide (NO)	Cd	MPK6	[117]
7	*Oryza sativa* (Root and Leaves)	Reactive oxygen species (ROS) and nitric oxide (NO)	As	OsMPK3, OsMPK4	[118]
8	*Arabidopsis thaliana*	Reactive oxygen species (ROS)	Cd	MAPK (MPK3 and MPK6)	[119]
9	*Oryza sativa* L. cv.TNG67	Reactive oxygen species (ROS)	Cd, Cu	OsMPK3, OsMPK6	[120]
10	*Medicago sativa* (seedling)	Reactive oxygen species (ROS)	Cu, Cd	SIMK, MMK2, MMK3 and SAMK	[121]

VARIOUS TRANSCRIPTIONAL FACTORS ARE INVOLVED IN THE REGULATION OF HEAVY METAL TOLERANCE

MAPK pathway plays a crucial role in a well-defined coordinated network with other signaling molecules to mediate growth and development-related processes and to activate stress-responsive genes to mitigate biotic and abiotic stress in plants. Thus, plants utilize multiple transcriptional factors to undergo HM tolerance, however, the activation of these TFs is regulated through phosphorylation *via* MAPKs under various stressors in plants [122, 123]. These TFs are known to have multiple phosphorylation sites and are efficient in controlling the expressions of multiple downstream genes under different stressful conditions. However, MAPK cascades involve multiple transcriptional factors to undergo the ameliorative and HMs detoxification processes, which include heat shock transcription factor (HSF), basic leucine zipper (bZIP), myeloblastosis protein (MYB), WRKY, and ethylene-responsive transcription factor (ERF) [108]. Among these, bZIPs family of TFs, includes 135 *bZIP* encoding genes in radish (*Raphanus Sativus*) which have been discovered in a whole genome study of the crop. *RsbZIP010* gets downregulated expressions under a variety of HMs stress conditions such as Pb, Cr, Cd, *etc*. This indicates the active roles of these genes and their specific involvement in multiple stress responses as well as in biological processes in plants [124]. Another study in rice concluded that the stress-related and ABA-dependent gene *LOC_Os02g52780* was found to have a significant contribution in maintaining the Cd stress tolerance and lowering Cd accumulation in rice grains [125]. On the other hand, MYB proteins are potential contributors and regulators to mediate biotic and biotic responses, metabolism, and developmental activities in plants. Studies also found that the *OsMYB45* gene in rice is effective in minimizing the Cd stresses, whereas its mutant undergoes low Catalase levels and high H_2O_2 concentrations in leaves with respect to their wild type [126, 127]. Also, MYB TFs are crucial in maintaining homeostasis as well as the absorption of some essential HMs like Fe and Zn other than Cd and As. However, *ArabidopsisMYB72* is well known to maintain homeostasis and its knockout mutant is highly sensitive to higher concentrations of Zn as well the Fe deficient conditions with respect to its wild type [128, 108]. WRKY is another TFs, known to get regulated under the HMs stress conditions. WRKY proteins show their composition with a zinc finger motif and a WRKY domain *(WRKYGQK),* which are quite effective in recognizing the downstream genes' cis-acting W-box elements (TTGACC/T). The radish genome database shows a total of 126 WRKY genes, among these RT-qPCR studies found that 36 *RsWRKY* genes undergo specific changes in the regulations under different HM stress conditions. Whereas Pb and Cd are found to mediate the induction of 20 and 24 *RsWRKY*gene transcripts [129]. On the other hand, comprehensive transcriptome studies in Soybeans concluded the presence of 29 Cd-responsive WRKY genes. However, *A. thaliana* and soybean show the overexpression of *GmWRKY142* and restrict the Cd uptake as well as help in the regulation of Cd stress tolerance [130]. HSF (heat shock transcription factors) is another TF known to get activated to respond the external heat stresses. *Sedum alfredii* is a Pb/Zn/Cd hyperaccumulator found to have 22 *Hsf* members and can be grouped into three different classes *SaHsfA, SaHsfB*, and *SaHsfC,* which are mainly based on its phylogenetic clustering. However, 18 *SaHsfs* are active against Cd toxicity, whereas the Class *HsfA* members mediate the HMs stress tolerance by the expression regulation of multiple genes like antioxidants and HM chelators [131, 108]. However, the mechanism of MAPKs cascade and the activation of transcription factors (TFs) to counter HM stresses is provided in Fig. (**5**).

CONCLUSION

This chapter discussed how HM stress can cause nutritional deficiency, decreased photosynthesis, oxidative stress, reduced development and yield, and ultimately plant mortality. Various signals are transduced during stressful conditions in order to induce physiological and biochemical changes necessary for survival. These responses in plants are regulated by the expression of multiple genes, the interaction between various molecules, and an increase in the activity of antioxidative enzymes. When plant roots encounter stress, multiple signalling molecules are activated, resulting in the activation of the MAPK cascade. In addition, the MAPK cascade is involved in the activation of transcription factors that augment the absorption, transport, and detoxification of HMs. Even though there is a vast amount of literature on signal transduction, there is still a knowledge gap regarding the main genes that are involved in the accumulation of HMs in plants. In addition, more research must be conducted on the various plant transduction pathways and their crosstalk with other signaling pathways that are involved in metal stress conditions in plants.

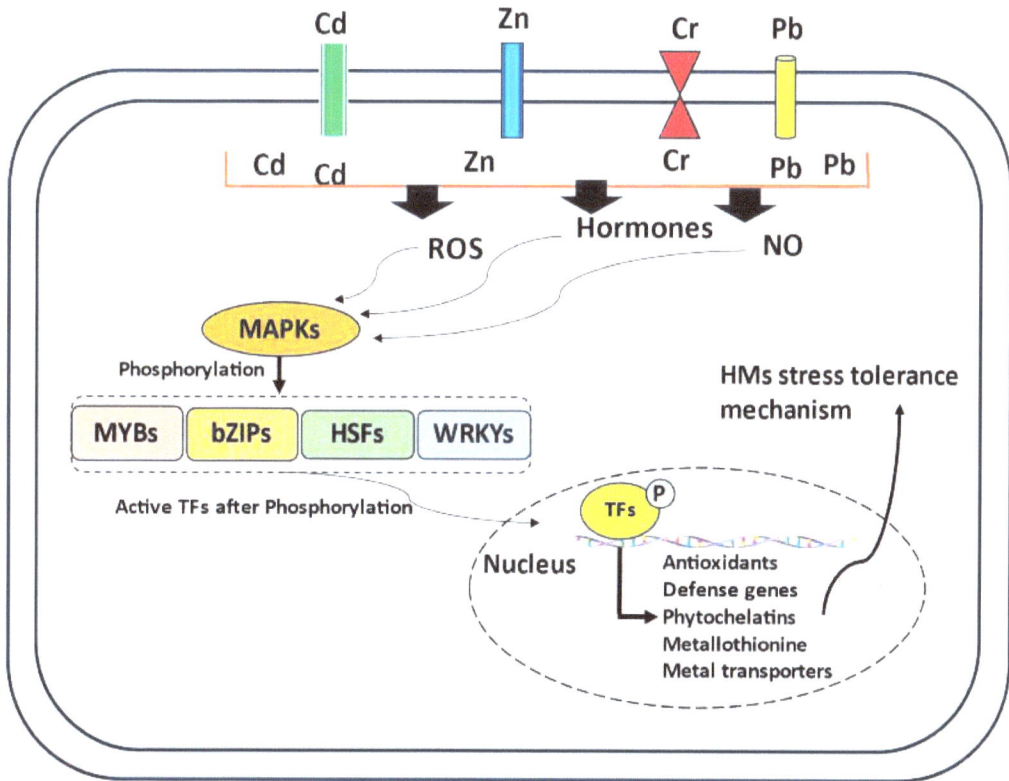

Fig. (5). Depicts the heavy metals exposure and the activation of signaling molecules like NO, ROS, and phytohormones. These signaling molecules further mediate the activation of MAPKs cascades. The activation of transcription factors (TFs) furthermore triggers the activation of defence-related genes, antioxidants, and other molecules associated with the HMs stress amelioration mechanisms.

REFERENCES

[1]   Ebrahimi M, Khalili N, Razi S, Keshavarz-Fathi M, Khalili N, Rezaei N. Effects of lead and cadmium on the immune system and cancer progression. J Environ Health Sci Eng 2020; 18(1): 335-43.

[http://dx.doi.org/10.1007/s40201-020-00455-2] [PMID: 32399244]

[2]   Husen A. The harsh environment and resilient plants: an overview. Harsh Environ Plant Resil: Mol Funct Asp 2021; 1-23.
[http://dx.doi.org/10.1007/978-3-030-65912-7_1]

[3]   Kumar K, Raina SK, Sultan SM. Arabidopsis MAPK signaling pathways and their cross talks in abiotic stress response. J Plant Biochem Biotechnol 2020; 29(4): 700-14.
[http://dx.doi.org/10.1007/s13562-020-00596-3]

[4]   Mondal S, Pramanik K, Ghosh SK, *et al.* Molecular insight into arsenic uptake, transport, phytotoxicity, and defense responses in plants: a critical review. Planta 2022; 255(4): 87.
[http://dx.doi.org/10.1007/s00425-022-03869-4] [PMID: 35303194]

[5]   Sharma D, Verma N, Pandey C, *et al.* MAP kinase as regulators for stress responses in plants: an overview. Protein Kinases Stress Signal Plants: Funct Genomic Perspect 2020; 369-92.

[6]   Sharma SS, Kumar V, Dietz KJ. Emerging trends in metalloid-dependent signaling in plants. Trends Plant Sci 2021; 26(5): 452-71.
[http://dx.doi.org/10.1016/j.tplants.2020.11.003] [PMID: 33257259]

[7]   Andreasson E, Ellis B. Convergence and specificity in the Arabidopsis MAPK nexus. Trends Plant Sci 2010; 15(2): 106-13.
[http://dx.doi.org/10.1016/j.tplants.2009.12.001] [PMID: 20047850]

[8]   Adedeji OH, Olayinka OO, Tope-Ajayi OO, Adekoya AS. Assessing spatial distribution, potential ecological and human health risks of soil heavy metals contamination around a Trailer Park in Nigeria. Sci Am 2020; 10: e00650.

[9]   Alengebawy A, Abdelkhalek ST, Qureshi SR, Wang MQ. Heavy metals and pesticides toxicity in agricultural soil and plants: Ecological risks and human health implications. Toxics 2021; 9(3): 42.
[http://dx.doi.org/10.3390/toxics9030042] [PMID: 33668829]

[10]   Wu Y, Li X, Yu L, Wang T, Wang J, Liu T. Review of soil heavy metal pollution in China: Spatial distribution, primary sources, and remediation alternatives. Resour Conserv Recycling 2022; 181: 106261.
[http://dx.doi.org/10.1016/j.resconrec.2022.106261]

[11]   Huang B, Yuan Z, Li D, Zheng M, Nie X, Liao Y. Effects of soil particle size on the adsorption, distribution, and migration behaviors of heavy metal(loid)s in soil: a review. Environ Sci Process Impacts 2020; 22(8): 1596-615.
[http://dx.doi.org/10.1039/D0EM00189A] [PMID: 32657283]

[12]   Dong Y, Liu S, Sun Y, Liu Y, Wang F. Effects of landscape features on the roadside soil heavy metal distribution in a tropical area in Southwest China. Appl Sci (Basel) 2021; 11(4): 1408.
[http://dx.doi.org/10.3390/app11041408]

[13]   Malunguja GK, Thakur B, Devi A. Heavy metal contamination of forest soils by vehicular emissions: ecological risks and effects on tree productivity. Environ Process 2022; 9(1): 11.
[http://dx.doi.org/10.1007/s40710-022-00567-x]

[14]   Oladoye PO, Olowe OM, Asemoloye MD. Phytoremediation technology and food security impacts of heavy metal contaminated soils: A review of literature. Chemosphere 2022; 288(Pt 2): 132555.
[http://dx.doi.org/10.1016/j.chemosphere.2021.132555] [PMID: 34653492]

[15]   Tang B, Xu H, Song F, Ge H, Yue S. Effects of heavy metals on microorganisms and enzymes in soils of lead–zinc tailing ponds. Environ Res 2022; 207: 112174.
[http://dx.doi.org/10.1016/j.envres.2021.112174] [PMID: 34637758]

[16]   Zhang Q, Wang C. Natural and human factors affect the distribution of soil heavy metal pollution: a review. Water Air Soil Pollut 2020; 231(7): 350.
[http://dx.doi.org/10.1007/s11270-020-04728-2]

[17] Aguilera A, Bautista F, Gutiérrez-Ruiz M, Ceniceros-Gómez AE, Cejudo R, Goguitchaichvili A. Heavy metal pollution of street dust in the largest city of Mexico, sources and health risk assessment. Environ Monit Assess 2021; 193: 1-6.

[18] Sidhu GS. Heavy metal toxicity in soils: sources, remediation technologies and challenges. Adv Plants Agric Res 2016; 5(1): 445-6.

[19] Li C, Zhou K, Qin W, et al. A review on heavy metals contamination in soil: effects, sources, and remediation techniques. Soil Sediment Contam 2019; 28(4): 380-94.
[http://dx.doi.org/10.1080/15320383.2019.1592108]

[20] Popovych V, Gapalo A. Monitoring of ground forest fire impact on heavy metals content in edafic horizons. J Ecol Eng 2021; 22(5): 96-103.
[http://dx.doi.org/10.12911/22998993/135872]

[21] Rybalova O, Bondarenko O, Korobkina K, Zolotarova S. Influence from forest fires on the environment. Sci Herit 2021; 74(1): 17-20.

[22] Yang J, Sun Y, Wang Z, et al. Heavy metal pollution in agricultural soils of a typical volcanic area: Risk assessment and source appointment. Chemosphere 2022; 304: 135340.
[http://dx.doi.org/10.1016/j.chemosphere.2022.135340] [PMID: 35709847]

[23] Sodango TH, Li X, Sha J, Bao Z. Review of the spatial distribution, source and extent of heavy metal pollution of soil in China: impacts and mitigation approaches. J Health Pollut 2018; 8(17): 53-70.
[http://dx.doi.org/10.5696/2156-9614-8.17.53] [PMID: 30524849]

[24] Vareda JP, Valente AJM, Durães L. Assessment of heavy metal pollution from anthropogenic activities and remediation strategies: A review. J Environ Manage 2019; 246: 101-18.
[http://dx.doi.org/10.1016/j.jenvman.2019.05.126] [PMID: 31176176]

[25] Sellami S, Zeghouan O, Dhahri F, Mechi L, Moussaoui Y, Kebabi B. Assessment of heavy metal pollution in urban and peri-urban soil of Setif city (High Plains, eastern Algeria). Environ Monit Assess 2022; 194(2): 126.
[http://dx.doi.org/10.1007/s10661-022-09781-4] [PMID: 35080670]

[26] Soltani-Gerdefaramarzi S, Ghasemi M, Ghanbarian B. Geogenic and anthropogenic sources identification and ecological risk assessment of heavy metals in the urban soil of Yazd, central Iran. PLoS One 2021; 16(11): e0260418.
[http://dx.doi.org/10.1371/journal.pone.0260418] [PMID: 34843585]

[27] Xu Y, Shi H, Fei Y, Wang C, Mo L, Shu M. Identification of soil heavy metal sources in a large-scale area affected by industry. Sustainability (Basel) 2021; 13(2): 511.
[http://dx.doi.org/10.3390/su13020511]

[28] Meng L, Alengebawy A, Ai P, Jin K, Chen M, Pan Y. Techno-economic assessment of three modes of large-scale crop residue utilization projects in china. Energies 2020; 13(14): 3729.
[http://dx.doi.org/10.3390/en13143729]

[29] Chen XX, Liu YM, Zhao QY, Cao WQ, Chen XP, Zou CQ. Health risk assessment associated with heavy metal accumulation in wheat after long-term phosphorus fertilizer application. Environ Pollut 2020; 262: 114348.
[http://dx.doi.org/10.1016/j.envpol.2020.114348] [PMID: 32182536]

[30] Ai P, Jin K, Alengebawy A, et al. Effect of application of different biogas fertilizer on eggplant production: Analysis of fertilizer value and risk assessment. Environ Technol Innov 2020; 19: 101019.
[http://dx.doi.org/10.1016/j.eti.2020.101019]

[31] Fan Y, Li Y, Li H, Cheng F. Evaluating heavy metal accumulation and potential risks in soil-plant systems applied with magnesium slag-based fertilizer. Chemosphere 2018; 197: 382-8.
[http://dx.doi.org/10.1016/j.chemosphere.2018.01.055] [PMID: 29366954]

[32] Komárek M, Čadková E, Chrastný V, Bordas F, Bollinger JC. Contamination of vineyard soils with

fungicides: A review of environmental and toxicological aspects. Environ Int 2010; 36(1): 138-51.
[http://dx.doi.org/10.1016/j.envint.2009.10.005] [PMID: 19913914]

[33] Wang X, Liu W, Li Z, Teng Y, Christie P, Luo Y. Effects of long-term fertilizer applications on peanut yield and quality and plant and soil heavy metal accumulation. Pedosphere 2020; 30(4): 555-62.
[http://dx.doi.org/10.1016/S1002-0160(17)60457-0]

[34] Chen R, Han L, Liu Z, *et al.* Assessment of soil-heavy metal pollution and the health risks in a mining area from Southern Shaanxi Province, China. Toxics 2022; 10(7): 385.
[http://dx.doi.org/10.3390/toxics10070385] [PMID: 35878290]

[35] Hossen MA, Chowdhury AIH, Mullick MRA, Hoque A. Heavy metal pollution status and health risk assessment vicinity to Barapukuria coal mine area of Bangladesh. Environ Nanotechnol Monit Manag 2021; 16: 100469.
[http://dx.doi.org/10.1016/j.enmm.2021.100469]

[36] Hui W, Hao Z, Hongyan T, Jiawei W, Anna L. Heavy metal pollution characteristics and health risk evaluation of soil around a tungsten-molybdenum mine in Luoyang, China. Environ Earth Sci 2021; 80(7): 293.
[http://dx.doi.org/10.1007/s12665-021-09539-0]

[37] Jeong H, Choi JY, Ra K. Heavy metal pollution assessment in stream sediments from urban and different types of industrial areas in South Korea. Soil Sediment Contam 2021; 30(7): 804-18.
[http://dx.doi.org/10.1080/15320383.2021.1893646]

[38] Hoque MM, Sarker A, Sarker ME, *et al.* Heavy metals in sediments of an urban river at the vicinity of tannery industries in Bangladesh: a preliminary study for ecological and human health risk. Int J Environ Anal Chem 2021; 103: 1-9.

[39] Zhou XY, Wang XR. Impact of industrial activities on heavy metal contamination in soils in three major urban agglomerations of China. J Clean Prod 2019; 230: 1-10.
[http://dx.doi.org/10.1016/j.jclepro.2019.05.098]

[40] Shammi SA, Salam A, Khan MAH. Assessment of heavy metal pollution in the agricultural soils, plants, and in the atmospheric particulate matter of a suburban industrial region in Dhaka, Bangladesh. Environ Monit Assess 2021; 193(2): 104.
[http://dx.doi.org/10.1007/s10661-021-08848-y] [PMID: 33521861]

[41] Obaideen K, Shehata N, Sayed ET, Abdelkareem MA, Mahmoud MS, Olabi AG. The role of wastewater treatment in achieving sustainable development goals (SDGs) and sustainability guideline. Energy Nexus 2022; 7: 100112.
[http://dx.doi.org/10.1016/j.nexus.2022.100112]

[42] Nahar N, Shahadat Hossen M. Influence of sewage sludge application on soil properties, carrot growth and heavy metal uptake. Commun Soil Sci Plant Anal 2021; 52(1): 1-10.
[http://dx.doi.org/10.1080/00103624.2020.1836201]

[43] Nahar N, Shahadat Hossen M. Influence of sewage sludge application on soil properties, carrot growth and heavy metal uptake. Commun Soil Sci Plant Anal 2021; 52(1): 1-10.
[http://dx.doi.org/10.1080/00103624.2020.1836201]

[44] Sayo S, Kiratu JM, Nyamato GS. Heavy metal concentrations in soil and vegetables irrigated with sewage effluent: A case study of Embu sewage treatment plant, Kenya. Sci Am 2020; 8: e00337.

[45] Swain A, Singh SK, Mohapatra KK, Patra A. Sewage sludge amendment affects spinach yield, heavy metal bioaccumulation, and soil pollution indexes. Arab J Geosci 2021; 14(8): 717.
[http://dx.doi.org/10.1007/s12517-021-07078-3]

[46] Chaoua S, Boussaa S, El Gharmali A, Boumezzough A. Impact of irrigation with wastewater on accumulation of heavy metals in soil and crops in the region of Marrakech in Morocco. J Saudi Soc Agric Sci 2019; 18(4): 429-36.

[http://dx.doi.org/10.1016/j.jssas.2018.02.003]

[47] Sulok KMT, Ahmed OH, Khew CY, *et al.* Chemical and biological characteristics of organic amendments produced from selected agro-wastes with potential for sustaining soil health: A laboratory assessment. Sustainability (Basel) 2021; 13(9): 4919.
[http://dx.doi.org/10.3390/su13094919]

[48] Rashad M, Hafez M, Popov AI, Gaber H. Toward sustainable agriculture using extracts of natural materials for transferring organic wastes to environmental-friendly ameliorants in Egypt. Int J Environ Sci Technol 2023; 20(7): 7417-32.
[http://dx.doi.org/10.1007/s13762-022-04438-8]

[49] Carpio MJ, Sánchez-Martín MJ, Rodríguez-Cruz MS, Marín-Benito JM. Effect of organic residues on pesticide behavior in soils: a review of laboratory research. Environments (Basel) 2021; 8(4): 32.
[http://dx.doi.org/10.3390/environments8040032]

[50] Abbott LK, Macdonald LM, Wong MTF, Webb MJ, Jenkins SN, Farrell M. Potential roles of biological amendments for profitable grain production – A review. Agric Ecosyst Environ 2018; 256: 34-50.
[http://dx.doi.org/10.1016/j.agee.2017.12.021]

[51] Mann RM, Vijver MG, Peijnenburg WJ. Metals and metalloids in terrestrial systems: bioaccumulation, biomagnification and subsequent adverse effects. Ecol Impacts Tox Chem 2011; 43-62.

[52] Liu WR, Zeng D, She L, *et al.* Comparisons of pollution characteristics, emission situations, and mass loads for heavy metals in the manures of different livestock and poultry in China. Sci Total Environ 2020; 734: 139023.
[http://dx.doi.org/10.1016/j.scitotenv.2020.139023] [PMID: 32460066]

[53] Li J, Xu Y, Wang L, Li F. Heavy metal occurrence and risk assessment in dairy feeds and manures from the typical intensive dairy farms in China. Environ Sci Pollut Res Int 2019; 26(7): 6348-58.
[http://dx.doi.org/10.1007/s11356-019-04125-1] [PMID: 30617882]

[54] Ma J, Chen Y, Antoniadis V, Wang K, Huang Y, Tian H. Assessment of heavy metal(loid)s contamination risk and grain nutritional quality in organic waste-amended soil. J Hazard Mater 2020; 399: 123095.
[http://dx.doi.org/10.1016/j.jhazmat.2020.123095] [PMID: 32534402]

[55] Khan A, Khan S, Khan MA, Qamar Z, Waqas M. The uptake and bioaccumulation of heavy metals by food plants, their effects on plants nutrients, and associated health risk: a review. Environ Sci Pollut Res Int 2015; 22(18): 13772-99.
[http://dx.doi.org/10.1007/s11356-015-4881-0] [PMID: 26194234]

[56] Zwolak A, Sarzyńska M, Szpyrka E, Stawarczyk K. Sources of soil pollution by heavy metals and their accumulation in vegetables: A review. Water Air Soil Pollut 2019; 230(7): 164.
[http://dx.doi.org/10.1007/s11270-019-4221-y]

[57] Ozaki H, Watanabe I, Kuno K. Investigation of the heavy metal sources in relation to automobiles. Water Air Soil Pollut 2004; 157(1-4): 209-23.
[http://dx.doi.org/10.1023/B:WATE.0000038897.63818.f7]

[58] Wang S, Cai LM, Wen HH, Luo J, Wang QS, Liu X. Spatial distribution and source apportionment of heavy metals in soil from a typical county-level city of Guangdong Province, China. Sci Total Environ 2019; 655: 92-101.
[http://dx.doi.org/10.1016/j.scitotenv.2018.11.244] [PMID: 30469072]

[59] Shahid M, Dumat C, Khalid S, Schreck E, Xiong T, Niazi NK. Foliar heavy metal uptake, toxicity and detoxification in plants: A comparison of foliar and root metal uptake. J Hazard Mater 2017; 325: 36-58.
[http://dx.doi.org/10.1016/j.jhazmat.2016.11.063] [PMID: 27915099]

[60] Yan A, Wang Y, Tan SN, Mohd Yusof ML, Ghosh S, Chen Z. Phytoremediation: a promising

approach for revegetation of heavy metal-polluted land. Front Plant Sci 2020; 11: 359.
[http://dx.doi.org/10.3389/fpls.2020.00359] [PMID: 32425957]

[61] Thakur S, Sharma SS. Characterization of seed germination, seedling growth, and associated metabolic responses of Brassica juncea L. cultivars to elevated nickel concentrations. Protoplasma 2016; 253(2): 571-80.
[http://dx.doi.org/10.1007/s00709-015-0835-0] [PMID: 26025262]

[62] Li L, Zhang Y, Ippolito JA, Xing W, Qiu K, Yang H. Lead smelting effects heavy metal concentrations in soils, wheat, and potentially humans. Environ Pollut 2020; 257: 113641.
[http://dx.doi.org/10.1016/j.envpol.2019.113641] [PMID: 31767230]

[63] Pasricha S, Mathur V, Garg A, Lenka S, Verma K, Agarwal S. Molecular mechanisms underlying heavy metal uptake, translocation and tolerance in hyperaccumulators-an analysis. Environ Chall (Amst) 2021; 4: 100197.
[http://dx.doi.org/10.1016/j.envc.2021.100197]

[64] Thakur S, Singh L, Wahid ZA, Siddiqui MF, Atnaw SM, Din MFM. Plant-driven removal of heavy metals from soil: uptake, translocation, tolerance mechanism, challenges, and future perspectives. Environ Monit Assess 2016; 188(4): 206.
[http://dx.doi.org/10.1007/s10661-016-5211-9] [PMID: 26940329]

[65] Shahid M, Dumat C, Pourrut B, Abbas G, Shahid N, Pinelli E. Role of metal speciation in lead-induced oxidative stress to *Vicia faba* roots. Russ J Plant Physiol 2015; 62(4): 448-54.
[http://dx.doi.org/10.1134/S1021443715040159]

[66] DalCorso G, Fasani E, Manara A, Visioli G, Furini A. Heavy metal pollutions: state of the art and innovation in phytoremediation. Int J Mol Sci 2019; 20(14): 3412.
[http://dx.doi.org/10.3390/ijms20143412] [PMID: 31336773]

[67] Dar FA, Pirzadah TB, Malik B. Accumulation of heavy metals in medicinal and aromatic plants. Plant Micronutr: Defic Tox Manag 2020; 113-27.
[http://dx.doi.org/10.1007/978-3-030-49856-6_6]

[68] Säumel I, Kotsyuk I, Hölscher M, Lenkereit C, Weber F, Kowarik I. How healthy is urban horticulture in high traffic areas? Trace metal concentrations in vegetable crops from plantings within inner city neighbourhoods in Berlin, Germany. Environ Pollut 2012; 165: 124-32.
[http://dx.doi.org/10.1016/j.envpol.2012.02.019] [PMID: 22445920]

[69] Pourrut B, Shahid M, Douay F, Dumat C, Pinelli E. Molecular mechanisms involved in lead uptake, toxicity and detoxification in higher plants. Heavy Met Stress Plants 2013; 121-47.
[http://dx.doi.org/10.1007/978-3-642-38469-1_7]

[70] Kumar R, Sharma S, Kaundal M, Sharma S, Thakur M. Response of damask rose (*Rosa damascena* Mill.) to foliar application of magnesium (Mg), copper (Cu) and zinc (Zn) sulphate under western Himalayas. Ind Crops Prod 2016; 83: 596-602.
[http://dx.doi.org/10.1016/j.indcrop.2015.12.065]

[71] Priyadarshanee M, Mahto U, Das S. Mechanism of toxicity and adverse health effects of environmental pollutants. Microb Biodegrad Bioremediat 2022; 33-53.

[72] Sweta SB. A review on heavy metal and metalloid contamination of vegetables: addressing the global safe food security concern. Int J Environ Anal Chem 2022; 104(3): 1-22.

[73] Feng Z, Ji S, Ping J, Cui D. Recent advances in metabolomics for studying heavy metal stress in plants. Trends Analyt Chem 2021; 143: 116402.
[http://dx.doi.org/10.1016/j.trac.2021.116402]

[74] Bungau S, Behl T, Aleya L, *et al.* Expatiating the impact of anthropogenic aspects and climatic factors on long-term soil monitoring and management. Environ Sci Pollut Res Int 2021; 28(24): 30528-50.
[http://dx.doi.org/10.1007/s11356-021-14127-7] [PMID: 33905061]

[75] Rai GK, Bhat BA, Mushtaq M, *et al.* Insights into decontamination of soils by phytoremediation: A

detailed account on heavy metal toxicity and mitigation strategies. Physiol Plant 2021; 173(1): ppl.13433.
[http://dx.doi.org/10.1111/ppl.13433] [PMID: 33864701]

[76]   Khalef RN, Hassan AI, Saleh HM. Heavy metal's environmental impact In: Saleh HM and Hassan AI Environmental Impact and Remediation of Heavy Metals IntechOpen; 2022
[http://dx.doi.org/10.5772/intechopen.103907]

[77]   Mukherjee S, Chatterjee N, Sircar A, *et al.* A comparative analysis of heavy metal effects on medicinal plants. Appl Biochem Biotechnol 2023; 195(4): 2483-518.
[http://dx.doi.org/10.1007/s12010-022-03938-0] [PMID: 35488955]

[78]   Saini S, Kaur N, Pati PK. Phytohormones: Key players in the modulation of heavy metal stress tolerance in plants. Ecotoxicol Environ Saf 2021; 223: 112578.
[http://dx.doi.org/10.1016/j.ecoenv.2021.112578] [PMID: 34352573]

[79]   Zakaria Z, Zulkafflee NS, Mohd Redzuan NA, *et al.* AbdullRazis AF. Understanding potential heavy metal contamination, absorption, translocation and accumulation in rice and human health risks. Plants 2021; 10(6): 1070.
[http://dx.doi.org/10.3390/plants10061070] [PMID: 34073642]

[80]   Naing AH, Kim CK. Abiotic stress-induced anthocyanins in plants: Their role in tolerance to abiotic stresses. Physiol Plant 2021; 172(3): 1711-23.
[http://dx.doi.org/10.1111/ppl.13373] [PMID: 33605458]

[81]   Verma S, Bhatt P, Verma A, Mudila H, Prasher P, Rene ER. Microbial technologies for heavy metal remediation: effect of process conditions and current practices. Clean Technol Environ Policy 2021; 1-23.

[82]   Kosakivska IV, Babenko LM, Romanenko KO, Korotka IY, Potters G. Molecular mechanisms of plant adaptive responses to heavy metals stress. Cell Biol Int 2021; 45(2): 258-72.
[http://dx.doi.org/10.1002/cbin.11503] [PMID: 33200493]

[83]   Saeed Q, Xiukang W, Haider FU, *et al.* Rhizosphere bacteria in plant growth promotion, biocontrol, and bioremediation of contaminated sites: A comprehensive review of effects and mechanisms. Int J Mol Sci 2021; 22(19): 10529.
[http://dx.doi.org/10.3390/ijms221910529] [PMID: 34638870]

[84]   Abdul Rahman NSN, Abdul Hamid NW, Nadarajah K. Effects of abiotic stress on soil microbiome. Int J Mol Sci 2021; 22(16): 9036.
[http://dx.doi.org/10.3390/ijms22169036] [PMID: 34445742]

[85]   Raklami A, Meddich A, Pajuelo E, Marschner B, Heinze S, Oufdou K. Combined application of marble waste and beneficial microorganisms: toward a cost-effective approach for restoration of heavy metals contaminated sites. Environ Sci Pollut Res Int 2022; 29(30): 45683-97.
[http://dx.doi.org/10.1007/s11356-022-19149-3] [PMID: 35147874]

[86]   Fakhar A, Gul B, Gurmani AR, *et al.* Heavy metal remediation and resistance mechanism of *Aeromonas*, *Bacillus*, and *Pseudomonas*: A review. Crit Rev Environ Sci Technol 2022; 52(11): 1868-914.
[http://dx.doi.org/10.1080/10643389.2020.1863112]

[87]   Shuaib M, Azam N, Bahadur S, Romman M, Yu Q, Xuexiu C. Variation and succession of microbial communities under the conditions of persistent heavy metal and their survival mechanism. Microb Pathog 2021; 150: 104713.
[http://dx.doi.org/10.1016/j.micpath.2020.104713] [PMID: 33387608]

[88]   Manzoor Z, Hassan Z, Ul-Allah S, *et al.* Transcription factors involved in plant responses to heavy metal stress adaptation. Plant Perspect Glob Clim Changes 2022; 221-31.
[http://dx.doi.org/10.1016/B978-0-323-85665-2.00021-2]

[89]   Tighe-Neira R, Gonzalez-Villagra J, Nunes-Nesi A, Inostroza-Blancheteau C. Impact of nanoparticles

and their ionic counterparts derived from heavy metals on the physiology of food crops. Plant Physiol Biochem 2022; 172: 14-23.
[http://dx.doi.org/10.1016/j.plaphy.2021.12.036] [PMID: 35007890]

[90]   Sychta K, Słomka A, Kuta E. Insights into plant programmed cell death induced by heavy metals—Discovering a terra incognita. Cells 2021; 10(1): 65.
[http://dx.doi.org/10.3390/cells10010065] [PMID: 33406697]

[91]   Pal A, Bhattacharjee S, Saha J, Sarkar M, Mandal P. Bacterial survival strategies and responses under heavy metal stress: a comprehensive overview. Crit Rev Microbiol 2022; 48(3): 327-55.
[http://dx.doi.org/10.1080/1040841X.2021.1970512] [PMID: 34473592]

[92]   Shen N, Xu C, Zhang J, *et al.* Molecular mechanism underlying cadmium tolerance differentiation in Lentinula edodes as revealed by mRNA and milRNA analyses. J Hazard Mater 2022; 440: 129841.
[http://dx.doi.org/10.1016/j.jhazmat.2022.129841]

[93]   Dutta S, Mitra M, Agarwal P, *et al.* Oxidative and genotoxic damages in plants in response to heavy metal stress and maintenance of genome stability. Plant Signal Behav 2018; 13(8): 1-49.
[http://dx.doi.org/10.1080/15592324.2018.1460048] [PMID: 29621424]

[94]   Gao H, Huang L, Gong Z, *et al.* Exogenous melatonin application improves resistance to high manganese stress through regulating reactive oxygen species scavenging and ion homeostasis in tobacco. Plant Growth Regul 2022; 98(2): 219-33.
[http://dx.doi.org/10.1007/s10725-022-00857-2]

[95]   Janeeshma E, Puthur JT. Physiological and metabolic dynamism in mycorrhizal and non-mycorrhizal *Oryza sativa* (var. Varsha) subjected to Zn and Cd toxicity: a comparative study. Environ Sci Pollut Res Int 2023; 30(2): 3668-87.
[http://dx.doi.org/10.1007/s11356-022-22478-y] [PMID: 35953749]

[96]   Ahmad S, Mfarrej MFB, El-Esawi MA, *et al.* Chromium-resistant *Staphylococcus aureus* alleviates chromium toxicity by developing synergistic relationships with zinc oxide nanoparticles in wheat. Ecotoxicol Environ Saf 2022; 230: 113142.
[http://dx.doi.org/10.1016/j.ecoenv.2021.113142] [PMID: 34990991]

[97]   Karmous I, Gammoudi N, Chaoui A. Assessing the potential role of zinc oxide nanoparticles for mitigating cadmium toxicity in *Capsicum annuum* L. under *in vitro* conditions. J Plant Growth Regul 2023; 42(2): 719-34.
[http://dx.doi.org/10.1007/s00344-022-10579-4]

[98]   Alharby HF, Ali S. Combined role of Fe nanoparticles (Fe NPs) and *Staphylococcus aureus* L. in the alleviation of chromium stress in rice plants. Life (Basel) 2022; 12(3): 338.
[http://dx.doi.org/10.3390/life12030338] [PMID: 35330089]

[99]   El-Shehawi AM, Arshi MJB, Elseehy MM, Kabir AH. Sugarcane bagasse acts as a metal absorber in the rhizosphere in mitigating arsenic toxicity in wheat. Rend Lincei Sci Fis Nat 2022; 33(3): 603-12.
[http://dx.doi.org/10.1007/s12210-022-01074-9]

[100]   Fatemi H, Esmaiel Pour B, Rizwan M. Foliar application of silicon nanoparticles affected the growth, vitamin C, flavonoid, and antioxidant enzyme activities of coriander (*Coriandrum sativum* L.) plants grown in lead (Pb)-spiked soil. Environ Sci Pollut Res Int 2021; 28(2): 1417-25.
[http://dx.doi.org/10.1007/s11356-020-10549-x] [PMID: 32839908]

[101]   Kaushik S, Sharma P, Kaur G, *et al.* Seed priming with methyl jasmonate mitigates copper and cadmium toxicity by modifying biochemical attributes and antioxidants in *Cajanus cajan.*. Saudi J Biol Sci 2022; 29(2): 721-9.
[http://dx.doi.org/10.1016/j.sjbs.2021.12.014] [PMID: 35197737]

[102]   Ren QQ, Huang ZR, Huang WL, *et al.* Physiological and molecular adaptations of *Citrus grandis* roots to long-term copper excess revealed by physiology, metabolome and transcriptome. Environ Exp Bot 2022; 203: 105049.
[http://dx.doi.org/10.1016/j.envexpbot.2022.105049]

[103]   Rehman S, Mansoora N, Al-Dhumri SA, *et al.* Associative effects of activated carbon biochar and arbuscular mycorrhizal fungi on wheat for reducing nickel food chain bioavailability. Environ Technol Innov 2022; 26: 102539.
[http://dx.doi.org/10.1016/j.eti.2022.102539]

[104]   Doria-Manzur A, Sharifan H, Tejeda-Benitez L. Application of zinc oxide nanoparticles to promote remediation of nickel by Sorghum bicolor: metal ecotoxic potency and plant response. Int J Phytoremediation 2023; 25(1): 98-105.
[http://dx.doi.org/10.1080/15226514.2022.2060934] [PMID: 35452585]

[105]   Najafi-Kakavand S, Karimi N, Ghasempour HR, Raza A, Chaichi M, Modarresi M. Role of jasmonic and salicylic acid on enzymatic changes in the root of two alyssum inflatumnáyr. populations exposed to nickel toxicity. J Plant Growth Regul 2023; 42(3): 1647-64.
[http://dx.doi.org/10.1007/s00344-022-10648-8]

[106]   Yang S, Ling G, Li Q, *et al.* Manganese toxicity-induced chlorosis in sugarcane seedlings involves inhibition of chlorophyll biosynthesis. Crop J 2022; 10(6): 1674-82.
[http://dx.doi.org/10.1016/j.cj.2022.04.008]

[107]   Giannakoula A, Therios I, Chatzissavvidis C. Effect of lead and copper on photosynthetic apparatus in citrus (*Citrus aurantium* L.) plants. The role of antioxidants in oxidative damage as a response to heavy metal stress. Plants 2021; 10(1): 155.
[http://dx.doi.org/10.3390/plants10010155] [PMID: 33466929]

[108]   Li S, Han X, Lu Z, *et al.* MAPK cascades and transcriptional factors: Regulation of heavy metal tolerance in plants. Int J Mol Sci 2022; 23(8): 4463.
[http://dx.doi.org/10.3390/ijms23084463] [PMID: 35457281]

[109]   Chmielowska-Bąk J, Gzyl J, Rucińska-Sobkowiak R, Arasimowicz-Jelonek M, Deckert J. The new insights into cadmium sensing. Front Plant Sci 2014; 5: 245.
[PMID: 24917871]

[110]   Jalmi SK, Sinha AK. ROS mediated MAPK signaling in abiotic and biotic stress- striking similarities and differences. Front Plant Sci 2015; 6: 769.
[http://dx.doi.org/10.3389/fpls.2015.00769] [PMID: 26442079]

[111]   Zhao FY, Wang K, Zhang SY, Ren J, Liu T, Wang X. Crosstalk between ABA, auxin, MAPK signaling, and the cell cycle in cadmium-stressed rice seedlings. Acta Physiol Plant 2014; 36(7): 1879-92.
[http://dx.doi.org/10.1007/s11738-014-1564-2]

[112]   Wu X, Liu Z, Liao W. The involvement of gaseous signaling molecules in plant MAPK cascades: function and signal transduction. Planta 2021; 254(6): 127.
[http://dx.doi.org/10.1007/s00425-021-03792-0] [PMID: 34812934]

[113]   Imran M, Khan AL, Mun BG, *et al.* Melatonin and nitric oxide: Dual players inhibiting hazardous metal toxicity in soybean plants *via* molecular and antioxidant signaling cascades. Chemosphere 2022; 308(Pt 3): 136575.
[http://dx.doi.org/10.1016/j.chemosphere.2022.136575] [PMID: 36155020]

[114]   Mumtaz MA, Hao Y, Mehmood S, *et al.* Physiological and Transcriptomic analysis provide molecular Insight into 24-epibrassinolide mediated Cr (VI)-Toxicity tolerance in pepper plants. Environ Pollut 2022; 306: 119375.
[http://dx.doi.org/10.1016/j.envpol.2022.119375] [PMID: 35500717]

[115]   Liu Y, Liu L, Qi J, Dang P, Xia T. Cadmium activates ZmMPK3-1 and ZmMPK6-1 *via* induction of reactive oxygen species in maize roots. Biochem Biophys Res Commun 2019; 516(3): 747-52.
[http://dx.doi.org/10.1016/j.bbrc.2019.06.116] [PMID: 31253404]

[116]   Ye Y, Li Z, Xing D. Nitric oxide promotes MPK6-mediated caspase-3-like activation in cadmium-induced *Arabidopsis thaliana* programmed cell death. Plant Cell Environ 2013; 36(1): 1-15.

[http://dx.doi.org/10.1111/j.1365-3040.2012.02543.x] [PMID: 22621159]

[117] Jin CW, Mao QQ, Luo BF, Lin XY, Du ST. Mutation of mpk6 enhances cadmium tolerance in Arabidopsis plants by alleviating oxidative stress. Plant Soil 2013; 371(1-2): 387-96. [http://dx.doi.org/10.1007/s11104-013-1699-8]

[118] Rao KP, Vani G, Kumar K, *et al.* Arsenic stress activates MAP kinase in rice roots and leaves. Arch Biochem Biophys 2011; 506(1): 73-82. [http://dx.doi.org/10.1016/j.abb.2010.11.006] [PMID: 21081102]

[119] Liu XM, Kim KE, Kim KC, *et al.* Cadmium activates Arabidopsis MPK3 and MPK6 *via* accumulation of reactive oxygen species. Phytochemistry 2010; 71(5-6): 614-8. [http://dx.doi.org/10.1016/j.phytochem.2010.01.005] [PMID: 20116811]

[120] Yeh CM, Chien PS, Huang HJ. Distinct signalling pathways for induction of MAP kinase activities by cadmium and copper in rice roots. J Exp Bot 2006; 58(3): 659-71. [http://dx.doi.org/10.1093/jxb/erl240] [PMID: 17259646]

[121] Jonak C, Nakagami H, Hirt H. Heavy metal stress. Activation of distinct mitogen-activated protein kinase pathways by copper and cadmium. Plant Physiol 2004; 136(2): 3276-83. [http://dx.doi.org/10.1104/pp.104.045724] [PMID: 15448198]

[122] Cornille A, Gladieux P, Smulders MJM, *et al.* New insight into the history of domesticated apple: secondary contribution of the European wild apple to the genome of cultivated varieties. PLoS Genet 2012; 8(5): e1002703. [http://dx.doi.org/10.1371/journal.pgen.1002703] [PMID: 22589740]

[123] Mao G, Meng X, Liu Y, Zheng Z, Chen Z, Zhang S. Phosphorylation of a WRKY transcription factor by two pathogen-responsive MAPKs drives phytoalexin biosynthesis in Arabidopsis. Plant Cell 2011; 23(4): 1639-53. [http://dx.doi.org/10.1105/tpc.111.084996] [PMID: 21498677]

[124] Fan L, Xu L, Wang Y, Tang M, Liu L. Genome-and transcriptome-wide characterization of bZIP gene family identifies potential members involved in abiotic stress response and anthocyanin biosynthesis in radish (*Raphanus sativus* L.). Int J Mol Sci 2019; 20(24): 6334. [http://dx.doi.org/10.3390/ijms20246334] [PMID: 31888167]

[125] Pan C, Ye H, Zhou W, *et al.* QTL mapping of candidate genes involved in Cd accumulation in rice grain. Zhiwu Xuebao 2021; 56(1): 25.

[126] Dubos C, Stracke R, Grotewold E, Weisshaar B, Martin C, Lepiniec L. MYB transcription factors in Arabidopsis. Trends Plant Sci 2010; 15(10): 573-81. [http://dx.doi.org/10.1016/j.tplants.2010.06.005] [PMID: 20674465]

[127] Hu S, Yu Y, Chen Q, Mu G, Shen Z, Zheng L. OsMYB45 plays an important role in rice resistance to cadmium stress. Plant Sci 2017; 264: 1-8. [http://dx.doi.org/10.1016/j.plantsci.2017.08.002] [PMID: 28969789]

[128] Van De Mortel J, Schat H, Moerland PD, *et al.* Expression differences for genes involved in lignin, glutathione and sulphate metabolism in response to cadmium in *Arabidopsis thaliana* and the related Zn/Cd☐hyperaccumulator *Thlaspi caerulescens*. Plant Cell Environ 2008; 31(3): 301-24. [http://dx.doi.org/10.1111/j.1365-3040.2007.01764.x] [PMID: 18088336]

[129] Karanja BK, Fan L, Xu L, *et al.* Genome-wide characterization of the WRKY gene family in radish (*Raphanus sativus* L.) reveals its critical functions under different abiotic stresses. Plant Cell Rep 2017; 36(11): 1757-73. [http://dx.doi.org/10.1007/s00299-017-2190-4] [PMID: 28819820]

[130] Cai Z, Xian P, Wang H, *et al.* Transcription factor GmWRKY142 confers cadmium resistance by up-regulating the cadmium tolerance 1-like genes. Front Plant Sci 2020; 11: 724.

[http://dx.doi.org/10.3389/fpls.2020.00724] [PMID: 32582254]

[131] Chen SS, Jiang J, Han XJ, Zhang YX, Zhuo RY. Identification, expression analysis of the Hsf family, and characterization of class A4 in *Sedum alfredii* hance under cadmium stress. Int J Mol Sci 2018; 19(4): 1216.
[http://dx.doi.org/10.3390/ijms19041216] [PMID: 29673186]

CHAPTER 4

Bioremediation Strategies for Heavy Metal Detoxification in Plants

Isha Madaan[1,7], Gurvarinder Kaur[1], Jaspreet Kour[2], Kulwinder Singh[3], Diksha Kalia[4,8], Komal Goel[4,8], Harpreet Kaur[5], Tanuja Verma[7], Manish Kumar[6], Parul Preet Gill[6], Rajesh Kumar Singh[4,8], Gaurav Zinta[4,8], Renu Bhardwaj[2] and Geetika Sirhindi[1,*]

[1] *Department of Botany, Punjabi University, Patiala, Punjab, India*

[2] *Department of Botanical and Environmental Sciences, Guru Nanak Dev University, Amritsar, 143001, India*

[3] *Sanmati Govt. College of Science Education and Research, Jagraon, Punjab, India*

[4] *Department of Biotechnology, CSIR-Institute of Himalayan Bioresource Technology, Palampur, Himachal Pradesh, India*

[5] *P.G. Department of Botany, Khalsa College Amritsar, Amritsar, Punjab, India*

[6] *Postgraduate Department of Botany, Dev Samaj College for Women, Ferozepur, Punjab, India*

[7] *Government College of Education, Jalandhar, Punjab, India*

[8] *Academy of Scientific and Innovative Research (AcSIR), Ghaziabad, Uttar Pradesh, India*

Abstract: Contamination of vital environmental components especially water and soil by heavy metals (HMs) due to effluents from various industries and the use of inorganic chemicals for farming is an issue of major concern these days. These heavy metals are non-biodegradable and cannot be vanished by any kind of natural biological or chemical decomposition. Therefore, eco-sustainable methods of heavy metal treatment need to be adopted to eliminate these harmful pollutants. Utilization of microorganisms and plants, living organisms for the eradication of HMs from soil and water has proven to be a highly efficient as well as cost-effective strategy. All types of microbes have the tendency to accumulate the optimum amount of heavy metals within them, however most common are bacteria and algae. Similarly, phytoremediation is a common detoxification mechanism, in which the plants absorb heavy metals from their surroundings and accumulate them in harvestable parts. By studying this chapter, students will become aware of the need for detoxification of heavy metals, various bioremediation strategies (phytoremediation, cyanoremediation, and mycoremediation), and biotechnological advances in these mechanisms for the removal of heavy metals from the ecosystem.

Corresponding author Geetika Sirhindi: Department of Botany, Punjabi University, Patiala, Punjab, India; E-mail: geetika.pbi.ac.in

Kanika Khanna, Sukhmeen Kaur Kohli & Renu Bhardwaj (Eds.)

Keywords: Bioremediation, Phytoremediation, Cyanoremediation, Mycoremediation.

INTRODUCTION

HMs are the concerning contaminants that are being added to the ecosystem on a large scale due to different reasons, both natural as well as anthropogenic. Uncontrolled accumulation of these HMs in soil and water leads to their entry into plant systems and further to the animal and human bodies through the food chain. Most of the HMs have no or little physiological function in living organisms and that too at very low concentrations. Therefore, they have been reported to pose serious threats to normal biological processes in living beings ultimately resulting in plant stress. In order to cope with the deleterious effects of these HMs, some sustainable, long-lasting, and cost-effective strategies must be chalked out. A vast number of research studies done in the past and recently have evidenced the implication of living organisms themselves to protect themselves from the noxious effects of HMs. This method of exploitation of living beings to get rid of HMs is termed Bio-remediation. The living organisms which are being utilised for this purpose include microbes and plants. Among microorganisms, algae, bacteria, and fungi have been reported to show potent efficiency in the elimination of HMs from agricultural soils. Various plants also possess the potential to collect the HMs within their roots and not transport them to the aerial parts, making them suitable remediation agents. This chapter mainly covers the details of processes of different bioremediation strategies along with the most common examples of living organisms being utilised for this purpose. In addition to this, the recent advances in molecular biology pertaining to the genetic engineering of plants for more effective utilisation as bio-remedial agents have also been discussed.

NEED FOR HEAVY METAL DETOXIFICATION

HMs are chemicals with a very high density as a result of which they are potentially toxic to all living organisms. A variety of natural as well as anthropogenic activities may result in HM accumulation in the environment (Fig. 1). One of the various natural processes that lead to the accumulation of Zn, Co, Ni, Cr, Fe, and Mn HMs in the biosphere is weathering of rocks. Waste from gasoline, paint industries, mining, coal combustion, pesticide usage, and leather tanning are the most common human-related causes of HM contamination [1, 2]. Some of the most common sources of various heavy metals have been enlisted briefly in Table 1.

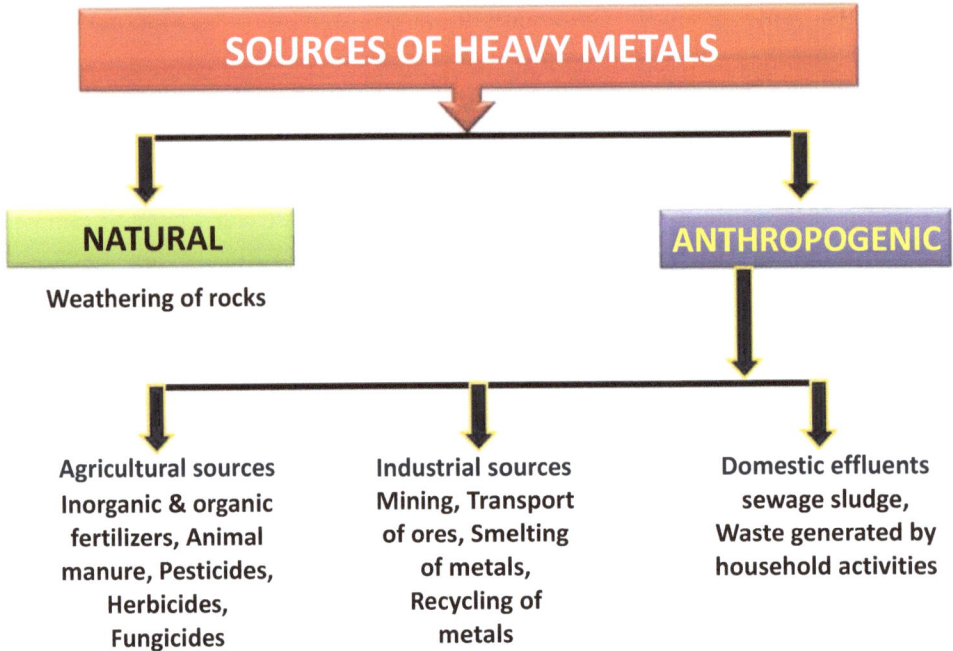

Fig. (1). Natural and anthropogenic sources of heavy metals.

Table 1. Various sources of heavy metals.

Sr. No.	Heavy Metals	Source of Heavy Metals	References
1.	Ar	Arsenical pesticides, Weathering and erosion of As-containing rocks.	[3 - 5]
2.	Hg	Industrial effluents, mining, agricultural waste, wastewater treatment and incineration; ore smelting, biogeochemical cycles.	[6 - 8]
3.	Pb	Fertilizers, Pesticides ore smelting, battery and paints industry effluents.	[9 - 11]
4.	Cr	Electroplating, metal processing, agricultural by-products, dye manufacturing, leather tanning, and paper manufacturing.	[12 - 14]
5.	Al	Acid rain, mining, natural deposition, and soil acidity from ore leaching.	[15, 16]
6.	Cd	Ore outcrops, smelting metal, sewage sludge, and home garbage, agricultural waste and pesticides.	[17 - 19]
7.	Au	Food manufacturing and packaging, jewelry production, pharmaceutical manufacturing, as well as ore leaching and smelting processes.	[20 - 23]

Enormous quantities of HMs pose serious threats to human health. Drainage of water from agricultural fields, mining, and industrial waste is mainly responsible for increasing levels of heavy metals in water bodies [24]. Therefore, the upsurging levels of heavy metals and their derivatives in landfills cause ecological damage with the highest damage to natural organisms with their toxic effects [25]. These toxic metals cause disastrous effects on human and animal bodies [26 - 30]. The first entry into the food chain for these chemicals is producers as many studies on higher plants reported that there are a number of plants that are hyperaccumulators of heavy metals [31 - 33].

Plants and animals bioaccumulate, the toxic metals they may or may not biotransform or detoxify the toxic forms of heavy metals, and when consumed by humans as their daily food source, they can cause metal intolerance [34]. Therefore, humans are at a greater risk of metal toxicity because being omnivores, they have diverse food sources because they do not secrete metals; instead, they store them in their bodies. Higher levels of As, Cd, Pb, Cr, Mn, Ni, Cu, and Zn have been reported in fruits, vegetables, and crops [35 - 37]. A study conducted in different districts of Tianjin, China [38] showed high deposition of Cd, Zn, Pb, Ni, Cu and Cr in locally produced vegetables and fish. Uncontrolled industrial revolution, improper sludge treatment, and other human activities have resulted in the accumulation of different HMs in the environment [39, 40]. These HMs have long persistence and high solubility therefore, may pose serious threats like cancer to humans and animals after entry into their bodies through the food chain [41, 42].

In developing countries like India, various agro-processing industries including tanneries, pulp-paper mills, olive mills, and distilleries have emerged as significant contributors to HM pollution [43, 44]. Their accumulation in tissues and cellular organelles ultimately results in enhanced toxicity by disrupting the absorption of essential elements within the body [45].

THE BASIC IDEA OF BIOREMEDIATION AND ITS VARIOUS STRATEGIES

Although many metals are necessary for plant growth, when their quantity exceeds safe levels, they become toxic. And of these metal ions, HMs like As, Cd, Pb, Hg, *etc.* are challenging to eliminate from the environment. Today, with the fast expansion of agriculture and metal industries, poor waste management practices, and excessive use of fertilizers and pesticides, HM pollution is increasing in our rivers, soil, and environment. In India, the Central Water Commission has reported that the samples from two-thirds of water quality stations are contaminated by one or more heavy metals. To solve this problem, the technique of bio-remediation is becoming increasingly popular because it is both affordable and efficient in nature. Bio-remediation is a technique mainly investigated for the treatment of metal-polluted waste. In this technique, microorganisms or plants are used in the biosorption of metals from contaminated water sources, streams, and soils. Heavy metals are of little importance to organisms but greatly harm living organisms and remain indestructible, unlike other pollutants, as they cannot be degraded chemically or biologically and continue to be a global threat [46]. They are present in the environment as trace elements, however, their elevated concentration above the threshold level is considered toxic [47]. Hence, the removal of these contaminants is crucial for sustainable development. Bioremediation is an effective method for reducing and eliminating HMs and appears to be a promising substitute for traditional methods which are often expensive, hazardous, and inefficient. As a result, employing microorganisms to remove heavy metals has become increasingly recognized as a cost-effective and eco-friendly method. Soil microbes, such as plant growth-promoting bacteria, (PGPB) possess metal-detoxifying traits and employ various strategies to evade metal toxicity [48]. Bioremediation cannot degrade pollutants but can transform them from a toxic state to a less toxic state or make them less bioavailable or volatile. When the process of bio-remediation is carried out specifically by plants, the process is called phytoremediation.

Various bio-remediation strategies include (Fig. **2**):

A. Phyto-remediation strategies
B. Microbial Bio-remediation
C. Combined strategies

Phytoextraction

Phytoextraction refers to the method of using plant roots to absorb or extract heavy metals from soil or water, which are then transported to the above-ground parts of the plant, such as the shoots.

Phytofiltration

Phytofiltration is the process through which plants eliminate soil pollutants from surface and wastewater.

Phytostabilization

In Phytostabilization, the movement, and presence of contaminants in the environment, as well as their infiltration into groundwater, are restricted.

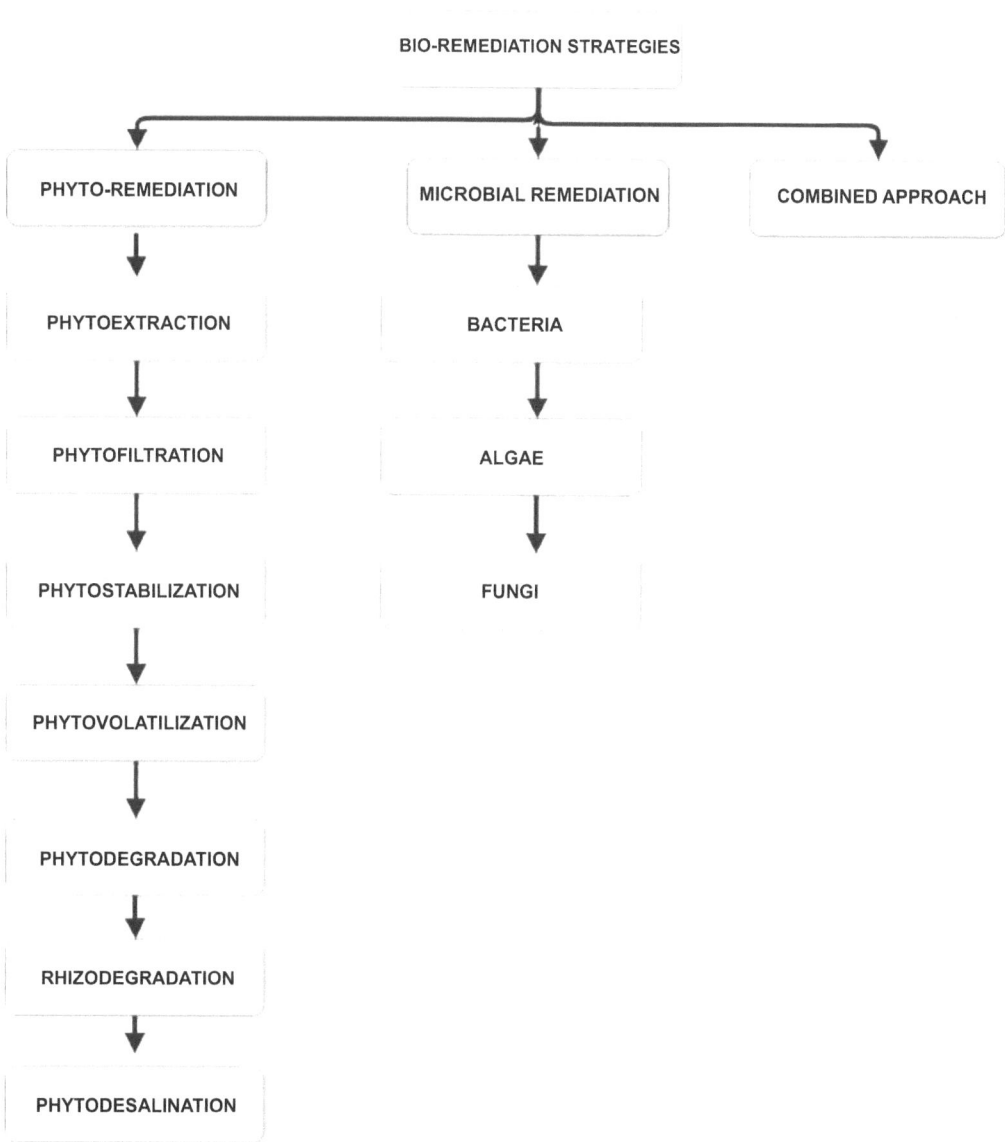

Fig. (2). A brief idea of Bio-remediation strategies.

Phytovolatilization

Phytovolatilization is the process by which plants absorb organic pollutants from soil and convert them into volatile form, for their easy removal into the atmosphere.

Phytodegradation

Phytodegradation is the technique mostly deployed for organic pollutants. In this technique, organic pollutants are removed enzymatically.

Rhizodegradation

Rhizodegradation is the decomposition of soil contaminants through microbial activity, which is promoted by the presence of the rhizosphere.

Phytodesalination

In Phytodesalination, halophytic plants are utilized to extract salts from soils affected by salinity.

MICROBES ASSISTED BIOREMEDIATION OF HEAVY METALS.

Various HMs like Cd, Cr, Pb, Hg, and Pt are released into the environment by anodizing cleaning, electrolyte deposition, and milling industries. Numerous organisms are utilised for HM eradication. The employment of microorganisms for heavy metal cleanup is more favourable, cost-effective, and simpler than other costly procedures. In addition, its availability and absorption capacity are also considerable. For heavy metal remediation, many bacteria (*Bacillus* sp., *Pseudomonas aeruginosa*, *Eichhornia* sp., *Ochrobactrum* sp., *Bacillus cereus*), fungi (*Aspergillus niger*, *Trichoderma*, *Aspergillus fumigates*, *Saccharomyces cerevisiae*), and algae (*Fucus vesiculosus*, *Sargassum*) are utilized [49]. This section addressed the function of several microorganisms in the elimination of HMs.

Cyanoremediation

Algae are utilized in cyanoremediation to remove pollutants, both organic and inorganic, derived from contaminated environments. Algal species such as *Chlorella, Spirulina, Oedogonium,* and *Spirogyra* are utilised for this purpose. As per research reports [50, 51], *Cladophora fascicularis* and *Spirogyra* are used for the remediation of Pb and Cu, respectively. Due to their potential to absorb HMs from their surrounding environment, micro and macro algae are seen as a viable option for heavy metal remediation. Additionally, they have the capacity to absorb heavy metals [52]. It has been discovered that microalgae utilise two processes for the elimination of HMs. These two processes are biosorption and decreased cellular intake of HMs. Microalgal species such as *Pseudochlorococcum typicum*, *Phormidiumambiguum*, and *Scenedesmus quadricauda* aid in the tolerance and eradication of Pb, Cd, and Hg from water [53]. In contrast, macroalgae are utilised for the removal and biomonitoring of HMs in saline aquatic ecosystems [54].

There are numerous algal species used for heavy metal cleanup in plants, as detailed in Table **2** below.

Table 2. Role of various algal species in bioremediation of heavy metals.

S. No.	Algal Species	Heavy Metal/s	References
1.	*Saccharina japonica and Sargassum fusiforme*	Cu, Cd, Zn	[55]
2.	*Chlorella pyrendoidosa*	U	[56]
3.	*Cladophora sp. and Spirogyra sp.*	Pb, Cu	[56]
4.	*Bacillus firmus, Chlorella fusca, Ascophyllum nodosum, Aspergillus niger,*	Cr, Ni, Cu, Cd, Pb, Zn	[56]
5.	*Spirullina sp*	Zn, Fe, Se, Mn, Cr, Cu,	[57]
6.	*Hydrodictylon sp., Oedogonium sp.*	As, V	[58]
7.	*Spirogyra sp.*	Cu, Pb	[59]
8.	*Desmodesmuspleiomorphus*	Cd	[60]

Mycoremediation

Mycoremediation is a bioremediation technique that uses fungus to remediate polluted soils and aqueous effluents from toxic pollutants. Fungi are multicellular microorganisms capable of remediating polluted soil and water *via* their biochemical and ecological potential to break down organic substances. Fungi are easy to grow, have robust growth, and can withstand fluctuating pH and temperature. They can also produce extracellular ligninolytic enzymes and metal-binding proteins, making them ideal candidates for bioremediation. Fungi are known to act on various harmful pollutants, including heavy metals (arsenic, nickel, copper, mercury, lead, chromium, cadmium, and silver), agricultural effluent pollutants (pesticides, herbicides, insecticides, antifungal drugs *etc*), industrial pollutants (polycyclic aromatic hydrocarbon, detergents, cyanotoxins, dyes) and medical and pharmaceuticals pollutants like antibiotics and phthalates, and help in their removal, degradation, or toxicity reduction. Fungi can not only be utilised for removing site contamination but also as bioreactors for the remediation of waste generated from industries. Additionally, fungi in association with plants (Arbuscular mycorrhiza, AM) not only provide them resistance to heavy metal toxicity but, *via* symbiosis, provide them stress tolerance.

Heavy metals are the micronutrients for most organisms, but increasing concentration from their limit causes toxicity. Fungi are highly tolerant to various heavy metals, and their tolerance and accumulation characteristics make them competent bioremediation agents against heavy metal pollution. They apply different biomechanics to survive highly toxic environments caused by heavy metals. Fungi-mediated removal of HMs depends on their interaction with each other, which further depends on several parameters such as the species of organism, type and quantity of the metal, and environmental conditions. Mechanisms involved in detoxifying and tolerating HMs using fungi may include intracellular or extracellular, active or passive acquisition, and valence modification [61]. Some extracellular mechanisms include chelation, precipitation, and binding to the cell wall, whereas intercellular mechanisms include transfer into intracellular compartments, and binding to sulphur compounds, organic acids, peptides, and polyphosphates. These different biomechanics are due to the presence of genes, proteins, and metabolites capable of surface sorption, precipitating, transporting, transforming, and sequestering heavy metals. These compounds are essential for the detoxification of metals [62].

Important Fungal Species involved in Heavy Metal Detoxification

There are numerous fungal species like Trichoderma, Fusarium, Penicillium, Aspergillus, Rhizopus, *etc.*, which can proficiently eradicate heavy metal pollutants from these fungi with filamentous (*Aspergillus sp.* and *Penicillium sp.*) are comparatively more efficient. Pharmaceutical xenobiotics, *e.g.*, acetaminophen [63] effectively eradicated by *Mucor hiemalis,* which is also used in the bioremediation of hydrocarbon pollutants [64].

Moreover, Mycorrhiza fungi not only removed the heavy metals from the soil but also improved the soil structure, which helps in plant nutrient uptake. Botrytis cinerea has the potential for lead removal [65]. The study [66] reviewed independently the comparative ability of keratinolytic fungi to other bacterial decomposers for hydrocarbons removal from different media and the ability to enhance plant growth under stressful conditions and found that Fungi are well suited relatively for PAH degradation [67]. On Earth, fungi are well-known as natural cleaners and decomposers [68]. The various biomechanical mechanisms that fungi use to tolerate heavy metals involve biosorption, bioaccumulation, chelation, and efflux transport. Biosorption refers to the process by which heavy metals are adsorbed onto the cell walls without their uptake into the cell. The specific nature of fungi cell walls *i.e.* their composition (around 90% polysaccharide) and the presence of functional groups (OH, CO, NH_2, and phosphoryl) groups with negative charges in their cell wall helps in non-chemical interaction and adsorption at the surface *via* electrostatic interactions, van der Waals forces, covalent bonding, *etc* [69]. It also acts as a ligand, allowing metal ions to attach. In bioaccumulation, fungi uptake the toxic heavy metals and utilize their machinery to generate the non-harmful variant of metal ions and precipitate/compartmentalise them in different cellular organelles, mainly vacuoles. Fungi also release chelating agents that attach to metal ions and form a non-toxic complex sequestered into different cellular organelles. Chelating molecules secreted by fungi mainly include organic acids, proteins, thiol-containing molecules, peroxidases, *etc.* The fungi, like other microorganisms, also have an efficient efflux system activated when there is a large quantity of HMs inside the cell. This active transport excretes the accumulated HMs in the cytosol out from the cell. It is one of the biomechanics that confers heavy metal resistance to fungi and is used in mycoremediation.

Mycoremediation, similar to most other bioremediation approaches, has some drawbacks: the process is usually slow, and the removal of contaminants is rarely close to 100%. The treatment outcome is influenced by the soil matrix and bioavailability of the specific pollutant. However, bioremediation is more affordable and environmentally friendly compared to physical or chemical treatment methods. Conversely, in many instances, the failure to completely degrade or only partially degrade pollutants results in the buildup of harmful metabolites, influenced by the type of pollutant involved. As a result, genetically engineered microorganisms (GEM) have been developed to alter microorganisms for targeted tasks such as degradation. Due to their simplicity in genetic engineering, scaling up, and transportation, fungi are a valuable solution for bioremediation [70]. Strain manipulation and targeted gene modifications are commonly employed to enhance the desired traits of enzymes and the associated metabolic pathways in microorganisms. By using induced mutation techniques, fungal mutants can be created to overproduce specific enzymes of interest. Using biotechnology methods, the white-rot fungus has also been used to clear polluted soil regions [71]. Genetic algorithms have contributed to the creation of experimental models for biosorption aimed at eliminating Cd (II) with the help of *Trichoderma sp* [72]. As the fungal mechanisms involved in genetic engineering for bioremediation remain poorly understood, there is significant potential for future research in this area [73].

Bacteria in Bioremediation of Heavy Metals

Microorganisms like bacteria are utilized to sequester, precipitate, or alter the oxidation state of HMs to endure their toxicity [74]. Studies reported that instead of a single bacterial strain, a mixture of bacteria is more efficient and provides greater resistance to HM remediation [75]. Microbial remediation uses various mechanisms of bioremediation, such as follows:

• Transformation of metals into harmless forms through enzymatic activity.
• Modulation of the biochemical pathway to block metal uptake.
• Efficient efflux systems reduce the intercellular metal concentration.

- Sequestration of metals by intracellular metal binding protein and peptides like metallothionein and bacterial siderophores.

Bacterial siderophores are iron-chelating substances that reduce the bioavailability of metals, enhance their mobility and ultimately remove them from the soil. For instance, *Azotobacter vinelandii* increases its siderophore production in the presence of Zn (II), leading to its enhanced metal mobility and hence extraction from soil [76]. *Desulfovibrio desulfuricans* transform sulfate into hydrogen sulfate, which interacts with heavy metals such as cadmium (Cd) and zinc (Zn), resulting in the formation of insoluble compounds, a process called bioprecipitation [77]. Bacteria use heavy metals as electron acceptors and generate energy for metal detoxification through both enzymatic and non-enzymatic mechanisms [78]. Biostimulation is the process that includes incorporating organic nutrients into the soil to make it more favorable for soil microbes. The same strategy can be used for the restoration of environments affected by heavy metals. As the organic nutrients reduce the soil pH, metals are more soluble or bioavailable and hence can be easily extracted from the soil [79].

Plant Growth-Promoting Bacteria (PGPB)

PGPB promotes plant growth by either facilitating minerals and elements or the production of phytohormones. They protect the plant from pathogen colonization by either attacking them or by the production of antifungal metabolites and inducing systemic resistance. PGPB also encourages plant growth by aiding in the fixation of nitrogen (N_2) or by lowering ethylene concentration. Rhizosperic bacteria enhance plant growth by improving the accessibility of nutrients present in soil and by producing phytohormones [80]. PGPR resides in the rhizosphere where they promote plant growth either by siderophore production or act as biocontrol where they eliminate harmful substances and pesticides [81]. *Flavobacterium, Micrococcus, Pseudomonas, Chromobacterium, Azospirillium, Azotobacter, and Agrobacterium* all promote plant growth by producing various mechanisms such as siderophore production, phosphate solubilization, ACC deaminase production, and IAA production [82].

Siderophores make nutrients available to plants by solubilizing the unavailable or limited metal or by reducing the pH of soil. Usually, siderophores solubilize the Fe in iron-limited soil and provide it to the plant [83]. Phosphorus is needed by the plants as a macronutrient but they tend to react with heavy metals and become precipitate, and hence is unavailable for plants. Few microbes like *Azotobacter, Bacillus, Microbacterium, Pseudomonas and Rhizobium, Enterobacter* solubilize phosphate and make it available as a soluble form. Similarly, ACC deaminase cleaves ACC thus lowering ethylene levels and developing an extensive root system that helps plants combat the conditions of abiotic stress [84].

PHYTOREMEDIATION AND ITS VARIOUS TECHNOLOGIES.

Phytoremediation of HMs referred to as green technology is a safe and sustainable strategy for HM elimination from contaminated lands [4]. Phytoremediation is an umbrella term comprising two Latin words '*phyto*' and '*remedium*', which means plant and remove/correct the evil, respectively. Since the discovery of technique and the term, the term has been used frequently with several definitions. It is an innovative, emerging, and cost-effective eco-friendly technology to improve the quality of contaminated soils using plants and their rhizospheric organisms [85, 86].

Phytoaccumulation occurs when the plants uptake and accumulate HMs from the soil in the root and shoot parts. On the basis of their ability to accumulate HMs, plants can be classified as hyperaccumulators, indicators, and tolerant plant species [87]. Plants suitable for phytoremediation must be able to accumulate high amounts of HMs along with displaying higher growth characteristics, higher biomass, extensive root system, and be harvested easily. Several plants have the tendency to accumulate HMs and they fall under the families like Brassicaceae, Fabaceae, Amaranthaceae, Lamiaceae Asteraceae, Euphorbiaceae, Cyperaceae, and Poaceae.

There are several mechanisms involved in phytoremediation phytovolatilization, phytodegradation, rhizodegradation, phytoextraction, phytofilteration, and phytostabilization [86, 88, 89]. The plants have the tendency to break or transform the toxic forms of HMs thereby rendering them non-toxic within their tissues and this process is termed phytodegradation. This degradation can take place inside the plant body when plants transform the HMs through various metabolic processes into ineffective form. However, the plants can render toxic from non-toxic ones outside the plant body. This can occur by the action of root exudates or the action of micro-organisms and this is rhizodegradation [90]. Well-developed plant root system helps enhance the uptake of water from the soil and precipitates toxins onto the root surface, which is referred to as rhizofilteration [86, 91]. The process of phytovolatilization helps in converting the HMs into their volatile compounds, which are then emitted into the atmosphere *via* the leaf surfaces [92]. The volatile forms are translocated from roots to shoots and then released through leaves. Phytofilteration is facilitated by the sequestration of HMs in water and phytoextraction helps in the accumulation of HMs in shoots by sequestration mechanisms (Fig. **3**).

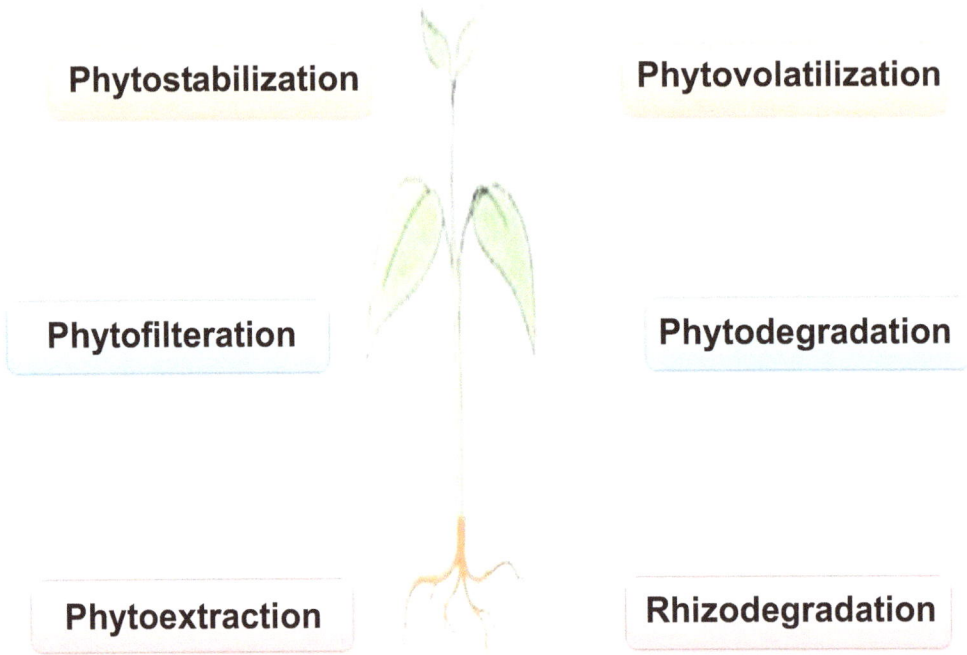

Fig. (3). Various techniques followed during the phytoremediation.

Phytoextraction is further of two types *viz.* induced and continuous. Induced phytoextraction is inducing enhanced phytoextraction using chemicals like chelating molecules and continuous phytoextraction is the natural ability of plants to accumulate HMs. The plants uptake HMs from soil and accumulate them in their roots thereby decreasing their availability to the shoots, this is phytostabilization. Also, the phytostabilised product is not allowed to enter into the water or soil system as it is bound to the root system of plants. Phytoextraction and phytostabilization hold several drawbacks as it is dependent upon the genotype and growth restoration capacity of plants as growth is affected by several factors. Therefore, the use of some genetic tools and other stimulants can help in enhancing growth [93]. A two-year project of phytoremediation introduced in HM-contaminated soil in China elucidated the cost of the project offset and the benefits for seven years [94]. It is a much more cost-effective approach as compared to other approaches like soil replacement, chemical leaching, precipitation, heat treatment and electro-remediation and others Phytoremediation is a successful method for eliminating heavy metals from polluted soils as several plants can act as accumulators of HMs, and potential candidates for HMs removal from soil and water sources.

GENO-REMEDIATION: A   BIOTECHNOLOGICAL   APPROACH TOWARDS HEAVY METAL DETOXIFICATION

As discussed in the chapter above, microorganisms have the capacity to metabolise heavy metals through biochemical processes related to the activity and growth of the organisms. Hazardous compounds in the soil can be degraded by microbes into innocuous byproducts generated during co-metabolism [95]. Certain bacteria and fungi are recognized for their ability to break down specific chemical compounds; however, the application of native microorganisms for cleaning contaminated sites has not yielded many successful results [96]. On the other hand, the genes responsible for bacterial inorganic transformation have been well-characterized, providing a molecular genetic foundation for improved metal tolerance [97 - 99]. Genetic engineering enables the creation of microorganisms designed to degrade specific pollutants. Common techniques include the manipulation of single genes or operons, constructing pathways, and modifying the sequences of existing genes [100]. This engineering approach has facilitated the development of bacteria capable of removing heavy metals such as Arsenic (As), Cadmium (Cd), Copper (Cu), Iron (Fe), Mercury (Hg), and Nickel (Ni) [101, 102]. There have been reports on the successful removal of Hg from polluted soil sediments using genetically modified *Escherichia coli* strain M109 and *Pseudomonas putida* carrying the merA gene [103 - 105]. A Mesorhizobium species transformed with a gene from *Arabidopsis thaliana* that encodes for phytochelatin synthase demonstrated an increased accumulation of Cd^{2+} [100]. Additionally, incorporating the mer operon from *Escherichia coli*, which is responsible for Hg^{2+} reduction, into the genetically engineered bacterium *Deinococcus geothermalis* enabled the organism to reduce mercury contamination at elevated temperatures through the expression of the mer genes [78]. Similarly, the comparative physiological and molecular investigations of related non hyperaccumulators and hyperaccumulator plants have revealed an intriguing finding: the majority of important hyperaccumulation processes depend on distinct regulatory mechanisms and gene expressions that are present in both types of plants. The constitutive overexpression of genes encoding transmembrane transporters, including those from the ZIP, HMA, MATE, YSL, and MTP families, has been observed in many accumulator plants [106, 107]. To enhance phytoremediation characteristics through transgenesis, a common approach is to introduce and overexpress genes that facilitate root uptake, root-to-shoot translocation, sequestration, and detoxification of toxic metals [108, 109].

The insertion of the arsM gene from the bacterium *Rhodopseudomonas palustris* into *Oryza sativa* cv. *Nipponbare* (Japonica rice) enhanced arsenic phytoremediation by promoting methylation and volatilization processes [110]. Tobacco (*Nicotiana tabacum*) plants were genetically modified to express a yeast metallothionein (MT) gene, which increased their tolerance to cadmium. In contrast, a mercury ion reductase gene was overexpressed in *Arabidopsis thaliana* (a type of mustard) to improve its mercury tolerance [111, 112]. Additionally, the algae Spirulina has been studied for their potential to biologically remove low concentrations of lead (below 50 mg/L) from wastewater. It was found that during the initial phase (0–12 minutes), the adsorption rate was extremely rapid, with 74% of the metal being biologically adsorbed, and the maximum biosorption capacity of live Spirulina was estimated to be 0.62 mg of lead per 10^5 algal cells [113]. Recently, researchers in phytoremediation have identified that Indian mustard (*Brassica juncea* (L.) Czern), AABB genome can accumulate high levels of metals, including zinc (Zn), cadmium (Cd), lead (Pb), and selenium (Se), while B. nigra (BB genome) has demonstrated excellent copper (Cu) accumulation capabilities [114 - 116]. These plants are ideal for phytoextraction due to their metal accumulation capacity and their ability to rapidly generate significant amounts of shoot biomass [106]. Consequently, these species have garnered substantial interest in the field of phytoremediation (Fig. **4**).

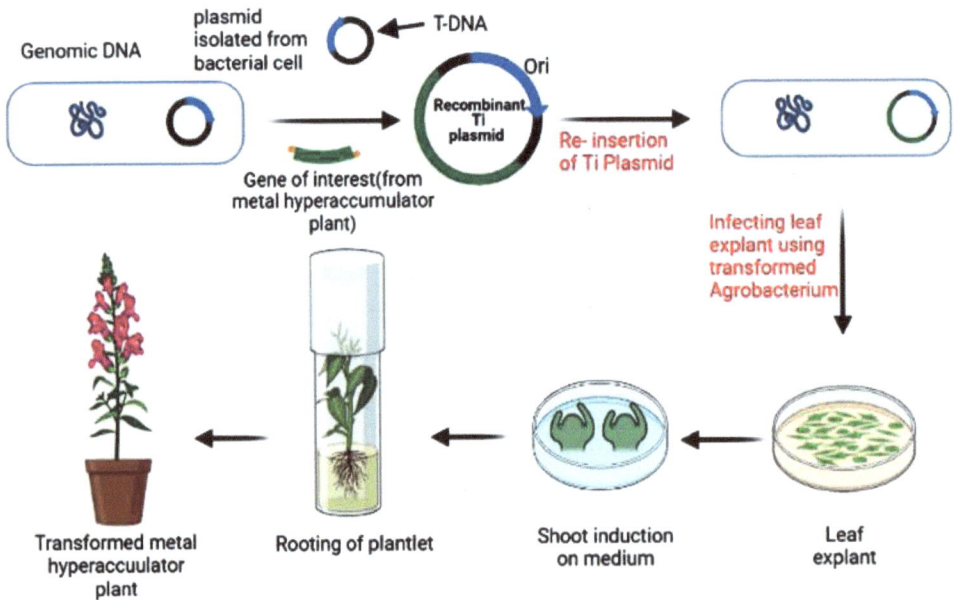

Fig. (4). Technique of Geno remediation for removal of heavy metals.

CONCLUSION AND FUTURE PERSPECTIVE

Removal of HMs through these strategies is a promising method to get rid of HM contamination of soil and water. These are green, sustainable, and economical strategies that provide long-lasting effects. However, further research on the biotechnological aspects of these strategies may pave the way for rapid clearance of HM toxicants from the environment.

REFERENCES

[1]   Tchounwou PB, Yedjou CG, Patlolla AK, Sutton DJ. Heavy metal toxicity and the environment. Molecular, clinical and environmental toxicology. Basel: Springer 2012; pp. 133-64.
[http://dx.doi.org/10.1007/978-3-7643-8340-4_6]

[2]   Guan Q, Wang F, Xu C, *et al.* Source apportionment of heavy metals in agricultural soil based on PMF: A case study in Hexi Corridor, northwest China. Chemosphere 2018; 193: 189-97.
[http://dx.doi.org/10.1016/j.chemosphere.2017.10.151] [PMID: 29131977]

[3]   Hossain MM, Khatun MA, Haque MN, *et al.* Silicon alleviates arsenic-induced toxicity in wheat through vacuolar sequestration and ROS scavenging. Int J Phytoremediation 2018; 20(8): 796-804.
[http://dx.doi.org/10.1080/15226514.2018.1425669] [PMID: 29775096]

[4]   Kumar, A., 2019. Phytoremediation: A Green Technology to clean the environment.

[5]   Vineetha RC, Binu P, Arathi P, Nair RH. L-ascorbic acid and α-tocopherol attenuate arsenic trioxide-induced toxicity in H9c2 cardiomyocytes by the activation of Nrf2 and Bcl2 transcription factors. Toxicol Mech Methods 2018; 28(5): 353-60.
[http://dx.doi.org/10.1080/15376516.2017.1422578] [PMID: 29297235]

[6]   Ahmad S, Mahmood R. Mercury chloride toxicity in human erythrocytes: enhanced generation of ROS and RNS, hemoglobin oxidation, impaired antioxidant power, and inhibition of plasma membrane redox system. Environ Sci Pollut Res Int 2019; 26(6): 5645-57.
[http://dx.doi.org/10.1007/s11356-018-04062-5] [PMID: 30612358]

[7]   Cabrita MT, Duarte B, Cesário R, *et al.* Mercury mobility and effects in the salt-marsh plant Halimione portulacoides: Uptake, transport, and toxicity and tolerance mechanisms. Sci Total Environ 2019; 650(Pt 1): 111-20.
[http://dx.doi.org/10.1016/j.scitotenv.2018.08.335] [PMID: 30196211]

[8]   León-Cañedo JA, Alarcón-Silvas SG, Fierro-Sañudo JF, *et al.* Mercury and other trace metals in lettuce (Lactuca sativa) grown with two low-salinity shrimp effluents: Accumulation and human health risk assessment. Sci Total Environ 2019; 650(Pt 2): 2535-44.
[http://dx.doi.org/10.1016/j.scitotenv.2018.10.003] [PMID: 30293006]

[9]   Sai Siva Ram AK, Pratap Reddy K, Girish BP, Supriya C, Sreenivasula Reddy P. Arsenic aggravated reproductive toxicity in male rats exposed to lead during the perinatal period. Toxicol Res (Camb) 2018; 7(6): 1191-204.
[http://dx.doi.org/10.1039/C8TX00146D] [PMID: 30510688]

[10]   Shi WG, Liu W, Yu W, *et al.* Abscisic acid enhances lead translocation from the roots to the leaves and alleviates its toxicity in Populus × canescens. J Hazard Mater 2019; 362: 275-85.
[http://dx.doi.org/10.1016/j.jhazmat.2018.09.024] [PMID: 30243250]

[11]   Minigalieva IA, Katsnelson BA, Panov VG, *et al.* In vivo toxicity of copper oxide, lead oxide and zinc oxide nanoparticles acting in different combinations and its attenuation with a complex of innocuous bio-protectors. Toxicology 2017; 380: 72-93.
[http://dx.doi.org/10.1016/j.tox.2017.02.007] [PMID: 28212817]

[12]   Antoniadis V, Zanni AA, Levizou E, *et al.* Modulation of hexavalent chromium toxicity on *Origanum vulgare* in an acidic soil amended with peat, lime, and zeolite. Chemosphere 2018; 195: 291-300.
[http://dx.doi.org/10.1016/j.chemosphere.2017.12.069] [PMID: 29272798]

[13]   Sihag S, Brar B, Joshi UN. Salicylic acid induces amelioration of chromium toxicity and affects antioxidant enzyme activity in *Sorghum bicolor* L. Int J Phytoremediation 2019; 21(4): 293-304.
[http://dx.doi.org/10.1080/15226514.2018.1524827] [PMID: 30873848]

[14]   Stambulska UY, Bayliak MM, Lushchak VI. Chromium(VI) toxicity in legume plants: modulation effects of rhizobial symbiosis. BioMed Res Int 2018; 2018: 1-13.
[http://dx.doi.org/10.1155/2018/8031213] [PMID: 29662899]

[15]   Maleki A, Hosseini MJ, Rahimi N, *et al.* Adjuvant potential of selegiline in treating acute toxicity of aluminium phosphide in rats. Basic Clin Pharmacol Toxicol 2019; 125(1): 62-74.
[http://dx.doi.org/10.1111/bcpt.13207] [PMID: 30712291]

[16]   Kura AU, Saifullah B, Cheah PS, Hussein MZ, Azmi N, Fakurazi S. Acute oral toxicity and biodistribution study of zinc-aluminium-levodopa nanocomposite. Nanoscale Res Lett 2015; 10(1): 105.
[http://dx.doi.org/10.1186/s11671-015-0742-5] [PMID: 25852400]

[17]   Irvine GW, Pinter TBJ, Stillman MJ. Defining the metal binding pathways of human metallothionein 1a: balancing zinc availability and cadmium seclusion. Metallomics 2016; 8(1): 71-81.
[http://dx.doi.org/10.1039/C5MT00225G] [PMID: 26583802]

[18]   Venter C, Oberholzer HM, Cummings FR, Bester MJ. Effects of metals cadmium and chromium alone and in combination on the liver and kidney tissue of male Spraque-Dawley rats: An ultrastructural and electron-energy-loss spectroscopy investigation. Microsc Res Tech 2017; 80(8): 878-88.
[http://dx.doi.org/10.1002/jemt.22877] [PMID: 28401733]

[19]   Bakti F, Sasse C, Heinekamp T, Pócsi I, Braus GH. Heavy metal-induced expression of PcaA provides cadmium tolerance to *Aspergillus fumigatus* and supports its virulence in the *Galleria mellonella* model. Front Microbiol 2018; 9: 744.
[http://dx.doi.org/10.3389/fmicb.2018.00744] [PMID: 29706948]

[20]   Hadrup N, Sharma AK, Loeschner K. Toxicity of silver ions, metallic silver, and silver nanoparticle materials after *in vivo* dermal and mucosal surface exposure: A review. Regul Toxicol Pharmacol 2018; 98: 257-67.
[http://dx.doi.org/10.1016/j.yrtph.2018.08.007] [PMID: 30125612]

[21]   Thummabancha K, Onparn N, Srisapoome P. Analysis of hematologic alterations, immune responses and metallothionein gene expression in Nile tilapia (*Oreochromis niloticus*) exposed to silver nanoparticles. J Immunotoxicol 2016; 13(6): 909-17.
[http://dx.doi.org/10.1080/1547691X.2016.1242673] [PMID: 27967301]

[22]   Lee WS, Kim E, Cho HJ, *et al.* The relationship between dissolution behavior and the toxicity of silver nanoparticles on zebrafish embryos in different ionic environments. Nanomaterials (Basel) 2018; 8(9): 652. b
[http://dx.doi.org/10.3390/nano8090652] [PMID: 30142912]

[23]   Konop M, Kłodzińska E, Borowiec J, *et al.* Application of micellar electrokinetic chromatography for detection of silver nanoparticles released from wound dressing. Electrophoresis 2019; 40(11): 1565-72.
[http://dx.doi.org/10.1002/elps.201900020] [PMID: 30848499]

[24]   Yin J, Wang AP, Li WF, Shi R, Jin HT, Wei JF. Time-response characteristic and potential biomarker identification of heavy metal induced toxicity in zebrafish. Fish Shellfish Immunol 2018; 72: 309-17.
[http://dx.doi.org/10.1016/j.fsi.2017.10.047] [PMID: 29111395]

[25]   Jadoon WA, Khpalwak W, Chidya RCG, *et al.* Evaluation of levels, sources and health hazards of road-dust associated toxic metals in Jalalabad and Kabul cities, Afghanistan. Arch Environ Contam Toxicol 2018; 74(1): 32-45.
[http://dx.doi.org/10.1007/s00244-017-0475-9] [PMID: 29159702]

[26]   Alho LOG, Gebara RC, Paina KA, Sarmento H, Melão MGG. Responses of *Raphidocelis subcapitata* exposed to Cd and Pb: Mechanisms of toxicity assessed by multiple endpoints. Ecotoxicol Environ Saf 2019; 169: 950-9.
[http://dx.doi.org/10.1016/j.ecoenv.2018.11.087] [PMID: 30597796]

[27] Chen H, Li Y, Ma X, *et al.* Analysis of potential strategies for cadmium stress tolerance revealed by transcriptome analysis of upland cotton. Sci Rep 2019; 9(1): 86.
[http://dx.doi.org/10.1038/s41598-018-36228-z] [PMID: 30643161]

[28] Zhang M, He P, Qiao G, Huang J, Yuan X, Li Q. Heavy metal contamination assessment of surface sediments of the Subei Shoal, China: Spatial distribution, source apportionment and ecological risk. Chemosphere 2019; 223: 211-22.
[http://dx.doi.org/10.1016/j.chemosphere.2019.02.058] [PMID: 30784728]

[29] Jaiswal A, Verma A, Jaiswal P. Detrimental effects of heavy metals in soil, plants, and aquatic ecosystems and in humans. J Environ Pathol Toxicol Oncol 2018; 37(3): 183-97.
[http://dx.doi.org/10.1615/JEnvironPatholToxicolOncol.2018025348] [PMID: 30317970]

[30] Mwakalapa EB, Simukoko CK, Mmochi AJ, *et al.* Heavy metals in farmed and wild milkfish (Chanos chanos) and wild mullet (*Mugil cephalus*) along the coasts of Tanzania and associated health risk for humans and fish. Chemosphere 2019; 224: 176-86.
[http://dx.doi.org/10.1016/j.chemosphere.2019.02.063] [PMID: 30822724]

[31] Lin L, Zheng F, Zhou H, Li S. Biomimetic gastrointestinal tract functions for metal absorption assessment in edible plants: comparison to *in vivo* absorption. J Agric Food Chem 2017; 65(30): 6282-7.
[http://dx.doi.org/10.1021/acs.jafc.7b02054] [PMID: 28685577]

[32] Akram S, Najam R, Rizwani GH, Abbas SA. Determination of heavy metal contents by atomic absorption spectroscopy (AAS) in some medicinal plants from Pakistani and Malaysian origin. Pak J Pharm Sci 2015; 28(5): 1781-7.
[PMID: 26408897]

[33] Xia H, Liang D, Chen F, *et al.* Effects of mutual intercropping on cadmium accumulation by the accumulator plants *Conyza canadensis, Cardamine hirsuta*, and *Cerastium glomeratum*. Int J Phytoremediation 2018; 20(9): 855-61.
[http://dx.doi.org/10.1080/15226514.2018.1438356] [PMID: 29873543]

[34] Boostani HR, Najafi-Ghiri M, Mirsoleimani A. The effect of biochars application on reducing the toxic effects of nickel and growth indices of spinach (*Spinacia oleracea* L.) in a calcareous soil. Environ Sci Pollut Res Int 2019; 26(2): 1751-60.
[http://dx.doi.org/10.1007/s11356-018-3760-x] [PMID: 30456609]

[35] Shaheen N, Irfan NM, Khan IN, Islam S, Islam MS, Ahmed MK. Presence of heavy metals in fruits and vegetables: Health risk implications in Bangladesh. Chemosphere 2016; 152: 431-8.
[http://dx.doi.org/10.1016/j.chemosphere.2016.02.060] [PMID: 27003365]

[36] Onakpa MM, Njan AA, Kalu OC. A review of heavy metal contamination of food crops in Nigeria. Ann Glob Health 2018; 84(3): 488-94.
[http://dx.doi.org/10.29024/aogh.2314] [PMID: 30835390]

[37] Sharma S, Nagpal AK, Kaur I. Heavy metal contamination in soil, food crops and associated health risks for residents of Ropar wetland, Punjab, India and its environs. Food Chem 2018; 255: 15-22.
[http://dx.doi.org/10.1016/j.foodchem.2018.02.037] [PMID: 29571461]

[38] Wang X, Sato T, Xing B, Tao S. Health risks of heavy metals to the general public in Tianjin, China *via* consumption of vegetables and fish. Sci Total Environ 2005; 350(1-3): 28-37.
[http://dx.doi.org/10.1016/j.scitotenv.2004.09.044] [PMID: 16227070]

[39] Ashraf S, Afzal M, Naveed M, Shahid M, Ahmad Zahir Z. Endophytic bacteria enhance remediation of tannery effluent in constructed wetlands vegetated with *Leptochloa fusca*. Int J Phytoremediation 2018; 20(2): 121-8.
[http://dx.doi.org/10.1080/15226514.2017.1337072] [PMID: 28621547]

[40] Rogowska J, Cieszynska-Semenowicz M, Ratajczyk W, Wolska L. Micropollutants in treated wastewater. Ambio 2020; 49(2): 487-503.

[http://dx.doi.org/10.1007/s13280-019-01219-5] [PMID: 31292910]

[41]   Rehman K, Fatima F, Waheed I, Akash MSH. Prevalence of exposure of heavy metals and their impact on health consequences. J Cell Biochem 2018; 119(1): 157-84.
[http://dx.doi.org/10.1002/jcb.26234] [PMID: 28643849]

[42]   Tripathi S, Sharma P, Singh K, Purchase D, Chandra R. Translocation of heavy metals in medicinally important herbal plants growing on complex organometallic sludge of sugarcane molasses-based distillery waste. Environmental Technology & Innovation 2021; 22: 101434. b
[http://dx.doi.org/10.1016/j.eti.2021.101434]

[43]   Juel MAI, Mizan A, Ahmed T. Sustainable use of tannery sludge in brick manufacturing in Bangladesh. Waste Manag 2017; 60: 259-69.
[http://dx.doi.org/10.1016/j.wasman.2016.12.041] [PMID: 28081994]

[44]   Sharma P, Ngo HH, Khanal S, Larroche C, Kim SH, Pandey A. Efficiency of transporter genes and proteins in hyperaccumulator plants for metals tolerance in wastewater treatment: Sustainable technique for metal detoxification. Environ. Technol. Innovathttps 2021.
[http://dx.doi.org/10.1016/j.eti.2021.101725]

[45]   Briffa J, Sinagra E, Blundell R. Heavy metal pollution in the environment and their toxicological effects on humans. Heliyon 2020; 6(9): e04691.
[http://dx.doi.org/10.1016/j.heliyon.2020.e04691] [PMID: 32964150]

[46]   Sharma P, Singh SP, Pandey S, Thanki A, Singh NK. 2020.Role of potential native weeds and grasses for phytoremediation of endocrine-disrupting pollutants discharged from pulp paper industry waste.
[http://dx.doi.org/10.1016/B978-0-12-819025-8.00002-8]

[47]   Mejáre M, Bülow L. Metal-binding proteins and peptides in bioremediation and phytoremediation of heavy metals. Trends Biotechnol 2001; 19(2): 67-73.
[http://dx.doi.org/10.1016/S0167-7799(00)01534-1] [PMID: 11164556]

[48]   Ahemad M, 2012. Implications of bacterial resistance against heavy metals in bioremediation: a review. Journal of Institute of Integrative Omics and Applied Biotechnology(IIOAB) 3.

[49]   Yin K, Wang Q, Lv M, Chen L. Microorganism remediation strategies towards heavy metals. Chem Eng J 2019; 360: 1553-63.
[http://dx.doi.org/10.1016/j.cej.2018.10.226]

[50]   Deng L, Su Y, Su H, Wang X, Zhu X. Sorption and desorption of lead (II) from wastewater by green algae *Cladophora fascicularis*. J Hazard Mater 2007; 143(1-2): 220-5.
[http://dx.doi.org/10.1016/j.jhazmat.2006.09.009] [PMID: 17049733]

[51]   Chekroun KB, Baghour M. The role of algae in phytoremediation of heavy metals: a review. J Mater Environ Sci 2013; 4(6): 873-80.

[52]   Mitra N, Rezvan Z, Seyed Ahmad M, Gharaie Mohamad Hosein M. Studies of water arsenic and boron pollutants and algae phytoremediation in three springs, Iran. International Journal of Ecosystem 2012; 2(3): 32-7.
[http://dx.doi.org/10.5923/j.ije.20120203.01]

[53]   Shanab S, Essa A, Shalaby E. Bioremoval capacity of three heavy metals by some microalgae species (Egyptian Isolates). Plant Signal Behav 2012; 7(3): 392-9.
[http://dx.doi.org/10.4161/psb.19173] [PMID: 22476461]

[54]   Gosavi K, Sammut J, Gifford S, Jankowski J. Macroalgal biomonitors of trace metal contamination in acid sulfate soil aquaculture ponds. Sci Total Environ 2004; 324(1-3): 25-39.
[http://dx.doi.org/10.1016/j.scitotenv.2003.11.002] [PMID: 15081694]

[55]   Poo KM, Son EB, Chang JS, Ren X, Choi YJ, Chae KJ. Biochars derived from wasted marine macro-algae (*Saccharina japonica* and *Sargassum fusiforme*) and their potential for heavy metal removal in aqueous solution. J Environ Manage 2018; 206: 364-72.
[http://dx.doi.org/10.1016/j.jenvman.2017.10.056] [PMID: 29101878]

[56]   Mani D, Kumar C. Biotechnological advances in bioremediation of heavy metals contaminated ecosystems: an overview with special reference to phytoremediation. Int J Environ Sci Technol 2014; 11(3): 843-72.
[http://dx.doi.org/10.1007/s13762-013-0299-8]

[57]   Mane, P. C., & Bhosle, A. B. (2012). Bioremoval of some metals by living algae Spirogyra sp. and Spirullina sp. from aqueous solution. Int J Journal of Environ Res 2012; 6(2): 571-6.

[58]   Saunders RJ, Paul NA, Hu Y, de Nys R. Sustainable sources of biomass for bioremediation of heavy metals in waste water derived from coal-fired power generation. PLoS One 2012; 7(5): e36470.
[http://dx.doi.org/10.1371/journal.pone.0036470] [PMID: 22590550]

[59]   Lee YC, Chang SP. The biosorption of heavy metals from aqueous solution by Spirogyra and Cladophora filamentous macroalgae. Bioresour Technol 2011; 102(9): 5297-304.
[http://dx.doi.org/10.1016/j.biortech.2010.12.103] [PMID: 21292478]

[60]   Monteiro CM, Marques APGC, Castro PML, Xavier Malcata F. Characterization of Desmodesmus pleiomorphus isolated from a heavy metal-contaminated site: biosorption of zinc. Biodegradation 2009; 20(5): 629-41.
[http://dx.doi.org/10.1007/s10532-009-9250-6] [PMID: 19225897]

[61]   Gupta VK, Nayak A, Agarwal S. Bioadsorbents for remediation of heavy metals: Current status and their future prospects. Environ Eng Res 2015; 20(1): 1-18.
[http://dx.doi.org/10.4491/eer.2015.018]

[62]   Bellion M, Courbot M, Jacob C, Blaudez D, Chalot M. Extracellular and cellular mechanisms sustaining metal tolerance in ectomycorrhizal fungi. FEMS Microbiol Lett 2006; 254(2): 173-81.
[http://dx.doi.org/10.1111/j.1574-6968.2005.00044.x] [PMID: 16445743]

[63]   Esterhuizen-Londt M, Schwartz K, Pflugmacher S. Using aquatic fungi for pharmaceutical bioremediation: Uptake of acetaminophen by Mucor hiemalis does not result in an enzymatic oxidative stress response. Fungal Biol 2016; 120(10): 1249-57.
[http://dx.doi.org/10.1016/j.funbio.2016.07.009] [PMID: 27647241]

[64]   Varjani SJ, Patel RK. Fungi: a remedy to eliminate environmental pollutants. Mycoremediation and Environmental Sustainability 2017; pp. 53-67.

[65]   Akar T, Tunali S, Kiran I. Botrytis cinerea as a new fungal biosorbent for removal of Pb(II) from aqueous solutions. Biochem Eng J 2005; 25(3): 227-35.
[http://dx.doi.org/10.1016/j.bej.2005.05.006]

[66]   Nadeem SM, Ahmad M, Zahir ZA, Javaid A, Ashraf M. The role of mycorrhizae and plant growth promoting rhizobacteria (PGPR) in improving crop productivity under stressful environments. Biotechnol Adv 2014; 32(2): 429-48.
[http://dx.doi.org/10.1016/j.biotechadv.2013.12.005] [PMID: 24380797]

[67]   Peng RH, Xiong AS, Xue Y, *et al.* Microbial biodegradation of polyaromatic hydrocarbons. FEMS Microbiol Rev 2008; 32(6): 927-55.
[http://dx.doi.org/10.1111/j.1574-6976.2008.00127.x] [PMID: 18662317]

[68]   Rhodes CJ. Applications of bioremediation and phytoremediation. Sci Prog 2013; 96(4): 417-27.
[http://dx.doi.org/10.3184/003685013X13818570960538] [PMID: 24547671]

[69]   Siddiquee S, Rovina K, Azad SA. Heavy metal contaminants removal from wastewater using the potential filamentous fungi biomass: a review. J Microb Biochem Technol 2015; 7(6): 384-95.
[http://dx.doi.org/10.4172/1948-5948.1000243]

[70]   Obire O, Putheti RR. Fungi in bioremediation of oil polluted environments. Sigma Xi Scientific Research Society 2009.

[71]   Gao D, Du L, Yang J, Wu WM, Liang H. A critical review of the application of white rot fungus to environmental pollution control. Crit Rev Biotechnol 2010; 30(1): 70-7.

[http://dx.doi.org/10.3109/07388550903427272] [PMID: 20099998]

[72]   Hlihor RM, Diaconu M, Leon F, Curteanu S, Tavares T, Gavrilescu M. Experimental analysis and mathematical prediction of Cd(II) removal by biosorption using support vector machines and genetic algorithms. N Biotechnol 2015; 32(3): 358-68.
[http://dx.doi.org/10.1016/j.nbt.2014.08.003] [PMID: 25224921]

[73]   Jafari M, *et al.* Bioremediation and genetically modified organisms. Fungi as bioremediators. Berlin, Heidelberg: Springer 2013; pp. 433-51.
[http://dx.doi.org/10.1007/978-3-642-33811-3_19]

[74]   Wang J, Chen C. Biosorbents for heavy metals removal and their future. Biotechnol Adv 2009; 27(2): 195-226.
[http://dx.doi.org/10.1016/j.biotechadv.2008.11.002] [PMID: 19103274]

[75]   Kang CH, Kwon YJ, So JS. Bioremediation of heavy metals by using bacterial mixtures. Ecol Eng 2016; 89: 64-9.
[http://dx.doi.org/10.1016/j.ecoleng.2016.01.023]

[76]   Huyer M, Page WJ. Zn2+ increases siderophore production in Azotobacter vinelandii. Appl Environ Microbiol 1988; 54(11): 2625-31.
[http://dx.doi.org/10.1128/aem.54.11.2625-2631.1988] [PMID: 16347766]

[77]   Chibuike GU, Obiora SC. Heavy metal polluted soils: effect on plants and bioremediation methods. Appl Environ Soil Sci 2014; 2014: 1-12.
[http://dx.doi.org/10.1155/2014/752708]

[78]   Dixit R, Wasiullah , Malaviya D, *et al.* Bioremediation of heavy metals from soil and aquatic environment: an overview of principles and criteria of fundamental processes. Sustainability (Basel) 2015; 7(2): 2189-212.
[http://dx.doi.org/10.3390/su7022189]

[79]   Karaca A. Effect of organic wastes on the extractability of cadmium, copper, nickel, and zinc in soil. Geoderma 2004; 122(2-4): 297-303.
[http://dx.doi.org/10.1016/j.geoderma.2004.01.016]

[80]   Nehra V, Choudhary M. A review on plant growth promoting rhizobacteria acting as bioinoculants and their biological approach towards the production of sustainable agriculture. J Appl Nat Sci 2015; 7(1): 540-56.
[http://dx.doi.org/10.31018/jans.v7i1.642]

[81]   Choudhary DK, Varma A. Microbial-mediated induced systemic resistance in plants. Springer 2016.
[http://dx.doi.org/10.1007/978-981-10-0388-2]

[82]   Bhattacharyya PN, Jha DK. Plant growth-promoting rhizobacteria (PGPR): emergence in agriculture. World J Microbiol Biotechnol 2012; 28(4): 1327-50.
[http://dx.doi.org/10.1007/s11274-011-0979-9] [PMID: 22805914]

[83]   Vejan P, Abdullah R, Khadiran T, Ismail S, Nasrulhaq Boyce A. Role of plant growth promoting rhizobacteria in agricultural sustainability—a review. Molecules 2016; 21(5): 573.
[http://dx.doi.org/10.3390/molecules21050573] [PMID: 27136521]

[84]   Chandra P, Singh E. 2016.Applications and mechanisms of plant growth-stimulating rhizobacteria.
[http://dx.doi.org/10.1007/978-981-10-2854-0_3] Chandra P, Singh E. Applications and Mechanisms of Plant Growth-Stimulating Rhizobacteria. In: Choudhary D, Varma A, Tuteja N (eds) Plant-Microbe Interaction: An Approach to Sustainable Agriculture. Springer, Singapore, 2016, pp. 37–62.
https://doi.org/10.1007/978-981-10-2854-0_3

[85]   Liu S, Yang B, Liang Y, Xiao Y, Fang J. Prospect of phytoremediation combined with other approaches for remediation of heavy metal-polluted soils. Environ Sci Pollut Res Int 2020; 27(14): 16069-85.
[http://dx.doi.org/10.1007/s11356-020-08282-6] [PMID: 32173779]

[86]   Sumiahadi A, Acar R. A review of phytoremediation technology: heavy metals uptake by plants. IOP Conf Ser Earth Environ Sci 2018; 142(1): 012023. []. IOP Publishing.]. [http://dx.doi.org/10.1088/1755-1315/142/1/012023]

[87]   Muthusaravanan S, Sivarajasekar N, Vivek JS, *et al.* Phytoremediation of heavy metals: mechanisms, methods and enhancements. Environ Chem Lett 2018; 16(4): 1339-59. [http://dx.doi.org/10.1007/s10311-018-0762-3]

[88]   Shen X, Dai M, Yang J, *et al.* A critical review on the phytoremediation of heavy metals from environment: Performance and challenges. Chemosphere 2022; 291(Pt 3): 132979. [http://dx.doi.org/10.1016/j.chemosphere.2021.132979] [PMID: 34801572]

[89]   Nedjimi B. Phytoremediation: a sustainable environmental technology for heavy metals decontamination. SN Applied Sciences 2021; 3(3): 286. [http://dx.doi.org/10.1007/s42452-021-04301-4]

[90]   Khan I, Iqbal M, Shafiq F. Phytomanagement of lead-contaminated soils: critical review of new trends and future prospects. Int J Environ Sci Technol 2019; 16(10): 6473-88. [http://dx.doi.org/10.1007/s13762-019-02431-2]

[91]   Woraharn S, Meeinkuirt W, Phusantisampan T, Avakul P. Potential of ornamental monocot plants for rhizofiltration of cadmium and zinc in hydroponic systems. Environ Sci Pollut Res Int 2021; 28(26): 35157-70. [http://dx.doi.org/10.1007/s11356-021-13151-x] [PMID: 33666846]

[92]   Kumar P, Kumar L. .Removal Of Fluoride Contamination In Waste Water By Using Eichhornia crassipes Nat Volatiles Essent Oils 2021; 8(4): 16624-33.

[93]   Fasani E, Manara A, Martini F, Furini A, DalCorso G. The potential of genetic engineering of plants for the remediation of soils contaminated with heavy metals. Plant Cell Environ 2018; 41(5): 1201-32. [http://dx.doi.org/10.1111/pce.12963] [PMID: 28386947]

[94]   Wan X, Lei M, Chen T. Cost–benefit calculation of phytoremediation technology for heavy-meta--contaminated soil. Sci Total Environ 2016; 563-564: 796-802. [http://dx.doi.org/10.1016/j.scitotenv.2015.12.080] [PMID: 26765508]

[95]   Ojuederie O, Babalola O. Microbial and plant-assisted bioremediation of heavy metal polluted environments: a review. Int J Environ Res Public Health 2017; 14(12): 1504. [http://dx.doi.org/10.3390/ijerph14121504] [PMID: 29207531]

[96]   Kumar NM, Muthukumaran C, Sharmila G, Gurunathan B. Genetically modified organisms and its impact on the enhancement of bioremediation. Bioremediation: Applications for Environmental Protection and Management. Singapore: Springer 2018; pp. 53-76. [http://dx.doi.org/10.1007/978-981-10-7485-1_4]

[97]   Singh JS, Abhilash PC, Singh HB, Singh RP, Singh DP. Genetically engineered bacteria: An emerging tool for environmental remediation and future research perspectives. Gene 2011; 480(1-2): 1-9. [http://dx.doi.org/10.1016/j.gene.2011.03.001] [PMID: 21402131]

[98]   Dhankher OP, Pilon-Smits EA, Meagher RB, Doty S. Biotechnological approaches for phytoremediation. Plant biotechnology and agriculture. Academic Press 2012; pp. 309-28. [http://dx.doi.org/10.1016/B978-0-12-381466-1.00020-1]

[99]   Kumar C, Mani D. Advances in bioremediation of heavy metals: a tool for environmental restoration. Lambert Academic Publishing 2012.

[100]   Kushwaha P, Kashyap PL. A review of advances in bioremediation of heavy metals by microbes and plants. Journal of Natural Resource Conservation and Management 2021; 2(1): 65-80. [http://dx.doi.org/10.51396/ANRCM.2.1.2021.65-80]

[101] Azad MAK, Amin L, Sidik NM. Genetically engineered organisms for bioremediation of pollutants in contaminated sites. Chin Sci Bull 2014; 59(8): 703-14.
[http://dx.doi.org/10.1007/s11434-013-0058-8]

[102] Kang JW. Removing environmental organic pollutants with bioremediation and phytoremediation. Biotechnol Lett 2014; 36(6): 1129-39.
[http://dx.doi.org/10.1007/s10529-014-1466-9] [PMID: 24563299]

[103] Chen S, Wilson DB. Genetic engineering of bacteria and their potential for Hg^{2+} bioremediation. Biodegradation 1997; 8(2): 97-103.
[http://dx.doi.org/10.1023/A:1008233704719] [PMID: 9342882]

[104] Barkay T, Miller SM, Summers AO. Bacterial mercury resistance from atoms to ecosystems. FEMS Microbiol Rev 2003; 27(2-3): 355-84.
[http://dx.doi.org/10.1016/S0168-6445(03)00046-9] [PMID: 12829275]

[105] Deckwer WD, Becker FU, Ledakowicz S, Wagner-Döbler I. Microbial removal of ionic mercury in a three-phase fluidized bed reactor. Environ Sci Technol 2004; 38(6): 1858-65.
[http://dx.doi.org/10.1021/es0300517] [PMID: 15074700]

[106] Memon AR. Metal hyperaccumulators: mechanisms of hyperaccumulation and metal tolerance. Phytoremediation. Management of Environmental Contaminants 2016; 3: 239-68.
[http://dx.doi.org/10.1007/978-3-319-40148-5_8]

[107] Memon AR, Schröder P. Implications of metal accumulation mechanisms to phytoremediation. Environ Sci Pollut Res Int 2009; 16(2): 162-75.
[http://dx.doi.org/10.1007/s11356-008-0079-z] [PMID: 19067014]

[108] Weber M, Trampczynska A, Clemens S. Comparative transcriptome analysis of toxic metal responses in *Arabidopsis thaliana* and the Cd $^{2+}$-hypertolerant facultative metallophyte *Arabidopsis halleri*. Plant Cell Environ 2006; 29(5): 950-63.
[http://dx.doi.org/10.1111/j.1365-3040.2005.01479.x] [PMID: 17087478]

[109] Ahsan N, Renaut J, Komatsu S. Recent developments in the application of proteomics to the analysis of plant responses to heavy metals. Proteomics 2009; 9(10): 2602-21.
[http://dx.doi.org/10.1002/pmic.200800935] [PMID: 19405030]

[110] Meng XY, Qin J, Wang LH, *et al.* Arsenic biotransformation and volatilization in transgenic rice. New Phytol 2011; 191(1): 49-56.
[http://dx.doi.org/10.1111/j.1469-8137.2011.03743.x] [PMID: 21517874]

[111] Misra S, Gedamu L. Heavy metal tolerant transgenic *Brassica napus* L. and *Nicotiana tabacum* L. plants. Theor Appl Genet 1989; 78(2): 161-8.
[http://dx.doi.org/10.1007/BF00288793] [PMID: 24227139]

[112] Rugh CL, Wilde HD, Stack NM, Thompson DM, Summers AO, Meagher RB. Mercuric ion reduction and resistance in transgenic *Arabidopsis thaliana* plants expressing a modified bacterial merA gene. Proc Natl Acad Sci USA 1996; 93(8): 3182-7.
[http://dx.doi.org/10.1073/pnas.93.8.3182] [PMID: 8622910]

[113] Chen F, Johns MR. Effect of C/N ratio and aeration on the fatty acid composition of heterotrophic *Chlorella sorokiniana*. J Appl Phycol 1991; 3(3): 203-9.
[http://dx.doi.org/10.1007/BF00003578]

[114] Cevher-Keskin B, Yıldızhan Y, Yüksel B, Dalyan E, Memon AR. Characterization of differentially expressed genes to Cu stress in *Brassica nigra* by Arabidopsis genome arrays. Environ Sci Pollut Res Int 2019; 26(1): 299-311.
[http://dx.doi.org/10.1007/s11356-018-3577-7] [PMID: 30397750]

[115] Dalyan E, Yüzbaşıoğlu E, Keskin BC, *et al.* The identification of genes associated with Pb and Cd response mechanism in *Brassica juncea* L. by using *Arabidopsis* expression array. Environ Exp Bot 2017; 139: 105-15.

[http://dx.doi.org/10.1016/j.envexpbot.2017.05.001]

[116] Memon AR. Genomics and transcriptomics analysis of Cu accumulator plant -*Brassica nigra* L. Journal of Applied Biological Sciences 2014; 8(2): 1-8.

CHAPTER 5

Probiotics as a New Means for Heavy Metal Detoxification

Shalini Dhiman[1], Jaspreet Kour[1], Arun Dev Singh[1], Tamanna Bhardwaj[1], Mohd Ali[2], Deepak Kumar[2], Parkiriti[2], Kanika Khanna[1,*], Puja Ohri[2] and Renu Bhardwaj[1]

[1] *Department of Botanical and Environmental Sciences, Guru Nanak Dev University, Amritsar, Punjab, India*

[2] *Department of Zoology, Guru Nanak Dev University, Amritsar, Punjab, India*

Abstract: Heavy metals can harm the bodies of all living organisms in terrible ways. This frightening situation will grow into a severe crisis when polluted living beings containing harmful heavy metals enter the human food chain. Probiotics have a great potential to minimize such kind of heavy metal poisoning in living beings. Probiotics have tremendous potential to effectively bioremediate various toxic heavy metals or metalloids (Cadmium (Cd), Mercury (Hg), Lead (Pb), Arsenic (As), *etc.*) from a variety of polluted environments. Endogenous presence of probiotics or exogenous application of probiotics can be proven to lessen the toxicity of these heavy metals inside living beings' bodies. Due to this reason, probiotics are also popularly known as biological detoxification tools. This book chapter provides comprehensive details related to probiotic strains, their mechanism of action, and their role in various heavy metal detoxification processes. This chapter also sheds light on recent progress in generating genetically engineered probiotics for treating HM toxicity.

Keywords: Detoxification, Heavy metals, Probiotics, Toxicity.

INTRODUCTION

Modern industrialization has caused major contamination inside soil, water and air ecosystems. Among the different types of contamination, heavy metals (HMs) are one of the most common and inescapable encounters a variety of organisms face either through direct or indirect sources, for example, HMs entering the food chain. Conventional ways to detox heavy metals from organism bodies, especially the organisms like fish, goats, *etc.*, that are an essential part of the human food chain, are an extremely expensive, risky, and tedious process. In contrast to this,

[*] **Corresponding author Kanika Khanna:** Department of Botanical and Environmental Sciences, Guru Nanak Dev University, Amritsar, Punjab, India; E-mail: kanika.27590@gmail.com

probiotic methods for heavy metal detoxification are very cheap, easy, specific, safe, effective, and green alternatives with lots of health-beneficial impacts on the organisms. Moreover, after examining the previous literature on probiotics' reactions to a variety of toxins, including heavy metals, we found that some strains and combinations improve health and detoxification processes *in vitro* and *in vivo*. There are diverse varieties of traditional and next generation probiotics strains like *Lactobacillus, Lactococcus, Bifidobacterium, Escherichia coli*, some species of *Bacillus, Streptococcus, Enterococcus, Pediococcus, Akkermansia, Bacteroides, Faecalibacterium, Eubacterium* and *Saccharomyces* yeast used for effective removal of toxic heavy metals from the organism bodies [1]. These probiotics remove heavy metals from the organisms' bodies mainly by the processes known as biosorption and bioaccumulation. The primary mechanism adopted by most probiotics especially *Lactobacillus* for heavy metal detoxification is the biosorption of heavy metal ions into bacterial cell walls, which is facilitated by the presence of teichoic acids and polysaccharides, which are helpful in attracting, and reducing oxidative stress, and sequestering heavy metals [1]. This biosorption process is then followed by bioaccumulation within the bacteria through the utilisation of cell membrane transition [1]. Further probiotics like *Bifidobacterium* actively transport or efflux heavy metals reducing circulating toxic heavy metal levels inside the organism exposed to contamination [1]. However, nowadays, genetically engineered microorganisms (GEMs) for heavy metal removal are popular due to their low cost, adaptability, and environmental friendliness. GEMs with high degradative potential are used in heavy metal bioremediation in plants, animals, groundwater, soil, and activated sludge. Two main methods are used to develop heavy metal-resistant GEMs: functional surface aggregates with a strong metal-binding ability for biosorption and metal ions carried into the cytoplasm and processed by storage mechanisms for improved intracellular bioaccumulation [2]. Therefore, the consequences of heavy metal toxicity are reversed inside the organism's body.

Potentially Effective Probiotic Strains (Traditional and Next Generation Probiotics) for Heavy Metals' Removal

According to the WHO (World Health Organization), probiotics are "live bacteria that, when supplied in suitable proportions, impart a health benefit to the host" [3]. Several probiotics have beneficial features such as the ability to bind to or resist heavy metals *in-vitro* that can be used for heavy metal detoxification and bioremediation (Fig. **1** and Table **1**). Previous studies have shown that probiotics have a strong ability to cling to the intestinal mucosa, to regulate the immune system, to tolerate gastrointestinal fluid, and to repress the growth of pathogens and to mitigate the heavy metal contamination [4 - 6]. Genera like *Lactobacillus, Lactococcus* and *Bifidobacterium, Escherichia coli*, some species of *Bacillus,*

Streptococcus, Enterococcus, Pediococcus, and *Saccharomyces* yeast, which are effective in the treatment of several gastrointestinal problems and remediating heavy metal toxicity, are traditional probiotic strains that are marketed [7]. Several studies have reported the role of traditional and next-generation probiotics in removing heavy metal toxicity. For instance, *Lactobacillus reuteri* P16 and *L. plantarum* CCFM8661 against Pb [8, 9] toxicity, *L. plantarum* TW1-1 employed against Cr toxicity [10], *L. brevis* 23017 for Hg toxicity, and *B. cereus* and *L. plantarum* CCFM8610 against Cd toxicity [11, 12]. Through intestinal heavy metal sequestration and intestinal peristalsis, these strains can encourage the fecal elimination of heavy metals, lowering heavy metal absorption in the gut and correcting heavy metal-induced alterations to the gut microbiota.

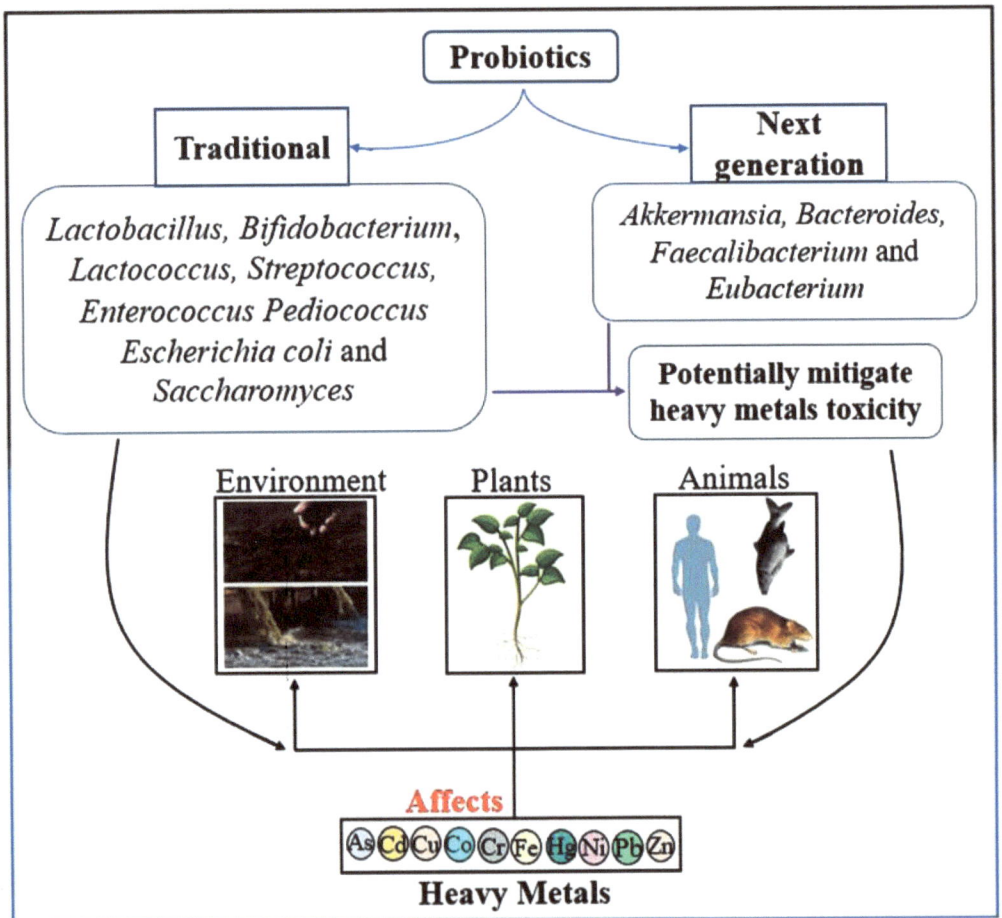

Fig. (1). Traditional and next-generation probiotics used for heavy metal removal.

Table 1. Different traditional and next-generation probiotics for heavy metal removal.

S.No.	Probiotic Strain used	Heavy Metal	Test Organism/Sample	Reported Effect	References
1	*Lactobacillus sporogenes*	As	Rat	Increased body weight, reduced all other As-induced morphological abnormalities, and improved hematological parameters.	[29]
2	*Lactobacillus* CLF9-1	Cd	Chickens	Decreased Cd accumulation in the liver and kidney of chicken.	[13]
3	*Enterococcus faecium*	Cd	Rice	Alleviated Cd-induced toxic effects.	[22]
4	*Pediococcusacidilactici GR-1*	Cu and Ni	Humans	Decreased Cu and Ni content in blood and enriched gut microbiota.	[30]
5	*L. casei*	Fe	*Cyprinus carpio*	Reduced hepatocellular pyknosis and necrosis.	[14]
6	*L. lactis*	Cr, Cd and Cu	*Cyprinus carpio*	Alleviated heavy metal-induced oxidative damage, tissue damage, and immunological responses.	[31]
7	*L. delbrueckii*subsp. *bulgaricus*	Pb	*Oncorhynchus mykiss*	Improved growth performance and activity of intestinal enzymes.	[32]
8	*L. casei*	Pb	Rat	Alleviated the toxic effects caused by As.	[33]
9	*Bacillus coagulans*	Cd	Common carp	Relieved Cd-induced oxidative stress.	[34]
10	Consortium of *Lactobacillus* spp.	Cu	*Oreochromis niloticus*	Maintained the histological structure of the gonads.	[35]
11	*L. fermentum* CQPC08	Pb	Rat	Decreased Pb in serum, kidney, liver, and brain of rats and reduced tumor necrosis factor and oxidative damage.	[36]
12	*E. faecium*	Cd, Cr and Ni	Fenugreek	Reduced toxicity of tannery sludge and increased plant growth parameters grew in treated tannery sludge.	[37]

(Table 1) cont.....

S.No.	Probiotic Strain used	Heavy Metal	Test Organism/Sample	Reported Effect	References
13	*L. rhamnosus*	Cd and Pb	Fish fillet	Reduced Cd and Pb by 72 and 84.3%.	[38]
14	*L. plantarum*	Cd	*Luciobarbuscapito*	Reduced accumulation of Cd in fish tissues, increased activity of antioxidants such as superoxide dismutase, glutathione peroxidase, and catalase.	[39]
15	*L. plantarum*	Cd	Soil and garlic bolt	Reduced available Cd in soil and increased biomass of garlic bolt.	[40]
16	*Saccharomyces cerevisiae* and *Lactobacillus* spp.	As	Humans and Cooked rice	Reduced As permeability and reduced As transport across the intestinal monolayer.	[41]
17	*P. pentosaceus*	Pb and Cd	Water	Exhibited Pd and Cd removal efficiency of more than 50%.	[42]
18	*L. ermentum*	Pb	*Danio rerio*	Lowered tissue damage such as lamella fusion, vein dilatation, and telangiectasia.	[43]
19	*Akkermansiamuciniphila*	Cd	Mice	Influenced gut microbiota and limited influence of chronic Cd exposure.	[26]
20	*S. cerevisiae*	Pb	Milk	Showed biosorbent potential and removed 705 of Pb.	[44]
21	*L. plantarum* CCFM8661	Pb	Mice	Increased Pb sequestration and biliary Pb elimination *via* enhanced production and excretion of bile acid.	[45]
22	*L. acidophilus*	Pb	*O. mykiss*	Increased activity of digestive enzymes.	[46]
	L. delbrueckii	Pb	Mice	Reduced mortality rate, alleviated tissue Pb accumulation, and elevated fecal Pb elimination.	[47]
23	*L. acidophilus* LA-5 and *B. lactis* BB-12	Pb	Rat	Reduced accumulation of Pb in rat brain.	[48]

(Table 1) cont.....

S.No.	Probiotic Strain used	Heavy Metal	Test Organism/Sample	Reported Effect	References
24	*Bifidobacterium angulatum*	Pb, As and Cd	Water	Enhanced removal of heavy metals.	[49]
25	*Escherichia coli* EcN-20	Cd and Hg	Rat	Mitigated Cd and Hg-induced liver and kidney damage.	[50]
26	*L. plantarum* CCFM8610	Cd	Mice	Increased Cd elimination.	[51]
27	*L. reuteri* Cd70-13 and Pb71-1	Cd and Lead	Fish	Increased uptake of heavy metals in the intestinal milieu.	[52]

Chen and his co-workers assessed the CLF9-1 strain of *Lactobacillus* isolated from chicken for its heavy metal removal potential [13]. They reported that chicken growth was enhanced by the Cd-tolerant CLF9-1 treatment, which also decreased the Cd-elevated liver and kidney coefficients in the chicken. The strain also resulted in the significant elimination of Cd in faeces leading to the reduced accumulation of Cd in the liver and kidney of chickens [13]. Similarly, *L. casei* alleviated the toxic effects induced by iron oxide nanoparticles in common carp, where reduced hepatocellular pyknosis and necrosis were observed in samples treated with nanoparticles and *L. casei* compared to the samples treated with nanoparticles alone [14]. Additionally, *L. plantarum* CCFM8661 and CCFM8610 in mice affected the excretion of heavy metals by boosting hepatic bile acid production and raising the excretion of bile acids in faeces, which included the heavy metal lead [15, 16]. Furthermore, the useful bacterium *L. acidophilus* was used to remove Pb, Cd, and Hg from milk, and its capacity to adsorb the specified heavy metals in milk was assessed. On the fourth day, with an initial ions' concentration of 100 g/L, the Cd, Hg, and Pd bio-absorption yields were at their highest levels at 75%, 78%, and 80%, respectively [17, 18]. Similarly, chromium (Cr) removal by *L. plantarum* at a starting Cr concentration of 100 ppm was studied by Ameen and his co-workers [19]. The maximal Cr uptake capacity in this investigation was 70% at 22 °C after 1 hour. Moreover, aminated *L. casei*, according to Halttunen (2007) may eliminate As from water [20]. Similar results were obtained while studying the heavy metal removal potential of *S. pyogenes* and *E. faecalis* from soil contaminated with used lubricating oils, where co-inoculation of both probiotics significantly enhanced the decontamination of soil [21]. In addition, *E. faecium* isolated from soil alleviated the toxic effects of Cd on rice. Inoculation of bacterium reduced the extractable and soluble Cd in the soil leading to reduced toxicity and reduced Cd concentration in rice leaves, stems, and roots [22] (Fig. **1**).

Genera *Akkermansia, Bacteroides, Eubacterium,* and *Faecalibacterium* make up the majority of probable next-generation probiotics [23]. The human gut commensal *F. prausnitzii* can produce short-chain fatty acids and has the arsM gene, which encodes a well-known methyltransferase implicated in microbial As detoxification [24]. Furthermore, in As-exposed mice, oral administration of 10^8 CFU of *F. prausnitzii* A2-165 mixed with *E.coli* significantly enhanced the survival rate of mice [25]. These findings demonstrated the protective effects of *F. prausnitzii* A2-165 against As toxicity, however, it is not obvious if this effect was brought on by some microbiome factor or *in-vivo* synthesis of As detoxification enzymes and short-chain fatty acids [25]. Similarly, after acute and chronic Cd exposure, 10^8 CFU of *A. muciniphila* supplementation enhanced the abundance of the gut microbiota in Cd-exposed mice, especially of several genera of Clostridiales, but did not significantly lower the levels of Cd in the blood and tissues [26]. Additionally, the inclusion of *F. prausnitzii* and *O. ruminantium* greatly decreased the absorption and buildup of lead and increased the expression of tight junction protein in the gut [27]. Furthermore, when exposed to Pb, some gut microbe abundances were noticeably reduced, raising the question of whether adding these intestine bacteria can lessen Pb toxicity. *A. muciniphila, F. prausnitzii,* or *Oscillibacter ruminantium*, three Pb-intolerant gut microorganisms, were given orally to mice exposed to Pb [28]. The findings demonstrated that treatment with *F. prausnitzii* or *O. ruminantium* greatly reduced Pb absorption and accumulation in host tissues and increased the expression of tight junction proteins and the synthesis of short-chain fatty acids by gut bacteria. These findings suggest that altering gut microbiota may be crucial for reducing the burden of Pb poisoning in the host [28]. Table **1**. shows the different traditional and next-generation probiotics for heavy metal removal.

Therefore, to lessen or reverse the consequences of heavy metal toxicity on humans, animals, plants, and the environment, potential probiotic strains should possess the following qualities: a) strong heavy metal binding, tolerance, or detoxification capacity; b) strong tolerance to acid and bile, allowing them to stay active in the digestive system; c) potent antioxidant or immunoregulatory skills to help them adapt to changes in the intestinal environment brought on by heavy metals and d) excellent intestinal adhesion or colonization ability to play positive functions in the gut.

Probiotics' Mechanism of Action Against Toxic Heavy Metals

The mechanism of heavy metal detoxification by probiotics involves the binding of metallic ions to bacterial cell walls, which is known as biosorption, followed by deposition inside the bacteria through the passage of the cell membrane, which is known as bioaccumulation [1]. Microbial cell surface characteristics and environmental factors have a significant impact on the biosorption and bioaccumulation mechanisms of probiotics heavy metal detoxification [1, 7]. The biosorption process occurs when negatively charged functional groups (*i.e.*, peptidoglycan and teichoic acids) on the probiotic cell wall interact electrostatically with heavy metal positive ions. This process is stained-specific and depends upon various factors like surface area, porosity, and surrounding pH [1]. Bioaccumulation, on the other hand, involves the active transport of heavy metals into microbial cells, and this process is primarily influenced by probiotics' metabolic activity and toxicity threshold [53]. For instance, methylated mercury is converted by *lactobacillus* bacteria into inorganic mercury (Hg^{2+}) and ultimately to Hg0, which is weakly absorbed in the gastrointestinal system [53]. Protein carboxyl groups and peptidoglycan hydroxyl groups are involved in the metal chelation of probiotics [28]. There is evidence that certain *L. rhamnosus* species and the yeast *Saccharomyces boulardii* can detoxify elements like mercury, lead, cadmium, and arsenic [54]. Probiotics are regarded as the most significant immune system modulators due to their natural anti-inflammatory properties. In addition, they have a clear impact on humeral and cell-regulated immunity. Substances that regulate the immune system are thought to be secreted by probiotics. Since probiotics are a natural anti-inflammatory regimen and clearly impact humeral and cell-regulated immunity, they are regarded as the most significant immune system modulator. Probiotics are believed to produce the substances necessary for the control of the immune system. According to earlier reports, L. factors released by the *reuteri* strain may block the expression of the NF-Kb gene, which would reduce cell development and affect protein kinases that are activated by nitrogen resulting in apoptosis [55].

Probiotic bacteria are one of the beneficial gram-positive bacteria that remove potentially toxic elements (PTEs) from foods . Despite speculation concerning the mechanism behind the removal of Potentially Toxic Elements (PTEs) through probiotic bacteria, they may be separated into two main processes: biosorption and bioaccumulation [56]. Bioaccumulation is a metabolically controlled and complicated response that occurs through transmembrane transport in live cells exposed to toxic heavy metals [57]. Additionally, biosorption of probiotic bacteria takes place at two distinct stages, *i.e.*, PTEs interaction with functional groups available up to attain the equilibrium, and second stage, *i.e.*, PTEs penetration into the cell and their intra-cellular uptake (as shown in Fig. **2**) [57]. Ion exchange, complexation, chelation, and metal precipitation are some processes that occur during toxic heavy metals binding to the extracellular functional groups feasible or inactivated probiotic bacteria [58]. Metal ion precipitation in the ion exchange mechanism was made more accessible by the molecules with negative charges that were present on the microbe's cell wall. Covalent or electrostatic interactions between organic molecules and PTE ions are also possible. These complex binding processes are typically influenced by the probiotic strain and the surrounding environment (initial metal ion concentration, acidity, and temperature) [59].

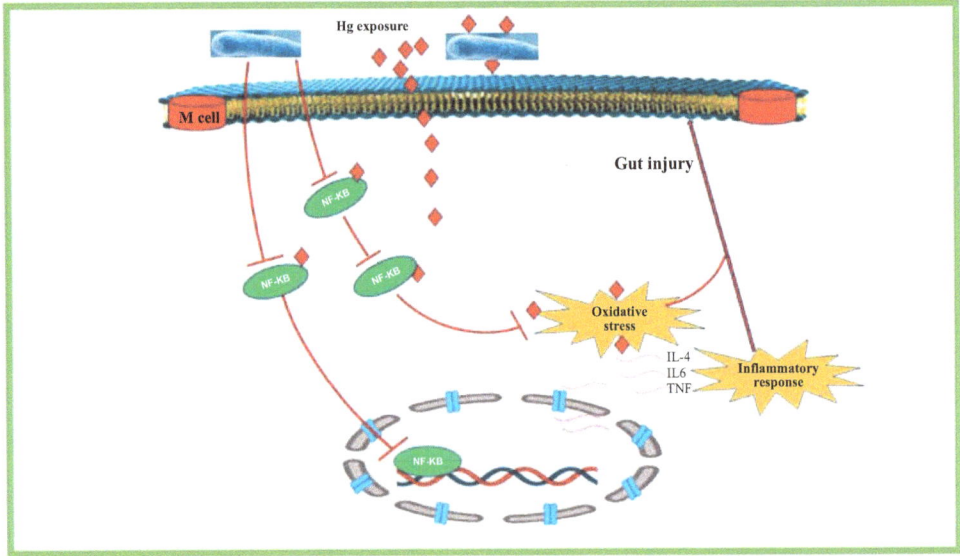

Fig. (2). Effects of *Lactobacillus brevis* 23017 treatment on the regulation of inflammation and oxidative stress induced by heavy metal mercury in mice.

Different proteins, polysaccharides, and lipids from the probiotic microorganisms's cell walls provide active sites, including sulfate, phosphate, carboxyl, and amino groups that can eliminate PTEs [58]. Gram-positive probiotic microbes are quite effective in removing PTE due to the presence of teichoic acid and peptidoglycan (PG) in their cell walls. Probiotic bacteria metabolites such as metal solubilization (organic acids), exopolysaccharides (EPs), bioabsorption through siderophore, and bioleaching *via* efflux pumps are among the key elements in PTE removal [60]. The functional hydroxyl groups of the PG and carboxyl group of proteins are also essential elements in PTE binding performance (as shown in Fig. **3**) [61]. Table **2** represents the mechanistic detoxification of heavy metals by probiotics

Probiotics-based Remediation of Heavy Metal (HM) Toxicity of Cadmium (Cd), Lead (Pb), Mercury (Hg), Arsenic (As), and Various other Heavy Metals

Probiotics are fermentative, nontoxigenic, and non-pathogenic bacteria present in various products, such as drugs, foods, and dietary supplements. Probiotics that are used most frequently are *Lactobacillus* and *Bifidobacterium* [1]. The toxicity of HMs such as Cd, Hg, Cr, As, and Pb has been successfully reduced in mammalians by supplementing with a single or a combination of probiotics [77]. *Lactobacilli* are the most common probiotics that are used to minimise the toxicity of HMs since they have a strong HM binding capacity, clearly decreasing the availability of HMs for the host [78]. Additionally, it has been suggested that the live probiotic strain *L. plantarum* CCFM8610 may increase the dissolution and uptake of important divalent ions, including Ca, Mg, and Fe, by inhibiting the intestinal absorption of Cd ions [79]. Probiotic strains may also promote peristalsis of the gastrointestinal tract, which facilitates HM excretion in stool. In addition, probiotic strains can also reduce the entry of HMs by improving intestinal barrier function and controlling the tight junction of the small intestine's epithelium. Intake of *L. plantarum* CCFM8610 decreased intestinal permeability and reduced Cd leakage into the systemic circulation [51]. Lead (Pb) is recognized as one of the most hazardous heavy metals. Lead removal abilities of the *Enterococcus faecium* strains EF031 and M74 from the water were first investigated by [80]. According to early findings, both *E. faecium* stains were capable of binding the Pb effectively [81]. According to some studies [82, 9], *Lactobacillus reuteri* P16, which was isolated from *Cyprinus carpio* fish intestinal contents, had effective Pb-binding capacity (>15% Pb removal), remarkable antioxidant activity (hydroxyl radical-scavenging, 42.18%), and good probiotic qualities. Mercury (Hg) poisoning in both humans and fish is a serious global issue. Due to its extreme toxicity, mercury is one of the most severe threats to aquatic ecology [83]. As a probiotic, *Bacillus* may generate antioxidant enzymes like SOD and glutathione to properly get rid of free radicals [84]. High concentrations of sulfhydryl sites in *Bacillus subtilis* bacterial biomasses can improve the removal of dissolved Hg (II), Cd (II), and Au (III) from water [85]. Probiotics elevated vitamin B12 and estradiol levels and lowered malondialdehyde (MDA) and conjugated dienes (CD) in rat uterine cells by enhancing the activity of antioxidative enzymes. Probiotics improved safety against mutagenic uterine DNA breakage, ovarian-uterine tissue damage, and necrosis in arsenic-exposed rats, in addition to their role in regulating the endogenous antioxidant system [86].

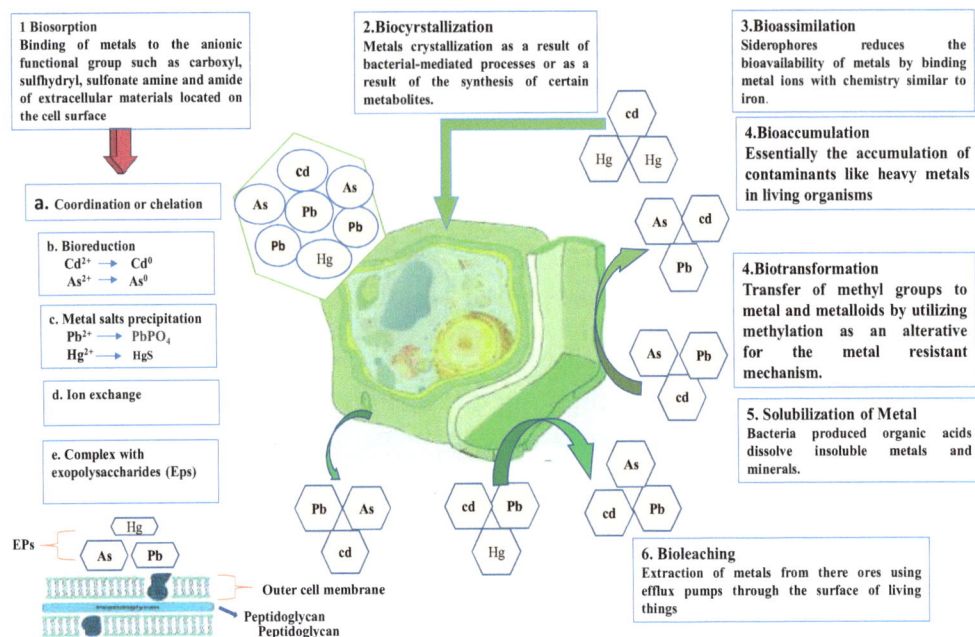

Fig. (3). Interactions between probiotic microorganisms and potentially toxic elements (PTEs) in food matrixes.

Probiotics Against Cadmium Toxicity

Cadmium (Cd) has damaging effects on organ systems that result in multisystem diseases. Cd poses a global hazard to human health and the environment. As such there is no definitive treatment for Cd toxicity in humans, but dietary products have come up in the prevention and mitigation of infirmities without resulting in any side effects. The probiotic bacteria act by various mechanisms in improving intestinal microbial balance and hence benefit the host [78]. Presently, many clinical studies are undertaken to demonstrate the potential action of probiotic microorganisms in protecting cell lines from chronic Cd toxicity symptoms. The probiotics show antioxidative effects and result in Cd sequestration. Also, the orally administrated probiotic agents have been shown to effectively decrease Cd absorption and accumulation in body tissues. It alleviated oxidative stress and amended histopathological changes in cadmium-exposed tissues. Many organisms are reported for exhibiting similar outcomes, but the most effective agent is *Lactobacilli.* The Cd ingested has a cell-damaging effect, making the gastrointestinal tract (GIT) its prime target. Cd causes harm by inflammation and disruption of tight cellular junctions, ultimately resulting in membrane permeability changes and more Cd absorption [51, 87]. Many studies have highlighted the binding capacity of microbes with Cd, decreasing its systemic absorption and accumulation in tissues facilitating Cd excretion from GIT by faeces [88]. Probiotics have been shown to have antioxidative properties in both *in vitro* and *in vivo* systems [88, 89]. This mechanism aids in the detoxification of Cd in organs like the liver and kidney (Table **1**). Hence, probiotics neutralize the harmful effects of Cd toxicity at local (intestinal) and systemic levels. Recent proteomic studies have highlighted the role of proteins generated from *Lactobacillus plantarum* strains and their action in the regulation of gene expression [90]. Studies conducted by Bisanz et al., 2014 stated the probable action of *L. rhamnosus* GR-1-supplemented yogurt in reducing Cd levels in children and pregnant women [91]. The strains having a potential role in ameliorating Cd toxicity are summarized in Table **3** [92]. The data procured from preclinical studies have shown that probiotic mixtures have the prospects for alleviating Cd toxicities. However, definitive clinical studies need to be planned and executed. Also, pre-clinically proven probiotics could be clinically tested along with putative chelating agents. Hence, probiotic supplementation can be regarded as a therapeutic strategy against cadmium toxicity along with having antioxidant, anti-inflammatory, and other supportive actions. More research and development in this direction may lead to the production of beneficial functional foods in the management of Cd toxicities in humans. (Fig. **4**).

Table 2. Detoxification of heavy metals by probiotics and their mechanism.

S.No.	Heavy Metals	Probiotics	Mechanism	References
1	Cd	*Lactobacillus plantarum (L. plantarum)* TW1-1	Reduce Cd's intestinal absorption by converting it to a less absorbable form.	[62]
2	Hg	*Comamonadaceae, Cloacibacterium, Xanthomonadaceae, Pirellula,* Deltaproteobacteria FAC87	To minimize its absorption, convert methylated Hg into Hg0.	[51, 63]
3	As, Cd, Fe	Sulfate-reducing bacteria (SRB), Fe-reducing bacteria methanogens, *Desulfovibrio* spp.	Methylation of HMs to change their chemical composition.	[64]
4	Cd, Pb	*Akkermansia muciniphila* (A. muciniphila) and *Lactobacillus rhamnosus (L. rhamnosus)* GR-1	Restoring the structural equilibrium by reversing the compositional alterations in the gut microbiota induced by HM.	[40]
5	Pb, Cd, Hg, Cr, As	*Pseudomonas, Oxalobacter formigens (O. formigens)*	Through siderophores and hydrogen sulfide, insoluble complexes are formed with HMs.	[65]
6	Pb	*L. plantarum* CCFM8610, CCFM 8611, and *Bacillus cereus.*	Enhance the HM excretion together with the formation of bile acid.	[45]

(Table 2) cont.....

S.No.	Heavy Metals	Probiotics	Mechanism	References
7	As	*Faecalibacterium prausnitzii* (*F. prausnitzii*), Bacteroides and *Faecalibacterium*	Produce enzymes that detoxify As.	[66]
8	Hg	*L. plantarum* CCFM639, CCFM8610, *L. brevis* 23017, Nocardia and Bacteroidales	By lowering the levels of proinflammatory cytokines, HM-induced inflammatory reactions are released.	[67]
9	As, Cd, Pb	*L. plantarum* CCFM639, *Bacillus coagulans* (*B. coagulans*)	Stimulate the expression of genes associated with antioxidants to produce antioxidant enzymes.	[68]
10	Pb, Cd	*L. plantarum* LC-705 and *Propionibacterium freudenreichii*	Decrease the PH of the intestine.	[69]
11	Pb	*L. plantarum* CCFM8661	Decrease the accumulation of Pb inside tissue.	[70]
12	Cd	*S. thermophilus*	Lacking visible absorption or sedimentation inside the cell, cadmium binding occurred at the surface of the *S. thermophilus* cell.	[71]
13	Cd	*L. acidophilus* Rosell-52, B. *longum* Rosell-175 and L. *rhamnosus* Rosell-11	With binding to the probiotics' cell wall, cadmium can excrete by faeces.	[72]
14	Pb	*L. reuteri* P16	Reduced the amount of lead that accumulated in all tissues (gills, spleen, liver, and kidney). excluding the gut.	[73]
15	Cd	*P. pentosaceus* GS4	Due to P. pentosaceus GS4's biosorption and complexation of Cd adsorption and deposition of Cd have been reduced.	[74]
16	Pb	*L. plantarum* KLDS 1.0344	Reduced the Pb amount through faecal elimination and restricted both capsulated and non-capsulated types of absorption into the blood.	[60]
17	Pb	*L. plantarum* CCFM8661	Both dietary supplements have a synergistic effect that increases Pb excretion and prevents Pb absorption.	[75]
18	Pb	*Leuc. mesenteroides* L-96	Effectively reduce lead toxicity by enhancing intestinal health and minimising Pb build-up in tissues.	[76]

Table 3. Shielding effects of probiotics against cadmium poisoning.

S.No.	Probiotic organisms	Cell line	Observations and proposed mechanisms	References
1. 1	*Lactobacillus plantarum* CCFM8610	Mice	Mitigates Cd toxicity by its sequestration and antioxidant effects.	[88]
2. 2	Mixture of *L. rhamnosus* Rosell-11, *L. acidophilus* Rosell-52, and *Bifidobacterium longum* Rosell-175	Rats and rat hepatocytes	Alleviate genotoxicity *in vitro* and *in vivo* by increasing Cd level in feces, lessening Cd concentration in blood and tissues.	[72]
3. 3	Genetically modified *Escherichia coli* Nissle 1917 (EcN-20)	Rats	Mitigates oxidative stress by the generation of antioxidant pyrroloquinoline quinone (PQQ).	[72]
4. 4	*Lactobacillus kefir*	Human hepatoma cell line (HepG2)	Improves cell viability by cadmium binding and adsorption.	[93]
5. 5	*Lactobacillus plantarum* CCFM8610	Human intestinal cancer cell line HT-29	Reduces Cd absorption by protecting the intestinal barrier through the alleviation of oxidative stress.	[51]
6. 6	*Lactobacillus acidophilus*	Mice	Lessens Cd-induced brain and testicular toxicity by excreting Cd in feces.	[94]
7. 7	Glycoprotein from *Lactobacillus plantarum* L67	Murine RAW 264.7 cells	Enhances cell viability by anti-inflammatory effects.	[95]
8. 8	*Saccharomyces cerevisiae*	Mice	Treats testicular toxicity	[96]
9. 9	*Streptococcus thermophilus*	Mice	Reduces blood Cd concentration through retraction of oxidative stress.	[71]
10	*Lactobacillus rhamnosus* GR-1 (LGR-1)	Human intestinal epithelial cell line (Caco-2)	Facilitates Cd absorption and immobilization across the intestinal epithelium.	[97]
11	*Lactobacillus* sp.	*in vitro* digestion model	Reduces Cd bio-accessibility.	[98]
12	*Lactobacillus plantarum* CNR273 *Bacillus coagulans*	Rat	Reduce Cd accumulation.	[99]
13	*Pediococcus pentosaceus* GS4	Mice	Decreases Cd deposition in tissues by increasing its excretion in feces.	[74]
14	*Lactobacillus plantarum* CCFM8610	Nile tilapia fish (*Oreochromis niloticus*)	Improves hematobiochemical parameters by mitigating oxidative stress and reduces Cd accumulation.	[100]

Probiotics Against Lead Toxicity

Lead, also known as *Plumbum* (Pb), is a soft and malleable heavy metal [101]. It is a very toxic environmental pollutant. In humans and animals, lead enters into the body *via* the oral route or by digestion through the gut lining from where it is absorbed and taken to target organs like the reproductive organs, nervous system, cardiovascular, gastrointestinal, hematopoietic system, and excretory system [102, 103]. It is a protoplasmic poison, and can bind to red blood cells and get deposited in the bones [104]. Even in plants, the higher concentration of Pb can cause severe negative impacts like the inhibition of photosynthesis, inhibition of seed germination, stunted growth, root blackening, less flowering, alteration in the structure of cell membrane, underdeveloped vascular system, and various changes in activities of enzymes and metabolic pathways [104]. Lead is also found to be present at toxically higher concentrations in different trophic levels of the food chain. Various food and fruit plants take up lead from the soil through roots and translocate it to aerial parts through the xylem [102]. The mechanism of toxicity induction by this heavy metal includes the weakening of antioxidant defense enzyme, oxidative stress, generation of reactive oxygen species, and inhibition of enzymatic activity [105]. The small traces of heavy metals present in aquatic as well as terrestrial food sources can cause a great threat to humanity. There should be a restricted maximum level of quantity of heavy metal in a foodstuff. The main causes for the accumulation of these heavy metals in the air, water, soil, and the food chain are urbanization, industrialization, long-term use of pesticides, fertilizers, mining, Pb-acid batteries, Pb-containing fuel, printing, exhaust from vehicles, *etc* [106, 107]. Considering the extreme toxicity of Pb, it becomes important to find an effective, safe, and specific method to compensate for its toxic effect. Out of the various methods present to overcome this effect, the use of probiotics is also an effective method. Probiotics are non-pathogenic living organisms. A substantial number of microorganisms are available as probiotics, but only some can lower the toxicity level of Pb. The principal mechanism behind the probiotic effect of microorganisms is their antioxidant effect. They have a tendency to reduce lead accumulation and oxidative stress, and prevent DNA, liver, and renal damage [108]. The heavy metals bound to the functional groups present on the surface of these probiotics once engulfed, are removed in the form of faecal matter in animals [109, 110]. In plants, various kinds of microorganisms are present that act as probiotics, like some plant endophytes and epiphytes, along with some soil probiotics well-known as probiotics; they all perform the function of adaptation of the stress of heavy metals in plants by lowering their accumulation and increasing biomass of plant [111]. PGPR is one of the examples that act as probiotics for plants [112]. In the aquatic environment, flora and fauna also deal with heavy content of heavy metal accumulation, which can be suppressed by the use of some microorganisms as probiotics like bacteria of bifidobacteria and lactobacilli, while other examples include *Bacillus subtilis, Enterococcus faecium, Escherichia coli* and *Saccharomyces boulardii* [52]. In animals, the effective probiotic microorganism can be bacteria or yeast. For example, bacteria available as probiotics include *Lactobacillus* sp., *Bacillus* sp., and *Pediococcus pentosaceus*, while *Saccharomyces cerevisiae* is the yeast effective as a probiotic. Instead, *L.plantarum* CCFM8661 is also found effective in ameliorating lead toxicity [75]. Some strains of lactic acid bacteria have high Pb binding sites and high antioxidant activities, and according to a study, the P16 *L. reuteri* strain is the best probiotic against Pb exposure [9]. Thus, many probiotics have been reported to be tolerant of heavy metals.

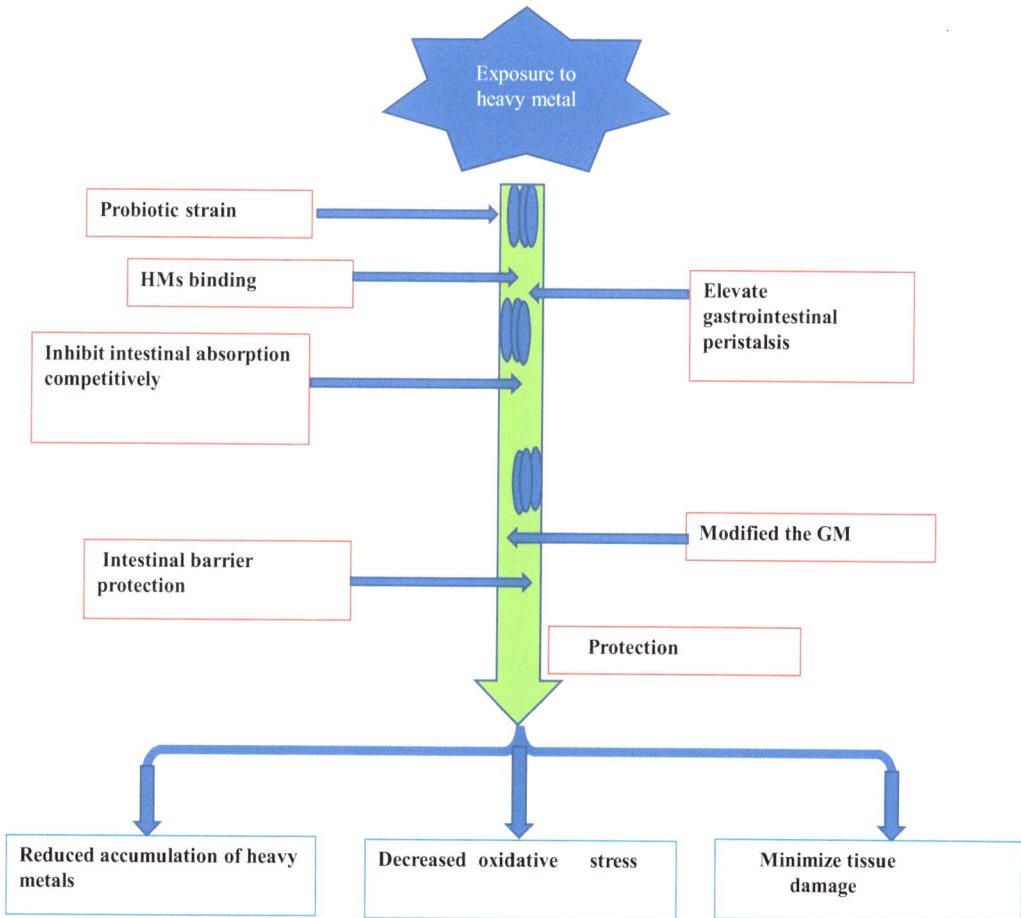

Fig. (4). Heavy metal remediation by probiotics

Probiotics Against Mercury Toxicity

Mercury (Hg) is another heavy metal known to exist in nature either in metallic, organic or inorganic forms and is consistently polluting aquatic and terrestrial ecosystems [83]. These heavy metals pose serious concerns in plants and humans as they are known to impair the functional activities of multiple proteins through modifications in functional groups like -SH and -SeH through specific modifications at their active sites [113]. However, the Hg (II) is functionally known to act corrosive and deteriorate the gastrointestinal layers, further allowing the entry and absorption of Hg. Mercury is well known to generate oxidative stress by the formation of mercaptans by undergoing reaction with thiol groups thereby leading to the inactivation of the antioxidant system by involving glutathione. Also, Hg is actively involved in the inactivation of certain antioxidative enzymes such as thioredoxin reductase, and glutathione peroxidase through specific modifications at the active sites in selenocysteine [114]. On the other hand, metallic mercury interactions and the formation of specific bonds with the sulfhydryl groups lead to their entry into the placenta and might further cross the blood-brain barriers. This metallic mercury form can be concentrated inside multiple organs either in the kidneys, lungs, breasts, liver, pancreas, brain and also in testes, therefore, leads to impaired functions of multiple tissues and organs. However, the human intestine is highly loaded with multiple bacterial cells with a high potential to sequester the heavy metals entering the human body. Whereas the beneficial microbes or probiotics like the strains of *Lactobacillus* are known to have HMs binding properties with respect to elements such as Hg, Cr, Cd, Pb, Ni, *etc* [19, 73]. The probiotics involve multiple reaction mechanisms to bins and sequestration of HMs, such as ion exchange reactions between teichoic acid and peptidoglycans [115], complex formation through ligands [116], and other nucleation reactions known to lead to precipitation [78]. *Pseudomonas* is among the gut microbiota that is effective in forming iron-chelating siderophores and which is known to show chelate formation with HMs such as Hg, Cr, Pb, and As, *etc* [117 - 119]. However, Hg methylation is another process induced by multiple bacteria such as methanogens, *Desulfovibrio* strains, *etc* [120]. Table **4**. shows the ameliorative potential of certain probiotics against mercury (Hg) toxicity.

Probiotics Against Arsenic Toxicity

Arsenic exists in many forms like crystal, sulfur, minerals or metals. It is also known as "the poison king" for its carcinogenic, mutagenic, and teratogenic nature [125]. It is released into the air, soil, and water *via* various processes like metal ore mining, smelting, industrial and domestic waste removal. It has toxic effects on plants, humans, and animals. Plants take this heavy metal from the soil through plant root cells, which is aided by the presence of various transporter channels or proteins in cells. From roots, it is translocated to other parts of the plant through the xylem resulting in the accumulation of arsenic in different parts of the plant [126]. At toxic concentrations, arsenic disturbs the normal metabolic and biochemical pathways of the plant. It inhibits the photosynthesis rate, lowers ATP production, damages cellular membranes, generates ROS, and alters DNA structures [127]. In humans and mammals, arsenic enters into the body *via* contaminated air and water. It is highly toxic even in low concentrations. Its toxicity results in haematological alterations, cytotoxic, genotoxic, or epidemic effects including skin cancer, cardiovascular disease, *etc.* The primary mechanism behind arsenic toxicity is their binding capacity with proteins. By interacting with zinc finger domains, cysteine clusters, and cysteine residues of proteins, it alters the normal genetic and cellular level structures [128, 29]. It competes for phosphate anion comprising transporters and inhibits ATP formation during oxidative phosphorylation. It can bind to Hb and accumulate in blood. It is a carcinogen and can cause cancer. It can inhibit Pyruvate Dehydrogenases Complex and disturb various metabolic pathways [129]. It generates free radicals that elevate oxidative stress and cellular damage like the hydroxyl radical produced by arsenic from radical lipid molecules. Various products of lipid peroxidations such as Melonaldehyde, act as stress biomarkers. Arsenic not only produces hydroxyl radicals but also peroxyl radicals, dimethylarsinic peroxyl radicals, dimethylarsinic radicals, singlet oxygen, and all these reactive oxygen species result in the elevation of oxidative stress [130]. Arsenic also participates in the formation of superoxides [1]. It also causes various autoimmune and metabolic disorders by altering the gut microbiota. To deal with the toxic effects of arsenic, various methods have been developed, of which one effective method is the use of probiotics to ameliorate the toxicity of heavy metals. Various microorganisms act as probiotics to deal with the toxicity of these heavy metals, like *Lactobacillus, Pediococcus, Bacillus, Bifidobacterium,* and yeast, but the main microorganisms effective in ameliorating the heavy metal arsenic involves bacteria like *Lactobacillus* strains, they can alter the toxicity of arsenic. These probiotic bacteria involve the removal of heavy metals *via* binding these metal ions on their surface by ion exchange method or biosorption or metabolism-independent uptake processes [1, 63]. Probiotics also help in lowering the absorption rate of these heavy metals in the gut lining by altering the expression of metal transporting channels, maintaining the gut-metal barrier, detoxifying heavy metals, and enhancing the sequestration of heavy metals present in the intestine [62]. Bacterial microorganisms like *Lactobacillus sporogenes* as probiotic cures haematological alterations [29]. Another bacteria *Bifidobacterium* sp. is found effective in alleviating arsenic toxicity, it induces high levels of SOD, and CAT, leading to the prevention of lipid peroxidation, reduction in glutathione, and prevention of DNA damage in the cells [131]. Arsenic-resistant strains of

Limosilactobacillus fermentum, a lactic acid bacterium is also resistant probiotic that has the potential to act as a bioremediation tool for fishes to survive in the high arsenic concentrations of aqueous environment [132]. Similarly, *Faecalibacteriumprausnitzii* is a marketed human probiotic that mitigates the toxic effects of arsenic [25]. In the same way, plant probiotic bacteria include rhizobial and non-rhizobial bacteria that help in heavy metal tolerance and plant growth [133, 134]. Some other strains of lactic acid bacteria *Pediococcus acidilactici* and *Pediococcus dextrinicus* can also help in the detoxification of arsenic [135]. All these microorganisms are effective in mitigating the toxicity of arsenic, but more study is required to isolate more isolates to different strains as probiotics.

Table 4. Ameliorative potentials of certain probiotics against mercury (Hg) toxicity.

S. No.	Heavy Metal	Harmful Effects	Probiotics	Ameliorative Impact of Probiotics	References
1.	Hg (10 ppm mercuric chloride)	Oxidative stress generation, and higher levels of creatinine, bilirubin, urea, alanine transaminase (ALT), and aspartate transaminase (AST) were observed in rats.	*Lactobacillus plantarum and Bacillus coagulans*	A decrease in the Hg level in the liver and kidney, with a reduction in the levels of creatinine, bilirubin, urea, ALT, and AST was observed (Study on rats).	[121]
2.	Hg	Oxidative stress and intestinal inflammation were observed.	*Lactobacillus brevis* 23017 (LAB, *L. brevis* 23017)	Ameliorates the injury of the small intestine by reducing intestinal inflammation and alleviating oxidative stress (Study on rats).	[117]
3.	Hg	Biochemical and histopathological alterations in rat serum and tissues were observed. Also, a decrease in body weight and an increase in kidney weight were observed, with histopathological changes in the brain and kidney.	High-fiber fermented mare's milk containing probiotics	Improves the histopathology of the kidney and brain; also restoration of the biochemical parameters in serum to almost normal values was observed.	[122]

(Table 4) cont.....

S. No.	Heavy Metal	Harmful Effects	Probiotics	Ameliorative Impact of Probiotics	References
4.	Hg (0.25 mg kg^{-1}mercuric chloride)	Mercuric chloride-induced oxidative stress and histopathological changes in the kidneys, alterations of enzymatic antioxidant defense system and histopathological changes in the kidneys of male rats were observed.	Protoxine, a probiotic (12 mg rat^{-1}) to mercuric chloride-treated group)	Ameliorates the levels of lipid peroxidation, improves the activity of enzymatic antioxidants and associated defense system. Also the histopathological changes in the kidneys of male rats were observed up to normal levels.	[123]
5.	1 mg/ml mercury [Hg(II) or methyl-Hg]	Impaired intestinal barrier, increased permeability, defective tight-junctions, and mucus hypersecretion in intestinal epithelial cells of rats due to Hg toxicity.	*Lactobacillus intestinalis* or *Lactobacillus johnsonii* cells	Alleviation of the damage.	[124]
6.	Hg	Oxidative stress-inducing potential, intestinal damage in *C. carpio* var. specularis (fish) was observed.	Se-rich *B. subtilis*	Significantly decreased the activity levels of CAT, SOD, and GSH-PX, while increasing the activity levels of GST, MDA, and GSH. Further alleviating the intestinal damage.	[83]

Probiotics Applications and their Effectiveness Against other Heavy Metals Toxicity

Multiple studies have demonstrated that in addition to the detoxifying effects, probiotics also have the ability to bind heavy metals. These probiotics are effective against numerous heavy metals, including Cr, Cu, Zn, Mn, Ni, Mo, and Ag. It has been reported that *Lactobacillus plantarum* TW1-1 can be utilised for Cr metal detoxification [15]. Similarly, other *Bacillus* species have been reported to function as metal-detoxifying agents, which have been tabulated in Table **5**.

Microorganisms have developed various strategies for the uptake and elimination of heavy metals. In *Bacillus subtilis*, Zn absorption is mediated by the Zur family and ZoSA, whereas Zn efflux is enhanced by the Cad-A CPx-type ATPase efflux pump [136]. Similarly, in the presence of high concentrations of Ni, Cu, Zn, and Co, CzcD, a cation diffusion transporter, protects the cell effectively [147]. According to research conducted by Gaballa *et al.* in 2003, CueR helps regulate

Cu in the cytosol. It regulates the Cop ZA operon, which encodes P-type ATPase and Cu chaperone, resulting in the export of copper from the cell [148].

Table 5. Role of microorganisms in heavy metal detoxification.

S.No.	Metal	Microorganisms	References
1	Zn	*Bacillus subtilis* *Bacillus cereus* *Bacillus licheniformis* *Bacillus jeotgali* *Bacillus firmus* *Bacillus altitudinis*	[136 - 138]
2	Ni	*Bacillus thuringiensis* *Bacillus megaterium*	[139, 140]
3	Cu	*Bacillus thuringiensis* *Bacillus cereus* *Bacillus licheniformis* *Bacillus sphaericus*	[139, 141, 142]
4	Cr	*Bacillus licheniformis* *Lactobacillus plantarum*	[143, 15]
5	Mn	*Bacillus subtilis*	[144]
6	Mo	*Bacillus subtilis*	[145]
7	Ag	*Bacillus licheniformis*	[146]

Probiotics offer a range of practical applications such as detoxification of various heavy metals, restoration of gut microbiota, and better nutrients bioavailability, thereby, playing a potential role in immune system modulations, improving aquatic animals' growth and well-being, and also helping in the production of various protective compounds (antimicrobial or antipathogenic compounds) [4, 52]. Furthermore, probiotics, such as *Lactobacillus* and *Bifidobacterium* play a crucial role in the processes of water purification and the enhancement of organisms' health [49]. Biofilm-forming probiotics, such as *Enterococcus faecium*, possess potential applications for the effective treatment of contaminated wastewater compared to conventional approaches [37]. Plants probiotics (*Bacillus* spp., *Pseudomonas* spp., Rhizobium spp., *etc.*) have a significant application in improving soil health as compared to traditional methods [111]. Therefore, probiotics' significance is highly evident as they demonstrate remarkable versatility on this earth planet, impacting environmental, ecological systems agricultural, and health-related domains in meaningful ways as compared to other conventional approaches [31, 49].

Promising Application of Engineered Probiotics for Heavy Metal Detoxification

Although the most naturally occurring probiotics do not exhibit infectivity or pathogenicity, there is still uncertainty about the safety of using probiotic-based bioagents. Unlike what was anticipated, some probiotic bacteria, particularly enterococci, have been found to promote a group of genes linked to communicable antibiotic resistance. Moreover, the ability of *Bacillus cereus* to create enterotoxins and emetic toxins was discovered [149]. Additionally, it was found that probiotic treatment resulted in certain typical side effects in vulnerable individuals, including harmful metabolic activities, gene transfer, systemic infections, and excessive immunological stimulation [150, 151]. The heavy metal-resistant characteristics are restricted to specific microbial species, and to ensure the therapeutic efficacy and better removal of heavy metals, it is necessary to mix several diverse strains. Therefore, due to the inherent flaws in naturally occurring probiotics, the development of engineered probiotic microorganisms with the ideal traits and functions has become a major emphasis [152]. Due to their low cost, versatility, and environmental friendliness, genetically engineered microbes (GEMs) have received a lot of attention in recent years when used for heavy metal removal. In plants, animals, groundwater, soil, and activated sludge bioremediation of heavy metals, GEMs with substantial degradative potential have been widely employed. The development of effective heavy metal-resistant GEMs typically relies on two major approaches: a) surface functional aggregates with a strong heavy metal-binding ability for biosorption; b) metal ions are carried into the cytoplasm and then processed by storage mechanisms for improved intracellular bioaccumulation [2].

A genetically engineered surface-displayed *E. coli* strain was created by Liu and his co-workers (2019) for the selective adsorption of Hg^{2+} [153]. By attaching to the N-terminal portion of the ice nucleation protein, a novel Hg^{2+}-binding peptide with the sequence, Cys-Lys-Cys-Lys-Cys-Lys-Cys, was surface-expressed on the *E. coli* BL21. The surface-engineered *E. coli* made it easier for mercury ions (Hg^{2+}) to be selectively adsorbed from a solution that contained other metal ions. The surface-engineered cell has a four-fold greater Hg^{2+} adsorption capability than the original *E. coli* cells. These whole-cell sorbents removed around 95% of the Hg^{2+} from the solution. Further, *Carassius auratus* was fed with the modified strains so that the bacteria may colonize the fish intestine. When compared to the control group, *C. auratus* that had been fed engineered bacteria displayed noticeably less (51.1%) accumulation of mercury. By adsorbing more Hg^{2+}, the surface-engineered *E. coli* successfully shielded fish against the toxic effects of Hg^{2+} in aquatic habitats. Additionally, the surface-engineered *E. coli* reduced the alterations in microbial diversity brought on by Hg^{2+} exposure in the colon, defending the intestinal microbial population [153]. Similar results were obtained by Liu *et al.* [153], where surface-engineered *E. coli* reduced the toxicity of methylmercury when fed to *C. auratus*. In another study, Xue *et al.* [154] built and improved a genetic circuit that can doprogrammable killing of cells with a cell suicide module after detecting Hg^{2+} in a waterbody and activating an Hg^{2+} adsorption module. Using this circuit, it is possible to instruct the altered *E. coli* cells to only express the Hg^{2+} adsorption protein when the Hg^{2+} concentration is higher than a predetermined threshold. The surviving cells are then set up to be programmed to be destroyed by the suicide unit when Hg^{2+} concentration goes below a threshold in the water body. Cells absorbed with Hg^{2+} can then be removed from natural habitats with a magnetically immobilized technique [154].

Additionally, with PbrR genetically modified on the cell surface of *E. coli* and lipoprotein (Lpp)-OmpA as the anchoring peptide, a whole-cell biosorbent for Pb was developed. *in vitro*, the PbrR-expressed cells showed improved hazardous Pb(II) ion immobilization abilities at both neutral and acidic pH levels. The integrity of the surface-displayed PbrR and the durability of engineered *E. coli* in the digestive tract of mice were both validated *in vivo*. Further, oral delivery of surface-engineered *E. coli* to mice was safe and had no impact on the blood levels of physiological metal ions. Furthermore, compared to the control, PbrR-displayed *E. coli* groups had considerably reduced blood and femur Pb levels [155]. In another experiment, *E. coli* was genetically altered by inserting a synthetic type VI secretory system cluster of *Pseudomonas putida* and a de novo synthetic heavy-metal-capturing gene, giving the synthetic cells a high ability to show the heavy-metal-capturing on the cell surface. Application of engineered *E. coli* with magnetic nanoparticles captured Pb and Cd with enhanced (more than 90%) removal effectiveness [156]. The study on the bioaccumulation of Cd was expanded by Deng *et al.* (2007) [157] using genetically modified *E. coli* (M4) that simultaneously expressed a Cd-transport system and metallothionein. The development of M4 demonstrated resistance to Cd presence. The ability of M4 to accumulate Cd^{2+} was increased more than one-fold when compared to the original host bacterial cells' capacity to do so. Though the application of genetically engineered probiotics is still in the testing phase, it has advanced significantly up to this point, thanks to the ongoing advancements in synthetic biology and recombinant DNA technology.

CONCLUSION AND FUTURE PERSPECTIVES

Heavy metals contaminate air, water, soil, and food, and harm diverse organisms in all parts of the world. According to research, the usage of probiotic bacteria may have the ability to give bio-protection or eradicate these hazardous heavy metals from organisms' bodies or ecosystems when provided alone or in combo at an adequate quantity and frequency. Probiotics are associated with a potent anti-carcinogenic, anti-immunomodulatory, and anti-mutagenic function. This is because of the fact that probiotics inhibit the absorption of toxins, which, in turn, lowers the toxicity of toxins like heavy metals. Probiotic intake also improves organisms' health which had previously been poisoned with the modest doses of heavy metals.

Most research has been performed on *Lactobacillus* probiotics, which are more likely to be able to bind heavy metals to the cell wall. Heavy metal absorption and binding vary between strains due to cell wall architectures and cell membranes. Moreover, the probiotic cell wall contains proteins, peptidoglycan, polysaccharides, and 1,3-β-glucan in the yeast cell wall, which are responsible for probiotic binding with heavy metals. Besides this, metabolites of probiotics, co-cultivation of different probiotics, or varying probiotic formulations (such as lactulose), gene expression, and maintaining faecal excretion are all linked to additional probiotic biodetoxification pathways. Despite several previous literature on probiotic biodetoxification, probiotics use is still limited in food and feed detoxification. Furthermore, although certain studies have enhanced our knowledge of interactions between probiotics and heavy metals, we still do not fully comprehend bacteria's vast and complicated involvement in these processes. Additionally, we should learn more about the genes of probiotic microorganisms that reduce or sequester heavy metals, so that, we can find the primary and most active probiotic microorganisms that act as bioprotectants.

REFERENCES

[1]   Abdel-Megeed RM. Probiotics: a promising generation of heavy metal detoxification. Biol Trace Elem Res 2021; 199(6): 2406-13.
[http://dx.doi.org/10.1007/s12011-020-02350-1] [PMID: 32821997]

[2]   Saravanan A, Kumar PS, Ramesh B, Srinivasan S. Removal of toxic heavy metals using genetically engineered microbes: Molecular tools, risk assessment and management strategies. Chemosphere 2022; 298: 134341.
[http://dx.doi.org/10.1016/j.chemosphere.2022.134341] [PMID: 35307383]

[3]   Hill C, Guarner F, Reid G, *et al.* The International Scientific Association for Probiotics and Prebiotics consensus statement on the scope and appropriate use of the term probiotic. Nat Rev Gastroenterol Hepatol 2014; 11(8): 506-14.
[http://dx.doi.org/10.1038/nrgastro.2014.66] [PMID: 24912386]

[4]   Chen R, Tu H, Chen T. Potential application of living microorganisms in the detoxification of heavy metals. Foods 2022; 11(13): 1905.
[http://dx.doi.org/10.3390/foods11131905] [PMID: 35804721]

[5]   Kim SJ, Shin JS, Park HS, *et al.* Production of antibacterial agents and genomic characteristics of probiotics strains for the foodborne pathogen control. Current Topic in Lactic Acid Bacteria and Probiotics 2022; 8(1): 1-16.
[http://dx.doi.org/10.35732/ctlabp.2022.8.1.1]

[6]   Singh EO. Probiotics: A fruitful approach to health improvement. Pharma Innovation Journal 2022; 11(9): 764-73.

[7]   Pop OL, Suharoschi R, Gabbianelli R. Biodetoxification and protective properties of probiotics. Microorganisms 2022; 10(7): 1278.
[http://dx.doi.org/10.3390/microorganisms10071278] [PMID: 35888997]

[8]   Tian F, Zhai Q, Zhao J, *et al. Lactobacillus plantarum* CCFM8661 alleviates lead toxicity in mice. Biol Trace Elem Res 2012; 150(1-3): 264-71.
[http://dx.doi.org/10.1007/s12011-012-9462-1] [PMID: 22684513]

[9]   Giri SS, Jun JW, Yun S, *et al.* Characterisation of lactic acid bacteria isolated from the gut of *Cyprinus carpio* that may be effective against lead toxicity. Probiotics Antimicrob Proteins 2019; 11(1): 65-73.
[http://dx.doi.org/10.1007/s12602-017-9367-6] [PMID: 29285742]

[10]   Wu G, Xiao X, Feng P, *et al.* Gut remediation: a potential approach to reducing chromium accumulation using *Lactobacillus plantarum* TW1-1. Sci Rep 2017; 7(1): 15000.
[http://dx.doi.org/10.1038/s41598-017-15216-9] [PMID: 29118411]

[11]   Zhai Q, Yu L, Li T, *et al.* Effect of dietary probiotic supplementation on intestinal microbiota and physiological conditions of Nile tilapia (*Oreochromis niloticus*) under waterborne cadmium exposure. Antonie van Leeuwenhoek 2017; 110(4): 501-13.
[http://dx.doi.org/10.1007/s10482-016-0819-x] [PMID: 28028640]

[12]   Wang N, Jiang M, Zhang P, *et al.* Amelioration of Cd-induced bioaccumulation, oxidative stress and intestinal microbiota by *Bacillus cereus* in *Carassius auratus* gibelio. Chemosphere 2020; 245: 125613.
[http://dx.doi.org/10.1016/j.chemosphere.2019.125613] [PMID: 31864061]

[13]   Chen DW, Li HJ, Liu Y, *et al.* Protective effects of fowl-origin cadmium-tolerant lactobacillus against sub-chronic cadmium-induced toxicity in chickens. Environ Sci Pollut Res Int 2022; 29(50): 76036-49.
[http://dx.doi.org/10.1007/s11356-022-19113-1] [PMID: 35665891]

[14]   Hedayati SA, Sheikh Veisi R, Hosseini Shekarabi SP, Shahbazi Naserabad S, Bagheri D, Ghafarifarsani H. Effect of dietary *Lactobacillus casei* on physiometabolic responses and liver histopathology in common carp (*Cyprinus carpio*) after exposure to iron oxide nanoparticles. Biol Trace Elem Res 2022; 200(7): 3346-54.
[http://dx.doi.org/10.1007/s12011-021-02906-9] [PMID: 34458957]

[15]   Breton J, Daniel C, Dewulf J, *et al.* Gut microbiota limits heavy metals burden caused by chronic oral exposure. Toxicol Lett 2013; 222(2): 132-8.
[http://dx.doi.org/10.1016/j.toxlet.2013.07.021] [PMID: 23916686]

[16]   Gao B, Chi L, Mahbub R, *et al.* Multi-omics reveals that lead exposure disturbs gut microbiome development, key metabolites, and metabolic pathways. Chem Res Toxicol 2017; 30(4): 996-1005.
[http://dx.doi.org/10.1021/acs.chemrestox.6b00401] [PMID: 28234468]

[17]   Massoud R, Khosravi-Darani K, Sharifan A, Asadi GH, Hadiani MR. Mercury Biodecontamination from Milk by using *L. acidophilus* ATCC 4356. J Pure Appl Microbiol 2020; 14(4): 2313-21.
[http://dx.doi.org/10.22207/JPAM.14.4.10]

[18]   Massoud R, Khosravi-Darani K, Sharifan A, Asadi G, Zoghi A. Lead and cadmium biosorption from milk by *Lactobacillus acidophilus* ATCC 4356. Food Sci Nutr 2020; 8(10): 5284-91.
[http://dx.doi.org/10.1002/fsn3.1825] [PMID: 33133531]

[19]   Ameen FA, Hamdan AM, El-Naggar MY. Assessment of the heavy metal bioremediation efficiency of the novel marine lactic acid bacterium, *Lactobacillus plantarum* MF042018. Sci Rep 2020; 10(1): 314.
[http://dx.doi.org/10.1038/s41598-019-57210-3] [PMID: 31941935]

[20]   Halttunen T, Finell M, Salminen S. Arsenic removal by native and chemically modified lactic acid bacteria. Int J Food Microbiol 2007; 120(1-2): 173-8.
[http://dx.doi.org/10.1016/j.ijfoodmicro.2007.06.002] [PMID: 17614152]

[21]   Adeleye AO, Yerima MB, Nkereuwem ME, Onokebhagbe VO, Daya MG. Effect of Bio-enhanced

Streptococcus pyogenes and *Enterococcus faecalis* Co-culture on Decontamination of Heavy Metals Content in Used Lubricating Oil Contaminated Soil. Journal of Soil. Plant and Environment 2022; 1(2): 1-5.

[22]   Cheng X, Sheng L, Peng S, Thorley E, Cao H, Li K. Integrated mechanism of heavy metal bioremediation from soil to rice (*Oryza sativa* L.) mediated by *Enterococcus faecium*. Plant Growth Regul 2022; 97(3): 523-35.
[http://dx.doi.org/10.1007/s10725-022-00811-2]

[23]   Wosinska L, Cotter PD, O'Sullivan O, Guinane C. The potential impact of probiotics on the gut microbiome of athletes. Nutrients 2019; 11(10): 2270.
[http://dx.doi.org/10.3390/nu11102270] [PMID: 31546638]

[24]   Qin J, Lehr CR, Yuan C, Le XC, McDermott TR, Rosen BP. Biotransformation of arsenic by a Yellowstone thermoacidophilic eukaryotic alga. Proc Natl Acad Sci USA 2009; 106(13): 5213-7.
[http://dx.doi.org/10.1073/pnas.0900238106] [PMID: 19276121]

[25]   Coryell M, McAlpine M, Pinkham NV, McDermott TR, Walk ST. The gut microbiome is required for full protection against acute arsenic toxicity in mouse models. Nat Commun 2018; 9(1): 5424.
[http://dx.doi.org/10.1038/s41467-018-07803-9] [PMID: 30575732]

[26]   Feng S, Liu Y, Huang Y, *et al.* Influence of oral administration of *Akkermansia muciniphila* on the tissue distribution and gut microbiota composition of acute and chronic cadmium exposure mice. FEMS Microbiol Lett 2019; 366(13): fnz160.
[http://dx.doi.org/10.1093/femsle/fnz160] [PMID: 31310663]

[27]   Lukovac S, Belzer C, Pellis L, *et al.* Differential modulation by *Akkermansia muciniphila* and *Faecalibacterium prausnitzii* of host peripheral lipid metabolism and histone acetylation in mouse gut organoids. MBio 2014; 5(4): e01438-14.
[http://dx.doi.org/10.1128/mBio.01438-14] [PMID: 25118238]

[28]   Zhai Q, Qu D, Feng S, *et al.* Oral supplementation of lead-intolerant intestinal microbes protects against lead (Pb) toxicity in mice. Front Microbiol 2020; 10: 3161.
[http://dx.doi.org/10.3389/fmicb.2019.03161] [PMID: 32038590]

[29]   Bora S, Lakshman M, Madhuri D, Kalakumar B, Udayakumar M. Protective Effect of *Lactobacillus sporogenes* against Arsenic-Induced Hematological Alterations in Male Albino Wistar Rats. Biol Trace Elem Res 2022; 200(11): 4744-9.
[http://dx.doi.org/10.1007/s12011-021-03055-9] [PMID: 34993908]

[30]   Feng P, Yang J, Zhao S, *et al.* Human supplementation with *Pediococcus acidilactici* GR-1 decreases heavy metals levels through modifying the gut microbiota and metabolome. NPJ Biofilms Microbiomes 2022; 8(1): 63-3.

[31]   Kakade A, Salama ES, Usman M, Arif M, Feng P, Li X. Dietary application of *Lactococcus lactis* alleviates toxicity and regulates gut microbiota in *Cyprinus carpio* on exposure to heavy metals mixture. Fish Shellfish Immunol 2022; 120: 190-201.
[http://dx.doi.org/10.1016/j.fsi.2021.11.038] [PMID: 34848303]

[32]   Mohammadian T, Ghanei-Motlagh R, Jalali M, *et al.* Protective effects of non-encapsulated and microencapsulated *Lactobacillus delbrueckii* subsp. bulgaricus in rainbow trout (*Oncorhynchus mykiss*) exposed to lead (Pb) *via* diet. Ann Anim Sci 2022; 22(1): 325-48.
[http://dx.doi.org/10.2478/aoas-2021-0026]

[33]   Ahmed SH, Badawi AS, Abdullah AK. Protective effects of *Lactobacillus casei* on lead-induced biochemical deteriorations in rats. 2021; 15(4): 1000-5.

[34]   Chang X, Chen Y, Feng J, Huang M, Zhang J. Amelioration of Cd-induced bioaccumulation, oxidative stress and immune damage by probiotic *Bacillus coagulans* in common carp (*Cyprinus carpio* L.). Aquacult Rep 2021; 20: 100678.
[http://dx.doi.org/10.1016/j.aqrep.2021.100678]

[35]   Hayati AL, Supriyanto AG, Suhargo LI, Hayaza SU, Ayubu A. Bioremediation potency of probiotics on cadmium pollution to improve fish reproductive health. Pollut Res 2020; 39(4): 980-4.

[36]   Long X, Sun F, Wang Z, *et al. Lactobacillus fermentum* CQPC08 protects rats from lead-induced oxidative damage by regulating the Keap1/Nrf2/ARE pathway. Food Funct 2021; 12(13): 6029-44.
[http://dx.doi.org/10.1039/D1FO00589H] [PMID: 34037025]

[37]   Maurya A, Kumar R, Singh A, Raj A. Investigation on biofilm formation activity of *Enterococcus faecium* under various physiological conditions and possible application in bioremediation of tannery effluent. Bioresour Technol 2021; 339: 125586.
[http://dx.doi.org/10.1016/j.biortech.2021.125586] [PMID: 34311409]

[38]   Samir O, Edris , Edris , Heikal G. Degradation effect of *Lactobacillus rhamnosus* on some heavy metals experimentally inoculated in fish fillet model. Benha Vet Med J 2021; 41(1): 132-6.
[http://dx.doi.org/10.21608/bvmj.2021.76500.1412]

[39]   Shang X, Xu W, Zhao Z, *et al.* Effects of sub-chronic exposure to cadmium (Cd) and selenium-enriched *Lactobacillus plantarum* in juvenile Luciobarbuscapito: bioaccumulation, antioxidant responses and intestinal microbes. SSRN Electronic Journal 2021

[40]   Zhu S, Xu T, Li Q, Deng X, Zhou X, Zhu G. Remediation effect of cadmium contaminated soil by the combination of *Lactobacillus plantarum* and red mud. IOP Conf Ser Earth Environ Sci. 821(1): 012021.
[http://dx.doi.org/10.1088/1755-1315/821/1/012021]

[41]   Clemente MJ, Vivó MÁ, Puig S, *et al. in vitro* evaluation of the efficacy of lactobacilli and yeasts in reducing bioavailability of inorganic arsenic. Lebensm Wiss Technol 2020; 126: 109272.
[http://dx.doi.org/10.1016/j.lwt.2020.109272]

[42]   Jaafar R. Bioremediation of lead and cadmium and the strive role of *Pediococcus pentosaceus* probiotic. Iraqi J Vet Sci 2020; 34(1): 51-7.
[http://dx.doi.org/10.33899/ijvs.2019.125581.1092]

[43]   Tehrani F, Shirazi H, Kazempoor R. Effect of lethal exposure of lead acetate on histopathology of gills of probiotic-treated zebra fish (Danio rerio). J Comp Pathol 2020; 17(1): 3033-44.

[44]   Massoud R, Khosravi-Darani K, Sharifan A, Asadi GH. Lead bioremoval from milk by Saccharomyces cerevisiae. Biocatal Agric Biotechnol 2019; 22: 101437.
[http://dx.doi.org/10.1016/j.bcab.2019.101437]

[45]   Zhai Q, Liu Y, Wang C, *et al. Lactobacillus plantarum* CCFM8661 modulates bile acid enterohepatic circulation and increases lead excretion in mice. Food Funct 2019; 10(3): 1455-64.
[http://dx.doi.org/10.1039/C8FO02554A] [PMID: 30768114]

[46]   Mohammadian T, Mohiseni M, Ahmadi Babadi B, Zeyaee Nezhad S. Effect of oral administration of different concentrations *Lactobacillus acidophilus* on growth performance and digestive enzyme rainbow trout fish (*Oncorhynchus mykiss*) in the face of lead toxicity in the diet. Indian Vet J 2018; 14(2): 68-79.

[47]   Li B, Jin D, Yu S, *et al.* EtareriEvivie S, Muhammad Z, Huo G, Liu F. *in vitro* and *in vivo* evaluation of *Lactobacillus delbrueckii* subsp. bulgaricus KLDS1. 0207 for the alleviative effect on lead toxicity. Nutrients 2017; 9(8): 845.
[http://dx.doi.org/10.3390/nu9080845] [PMID: 28786945]

[48]   Zanjani SY, Eskandari MR, Kamali K, Mohseni M. The effect of probiotic bacteria (*Lactobacillus acidophilus* and *Bifidobacterium lactis*) on the accumulation of lead in rat brains. Environ Sci Pollut Res Int 2017; 24(2): 1700-5.
[http://dx.doi.org/10.1007/s11356-016-7946-9] [PMID: 27796979]

[49]   Elsanhoty RM, Al-Turki IA, Ramadan MF. Application of lactic acid bacteria in removing heavy metals and aflatoxin B1 from contaminated water. Water Sci Technol 2016; 74(3): 625-38.
[http://dx.doi.org/10.2166/wst.2016.255] [PMID: 27508367]

[50]   Raghuvanshi R, Chaudhari A, Kumar GN. Amelioration of cadmium- and mercury-induced liver and kidney damage in rats by genetically engineered probiotic *Escherichia coli* Nissle 1917 producing pyrroloquinoline quinone with oral supplementation of citric acid. Nutrition 2016; 32(11-12): 1285-94. [http://dx.doi.org/10.1016/j.nut.2016.03.009] [PMID: 27209211]

[51]   Zhai Q, Tian F, Zhao J, Zhang H, Narbad A, Chen W. Oral administration of probiotics inhibits absorption of the heavy metal cadmium by protecting the intestinal barrier. Appl Environ Microbiol 2016; 82(14): 4429-40. [http://dx.doi.org/10.1128/AEM.00695-16] [PMID: 27208136]

[52]   Bhakta JN, Munekage Y, Ohnishi K, Jana BB. Isolation and identification of cadmium- and lead-resistant lactic acid bacteria for application as metal removing probiotic. Int J Environ Sci Technol 2012; 9(3): 433-40. [http://dx.doi.org/10.1007/s13762-012-0049-3]

[53]   Ninkov M, Popov Aleksandrov A, Demenesku J, *et al.* Toxicity of oral cadmium intake: Impact on gut immunity. Toxicol Lett 2015; 237(2): 89-99. [http://dx.doi.org/10.1016/j.toxlet.2015.06.002] [PMID: 26051590]

[54]   Wang J, Hu W, Yang H, *et al.* Arsenic concentrations, diversity and co-occurrence patterns of bacterial and fungal communities in the feces of mice under sub-chronic arsenic exposure through food. Environ Int 2020; 138: 105600. [http://dx.doi.org/10.1016/j.envint.2020.105600] [PMID: 32120061]

[55]   Trapecar M, Goropevsek A, Gorenjak M, Gradisnik L, Slak Rupnik M. A co-culture model of the developing small intestine offers new insight in the early immunomodulation of enterocytes and macrophages by Lactobacillus spp. through STAT1 and NF-kB p65 translocation. PLoS One 2014; 9(1): e86297. [http://dx.doi.org/10.1371/journal.pone.0086297] [PMID: 24454965]

[56]   Ahemad M, Kibret M. Recent trends in microbial biosorption of heavy metals: a review. Biochem Mol Biol (N Y) 2013; 1(1): 19-26. [http://dx.doi.org/10.12966/bmb.06.02.2013] [PMID: 23466392]

[57]   Mirza Alizadeh A, Hosseini H, Mollakhalili Meybodi N, *et al.* Mitigation of potentially toxic elements in food products by probiotic bacteria: A comprehensive review. Food Res Int 2022; 152: 110324. [http://dx.doi.org/10.1016/j.foodres.2021.110324] [PMID: 35181105]

[58]   Javanbakht V, Alavi SA, Zilouei H. Mechanisms of heavy metal removal using microorganisms as biosorbent. Water Sci Technol 2014; 69(9): 1775-87. [http://dx.doi.org/10.2166/wst.2013.718] [PMID: 24804650]

[59]   Pakdel M, Soleimanian-Zad S, Akbari-Alavijeh S. Screening of lactic acid bacteria to detect potent biosorbents of lead and cadmium. Food Control 2019; 100: 144-50. [http://dx.doi.org/10.1016/j.foodcont.2018.12.044]

[60]   Muhammad Z, Ramzan R, Zhang S, *et al.* Comparative assessment of the bioremedial potentials of potato resistant starch-based microencapsulated and non-encapsulated *Lactobacillus plantarum* to alleviate the effects of chronic lead toxicity. Front Microbiol 2018; 9: 1306. [http://dx.doi.org/10.3389/fmicb.2018.01306] [PMID: 29971052]

[61]   Lin D, Ji R, Wang D, *et al.* The research progress in mechanism and influence of biosorption between lactic acid bacteria and Pb(II): A review. Crit Rev Food Sci Nutr 2019; 59(3): 395-410. [http://dx.doi.org/10.1080/10408398.2017.1374241] [PMID: 28886254]

[62]   Duan H, Yu L, Tian F, Zhai Q, Fan L, Chen W. Gut microbiota: A target for heavy metal toxicity and a probiotic protective strategy. Sci Total Environ 2020; 742: 140429. [http://dx.doi.org/10.1016/j.scitotenv.2020.140429] [PMID: 32629250]

[63]   Bist P, Choudhary S. Impact of heavy metal toxicity on the gut microbiota and its relationship with metabolites and future probiotics strategy: a review. Biol Trace Elem Res 2022; 200(12): 5328-50.

[http://dx.doi.org/10.1007/s12011-021-03092-4] [PMID: 34994948]

[64]   Wani AL, Ara A, Usmani JA. Lead toxicity: a review. Interdiscip Toxicol 2015; 8(2): 55-64.
       [http://dx.doi.org/10.1515/intox-2015-0009] [PMID: 27486361]

[65]   He X, Qi Z, Hou H, Qian L, Gao J, Zhang XX. Structural and functional alterations of gut microbiome
       in mice induced by chronic cadmium exposure. Chemosphere 2020; 246: 125747.
       [http://dx.doi.org/10.1016/j.chemosphere.2019.125747] [PMID: 31891852]

[66]   Nielsen KM, Zhang Y, Curran TE, *et al.* Alterations to the intestinal microbiome and metabolome of
       Pimephales promelas and Mus musculus following exposure to dietary methylmercury. Environ Sci
       Technol 2018; 52(15): 8774-84.
       [http://dx.doi.org/10.1021/acs.est.8b01150] [PMID: 29943971]

[67]   Chen Y, Xu J, Chen Y. Regulation of neurotransmitters by the gut microbiota and effects on cognition
       in neurological disorders. Nutrients 2021; 13(6): 2099.
       [http://dx.doi.org/10.3390/nu13062099] [PMID: 34205336]

[68]   Honda K, Littman DR. The microbiota in adaptive immune homeostasis and disease. Nature 2016;
       535(7610): 75-84.
       [http://dx.doi.org/10.1038/nature18848] [PMID: 27383982]

[69]   Dheer R, Patterson J, Dudash M, *et al.* Arsenic induces structural and compositional colonic
       microbiome change and promotes host nitrogen and amino acid metabolism. Toxicol Appl Pharmacol
       2015; 289(3): 397-408.
       [http://dx.doi.org/10.1016/j.taap.2015.10.020] [PMID: 26529668]

[70]   Zhai Q, Wang H, Tian F, Zhao J, Zhang H, Chen W. Dietary *Lactobacillus plantarum*
       supplementation decreases tissue lead accumulation and alleviates lead toxicity in Nile tilapia
       (*Oreochromis niloticus*). Aquacult Res 2017; 48(9): 5094-103.
       [http://dx.doi.org/10.1111/are.13326]

[71]   G Allam N, M Ali EM, Shabanna S, Abd-Elrahman E. Protective efficacy of *Streptococcus
       thermophilus* against acute cadmium toxicity in mice. Iranian journal of pharmaceutical research. Iran
       J Pharm Res 2018; 17(2): 695-707.
       [PMID: 29881427]

[72]   Djurasevic S, Jama A, Jasnic N, *et al.* The protective effects of probiotic bacteria on cadmium toxicity
       in rats. J Med Food 2017; 20(2): 189-96.
       [http://dx.doi.org/10.1089/jmf.2016.0090] [PMID: 27976972]

[73]   Giri SS, Yun S, Jun JW, *et al.* Therapeutic effect of intestinal autochthonous *Lactobacillus reuteri* P16
       against waterborne lead toxicity in *Cyprinus carpio*. Front Immunol 2018; 9: 1824.
       [http://dx.doi.org/10.3389/fimmu.2018.01824] [PMID: 30131809]

[74]   Dubey V, Mishra AK, Ghosh AR, Mandal BK. Probiotic *Pediococcus pentosaceus*GS 4 shields brush
       border membrane and alleviates liver toxicity imposed by chronic cadmium exposure in Swiss albino
       mice. J Appl Microbiol 2019; 126(4): 1233-44.
       [http://dx.doi.org/10.1111/jam.14195] [PMID: 30614180]

[75]   Zhai Q, Yang L, Zhao J, Zhang H, Tian F, Chen W. Protective effects of dietary supplements
       containing probiotics, micronutrients, and plant extracts against lead toxicity in mice. Front Microbiol
       2018; 9: 2134.
       [http://dx.doi.org/10.3389/fmicb.2018.02134] [PMID: 30254621]

[76]   Yi YJ, Lim JM, Gu S, *et al.* Potential use of lactic acid bacteria Leuconostoc mesenteroides as a
       probiotic for the removal of Pb(II) toxicity. J Microbiol 2017; 55(4): 296-303.
       [http://dx.doi.org/10.1007/s12275-017-6642-x] [PMID: 28361342]

[77]   Feng P, Ye Z, Kakade A, Virk AK, Li X, Liu P. A review on gut remediation of selected
       environmental contaminants: possible roles of probiotics and gut microbiota. Nutrients 2018; 11(1):
       22.

[http://dx.doi.org/10.3390/nu11010022] [PMID: 30577661]

[78]   Monachese M, Burton JP, Reid G. Bioremediation and tolerance of humans to heavy metals through microbial processes: a potential role for probiotics?. Appl Environ Microbiol 2012; 78(18): 6397-404.
[http://dx.doi.org/10.1128/AEM.01665-12] [PMID: 22798364]

[79]   Zhai Q, Wang G, Zhao J, *et al.* Protective effects of *Lactobacillus plantarum* CCFM8610 against acute cadmium toxicity in mice. Appl Environ Microbiol 2013; 79(5): 1508-15.
[http://dx.doi.org/10.1128/AEM.03417-12] [PMID: 23263961]

[80]   Topcu A, Bulat T. Removal of cadmium and lead from aqueous solution by *Enterococcus faecium* strains. J Food Sci 2010; 75(1): T13-7.
[http://dx.doi.org/10.1111/j.1750-3841.2009.01429.x] [PMID: 20492209]

[81]   Liu W, Feng H, Zheng S, *et al.* Pb toxicity on gut physiology and microbiota. Front Physiol 2021; 12: 574913.
[http://dx.doi.org/10.3389/fphys.2021.574913] [PMID: 33746764]

[82]   Giri SS, Jun JW, Yun S, *et al.* Characterisation of lactic acid bacteria isolated from the gut of *Cyprinus carpio* that may be effective against lead toxicity. Probiotics Antimicrob Proteins 2019; 11(1): 65-73.
[http://dx.doi.org/10.1007/s12602-017-9367-6] [PMID: 29285742]

[83]   Shang X, Yu P, Yin Y, *et al.* Effect of selenium-rich Bacillus subtilis against mercury-induced intestinal damage repair and oxidative stress in common carp. Comp Biochem Physiol C Toxicol Pharmacol 2021; 239: 108851.
[http://dx.doi.org/10.1016/j.cbpc.2020.108851] [PMID: 32777471]

[84]   Vidhya Hindu S, Thanigaivel S, Vijayakumar S, Chandrasekaran N, Mukherjee A, Thomas J. Effect of microencapsulated probiotic *Bacillus vireti* 01-polysaccharide extract of *Gracilaria folifera* with alginate-chitosan on immunity, antioxidant activity and disease resistance of *Macrobrachium rosenbergii* against *Aeromonas hydrophila* infection. Fish Shellfish Immunol 2018; 73: 112-20.
[http://dx.doi.org/10.1016/j.fsi.2017.12.007] [PMID: 29208500]

[85]   Yu Q, Fein JB. Enhanced removal of dissolved Hg (II), Cd (II), and Au (III) from water by *Bacillus subtilis* bacterial biomass containing an elevated concentration of sulfhydryl sites. Environ Sci Technol 2017; 51(24): 14360-7.
[http://dx.doi.org/10.1021/acs.est.7b04784] [PMID: 29154538]

[86]   Chattopadhyay S, Khatun S, Maity M, *et al.* Association of vitamin B 12, lactate dehydrogenase, and regulation of NF-κB in the mitigation of sodium arsenite-induced ROS generation in uterine tissue by commercially available probiotics. Probiotics Antimicrob Proteins 2019; 11(1): 30-42.
[http://dx.doi.org/10.1007/s12602-017-9333-3] [PMID: 28994024]

[87]   Tinkov AA, Gritsenko VA, Skalnaya MG, Cherkasov SV, Aaseth J, Skalny AV. Gut as a target for cadmium toxicity. Environ Pollut 2018; 235: 429-34.
[http://dx.doi.org/10.1016/j.envpol.2017.12.114] [PMID: 29310086]

[88]   Zhai Q, Yin R, Yu L, *et al.* Screening of lactic acid bacteria with potential protective effects against cadmium toxicity. Food Control 2015; 54: 23-30.
[http://dx.doi.org/10.1016/j.foodcont.2015.01.037]

[89]   Wang Y, Wu Y, Wang Y, *et al.* Antioxidant properties of probiotic bacteria. Nutrients 2017; 9(5): 521.
[http://dx.doi.org/10.3390/nu9050521] [PMID: 28534820]

[90]   Zhai Q, Xiao Y, Zhao J, *et al.* Identification of key proteins and pathways in cadmium tolerance of *Lactobacillus plantarum* strains by proteomic analysis. Sci Rep 2017; 7(1): 1182.
[http://dx.doi.org/10.1038/s41598-017-01180-x] [PMID: 28446769]

[91]   Bisanz JE, Enos MK, Mwanga JR, *et al.* Randomized open-label pilot study of the influence of probiotics and the gut microbiome on toxic metal levels in Tanzanian pregnant women and school children. MBio 2014; 5(5): e01580-14.
[http://dx.doi.org/10.1128/mBio.01580-14] [PMID: 25293764]

[92]   Bhattacharya S. The role of probiotics in the amelioration of cadmium toxicity. Biol Trace Elem Res 2020; 197(2): 440-4.
[http://dx.doi.org/10.1007/s12011-020-02025-x] [PMID: 31933279]

[93]   Gerbino E, Carasi P, Tymczyszyn EE, Gómez-Zavaglia A. Removal of cadmium by *Lactobacillus kefir* as a protective tool against toxicity. J Dairy Res 2014; 81(3): 280-7.
[http://dx.doi.org/10.1017/S0022029914000314] [PMID: 24960206]

[94]   Kadry MO, Megeed RA. Probiotics as a complementary therapy in the model of cadmium chloride toxicity: crosstalk of β-catenin, BDNF, and StAR signaling pathways. Biol Trace Elem Res 2018; 185(2): 404-13.
[http://dx.doi.org/10.1007/s12011-018-1261-x] [PMID: 29427035]

[95]   Song S, Oh S, Lim KT. Lactobacillus plantarum L67 glycoprotein protects against cadmium chloride toxicity in RAW 264.7 cells. J Dairy Sci 2016; 99(3): 1812-21.
[http://dx.doi.org/10.3168/jds.2015-10121] [PMID: 26774722]

[96]   Farag IM, Abd-El-Moneim OM, Abd El-Kader HAM, Abd El-Rahim AH, Radwan HA, Fadel M. Modulatory role of *Saccharomyces cerevisiae* against cadmium-induced genotoxicity in mice. J Arab Soc Med Res 2017; 12(1): 27-38.
[http://dx.doi.org/10.4103/jasmr.jasmr_2_17]

[97]   Daisley BA, Monachese M, Trinder M, *et al.* Immobilization of cadmium and lead by *Lactobacillus rhamnosus* GR-1 mitigates apical-to-basolateral heavy metal translocation in a Caco-2 model of the intestinal epithelium. Gut Microbes 2019; 10(3): 321-33.
[http://dx.doi.org/10.1080/19490976.2018.1526581] [PMID: 30426826]

[98]   Kumar N, Kumar V, Panwar R, Ram C. Efficacy of indigenous probiotic Lactobacillus strains to reduce cadmium bioaccessibility - An *in vitro* digestion model. Environ Sci Pollut Res Int 2017; 24(2): 1241-50.
[http://dx.doi.org/10.1007/s11356-016-7779-6] [PMID: 27770327]

[99]   Jafarpour D, Shekarforoush SS, Ghaisari HR, Nazifi S, Sajedianfard J, Eskandari MH. Protective effects of synbiotic diets of *Bacillus coagulans, Lactobacillus plantarum* and inulin against acute cadmium toxicity in rats. BMC Complement Altern Med 2017; 17(1): 291.
[http://dx.doi.org/10.1186/s12906-017-1803-3] [PMID: 28583137]

[100] Zhai Q, Yu L, Li T, *et al.* Effect of dietary probiotic supplementation on intestinal microbiota and physiological conditions of Nile tilapia (*Oreochromis niloticus*) under waterborne cadmium exposure. Antonie van Leeuwenhoek 2017; 110(4): 501-13.
[http://dx.doi.org/10.1007/s10482-016-0819-x] [PMID: 28028640]

[101] Mason LH, Harp JP, Han DY. Pb neurotoxicity: neuropsychological effects of lead toxicity. Biomed Res Int 2014; 2014-4.
[http://dx.doi.org/10.1155/2014/840547]

[102] Kumar A, Kumar A, M M S CP, *et al.* Lead toxicity: health hazards, influence on food chain, and sustainable remediation approaches. Int J Environ Res Public Health 2020; 17(7): 2179.
[http://dx.doi.org/10.3390/ijerph17072179] [PMID: 32218253]

[103] Assi MA, Hezmee MNM, Haron AW, Sabri MY, Rajion MA. The detrimental effects of lead on human and animal health. Vet World 2016; 9(6): 660-71.
[http://dx.doi.org/10.14202/vetworld.2016.660-671] [PMID: 27397992]

[104] Hadi F, Aziz T. A mini review on lead (Pb) toxicity in plants. J Biol Life Sci 2015; 6(2): 91-101.
[http://dx.doi.org/10.5296/jbls.v6i2.7152]

[105] Balali-Mood M, Naseri K, Tahergorabi Z, Khazdair MR, Sadeghi M. Toxic mechanisms of five heavy metals: mercury, lead, chromium, cadmium, and arsenic. Front Pharmacol 2021; 12: 643972.
[http://dx.doi.org/10.3389/fphar.2021.643972] [PMID: 33927623]

[106] Zwolak A, Sarzyńska M, Szpyrka E, Stawarczyk K. Sources of soil pollution by heavy metals and

their accumulation in vegetables: A review. Water Air Soil Pollut 2019; 230(7): 164.
[http://dx.doi.org/10.1007/s11270-019-4221-y]

[107] Shabaan M, Asghar HN, Akhtar MJ, Ali Q, Ejaz M. Role of plant growth promoting rhizobacteria in the alleviation of lead toxicity to *Pisum sativum* L. Int J Phytoremediation 2021; 23(8): 837-45.
[http://dx.doi.org/10.1080/15226514.2020.1859988] [PMID: 33372547]

[108] Bhattacharya S. Probiotics against alleviation of lead toxicity: recent advances. Interdiscip Toxicol 2019; 12(2): 89-92.
[http://dx.doi.org/10.2478/intox-2019-0010] [PMID: 32206029]

[109] Banwo K, Alonge Z, Sanni AI. Binding capacities and antioxidant activities of *Lactobacillus plantarum* and *Pichia kudriavzevii* against cadmium and lead toxicities. Biol Trace Elem Res 2021; 199(2): 779-91.
[http://dx.doi.org/10.1007/s12011-020-02164-1] [PMID: 32436065]

[110] Teemu H, Seppo S, Jussi M, Raija T, Kalle L. Reversible surface binding of cadmium and lead by lactic acid and bifidobacteria. Int J Food Microbiol 2008; 125(2): 170-5.
[http://dx.doi.org/10.1016/j.ijfoodmicro.2008.03.041] [PMID: 18471917]

[111] Walker R, Otto-Pille C, Gupta S, Schillaci M, Roessner U. Current perspectives and applications in plant probiotics. Microbiol Aust 2020; 41(2): 95-9.
[http://dx.doi.org/10.1071/MA20024]

[112] Arif I, Batool M, Schenk PM. Plant microbiome engineering: expected benefits for improved crop growth and resilience. Trends Biotechnol 2020; 38(12): 1385-96.
[http://dx.doi.org/10.1016/j.tibtech.2020.04.015] [PMID: 32451122]

[113] Bernhoft RA. Mercury toxicity and treatment: a review of the literature. J Environ Public Health 2012; 2012-2.
[http://dx.doi.org/10.1155/2012/460508]

[114] Zefferino R, Piccoli C, Ricciardi N, Scrima R, Capitanio N. Possible mechanisms of mercury toxicity and cancer promotion: Involvement of gap junction intercellular communications and inflammatory cytokines. Oxid Med Cell Longev 2017; 2017(1): 7028583.
[http://dx.doi.org/10.1155/2017/7028583] [PMID: 29430283]

[115] Zoghi A, Khosravi-Darani K, Sohrabvandi S. Surface binding of toxins and heavy metals by probiotics. Mini Rev Med Chem 2014; 14(1): 84-98.
[http://dx.doi.org/10.2174/1389557513666131211105554] [PMID: 24329992]

[116] Landersjö C, Yang Z, Huttunen E, Widmalm G. Structural Studies of the Exopolysaccharide Produced by *Lactobacillus r hamnosus* strain GG (ATCC 53103). Biomacromolecules 2002; 3(4): 880-4.
[http://dx.doi.org/10.1021/bm020040q] [PMID: 12099838]

[117] Jiang X, Gu S, Liu D, *et al. Lactobacillus brevis* 23017 relieves mercury toxicity in the colon by modulation of oxidative stress and inflammation through the interplay of MAPK and NF-κBsignaling cascades. Front Microbiol 2018; 9: 2425.
[http://dx.doi.org/10.3389/fmicb.2018.02425] [PMID: 30369917]

[118] Ahmed E, Holmström SJM. Siderophores in environmental research: roles and applications. Microb Biotechnol 2014; 7(3): 196-208.
[http://dx.doi.org/10.1111/1751-7915.12117] [PMID: 24576157]

[119] Hassan ZU, Ali S, Rizwan M, Ibrahim M, Nafees M, Waseem M. Role of bioremediation agents (bacteria, fungi, and algae) in alleviating heavy metal toxicity. probiotics in agroecosystem. Environ Sci Pollut Res 2017; 517-37.

[120] Gilmour CC, Podar M, Bullock AL, *et al.* Mercury methylation by novel microorganisms from new environments. Environ Sci Technol 2013; 47(20): 11810-20.
[http://dx.doi.org/10.1021/es403075t] [PMID: 24024607]

[121] Majlesi M, Shekarforoush SS, Ghaisari HR, Nazifi S, Sajedianfard J, Eskandari MH. Effect of

probiotic *Bacillus coagulans* and *Lactobacillus plantarum* on alleviation of mercury toxicity in rat. Probiotics Antimicrob Proteins 2017; 9(3): 300-9.
[http://dx.doi.org/10.1007/s12602-016-9250-x] [PMID: 28084611]

[122] Abdel-Salam A, Al-Dekheil A, Babkr A, Farahna M, Mousa H. High fiber probiotic fermented mare's milk reduces the toxic effects of mercury in rats. N Am J Med Sci 2010; 2(12): 569-75.
[http://dx.doi.org/10.4297/najms.2010.2569] [PMID: 22558569]

[123] Assumaidaee AA, Ali NM, Akutbi SH, Fadhil AA. Efficacy of probiotic (protoxine) on mercury-induced nephrotoxicity and lipid peroxidation in rats. Magallat Diyalá Li-l-'ulum al-Zira'iyyat 2018; 10(special Issue): 114-26.

[124] Rodríguez-Viso P, Domene A, Vélez D, Devesa V, Zúñiga M, Monedero V. Lactic acid bacteria strains reduce *in vitro* mercury toxicity on the intestinal mucosa. Food Chem Toxicol 2023; 173: 113631.
[http://dx.doi.org/10.1016/j.fct.2023.113631] [PMID: 36690269]

[125] Shaji E, Santosh M, Sarath KV, Prakash P, Deepchand V, Divya BV. Arsenic contamination of groundwater: A global synopsis with focus on the Indian Peninsula. Geoscience Frontiers 2021; 12(3): 101079.
[http://dx.doi.org/10.1016/j.gsf.2020.08.015]

[126] Khan I, Awan SA, Rizwan M, Ali S, Zhang X, Huang L. Arsenic behavior in soil-plant system and its detoxification mechanisms in plants: A review. Environ Pollut 2021; 286: 117389.
[http://dx.doi.org/10.1016/j.envpol.2021.117389] [PMID: 34058445]

[127] Abbas G, Murtaza B, Bibi I, *et al.* Natasha. Arsenic uptake, toxicity, detoxification, and speciation in plants: physiological, biochemical, and molecular aspects. Int J Environ Res Public Health 2018; 15(1): 59.
[http://dx.doi.org/10.3390/ijerph15010059] [PMID: 29301332]

[128] Vergara-Gerónimo CA, León Del Río A, Rodríguez-Dorantes M, Ostrosky-Wegman P, Salazar AM. Arsenic-protein interactions as a mechanism of arsenic toxicity. Toxicol Appl Pharmacol 2021; 431: 115738.
[http://dx.doi.org/10.1016/j.taap.2021.115738] [PMID: 34619159]

[129] Shen S, Li XF, Cullen WR, Weinfeld M, Le XC. Arsenic binding to proteins. Chem Rev 2013; 113(10): 7769-92.
[http://dx.doi.org/10.1021/cr300015c] [PMID: 23808632]

[130] Flora SJ, Mittal M, Mehta A. Heavy metal induced oxidative stress & its possible reversal by chelation therapy. Indian J Med Res 2008; 128(4): 501-23.
[PMID: 19106443]

[131] Ray M, Hor P, Singh SN, Mondal KC. Multipotent antioxidant and antitoxicant potentiality of an indigenous probiotic *Bifidobacterium* sp. MKK4. J Food Sci Technol 2021; 58(12): 4795-804.
[http://dx.doi.org/10.1007/s13197-021-04975-z] [PMID: 34629544]

[132] Bhakta JN, Bhattacharya S, Lahiri S, Panigrahi AK. Probiotic characterization of arsenic-resistant lactic acid bacteria for possible application as arsenic bioremediation tool in fish for safe fish food production. Probiotics Antimicrob Proteins 2023; 15(4): 889-902.
[http://dx.doi.org/10.1007/s12602-022-09921-9] [PMID: 35119613]

[133] Menéndez E, Paço A. Is the application of plant probiotic bacterial consortia always beneficial for plants? Exploring synergies between rhizobial and non-rhizobial bacteria and their effects on agro-economically valuable crops. Life (Basel) 2020; 10(3): 24.
[http://dx.doi.org/10.3390/life10030024] [PMID: 32178383]

[134] Chiboub M, Jebara SH, Abid G, Jebara M. Co-inoculation effects of Rhizobium sullae and Pseudomonas sp. on growth, antioxidant status, and expression pattern of genes associated with heavy metal tolerance and accumulation of cadmium in Sulla coronaria. J Plant Growth Regul 2020; 39(1): 216-28.

[http://dx.doi.org/10.1007/s00344-019-09976-z]

[135] Bhakta JN, Ohnishi K, Munekage Y, Iwasaki K. Isolation and probiotic characterization of arsenic-resistant lactic acid bacteria for uptaking arsenic. Int J Bioeng Life Sci 2010; 4(11): 831-8.

[136] Gaballa A, Wang T, Ye RW, Helmann JD. Functional analysis of the Bacillus subtilis Zur regulon. J Bacteriol 2002; 184(23): 6508-14.
[http://dx.doi.org/10.1128/JB.184.23.6508-6514.2002] [PMID: 12426338]

[137] Wierzba S. Biosorption of lead(II), zinc(II) and nickel(II) from industrial wastewater by Stenotrophomonas maltophilia and Bacillus subtilis. Pol J Chem Technol 2015; 17(1): 79-87.
[http://dx.doi.org/10.1515/pjct-2015-0012]

[138] Khan M, Ijaz M, Chotana GA, Murtaza G, Malik A, Shamim S. *Bacillus altitudinis* MT422188: a potential agent for zinc bioremediation. Bioremediat J 2022; 26(3): 228-48.
[http://dx.doi.org/10.1080/10889868.2021.1927973]

[139] Oves M, Khan MS, Zaidi A. Biosorption of heavy metals by Bacillus thuringiensis strain OSM29 originating from industrial effluent contaminated north Indian soil. Saudi J Biol Sci 2013; 20(2): 121-9.
[http://dx.doi.org/10.1016/j.sjbs.2012.11.006] [PMID: 24115905]

[140] Njoku KL, Akinyede OR, Obidi OF. Microbial remediation of heavy metals contaminated media by *Bacillus megaterium* and *Rhizopus stolonifer*. Sci Am 2020; 10: e00545.

[141] Raj A S, Muthukumar P V, B B, M P. Comparative biosorption capacity of copper and chromium by *Bacillus cereus*. IACSIT Int J Eng Technol 2018; 7(3.34): 442-4.
[http://dx.doi.org/10.14419/ijet.v7i3.34.19355]

[142] Liu Y, Liao T, He Z, *et al.* Biosorption of copper(II) from aqueous solution by Bacillus subtilis cells immobilized into chitosan beads. Trans Nonferrous Met Soc China 2013; 23(6): 1804-14.
[http://dx.doi.org/10.1016/S1003-6326(13)62664-3]

[143] Samarth DP, Chandekar CJ, Bhadekar RK. Biosorption of heavy metals from aqueous solution using *Bacillus licheniformis*. Int J Pure Appl Sci Technol 2012; 10(2): 12.

[144] Guedon E, Helmann JD. Origins of metal ion selectivity in the DtxR/MntR family of metalloregulators. Mol Microbiol 2003; 48(2): 495-506.
[http://dx.doi.org/10.1046/j.1365-2958.2003.03445.x] [PMID: 12675807]

[145] Zhong Q, Kobe B, Kappler U. Molybdenum enzymes and how they support virulence in pathogenic bacteria. Front Microbiol 2020; 11: 615860.
[http://dx.doi.org/10.3389/fmicb.2020.615860] [PMID: 33362753]

[146] Sun DH, Li XL, Zhang GL. Biosorption of Ag (I) from aqueous solution by *Bacillus licheniformis* strain R08. Appl Mech Mater 2013; 295-298: 129-34.
[http://dx.doi.org/10.4028/www.scientific.net/AMM.295-298.129]

[147] Moore CM, Gaballa A, Hui M, Ye RW, Helmann JD. Genetic and physiological responses of *Bacillus subtilis* to metal ion stress. Mol Microbiol 2005; 57(1): 27-40.
[http://dx.doi.org/10.1111/j.1365-2958.2005.04642.x] [PMID: 15948947]

[148] Gaballa A, Helmann JD. Bacillus subtilis CPx-type ATPases: Characterization of Cd, Zn, Co and Cu efflux systems. Biometals 2003; 16(4): 497-505.
[http://dx.doi.org/10.1023/A:1023425321617] [PMID: 12779235]

[149] McKillip JL. Prevalence and expression of enterotoxins in *Bacillus cereus* and other Bacillus spp., a literature review. Antonie van Leeuwenhoek 2000; 77(4): 393-9.
[http://dx.doi.org/10.1023/A:1002706906154] [PMID: 10959569]

[150] Dore MP, Bibbò S, Fresi G, Bassotti G, Pes GM. Side effects associated with probiotic use in adult patients with inflammatory bowel disease: a systematic review and meta-analysis of randomized controlled trials. Nutrients 2019; 11(12): 2913.

[http://dx.doi.org/10.3390/nu11122913] [PMID: 31810233]

[151]   Rodriguez-Arrastia M, Martinez-Ortigosa A, Rueda-Ruzafa L, Folch Ayora A, Ropero-Padilla C. Probiotic supplements on oncology patients' treatment-related side effects: a systematic review of randomized controlled trials. Int J Environ Res Public Health 2021; 18(8): 4265.
[http://dx.doi.org/10.3390/ijerph18084265] [PMID: 33920572]

[152]   Mazhar SF, Afzal M, Almatroudi A, *et al.* The prospects for the therapeutic implications of genetically engineered probiotics. J Food Qual 2020; 2020: 1-11.
[http://dx.doi.org/10.1155/2020/9676452]

[153]   Liu M, Kakade A, Liu P, Wang P, Tang Y, Li X. Hg^{2+}-binding peptide decreases mercury ion accumulation in fish through a cell surface display system. Sci Total Environ 2019; 659: 540-7.
[http://dx.doi.org/10.1016/j.scitotenv.2018.12.406] [PMID: 31096383]

[154]   Xue Y, Du P, Ibrahim Shendi AA, Yu B. Mercury bioremediation in aquatic environment by genetically modified bacteria with self-controlled biosecurity circuit. J Clean Prod 2022; 337: 130524.
[http://dx.doi.org/10.1016/j.jclepro.2022.130524]

[155]   Hui C, Guo Y, Zhang W, *et al.* Surface display of PbrR on *Escherichia coli* and evaluation of the bioavailability of lead associated with engineered cells in mice. Sci Rep 2018; 8(1): 5685.
[http://dx.doi.org/10.1038/s41598-018-24134-3] [PMID: 29632327]

[156]   Zhu N, Zhang B, Yu Q. Genetic engineering-facilitated coassembly of synthetic bacterial cells and magnetic nanoparticles for efficient heavy metal removal. ACS Appl Mater Interfaces 2020; 12(20): 22948-57.
[http://dx.doi.org/10.1021/acsami.0c04512] [PMID: 32338492]

[157]   Deng X, Yi XE, Liu G. Cadmium removal from aqueous solution by gene-modified *Escherichia coli* JM109. J Hazard Mater 2007; 139(2): 340-4.
[http://dx.doi.org/10.1016/j.jhazmat.2006.06.043] [PMID: 16890348]

Plant Growth Promoting Bacteria as Promising Candidates for Heavy Metal Detoxification

Ravinderjit Kaur[1], Nandni Sharma[4], Deepak Kumar[2], Sandeep Kour[2], Roohi Sharma[2], Puja Ohri[2,*] and Renu Bhardwaj[3]

[1] *Department of Zoology, S.R. Government College (Women), Amritsar, Punjab, India*

[2] *Department of Zoology, Guru Nanak Dev University, Amritsar, Punjab, India*

[3] *Department of Botanical and Environmental Sciences, Guru Nanak Dev University, Amritsar, Punjab, India*

[4] *Department of Zoology, DAV University, Jalandhar, Punjab, India*

Abstract: Heavy metals are among the primary contaminants of the environment, which are released due to geological and anthropological activities. Although some heavy metals are vital for life processes, however, higher concentrations of certain metallic ions terribly disturb the equilibrium and remarkably become risky for human health. Owing to their toxicity and accumulation, the conventional means of their management are pricey, unfeasible, and trigger secondary pollution concerns. Hence, a novel, practical approach like bioremediation has gained importance where innate biological mechanisms in the microbes are utilized to degrade and detoxify metal ions. Being ubiquitous, eco-friendly, and cost-effective, the microorganisms alter soil characteristics including pH, valency, adsorption, chelation, and precipitation of pollutants. Moreover, they utilize direct and indirect mechanisms to suppress the toxic effects of metalloids and other combative in the environment. Also, rhizosphere-plant consortium in polluted soils plays a crucial role in improving plant tolerance by driving crucial nutrient cycles thereby, facilitating their survival in harsh circumstances. In addition to bacteria, fungi, and algae have also been utilized over the past years to nullify the danger caused by heavy metals. The present book chapter highlights the toxicity of heavy metals on plants and the mechanisms employed by plant growth-promoting bacteria to detoxify these heavy metals.

Keywords: Bioremediation, Heavy metals, Microbes, Plant growth-promoting bacteria, Rhizosphere-plant consortium.

* **Corresponding author Puja Ohri:** Department of Zoology, Guru Nanak Dev University, Amritsar, Punjab, India; E-mail: ohri11puja@gmail.com

Kanika Khanna, Sukhmeen Kaur Kohli & Renu Bhardwaj (Eds.)

INTRODUCTION

Exponential growth in urbanization and industrialization has led to one of the major environmental pollutions known as metal pollution. The escalated concentrations of some of the inorganic elements pose a greater liability for human health and their surroundings owing to higher toxicity [1 - 4]. As compared to the biodegradable entities, these metalloids do not undergo chemical/ microbial degradation or be removed but are known to magnify into the major sinks such as soil or water over the years. Their bioaccumulation in organisms both on land and in water occurs through the food chains/ food webs eventually reaching the upper trophic levels [4 - 6]. Some of the well-known heavy elements include Lead (Pb), Mercury (Hg), Zinc (Zn), Iron (Fe), Cadmium (Cd), Cobalt (Co), Copper (Cu), Nickel (Ni), Arsenic (As), Chromium (Cr), and Selenium (Se). Though at lower concentrations, some of these elements- Fe, Zn, Cu, N, Mn, and Co are vital for the normal functioning and growth of living organisms, while others prove to be toxic. Thus, both the deficiency and sufficiency of heavy metals (HMs) can influence the plant systems [1, 7 - 9, 321].

In order to reduce metal pollution, various conventional methods have been adopted over the past years. These include physicochemical remedies such as excavation, electrochemical treatment, ion exchange, soil washing, in situ fixation, precipitation, adsorption, and evaporation technology [10 - 12]. However, these procedures are expensive, intrusive, demands high labour, are not apt for large-scale field applications, and raise serious concerns by disturbing both soil structure and its native microflora [6, 13 - 15]. Therefore, a more feasible, ecologically sound, non-invasive approach of bioremediation is being adopted for the restoration of a healthy environment free from heavy metal pollution [16 - 18].

Bioremediation is a surging, remarkable, cost-effective, and modest practice for restoring soils attenuated with heavy metals. It utilizes green plants, micro-fauna such as fungi, bacteria, yeast, and algae or their enzymes to restore the polluted grounds to their primary states [16, 19 - 21, 326]. The whole process solely relies on the native biological vigor, which in addition is dependent on several soil-related parameters including structure, pH, temperature, oxygen availability, moisture content, nature of contaminants, nutritional value, microbial heterogeneity, *etc* [16, 22, 23]. Detoxification *via* plant growth-promoting bacteria (PGPB) is the most crucial method that can alleviate heavy metal (HM) toxicity through a variety of processes including mobilizing/ immobilizing, uptaking, and transforming heavy metals and acclimatizing plants with the metal-polluted environment by producing siderophores, phytohormones, and antibiotics, and causing chelation, phosphate solubilization, biological nitrogen fixation and synthesis of lytic enzymes [2, 17, 24, 25, 322 - 326]. They can be free-living (communicate with plants under apt conditions), symbiotic (live in rhizospheric zones), or endophytic (form substantial associations with plant tissues/organs). They positively alter growth and productivity by producing plant growth regulators, driving essential nutrient cycles, and altering translocation and accumulation by modifying the photo-availability of HMs in attenuated soils [26 - 28]. Some commonly known genera of PGPB include: *Agrobacterium* spp., *Acinetobacter* spp., *Arthrobacter* spp., *Azospirillum* spp., *Azotobacter* spp., *Bacillus* spp., *Delfitia* spp., *Paenobacillus* spp., *Pantoea* spp., *Pseudomonas* spp., *Streptomyces* spp. and *Rhizobia* spp [28, 29].

Naturally, microbial-heavy metal detoxification in polluted spots is known as natural attenuation. However, depending on whether excavation is required or not, the process is mainly divided into two approaches- "*in-situ*" and "*ex-situ*" (Fig. 1). Of the two reclamation processes, *in-situ* (on-site) is usually preferred with minimal health risks, no site disturbance and transport of attenuated soil, with low cost as the soil is confined to their actual place throughout the process [18, 30, 328, 329]. It is further categorized into two types of remediation- intrinsic and engineered bioremediation depending upon the kind of microorganisms used. For intrinsic remediation, the metabolic activity of indigenous microbes is boosted while in the engineered process, specific genetically engineered microorganisms are used [31, 32]. In the case of *ex-situ* bioremediation, polluted soil, and water are excavated from the actual location and are further divided into solid-phase, slurry-phase, and vapour-phase systems (Fig. **1**) depending upon the type of contaminated samples [18, 32 - 34, 327, 328]. The current chapter summarizes the deleterious effects of heavy metals on plants and highlights the role and mechanisms followed by plant growth-promoting bacteria for the detoxification of heavy metals.

Sources of Heavy Metals

In the environment, weathering of rocks and parental chemicals normally occurs in trivial quantities (1000 mg/kg) and is rarely harmful [35, 36]. Most soils could build up heavy metals over the allowed average levels, significant enough to cause health concerns to people, flora and fauna, ecosystems, or other media in remote or urbanized settings [37, 38]. Various studies have recognized heavy metals as potential contaminants in the soil ecology for varied reasons; a) Heavy metals serve as soil pests because their rate of proliferation through synthetic phases is faster than those of natural sources; b) They are transported from mines to arbitrary ecological regions with increasing chances of direct exposure; c) Redundant products have higher heavy metal content than the receiving area; and d) In comparison to pedogenic or granular metallurgical soils, normal soils hold more movable and bioavailable heavy metals [39]. Therefore, heavy metals get into the atmosphere both by natural phenomena and anthropogenic behaviour (Fig. **2**).

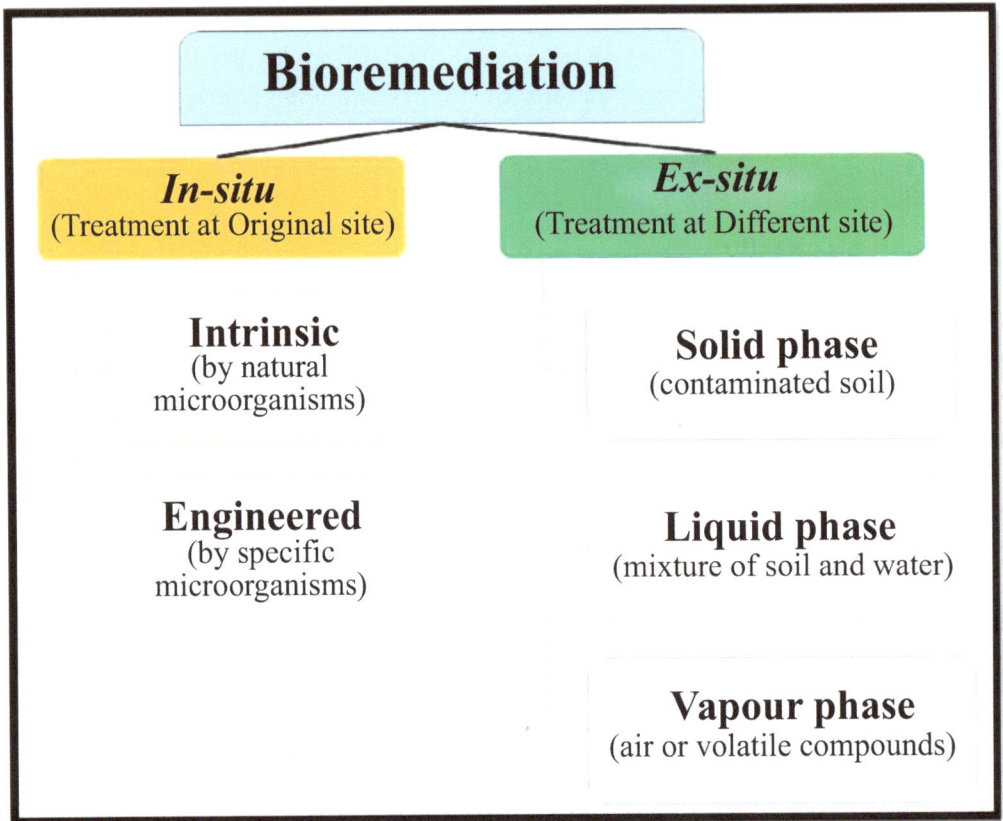

Bioremediation

In-situ
(Treatment at Original site)

Ex-situ
(Treatment at Different site)

Intrinsic
(by natural microorganisms)

Solid phase
(contaminated soil)

Engineered
(by specific microorganisms)

Liquid phase
(mixture of soil and water)

Vapour phase
(air or volatile compounds)

Fig. (1). Strategies employed for heavy metal detoxification.

Natural Processes

The primary source of heavy metals is the parent material from which they are derived [40]. The discharge of HMs in soil involves natural processes like erosion and surface winds that blow dust particles, as well as biological sources like damaging forest fires, earthy weathering of metal-bearing rocks, and other geological matter and volcanic events [40, 41]. The gaseous exchange and bubbles creating bursts in the sea are also a source for HM flow in nature. However, natural sources are less significant than the anthropogenic ones [42].

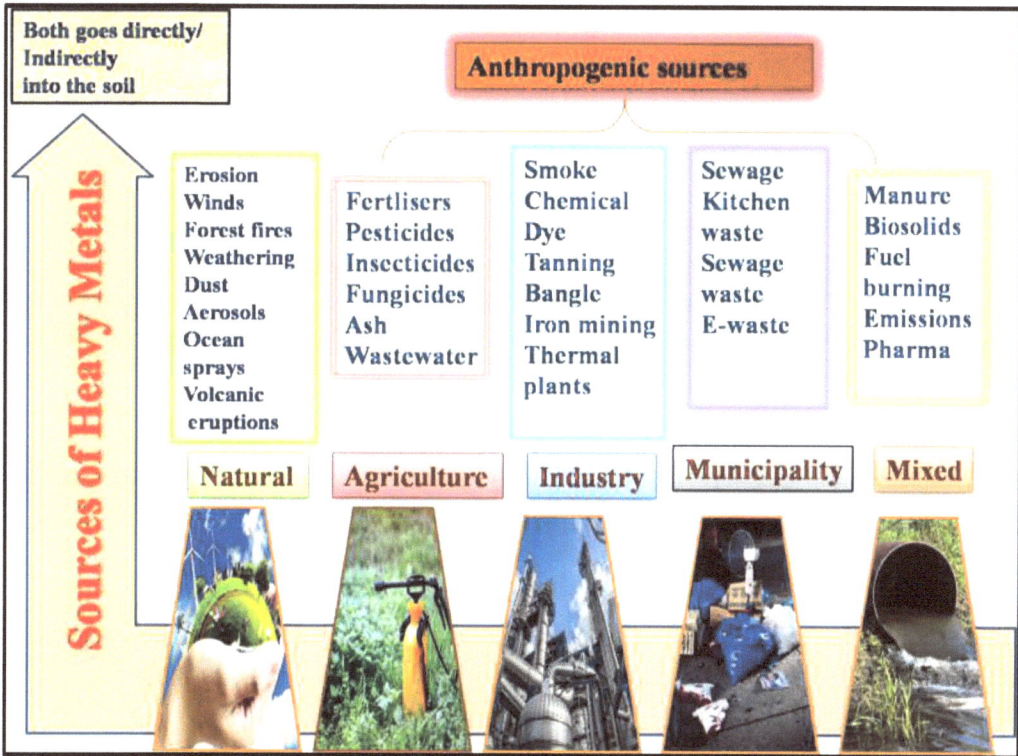

Fig. (2). Varied sources of heavy metals contamination in the soil environment.

Anthropogenic Sources

Industrial processes cause an excessive build-up of metals in the environment and are hazardous. These include mining and extraction of metalliferous ore deposits, metal making and electroplating, fossil fuel combustion and their by-products, sewage sludge, tannery and other dye works, insecticides, fertilizers, lead-based products like gasoline and paints, auto-motors, electronic gadgets like batteries and various agricultural practices [43 - 47]. Some of these sources are explained in the following sections.

Insecticides and Pesticides

The last several years have witnessed a rise in the usage of conventional insecticides and pesticides in agriculture and horticultural techniques that have considerable metal concentrations. Recently, 10% of the pesticides licensed for use as antimicrobials (for fungus) and insecticides in the UK were centered on substances that contained Cu, Hg, Pb, Mn, As, or Zn [48]. Administering phytosanitary chemical preparations incorrectly and frequently leads to soil contamination with copper. For example, in the 19th century, a serious Cu build-up arose in the soil of grapevine orchards in countries like France, Brazil, Croatia, and Spain as a result of the frequent use of copper-based antifungal agents in viniculture. The Bordeaux mixture of $CuSO_4$ and $Ca(OH)_2$ and other compounds like $Cu(OH)_2$, $CuCl_2$, $CuSO_4$, Cu_2O, and $Cu(OH)_2$ are examples of such pesticides that contained copper in the fungicidal spray [49]. In orchards, lead arsenate $(PbHAsO_4)$ was used to eradicate insects for a long time [50, 51]. Such insecticides containing arsenic had been routinely employed in Australia and New Zealand to combat banana diseases and manage livestock ticks [7]. In a particular area, formulations of compounds comprising Cr, Cu, and As play a vital role in maintaining timber. However, if surrounding areas are remodeled for farming or non-farming activities, then they become vulnerable to elevated levels of such heavy metals, resulting in problematic outcomes [50, 52]. Therefore, under such circumstances, these compounds are more hazardous than actual fertilizers, and they should indeed be confined to specific areas and types of plants and places only [51, 53, 54].

Sewage Effluents

In many regions of the world, the tradition of discharging municipal, industrial, and associated wastewater onto land has been practiced for ages [55]. According to estimates, wastewater has been used to feed 20 million hectares of ploughland worldwide [56]. Sewage irrigation-based cultivation provides 50% of the city's supply of vegetables [57] by providing a substitute for fertilizers but also promotes the circulation and retention of heavy metals in soils [58]. Sewage sludge is a potent amendment to soil quality that changes its properties and produces more crop yield as compared to fertilizers [59]. However, numerous studies indicated a tendency in the opposite direction for the amounts of Zn, Mn, Fe, Cu, Pb, Ni, and Cd in soils supplied *via* sewage or well water, which means that after a certain concentration, sewage effluent has a mordant effect on soil quality and subsequently affects plant growth [60, 61]. The risk sources include the application of sewage sludge, wastewater, chemical fertilizers, insecticides, sprays, and liquid waste irrigation, due to which the proportion of heavy metals in the soil builds up [62, 63]. Furthermore, farmers focus on elevating their harvests and earnings rather than on environmental issues [64]. Also, the discharges typically have low metal concentrations (99% water and 1% dissolved, suspended, and colloidal matter); however, protracted irrigation with them can ultimately lead to substantial metal accumulation in the soil [65, 66].

Aerial Sources

In the atmosphere, metallic particles are released by fugitive emissions such as dust from storerooms, garbage piles, and tunnel or pile emissions of air, gas, or condensed streams. Processing at extreme heat, metals including Cd, As, and Pb vaporize and transit into oxides and condense as small particles in the absence of a reducing agent [67]. Depending upon site-specific conditions, total pollutant emissions are of two types: stack and fugitive emissions [68]. Although drying and/or moist precipitation techniques clear pollutants from the gas stream, stack emissions are frequently dispersed over a large region by naturally occurring air currents. On the other side, fugitive emissions are typically scattered over a narrow zone since they are produced near the bottomland [68]. Fumes from factory fireplaces, burning, and other emissions add solids that are ultimately placed on land or the ocean [69]. The onset of industrialization led to heavy metal contamination with various types of fossil fuels, thus polluting the environment on a large scale [70]. For illustration, plants and soils close to metallurgical operations have extremely high amounts of Pb, Cd, and Zn. The effuse from industrial sources, combustion of waste and open fields or lands, traffic and vehicular emissions like the wearing of car brake and clutch liners, road conditions, other road amenities, and construction materials are the main reservoirs of heavy metals [71 - 73]. Similarly, Pb is consumed in automobile engines, resulting in the formation of lead salts (bromines, chlorines, and oxides) [74]. Also, tetraethyl Pb-containing gasoline ignition releases Pb into the air, thereby adding a disproportionate amount of Pb to the land in cities and areas close to leading roads [75]. Cd, Zn, and Cu are released in tyre dirt, diesel engine exhausts, and car washes [53, 76 - 78].

Fertilizers

Agriculture originally had a major impact on the soil [79]. In order to flourish quickly and accomplish the life span at the same time, plants can absorb macronutrients like N, P, K, S, Ca, and Mg but also need and gather micronutrients [80]. Some soils lack certain heavy metals, including Co, Cu, Fe, Mn, Mo, Ni, and Zn required for optimal plant growth; hence, crops may be supplemented with these metals as a foliar solution [81]. In current agricultural systems, considerable sums of fertilizers are routinely added to the soil to supply enough Nitrogen, Phosphorus, and Potassium (NPK) for crop production. Heavy metals (such as Cu, Cd, and Pb) are contaminants present in trace concentrations in the chemicals used to provide the above nutrients which, following repeated applications, may considerably augment their amount in the soil [46, 82, 83]. For instance, the application of phosphorous and nitrogen fertilizers to the soil introduces Cd, Cr, and other hazardous ions like Hg, Fe, and Pb, of which Cd and Pb have no physiological activity [46, 84 - 86]. Also, inorganic fertilizers largely cause heavy metal accumulation in soils *via* drainage, sludge, and lining [87]. Fungicides, phosphate, and inorganic fertilizers have varying concentrations of Cd, Cr, Ni, Pb, and Zn [88].

Manure and Organic Wastes

A popular practice in agriculture is to administer animal wastes, such as chicken, cattle, and pig manures to crops and grazing land as solids or slurries. Hence, these wastes, along with a variety of other organic wastes such as livestock organic manure, compost, municipal trash, and sewage waste, cause heavy metals (Cd, Cr, Cu, Pb, Hg, Ni, Se, Mo, Zn, Ti, and Sb) deposition in soil [60, 89]. In metropolitans, organic material, and nutrients are recycled by applying biological materials to the land [90]. However, organic wastes are employed in every region of the globe, but only half of the total amount of sewage sludge is used or disposed of annually, and around 30 percent of wastewater sludge is applied only for fertilization [91, 92]. Garden waste is also a potent biosolid used for biofertilizers [93]. Penetration of heavy metals in soils has the ability to pollute water, especially with Cd, Zn, and Ni of drainage leachates [94, 95].

Although long-term potent biosolid application in agriculture has coarsely influenced soil health by supplementing it with heavy metals like Pb, Ni, Cd, Zn, and Cu, the method applied for treating different biosolids has a remediation effect as well [63, 96]. For example, in extremely mineralized soil with little organic matter, possibilities of lowering the amount of hexavalent chromium using various organic modifications, such as solid biomass, compost, fish and horse manure, leftover mushroom or animal farm manure, and the duration of soil absorption by adsorption experiments reduced the proportions of Cr (III) and Cr (VI) in a soil-solution equilibrium [97]. Also, the addition of an elevated Fe compost reduced soil Pb bioavailability in lead-polluted soil [98]. Moreover, sewage sludge, compost, and limes are big contributors that enrich the soil with Cd and As, which further enter humans and animals by consuming crops grown on these contaminated lands [99, 100].

Metal Excavation, Manufacturing, and Industry

The inheritance of widespread distribution of metal pollutants in soil has left many nations with the extraction and processing of metal ores in conjunction with industries [7]. In India, major industries (chemical, bangle, automobile, thermal, textile, dye, *etc.*) in towns are known to accumulate heavy metals in the soil [101]. Dry depositing (wind) and wet depositing (rainfall) are the two methods by which heavy metals enter the environment. The direct release of mining and tailings with heavier and larger particles that collect at the bottom of the flotation process during mining in on-site wetlands resulted in higher concentrations of contaminants [102]. The comprehensive metallurgical treatment of Pb and Zn ore has tainted the soil, endangering both man and ecological health [103]. For instance, the exploitation of Hg in gold mines has significantly emitted increased amounts of this element into the atmosphere [104].

Heavy metals pose a significant environmental threat to people primarily due to their bioavailability, where the food web or direct consumption of polluted soil are the main modes of their absorption [46, 105, 106]. The components from pharmaceutics, pesticides, or hydrocarbon-based products are produced by a number of diverse sources, including petrochemicals (unintentional oil spillage), textiles, and dyeing. In contrast, certain sources have the ability to dispose of waste on the land, while others are just helpful for agriculture or forests. However, many of these sources are potentially harmful when applied to land due to the presence of toxic organics or heavy metals like Pb, Zn, and Cr [89, 107]. For example, coal mines serve as the main root of Cd, Fe, and As for contaminating the nearby land. Vapour forms of heavy metals (Zn, Pb, Cd, and Cu), when mixed with water, make aerosols on condensation. Although numerous rehabilitation techniques are applied for their amelioration at the place of their origin, they are time- and money-consuming and oftentimes do not improve soil output [53, 108]. Thus, due to massive sources, frequent prevalence, and acute and long-term toxic effects on humans, flora, and fauna, heavy metal contamination has become a serious environmental threat [109 - 111].

Toxic Effects of Heavy Metals

Plants acquire heavy metals through the roots for their growth and development; however, their excessive concentration can have negative impacts and lead to toxicity in plants [112]. As they cannot be broken down, their accumulation within the plants inhibits plant growth, enzymatic and photosynthetic activities, stomatal functions, root systems, absorption and accumulation of other nutrients [111]. HMs also induce oxidative stress in plants, damage cell structures, and inhibit cytoplasmic enzymes [112]. Different heavy metals have varied toxicity effects on plants. For instance, Cd accumulation results in oxidative stress that can negatively impact seed germination and ion and water balance, reduce photosynthesis, disintegrate cell organelles and membranes, and decrease crop yield [113]. Pb toxicity impairs cell permeability, germination and growth of seedlings, elongation of roots, transpiration, and biosynthesis of chlorophyll [114]. As is another non-essential metal that interferes with various metabolic processes in plants and inhibits biomass production and root growth [115]. Other metals like Zn, Cu, Fe, Mn, *etc.*, can improve plant growth at low concentrations, but at higher concentrations, they alter the plant morphology and physiology and affect various metabolic processes in plants [116 - 118]. Heavy metal toxicity leads to decreased plant growth, yield impairment, chlorosis, reduction in nutrient uptake, metabolic disorders, *etc* [8]. One of the primary targets of heavy metals is proteins. They affect physiological processes by either dislodging necessary ions from metalloproteins or by forming a compound with functional side chain groups of proteins. Furthermore, heavy metals disrupt the native conformation of proteins by preventing the folding of nascent or alien proteins, which results in an absolute shortage of the impacted proteins or the production of proteotoxic complexes [330]. By attaching to the active sites of cysteine residues, Cd disrupted thiol transferase activity in *Brassica juncea*, resulting in oxidative damage [331]. The transformations caused by Cd toxicity damaged the stabilizing reactions linked to modifications in the tertiary structure of the protein, resulting in the loss of protein function [332]. To combat these adverse consequences of heavy metal toxicity, plants have evolved certain detoxifying mechanisms. Cell wall binding, subcellular metal compartmentation, metal chelation, uptake reduction or metal pumping, and protein damage repair are some examples of these mechanisms [333]. Plant development and stress tolerance are molecularly controlled by adaptive gene expression. Heavy metal exposure triggers a large number of genes and resistance proteins in plants to help them overcome metal toxicity [334]. For instance, the Cd resistance gene *AtNramp* cDNAs, extracted from *Arabidopsis*, encrypted metal transporters that lead to improvement in accumulation and sensitivity to Cd [335]. An investigation of the effect of various concentrations of Cd, Cr, and Pb on seed germination of *Trigonella foenum-graecum* L. found alterations in defense gene expression patterns as compared to the control [334]. Plants use a variety of physiological and biochemical processes to withstand stress, such as cell receptors on the cell surface identifying the external heavy metal stress, signal transduction cascades within the cell, and initiating various countermeasures to mitigate the adverse effects of heavy metal toxicity [336]. For example, alterations in gene expression result in the stimulation of certain antioxidative enzymes, xylem translocation, sequestration, detoxification, chelation, and regulation of metal transport *via* plasma membrane [337]. Table **1** summarizes the toxic effects of certain metals on the growth, physiology, and biochemistry of some plant species.

Plant Growth Promoting Bacteria

Soil is the major hub where different microbial life forms, including bacteria, fungi, algae, and protozoa, exist. Bacteria make up the vast majority (approximately 95%) of these microbial forms, although they are not always evenly dispersed. In other words, the number of bacteria near plant roots is typically much larger than the number of bacteria in the whole soil. This concentration is attributed to the fact that bacteria frequently utilize about 3-30% of the carbon fixed during photosynthesis in the form of root exudates (organic acids, sugars, and amino acids) as sustenance [156, 157]. Moreover, no matter how many bacteria are present in a given soil sample, there are three different ways that microbes might impact plants. Depending on the plant's perspective, the interaction between soil bacteria and plants may be advantageous, damaging, or neutral [158]. Advantageous interactions include the plant-growth promoting and stress-alleviating nature of bacteria. Bacterial strains that exhibit the potential for plant growth promotion are categorized as PGPB, which include a group of bacteria that reside in the rhizospheric area around plant roots. Additionally, they also include rhizobial strains that can develop nodules on the roots of particular plants (legumes) and endophytes that can live inside plants' internal tissues [159]. Keeping into consideration all these properties, a number of PGPBs have currently been commercialized as biofertilizers or as biological control agents against a wide variety of stressors [160, 161]. For instance, various PGPBs like *Agrobacterium*, *Rhizobia,* and *Azospirilla* frequently boost the yield of various agricultural crops when applied as seed inoculants in fields, though the application of some PGPBs like *Pseudomonads* shows less satisfactory results. So, in order to overcome this, researchers have now started working on genetic modification of the beneficial traits of bacteria [162]. All these three categories, *i.e.*, rhizospheric, endophytic, and genetically modified PGPB, are discussed in the following sections.

Table 1. Toxic effects of various heavy metals on plants.

Heavy Metal	Plant Species	Toxic effect of Heavy Metal	References
Arsenic	*Triticum aestivum* L.	Reduced nitrate reductase activity, decreased carotenoids, total chlorophyll and soluble protein content, increased accumulation of proline, decreased grain yield.	119
	Glycine max	Leaf area and plant height were reduced; a decrease in stomatal conductance and net photosynthetic rate was observed.	120
	Brassica napus	Considerable reduction in growth parameters, decreased chlorophyll content, and RUBISCO activity, enhanced ROS accumulation, and oxidative stress in leaves.	121
	Oryza sativa L.	Reduced plant dry weight, and height, decreased chlorophyll a, b, carotenoid content, enhanced hydrogen peroxide (H_2O_2), and leaf malondialdehyde (MDA) content, increased electrolyte leakage in leaves.	122

(Table 1) cont.....

Heavy Metal	Plant Species	Toxic effect of Heavy Metal	References
Iron	*O. sativa* L.	Decreased growth, and yield of the crop, induced oxidative damage, and reduced nutritional content of some rice varieties.	123
	Rice	Reduced growth, and biomass of plant, leaf bronzing.	124
	O. sativa	Reduced growth attributes, decreased photosynthesis, stomatal conductance, and transpiration.	125
	O. sativa L.	Reduced number, and width of leaves, decreased plant height, and shoot and root biomass, leaf bronzing was observed.	118
Zinc	*Arabidopsis halleri* and *A. arenosa*	Decreased root length, photosynthetic pigments, and photosynthetic rate.	126
	Zea mays	Decreased plant growth parameters, and chlorophyll content, reduced antioxidant enzyme activities.	117
	Carthamus tinctorius L.	Reduced production of biomass, decreased chlorophyll content, increased lipoxygenase activity, and MDA and H_2O_2 content.	127
	T. aestivum	Decreased plant biomass, photosynthetic ability, and chlorophyll content, enhanced MDA content, and lipid peroxidation.	128
Copper	Cucumber	Decreased biomass, inhibited root growth, and caused oxidative damage.	116
	Linum usitatissimum L.	Reduced plant growth, and yield, enhanced the accumulation of secondary metabolites, enzymatic, and non-enzymatic antioxidants, and induced oxidative damage.	129
	Corchorus capsularis L.	Reduced plant height, fresh, and dry weight, decreased total chlorophyll content, enhanced MDA content, and oxidative stress in seedlings.	130
	Citrus sinensis	Affected the growth of seedlings, yellowing of leaves, lowered transpiration, stomatal conductance, water use efficiency, enhanced MDA content, and leaf superoxide production.	131

(Table 1) cont.....

Heavy Metal	Plant Species	Toxic effect of Heavy Metal	References
Cadmium	*Momordica charantia*	Declined root length, chlorophyll a, total chlorophyll content, and fresh weight of the plant.	132
	Sorghum bicolor L.	Decreased plant growth, enhanced electrolyte leakage, H_2O_2, and MDA content.	133
	O. sativa	Reduced plant growth parameters, decreased stomatal conductance, transpiration, and photosynthesis rate, and enhanced MDA content.	134
	Aloe ferox Mill.	Decreased length, width, and thickness of leaves, increased number of stomata in leaf epidermis, reduced Mg accumulation in leaves, enhanced lipid peroxidation, and affected cuticle deposition.	135
Chromium	*B. juncea* L.	Decreased root, and shoot length, enhanced ROS, and H_2O_2 content, reduced photosynthetic pigments, declined contents of non-enzymatic antioxidant enzymes.	136
	B. napus L.	Negatively affected plant morphology, decreased total chlorophyll, and carotenoid content, reduced net photosynthesis, stomatal conductance, transpiration rate, and water use efficiency, and enhanced oxidative stress.	137
	O. sativa	Decreased shoot height, root length, dry biomass, chlorophyll, and carotenoid contents, reduced water-soluble proteins, soluble sugars, and ascorbate but increased proline, and glutathione content, and decreased mineral content in roots.	138
	T. aestivum L.	Reduced growth, decreased chlorophyll, and carotenoid contents, and declined antioxidant enzyme activities.	139
Lead	*Pisum sativum* L.	Decreased growth, biomass, and yield of pea, reduced chlorophyll, and relative water content.	140
	Cynodondactylon	Reduced plant growth, seed germination, chlorophyll a and b content, enhanced proline, and MDA content.	141
	Coriandrum sativum L.	Deceased plant biomass, and vitamin C content, and enhanced flavonoid, MDA, and antioxidant enzyme activities.	142
	Vigna radiate	Reduced plant height, fresh weight, total chlorophyll content, and water use efficiency, suppressed photosynthetic activity, and enhanced MDA and H_2O_2 content.	143

(Table 1) cont.....

Heavy Metal	Plant Species	Toxic effect of Heavy Metal	References
Mercury	*B. napus* L.	Inhibited seedling growth, decreased plant biomass, and caused oxidative stress by increasing MDA, and H_2O_2 content.	144
	Soybean	Reduced plant growth, biomass, and yield, decreased gas exchange parameters, enhanced electrolyte leakage, and MDA, and H_2O_2 levels.	145
	Cajanus cajan L.	Reduced seed germination, decreased root, and shoot length, and root and shoot dry weight.	146
	Lactuca sativa L.	Decreased seed germination.	147
Nickel	*Eruca sativa*	Reduction in fresh and dry weight of plant, reduced shoot and root length, decreased chlorophyll levels, and enhanced proline accumulation.	148
	Eleusine coracana L.	Decreased shoot and root length, plant dry mass, chlorophyll, mineral content, and enhanced oxidative stress.	149
	O. sativa L.	Increased oxidative stress in leaves, enhanced lipid peroxidation, and membrane damage, decreased growth parameters, water balance, and chlorophyll content.	150
	Sesame	Reduced plant growth, and biomass, decreased crop yield.	151
Selenium	*O. sativa*	Retarded plant growth, and biomass, decreased photosynthetic pigments, reduced water status, and induced osmotic stress.	152
	O. sativa	Reduced chlorophyll, carotenoids, and protein contents, caused leaf chlorosis, and necrosis, decreased plant biomass.	153
	O. sativa	Inhibited growth, increased oxidative stress, and caused burning of leaves, and chlorosis.	154
-	*Solanum lycopersicum* L.	Increased ROS, and MDA content, caused cell death, stunted growth, and chlorosis.	155

Rhizospheric Bacteria

The constrictive area of soil that surrounds the plant roots is termed "rhizosphere/phyllosphere". This region is rich in nutrients, which makes it a vigorous spot for biological and chemical activities. The rhizospheric region, being the home to a broad variety of organisms including bacteria, fungi, protozoa, viruses, nematodes, algae, *etc.*, is considered an area where these organisms interact with one another and also with plants in a variety of ways. Furthermore, with a significant energy flux, the rhizosphere is considered the greatest ecosystem on earth [163].

Rhizosphere-dwelling bacteria are friendly to plants and have a positive impact on them; hence, they are known as rhizospheric plant growth-promoting bacteria [164]. This diverse group of PGPB is an essential part of the soil microbiome and is intended to secrete several regulatory compounds close to plant roots that induce growth promotion in plants [165, 166]. Rhizospheric PGPBs have an impact on the general health of plants. They aid in improved nutrient uptake by host plants, defend plants against phytopathogenic organisms, and foster tolerance to a variety of abiotic stresses [167, 168]. A variety of rhizospheric PGPB strains have been reviewed in the literature that can boost crop yield, display biological control potential, improve foliar disease resistance, encourage nodulation in legumes, and improve seedling emergence [169 - 174].

For instance, a study by Becze *et al.* [173] investigated the effects of two specific PGPR strains, *Viridibacillus* sp. (BP13) and *Delftia acidovorans* (BP12), on the growth and development of *Zea mays* under Cd stress. The study demonstrated that *Z. mays* plants inoculated with these strains exhibited superior growth performance compared to non-inoculated controls under heavy metal toxicity. Both PGPR strains facilitated Cd immobilization in the rhizosphere through exopolysaccharide production, thereby reducing the bioavailability and toxicity of Cd. Additionally, the strains significantly enhanced antioxidant enzyme activity in the plants, which mitigated oxidative stress caused by the accumulation of reactive oxygen species (ROS). The production of phytohormones, particularly indole-3-acetic acid (IAA), further supported improved root and shoot growth by enhancing nutrient uptake and stabilizing plant physiological processes.

Plant Growth Promoting Endophytic Bacteria

Plant endophytes are microorganisms that can colonize and go through all the stages of life inside various plant parts, such as roots, stems, leaves, flowers, seeds, and pollen, without harming the host plant [175]. Different endophytic bacteria coexist with plants in symbiotic relationships. In these relationships, both partners change their metabolism in accordance with the symbiotic settings, which has an impact on each other's biochemical characteristics [176]. Under both favorable and unfavorable conditions, this association results in enhanced plant development [177]. *Azospirillum, Pseudomonas, Bradyrhizobium, Bacillus, Pantoea, Burkholderia, Rhizobacter, etc.*, are a few genera that fall under this category [178 - 181].

Endophytic bacterial colonization of plants is a complicated occurrence that requires cooperation between two parties. These bacteria can spread to plants either by seeds and vegetative planting material or from the area around the roots [182]. Plants secrete root exudates that include amino acids, proteins, and organic acids, which attract these bacteria [183]. These exudates serve as signalling molecules that draw bacteria, which then travel in the direction of plant roots [184]. When endophytic bacteria reach the root surface, they first communicate with the plant cell wall through weak electrostatic junctions [185]. The endophytes enter root cells by secreting various enzymes that degrade the cell wall [186]. These bacteria then penetrate the roots either through the openings created by any mechanical injury or through the sites from where lateral roots originate [184]. The root hairs, zone of secondary root emergence, and apex being the major producers of exudates, are among the areas where bacteria are most likely to invade [187]. Bacteria attach to the surface of plants through a variety of structures, including flagella, fimbriae, lipopolysaccharides, and bacterial surface polysaccharides [186]. Using bacterial flagella or the transpiration stream, they can spread to other plant components after entering the roots [188]. For instance, the study conducted by Shahjan and co-workers [173] on *Bacillus amyloliquefaciens* RWL-1 demonstrates its potential for alleviating heavy metal stress in rice plants through regulating metabolic changes, offering promising insights for bacterial bioremediation. The bacterium enhances rice plant growth under conditions of metal toxicity, particularly in the presence of metals like cadmium and lead. This bioremediation potential is attributed to the bacteria's ability to modulate key metabolic pathways, such as increasing antioxidant enzyme activities and promoting the production of osmoprotectants, which help mitigate oxidative stress caused by heavy metals. Additionally, *B. amyloliquefaciens* RWL-1 contributes to improved nutrient uptake, further supporting plant resilience. By influencing the plant's metabolism, the bacterium helps to maintain cellular integrity and promote overall plant health under stressful environmental conditions. This research supports the potential application of *B. amyloliquefaciens* in sustainable agricultural practices aimed at heavy metal detoxification, particularly in polluted soils, aligning with other studies that highlight the role of plant growth-promoting bacteria in environmental remediation.

Genetically Modified Plant Growth Promoting Bacteria

Despite the existence of a wide variety of plant growth-promoting rhizospheric and endophytic bacteria, in some cases, their direct application to the field does not impose a growth promotion impact, which makes them unsuitable for commercialization. So, in order to overcome all these drawbacks, researchers are working on genetic modification of these PGPB strains [317]. However, a lot of problems need to be resolved in order to commercialize PGPB strains more widely. These include: (a) identification of traits that are crucial for effective functioning and subsequently choosing PGPB strains with appropriate biological activities; (b) ensuring consistency among regulatory agencies in various nations regarding which strains can be released into the environment and under what circumstances genetically engineered strains are suitable for environmental use; (c) gaining a better understanding of the benefits and drawbacks of using rhizospheric and endophytic PGPB; (d) development of more effective means of applying PGPB to plants in specific environmental conditions (viz. under in-vitro and in-vivo conditions); (e) developing a greater awareness of the potential communications between PGPB and mycorrhizae and other soil fungi [162, 189]. One notable example is a genetically modified strain of *Pseudomonas fluorescens*, which was engineered to express metallothioneins and phytochelatins-proteins involved in heavy metal sequestration. This modification enabled the bacteria to tolerate higher concentrations of metals like cadmium, lead, and mercury while simultaneously promoting plant growth through phosphate solubilization and the production of plant growth regulators. Field trials demonstrated that plants inoculated with these modified bacteria exhibited increased biomass and lower heavy metal accumulation in edible parts, showcasing their potential for use in phytoremediation. The study underscores the feasibility of integrating genetically modified plant growth-promoting bacteria into remediation strategies for contaminated soils, offering a promising avenue for sustainable agricultural practices in polluted environments [317]. The following (Table **2**) highlights some of the known groups and species of PGPB and their effect on different plants.

Table 2. Plant growth promoting bacteria and their positive impact on plants.

Group of PGPB	Bacterial Species	Exerted Mechanism	Beneficial Plants	References
Plant Growth Promoting Rhizobacteria	*Azospirillum brasilens*	Nitrogen fixation	Rice	[190]
	A. brasilens	Production of Indole acetic acid	Cereals	[191]
	Bacillus filamentosus	Zinc solubilization	*Medicago sativa* L.	[192]
	B. pseudomycoides	Potassium solubilization	*Camellia sinensis*	[193]
	B. siamensis	Gibberellin production	Arabidopsis mutants	[194]
	Burkholderia cenocepacia	Phosphate solubilization	*Nicotiana tabacum* L.	[195]
	Enterobacter oryzae	Nitrogen fixation	*Acacia acuminate*	[196]

(Table 2) cont.....

Group of PGPB	Bacterial Species	Exerted Mechanism	Beneficial Plants	References
-	*Sphingomonas trueperi*	Nitrogen fixation	*A. acuminate*	[196]
	Stenotrophomonas maltophilia	Nitrogen fixation	*Cicer arietinum* L.	[197]
	Acetobacter diazotrophicus	Nitrogen fixation	-	[198]
	Rhizobium phaseoli	Indole acetic acid production	Common bean	[199]
	Achromobacterxylosoxidase	Ammonia production	Agricultural fields	[200]
	B. amyloliquefaciens	Production of gibberellin	Rice	[201]
	B. circulans	Potassium solubilization	*Solanum lycopersicum* L.	[202]
	A. brasilens	Phosphate solubilization	Wheat, maize and rice	[203]
	B. atrophaeus	Production of Indole acetic acid	*Glycine max* L.	[204]
	Bradyrhizobium japonicum	Nitrogen fixation	*G. max* L.	[205]
	Azotobacter chroococcum	Production of siderophore and gibberellin	Cereals	[206]
Plant Growth Promoting Endophytic Bacteria	*Acinetobacter guillouiae*	Phosphorus solubilization and indole acetic production	Wheat	[207]
	Pseudomonas aeruginosa	Phosphorus solubilization; indole acetic acid and siderophore production	Soybean	[208]
	A. calcoaceticus	Phosphorus solubilization; indole acetic acid and siderophore production	Soybean	[209]
	A. braumalli	Production of indole acetic acid and siderophores	Maize	[210]
	B. subtilis	Siderophore production	Soybean	[211]

(Table 2) cont.....

Group of PGPB	Bacterial Species	Exerted Mechanism	Beneficial Plants	References
-	*Rhizobium leguminosurum*	Zinc and Phosphorus solubilization; production of indole acetic acid and siderophoes	Beans	[212]
	Paenibacilluspolymyrea	Nitrogen fixation	Maize	[213]
	B.amyloliquefaciens	Nitrogen fixation; phosphorus and zinc solubilization; production of indole acetic acid and siderophores	Wheat	[214]
	Methylobacterium oryzae	Nitrogen fixation and indole acetic acid production	Soybean	[215]
	P. fluorescence	Nitrogen fixation and indole acetic acid production	Rice	[216]
	R. endophytum	Phosphorus solubilization	Beans	[217]
	Burkholderia vietnamiensis	Nitrogen fixation	Rice	[218]
	Klebsiella pneumonias	Nitrogen fixation	Maize	[219]
Genetically Modified Plant Growth PromotingBacteria	*P. stutzeri* DSM4166	Enhanced nitrogen fixation	*Arabidopsis*	[200]
	Ensifermedicae MA11	Increased copper rhizo-stabilization	*Medicago truncatula*	[221]
	Sinorhizobium meliloti GR4	Enhanced root length	Lettuce	[222]
	S. meliloti RMBPL-2	Increased yield and biomass	*Alfa alfa*	[223]
	Agrobacterium radiobacter K84	Control of crown gall disease caused by *A. tumefaciens*	*Rubus* sp., grape vines	[224]

Physiology/Bioremediation of Heavy Metal Toxicity

Heavy metal remediation is vital for the preservation and protection of the environment [225]. Various physicochemical and biological methods have been employed to remove HMs from the environment. Physicochemical methods are quick but still challenging because of the expense and technical difficulty. They impart negative effects on the biological, chemical, and physical characteristics of soil and result in secondary contamination [226 - 229]. In contrast, biological remediation is thought to be the most efficient way to remove harmful metals since it uses natural, environmentally benign, inexpensive, and widely accepted processes [230]. One such strategy is the employment of PGPB for the bioremediation of HMs [231, 232]. For example, it has been determined that PGPR such as *P. fluorescens*, *P. stutzeri*, and *P. gessardii* can be used for bioremediation of Pb stress and also significantly improve the physiological traits and plant growth of cowpea in Pb-contaminated areas [338]. Another study reported that the application of *P. aeruginosa* increased the growth and biomass of ryegrass and its ability to absorb and accumulate heavy metals in soil; the bioconcentration factors of Cu, Pb, and Cd in ryegrass also rose by 35.9%, 55.6%, and 283.5%, respectively. Cu, Pb, and Cd accumulation in ryegrass shoots increased significantly due to the external microbe; however, only Pb accumulation in roots increased by 16.3%, and that of both Cu and Cd reduced. Furthermore, the transfer factor of Pb dropped in the *P. aeruginosa* inoculated system, whereas the transfer factor of Cu and Cd in plants rose noticeably [339]. Additionally, the application of *Pseudomonas* sp. OBA 2.4.1 enhanced plant growth, siderophore activity, and biofilm production [340]. Similarly, the application of *Bacillus subtilis* NA2, and *Aspergillus niger* PMI-118 enhanced plant biomass, shoot length, root length (25.06%), chlorophyll, total sugars, total proteins, and cadmium tolerance in wheat plants [341]. Additionally, treatment of *Pseudomonas putida*, *Bacillus pumilus*, *Lysinibacillus sphaericus*, and *Exiguobacterium aurantiacum* significantly enhanced bioremediation of heavy metals and increased Zn and Fe accumulation in leaves in *Zea mays* under saline sodic soil [342]. However, the application of PGPR such as *Bacillus gibsonii* and *B. xiamenensis* enhanced root length, shoot length, root weight, and shoot weight of *Sesbania sesban* plants under industrially contaminated soils, and it was also found that the inoculation of both PGPR strains enhanced the phytoextraction capacity of *S. sesban* plants. The growth of the *S. sesban* plant and the phytoextraction of heavy metals from the soil were both enhanced by heavy metal-tolerant bacterial isolates, such as *B. xiamenensis* and *B. gibsonii*. The immunization of *B. gibsonii* resulted in a greater absorption of heavy metals Zn (15.74 ppm), Mn (16.4 ppm), Cd (20.5 ppm), Ni (36 ppm), and Cu (25 ppm) than that of *B. xiamenensis*. *B. xiamenensis* vaccination enhanced Pb uptake (64 ppm) and Cr accumulation in *S. sesban* plants [343].

Mode of Action

Despite the fact that heavy metals are typically poisonous, PGPBs have evolved specific defense mechanisms and metabolic pathways to utilize HMs for cellular benefits [233]. Detoxification strategies comprise extracellular barrier exclusion (which limits the entry of HM into the cell), extracellular and intracellular sequestration (which traps metal within the cytoplasm), active transport efflux systems (which export metal away from the intracellular compartment with the help of efflux systems), and enzymatic detoxification (which reduces the toxicity of metal using enzymes) [234 - 237]. PGPB-mediated bioremediation of HM toxicity generally embraces biosorption, bioaccumulation, bioleaching, biovolatilization, bioprecipitation, biotransformation, and biosurfactant technology (Fig. **3**) [27, 238]. All these strategies have been discussed in the following sections.

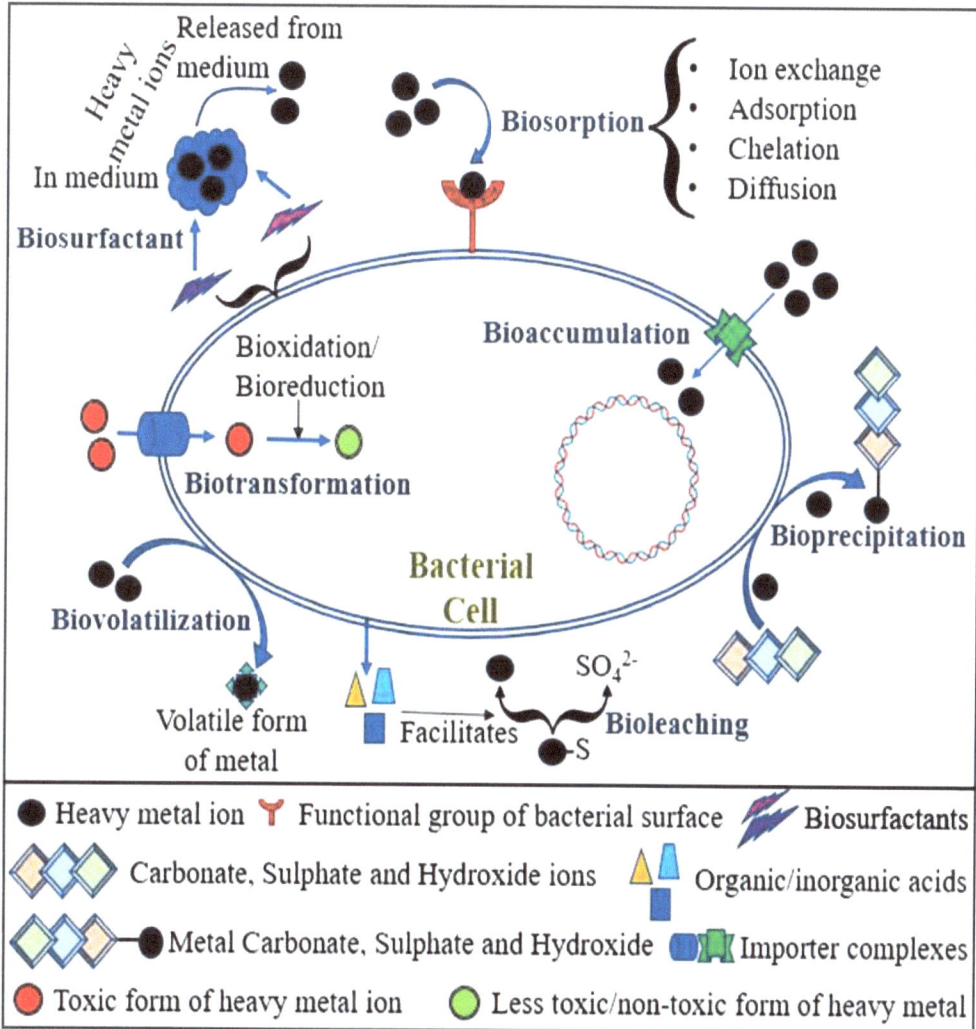

Fig. (3). Mode of action of PGPB during bioremediation of heavy metal toxicity.

Biosorption and Bioaccumulation

Biosorption is the collective term for all mechanisms in which living or dead biomass extracts HMs or other contaminants from fluids. It includes ion exchange, adsorption, chelation, and diffusion across cell walls and membranes. These methods vary and are dependent on the species used, the source, the preparation of the biomass, and the solution's chemistry. The uptake of metal can be either active or passive, and both processes can take place separately or simultaneously. The passive mechanism is the least specific of these and does not include cellular metabolism [239, 240]. Through ion exchange, metal binds to poly-ionic cell walls and is independent of physical conditions like ionic strength and pH. It is a quick and reversible process that takes about 5 to 10 minutes to completely biosorb HMs [241]. However, temperature, uncouplers, and metabolic inhibitors have an impact on the sluggish, cellular metabolism-dependent active process. Metal may form complexes with certain proteins, such as metallothioneins, located in vacuoles during the active process [242, 243]. Scanning electron microscopy revealed that due to their high binding affinities to HMs and anionic functional groups (amine, carboxyl, hydroxyl, phosphonate, sulfhydryl, and sulfonate) present on the surface of bacteria as well as extracellular polymers (such as humic substances, polysaccharides, and proteins), they are responsible for bacterial biosorption [244, 245]. For instance, living and dead biomass of *B. xiamenensis* PbRPSD202 bio absorbs Pb both actively and passively from the polluted environment at 35°C, pH 6, and 140 rpm and showed resistance to other HMs such as As, Cd, Cr, Cu, Ni, and Zn [246]. Similarly, *P. aeruginosa*, *A. chroococcum, B. subtilis*, O*chrobactrum* MT180101, *B. cereus*, and *Staphylococcus hominis* AMB-2 showed biosorption capability for Cd, Cr, Pd, Ni, Pb, and Cu [244, 247 - 249].

Further, bioaccumulation is a metabolically active process in which microbes absorb HMs into their internal space by employing importer complexes that make a translocation pathway in the lipid bilayer [250]. Proteins and peptide ligands may sequester HMs once they have entered the intracellular space. The performance indicator known as μmol x or mg x /g dry weight, where x is the target HM, refers to the bioaccumulative capacity of a bacterial biomass for that target HM [251]. For example, *Pseudomonas* spp. from wastewater exhibited bioaccumulative properties for Cu and Zn and showed resistance to Hg, Pb, Ni, and Cr toxicity [252]. Likewise, HM-resistant *B. cereus* from the rhizospheric soil of *Tagetes minuta* can accumulate significant amounts of different heavy metals, such as Cd, Cr, and Ni [253].

Moreover, HM-resistant plant endophytes have been isolated, which can promote the growth of plants and accumulate various HMs, leading to the remediation of

HM toxicity. Additionally, bacterial strains *viz.A. baumannii, Enterobacter* sp., *Klebsiella pneumoniae, Pseudomonas* sp., *Serratia marcescen, B. amyloliquefaciens*, and *Jeotgalicoccus huakuii* showed HM bioaccumulation [254, 255]. Furthermore, genetically modified bacteria also increase the bioaccumulation of HMs in order to remediate the toxicity of HMs. For example, the genetic modification of *Escherichia coli* by introducing a synthetic type VI secretory system cluster of *P. putida* and a *de novo* synthetic HM-capturing gene increased its metal bioaccumulative efficiency by modulating the ability to present HM-capturing protein on the bacterial surface [256].

Bioleaching and Biovolatilization

Bioleaching is a quick and efficient method for extracting metals from low-grade ores and mineral concentrates, which uses the metabolic activity of microbes such as bacteria and fungi to break down organic molecules. It relies on the notion of converting sparingly soluble metal compounds to forms that can be extracted simply [257]. The formation of inorganic or organic acids by bacteria in acidic environments (pH ranging from 1.5-3) can result in bioleaching [258]. On the other hand, bio volatilization is the use of the catalytic activity of microorganisms to transform HMs into their volatile derivatives. Recently, a new bacterial strain, *Providencia* sp. LLDRA6, was found to have bioleaching ability for Mn, Zn, Pb, Cr, Cd, and Cu [259]. Similarly, *Acidithiobacillus ferrooxidans, A. thiooxidans*, and *Candidatus* sp. exhibited bioleaching affinity for Ba, Li, Ni, Zn, and Cu [260, 261]. Further, bioleaching and biovolatilization methods can be used effectively with advanced technical features and genetically modified microorganisms [262]. For instance, around 2.2-4.5% As was extracted from contaminated soil by biovolatilization using genetically modified bacterial strains *B. idriensis* and *Sphingomonas desiccabilis* [263]. Similarly, *Arsenici bacterrosenii* (isolated from paddy fields), *B. flexus*, and *Acinetobacter junii* exhibited biovolatilization efficiency for As. Further, the arsenite methyltransferase gene cloned from *A. rosenii* provided arsenite resistance to *E. coli* [264, 265].

Bioprecipitation and Biotransformation

Microbiological activity, known as bioprecipitation, transforms soluble metal species into insoluble forms like carbonates, hydroxides, phosphates, and sulphides. These processes can occur singly or in combination [266]. It is important to note that microbial precipitation does not always depend on microbial metabolism, as this can happen with either active or dead microbes. Additionally, metal precipitates produced by bioprecipitation may adhere directly to microbial cells. Environmental factors like pH and redox potentials have an impact on the occurrence of bioprecipitation and the longevity of bioprecipitates

[267]. For example, *Serratia liquefaciens* and *B. megaterium* bio precipitated Cd and Pb and reduced uptake of these metals, producing abscisic acid, siderophores, and indole-3-acetic acid which might be responsible for the protective role of bacterial strains against HM toxicity in pakchoi plants [268]. Similarly, *Cupriavidus* sp. Cd02, *S. marcescens*, *Stenotrophomonas rhizophila*, *S. pasteurii*, and *Variovorax boronicumulans* increased the pH of culture media, leading to bioprecipitation of Cd, Cu, Pb, and Zn, resulting in decreased availability and toxicity of HMs [269 - 271].

A modern and advantageous technology for the removal of HMs is biotransformation, where they are changed from highly toxic to less/non-toxic forms to lessen their permanence and lethality [272, 273]. The process is mediated by enzymatic complexes of microbes. Typically, bacterial and fungal structures, including spores, enzymes, vegetative cells, and resting cells, are used to carry out the microbial transformation. HM biotransformation reactions can occur on the surface of cells, in the extracellular space, or in vacuoles and are controlled by the metal being recovered or removed [274]. Through diverse metabolic processes such as oxidation, reduction, methylation, demethylation, and complexation, microorganisms can interact with HMs and regulate their biotransformation. For instance, *Bacillus* sp. CRB-7 totally eliminated 120 mg/L Cr(VI) in 48 hours under ideal conditions (pH 7 and 37° C). Further, FTIR analysis showed the elimination mechanism was dominated by the bioreduction of Cr(VI) along with a small amount of bioadsorption [275]. *Enterobacter cloacae*, *Comamonas testosterone*, and *Geobacter sulfurreducens* removed toxic Se and Cr(VI), respectively by their direct and indirect reduction to less toxic forms [276 - 278].

Biosurfactant Technology

Biosurfactants are amphiphilic molecules that possess surface activity and can form complexes with metals. These molecules are produced by a range of microorganisms, *viz.*, bacteria and fungi, and have drawn significant attention as an alternative strategy to chemical leaching agents, such as synthetic surfactants, due to their low toxicity, high biodegradability, and promise of environmental compatibility [279 - 281]. For instance, biosurfactant rhamnolipid belonging to the family of glycolipids obtained from *P. aeruginosa* significantly enhanced As removal along with Cu, Zn, and Pb in mine tailings at a concentration of 0.1% [282]. While glycolipid biosurfactant obtained from *Burkholderia* sp. Z-90 eliminated Cd, Mn, and Zn from a site contaminated with a mixture of hazardous HMs [283]. Similarly, lipopeptide biosurfactant produced by *Paenibacillus* sp. D9 and extracellular biosurfactant of *P. fluorescence* exhibited bioremediating potential for Ni and Zn, respectively [119, 284].

A Direct Mechanism for Remediation

PGPBs impact the growth of plants both directly and indirectly under stress [285]. The former involves facilitating resource acquisition and producing phytohormones that have a direct impact on plant growth and help the plant withstand the toxic effects of HMs [286]. Bacteria can supply mineral resources that plants lack, such as nitrogen (N), iron (Fe), and phosphorus (P) [287]. The atmospheric nitrogen (N_2) is crucial for plant growth and development but is unavailable to plants directly. Hence, through the complex enzyme nitrogenase, nitrogen-fixing bacteria continuously transform atmospheric N_2 into phyto-available molecules such as ammonia and nitrate (Fig. 4) [288]. In maize, nitrogen-fixing bacteria *A. chroococcum* enhance root biomass by 20% and 28% under 585 mg Pb/kg and 2007 mg Cu/kg respectively along with an increased number, yield, and protein of kernels [289]. Iron is the fourth most prevalent metal on Earth, but it is not easily digested either by bacteria or plants in aerobic soils because ferric ion (Fe^{3+}), the most common form in nature, is sparingly soluble, leaving very little iron available for digestion by living things [290]. High levels of iron are necessary for both microorganisms and plants, and getting enough iron is difficult because bacteria, fungi, and plants compete for it in the rhizosphere. Therefore, bacteria create low-molecular-mass siderophores with an extraordinarily high affinity for Fe^{3+} and membrane receptors that can bind the Fe-siderophore complex, boosting iron uptake by microbes in order to survive under a restricted supply of iron [291].

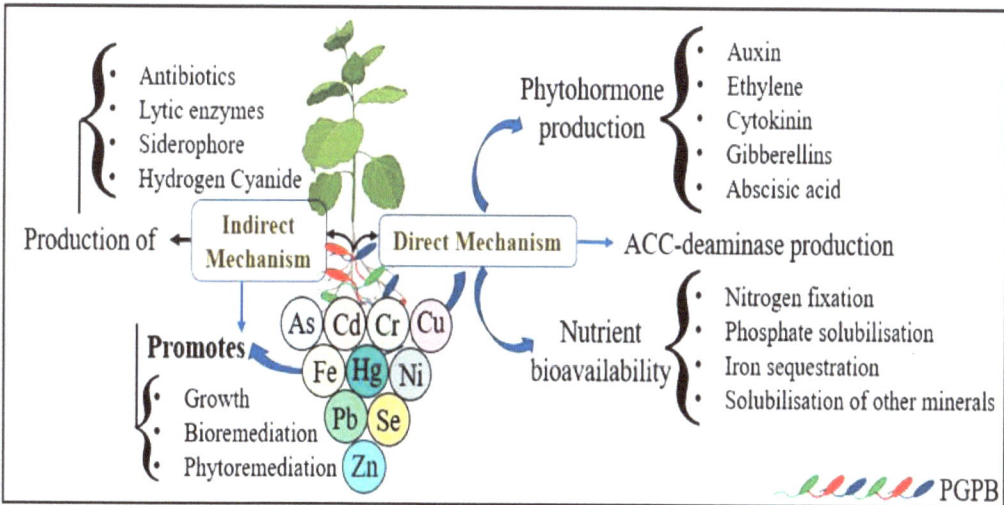

Fig. (4). Direct and indirect mechanisms employed by PGPB during heavy metal toxicity.

Phosphorus is rare in many soils across the globe, hence restricting the amount that plants may consume. Both monobasic and dibasic forms of soluble phosphorus are typically obtained by plants [165]. For example, Apatite is an inorganic phosphorus-containing mineral, while organic forms include phosphomonoesters, phosphotriesters, and inositol phosphate [292]. In fields, inorganic phosphorus is used as a chemical fertilizer together with other elements like nitrogen. However, because phosphorus is largely insoluble and used by plants, it leaches into the ground, poisoning the water supplies [293]. As a result, the usage of phosphate-solubilizing PGPB plays a crucial role in the solubilization of insoluble forms of phosphorus, mostly through mechanisms including the generation of acid phosphatases, which aid in the mineralization of organic phosphorus in the soil [294]. Similar to this, PGPB produces organic acids like citric and gluconic acids that aid in the solubilization of phosphorus. When plants consume these mineralized or solubilized molecules, their growth, development, and resistance to HM toxicity increase [295].

PGPBs also solubilize other essential nutrients such as potassium and zinc that directly influence the growth of plants and provide tolerance to HM toxicity [172, 296, 297]. PGPB supplies plants with phytohormones such as auxins, abscisic acid, cytokinins, gibberellins, and ethylene, which are responsible for enhancing plant growth and may also prevent metal phytotoxicity [298]. Naturally synthesized auxins act as potent compounds that are involved in nearly every aspect of plant physiology, such as regulating cell division, differentiation, expansion, and mitigation of abiotic stress [299]. Depending upon the strain, over 80% of the bacteria in the rhizosphere produce and excrete auxins along with their direct role in mitigating HM stress such as Cd [286, 300].

Similarly, PGPB secretes other hormones that are dependent on their type and promote cell division, differentiation, and elongation; control seed germination, root growth, shoot growth, leaf expansion, fruit maturation, and other beneficial effects; and provide direct resistance to HMs [6]. For example, cucumber plants after *P. psychrotolerans* CS51 inoculation produced more endogenous indole--acetic acid and gibberellins, which considerably boosted plant growth (measured by root shoot length) and increased their resistance to HMs [301]. Similarly, the effects of cytokinin, abscisic acid, and ethylene-producing PGPB in alleviating HM toxicity and modulating plant growth have been extensively studied [189, 302 - 304]. Additionally, certain PGPBs such as *B. gibsonii* PM11, *B. xiamenensis*, *P. thivervalensis* Y1-3-9, *Pantoea agglomerans* Jp3-3, and *Ralstonia* sp. J1-22-2, produce ACC-deaminase (aminocyclopropane-1-carboxylate deaminase) that directly promotes plant growth by increasing fresh and dry biomass, root and shoot length, and chlorophyll content and also mitigates HM toxicity through improved antioxidant activity and proline content [305, 306].

Indirect Mechanisms for Remediation

The creation of antagonistic chemicals like antibiotics, siderophores, lytic enzymes, and hydrogen cyanide for biocontrol and the generation of systemic resistance are examples of indirect processes, which indirectly shift the innate defense system of plants to combat HM [233, 307]. PGPB is attributed most frequently to its capacity to inhibit the spread of plant diseases and its ability to synthesize a variety of antibiotics [308 - 310]. Some bacterial strains produce siderophores which indirectly promote the growth of plants by acting as biocontrol agents or facilitating HM sequestration (Fig. **4**) [311 - 313]. For instance, in mung bean siderophore-producing *P. putida* KNP9, reduced accumulation of Cd in the tissues led to enhanced growth of Cd-stressed plants [314]. In other plants also, soil inoculation with *B. thuringiensis, Alcaligenes faecalis, B. subtilis,* and *P. aeruginosa* increased biomass, and content of soluble HMs and produced hydrogen cyanide and indole-3-acetic acid, which protected the plants from fungal pathogens, aiding in the indirect growth promotion of plants [315, 316].

Problems Encountered

Bacterial-mediated bioremediation of HM toxicity has gained significant importance over time, but it encounters challenges that make it difficult to be used on a wide scale. The majority of research on HM bioremediation has been conducted in laboratories. While, in nature, the detoxification process involves multiple elements; understanding the decontamination of polluted areas is extremely complex [317]. Another problem is the requirement for intricate laboratory equipment. Further, microbial metabolic processes are typically slower and more complex than chemical processes; hence, it is necessary to optimize all factors involved, including pH, temperature, urea concentration, salts present, bacterial strains to be employed, and appropriate conditions to favour their metabolism [318]. Furthermore, when ureolytic bacteria are used, byproducts of the metabolism, such as ammonium and nitrate are produced, which can be poisonous and hazardous to human health in high concentrations [319]. Although the establishment and growth of genetically modified bacteria under provided conditions play a significant role in their successful use in field conditions, growth rate, inoculum size, environmental factors, particularly spatial dispersion, and the existence of competing microorganisms are also significant factors (Fig. **5**) [320].

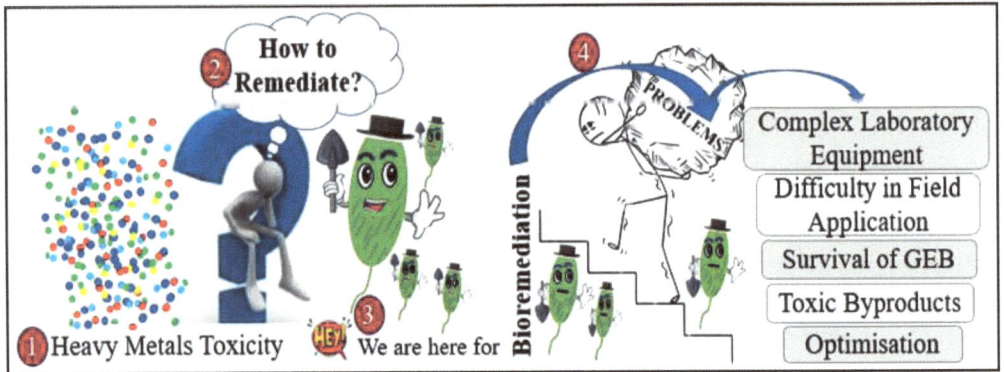

Fig. (5). Problems encountered during PGPB-mediated bioremediation of heavy metal toxicity. GEB: Genetically engineered bacteria.

FUTURE CHALLENGES AND CONCLUSION

Escalating heavy metal contamination poses a significant threat to the environment and human health. These persistent entities accumulate over time in the soils and aquatic bodies, eventually infiltrating food chains and food webs and causing serious ecological and health problems. Therefore, microbial bioremediation offers a promising and sustainable solution to this pressing issue. While significant progress has been made, several research gaps remain. By harnessing the metabolic capabilities of microorganisms, heavy metals can be effectively degraded and detoxified. However, there is a need to explore innovative approaches to optimize bioremediation processes, such as identifying novel microbial strains, investigating synergistic interactions between microbes, and considering the potential of genetic engineering. Furthermore, integrating advanced technologies like nanotechnology, metagenomics, and systems biology can revolutionize the field. This can be achieved by fostering collaboration between microbiologists, environmental engineers, chemists, and other relevant disciplines to develop cost-effective and scalable bioremediation technologies. However, rigorous risk assessments and supportive regulatory frameworks are essential to ensure the safe and sustainable application of these technologies. By prioritizing research, innovation, and ethical implementation, the power of microbial bioremediation can be harnessed to protect our environment and safeguard human health.

REFERENCES

[1]   Ali H, Khan E, Sajad MA. Phytoremediation of heavy metals—Concepts and applications. Chemosphere 2013; 91(7): 869-81.
[http://dx.doi.org/10.1016/j.chemosphere.2013.01.075] [PMID: 23466085]

[2]   Hassan TU, Bano A, Naz I. Alleviation of heavy metals toxicity by the application of plant growth promoting rhizobacteria and effects on wheat grown in saline sodic field. Int J Phytoremediation 2017;

19(6): 522-9.
[http://dx.doi.org/10.1080/15226514.2016.1267696] [PMID: 27936865]

[3]   Briffa J, Sinagra E, Blundell R. Heavy metal pollution in the environment and their toxicological effects on humans. Heliyon 2020; 6(9): e04691.
[http://dx.doi.org/10.1016/j.heliyon.2020.e04691] [PMID: 32964150]

[4]   Mushtaq Z, Liaquat M, Nazir A, *et al.* Potential of plant growth promoting rhizobacteria to mitigate chromium contamination. Environ Technol Innov 2022; 28(102826): 102826.
[http://dx.doi.org/10.1016/j.eti.2022.102826]

[5]   Ali H, Khan E. Trophic transfer, bioaccumulation, and biomagnification of non-essential hazardous heavy metals and metalloids in food chains/webs—Concepts and implications for wildlife and human health. Hum Ecol Risk Assess 2019; 25(6): 1353-76.
[http://dx.doi.org/10.1080/10807039.2018.1469398]

[6]   Nazli F, Mustafa A, Ahmad M, *et al.* A review on practical application and potentials of phytohormone-producing plant growth-promoting rhizobacteria for inducing heavy metal tolerance in crops. Sustainability (Basel) 2020; 12(21): 9056.
[http://dx.doi.org/10.3390/su12219056]

[7]   Wuana RA, Okieimen FE. Heavy metals in contaminated soils: A review of sources, chemistry, risks and best available strategies for remediation. ISRN Ecol 2011; 2011: 1-20.
[http://dx.doi.org/10.5402/2011/402647]

[8]   Jaiswal A, Verma A, Jaiswal P. Detrimental effects of heavy metals in soil, plants, and aquatic ecosystems and in humans. J Environ Pathol Toxicol Oncol 2018; 37(3): 183-97.
[http://dx.doi.org/10.1615/JEnvironPatholToxicolOncol.2018025348] [PMID: 30317970]

[9]   Hou D, O'Connor D, Igalavithana AD, *et al.* Metal contamination and bioremediation of agricultural soils for food safety and sustainability. Nat Rev Earth Environ 2020; 1(7): 366-81.
[http://dx.doi.org/10.1038/s43017-020-0061-y]

[10]   Rajasulochana P, Preethy V. Comparison on efficiency of various techniques in treatment of waste and sewage water – A comprehensive review. Resource-Efficient Technologies 2016; 2(4): 175-84.
[http://dx.doi.org/10.1016/j.reffit.2016.09.004]

[11]   Mishra J, Singh R, Arora NK. Alleviation of heavy metal stress in plants and remediation of soil by rhizosphere microorganisms. Front Microbiol 2017; 8: 1706.
[http://dx.doi.org/10.3389/fmicb.2017.01706] [PMID: 28932218]

[12]   Mushtaq Z, Asghar HN, Zahir ZA. Comparative growth analysis of okra (*Abelmoschus esculentus*) in the presence of PGPR and press mud in chromium contaminated soil. Chemosphere 2021; 262(127865): 127865.
[http://dx.doi.org/10.1016/j.chemosphere.2020.127865] [PMID: 32791369]

[13]   Lasat MM. Phytoextraction of toxic metals: a review of biological mechanisms. J Environ Qual 2002; 31(1): 109-20.
[http://dx.doi.org/10.2134/jeq2002.1090] [PMID: 11837415]

[14]   Garbisu C, Alkorta I, Kidd P, Epelde L, Mench M. Keep and promote biodiversity at polluted sites under phytomanagement. Environ Sci Pollut Res Int 2020; 27(36): 44820-34.
[http://dx.doi.org/10.1007/s11356-020-10854-5] [PMID: 32975751]

[15]   Raffa CM, Chiampo F, Shanthakumar S. Remediation of metal/metalloid-polluted soils: A short review. Appl Sci (Basel) 2021; 11(9): 4134.
[http://dx.doi.org/10.3390/app11094134]

[16]   Mani D, Kumar C. Biotechnological advances in bioremediation of heavy metals contaminated ecosystems: an overview with special reference to phytoremediation. Int J Environ Sci Technol 2014; 11(3): 843-72.
[http://dx.doi.org/10.1007/s13762-013-0299-8]

[17]   Tiwari S, Lata C. Heavy metal stress, signaling, and tolerance due to plant-associated microbes: An overview. Front Plant Sci 2018; 9: 452.
[http://dx.doi.org/10.3389/fpls.2018.00452] [PMID: 29681916]

[18]   Saha L, Tiwari J, Bauddh K, Ma Y. Recent developments in microbe-plant-based bioremediation for tackling heavy metal-polluted soils. Front Microbiol 2021; 12: 731723.
[http://dx.doi.org/10.3389/fmicb.2021.731723] [PMID: 35002995]

[19]   Chakraborty R, Wu CH, Hazen TC. Systems biology approach to bioremediation. Curr Opin Biotechnol 2012; 23(3): 483-90.
[http://dx.doi.org/10.1016/j.copbio.2012.01.015] [PMID: 22342400]

[20]   Ekperusi OA, Aigbodion FI. Bioremediation of petroleum hydrocarbons from crude oil-contaminated soil with the earthworm: Hyperiodrilus africanus. 3 Biotech 2015; 5(6): 957-65.

[21]   Ayangbenro A, Babalola O. A new strategy for heavy metal polluted environments: A review of microbial biosorbents. Int J Environ Res Public Health 2017; 14(1): 94.
[http://dx.doi.org/10.3390/ijerph14010094] [PMID: 28106848]

[22]   Thapa B, Kc AK, Ghimire A. KC AK, Ghimire A. A review on bioremediation of petroleum hydrocarbon contaminants in soil. J Sci Eng Technol 1970; 8(1): 164-70.
[http://dx.doi.org/10.3126/kuset.v8i1.6056]

[23]   Mangunwardoyo W, Sudjarwo T, Patria MP. Bioremediation of effluent wastewater treatment plant Bojongsoang Bandung Indonesia using consortium aquatic plants and animals. Int J Res Rev Appl Sci 2013; 14: 150-60.

[24]   Chen M, Xu P, Zeng G, Yang C, Huang D, Zhang J. Bioremediation of soils contaminated with polycyclic aromatic hydrocarbons, petroleum, pesticides, chlorophenols and heavy metals by composting: Applications, microbes and future research needs. Biotechnol Adv 2015; 33(6): 745-55.
[http://dx.doi.org/10.1016/j.biotechadv.2015.05.003] [PMID: 26008965]

[25]   Gami B, Limbasiya J, Bloch K. Bioremediation by plant growth promoting rhizobacteria. Int J Sci Res Rev 2018; 7(1): 285-94.

[26]   Nadeem SM, Ahmad M, Zahir ZA, Javaid A, Ashraf M. The role of mycorrhizae and plant growth promoting rhizobacteria (PGPR) in improving crop productivity under stressful environments. Biotechnol Adv 2014; 32(2): 429-48.
[http://dx.doi.org/10.1016/j.biotechadv.2013.12.005] [PMID: 24380797]

[27]   Ma Y, Rajkumar M, Zhang C, Freitas H. Beneficial role of bacterial endophytes in heavy metal phytoremediation. J Environ Manage 2016; 174: 14-25.
[http://dx.doi.org/10.1016/j.jenvman.2016.02.047] [PMID: 26989941]

[28]   Saravanakumar D, Thomas A, Banwarie N. Antagonistic potential of lipopeptide producing *Bacillus amyloliquefaciens* against major vegetable pathogens. Eur J Plant Pathol 2019; 154(2): 319-35.
[http://dx.doi.org/10.1007/s10658-018-01658-y]

[29]   Saravanakumar D, Samiyappan R. ACC deaminase from *Pseudomonas fluorescens* mediated saline resistance in groundnut (*Arachis hypogea*) plants. J Appl Microbiol 2007; 102(5): 1283-92.
[http://dx.doi.org/10.1111/j.1365-2672.2006.03179.x] [PMID: 17448163]

[30]   Lombi E, Hamon RE. Remediation of polluted soils. Encycl Soils Environ 2005; pp. 379-85.

[31]   Hazen TC. In situ: groundwater bioremediation. In: Timmis KN, Ed. Handbook of Hydrocarbon and Lipid Microbiology. Berlin: Springer 2010; pp. 2583-94.
[http://dx.doi.org/10.1007/978-3-540-77587-4_191]

[32]   Kumar A, Bisht BS, Joshi VD, Dhewa T. Review on bioremediation of polluted environment: a management tool. Int J Environ Sci 2011; 1(6): 1079-93.

[33]   Rayu S, Karpouzas DG, Singh BK. Emerging technologies in bioremediation: constraints and opportunities. Biodegradation 2012; 23(6): 917-26.

[http://dx.doi.org/10.1007/s10532-012-9576-3] [PMID: 22836784]

[34] Tomei MC, Daugulis AJ. Ex situ bioremediation of contaminated soils: an overview of conventional and innovative technologies. Crit Rev Environ Sci Technol 2013; 43(20): 2107-39.
[http://dx.doi.org/10.1080/10643389.2012.672056]

[35] Kabata-Pendias A, Pendias H. Trace metals in soils and plants. Boca Raton, Fla, USA: CRC Press 2010.
[http://dx.doi.org/10.1201/b10158]

[36] Kumar A, Mishra S, Kumar A, Singhal S. Environmental quantification of soil elements in the catchment of hydroelectric reservoirs in India. Hum Ecol Risk Assess 2017; 23(5): 1202-18.
[http://dx.doi.org/10.1080/10807039.2017.1309266]

[37] D'Amore JJ, Al-Abed SR, Scheckel KG, Ryan JA. Methods for speciation of metals in soils: a review. J Environ Qual 2005; 34(5): 1707-45.
[http://dx.doi.org/10.2134/jeq2004.0014] [PMID: 16151225]

[38] Kumar A, Kumar A, M M S CP, *et al.* Lead toxicity: Health hazards, influence on food chain, and sustainable remediation approaches. Int J Environ Res Public Health 2020; 17(7): 2179.
[http://dx.doi.org/10.3390/ijerph17072179] [PMID: 32218253]

[39] Kaasalainen M, Yli-Halla M. Use of sequential extraction to assess metal partitioning in soils. Environ Pollut 2003; 126(2): 225-33.
[http://dx.doi.org/10.1016/S0269-7491(03)00191-X] [PMID: 12927493]

[40] Jayakumar M, Surendran U, Raja P, Kumar A, Senapathi V. A review of heavy metals accumulation pathways, sources and management in soils. Arab J Geosci 2021; 14(20): 2156.
[http://dx.doi.org/10.1007/s12517-021-08543-9]

[41] Muradoglu F, Gundogdu M, Ercisli S, *et al.* Cadmium toxicity affects chlorophyll a and b content, antioxidant enzyme activities and mineral nutrient accumulation in strawberry. Biol Res 2015; 48(1): 11.
[http://dx.doi.org/10.1186/s40659-015-0001-3] [PMID: 25762051]

[42] Dixit R, Wasiullah , Malaviya D, *et al.* Bioremediation of heavy metals from soil and aquatic environment: An overview of principles and criteria of fundamental processes. Sustainability (Basel) 2015; 7(2): 2189-212.
[http://dx.doi.org/10.3390/su7022189]

[43] Khan S, Cao Q, Zheng YM, Huang YZ, Zhu YG. Health risks of heavy metals in contaminated soils and food crops irrigated with wastewater in Beijing, China. Environ Pollut 2008; 152(3): 686-92.
[http://dx.doi.org/10.1016/j.envpol.2007.06.056] [PMID: 17720286]

[44] Zhang MK, Liu ZY, Wang H. Use of single extraction methods to predict bioavailability of heavy metals in polluted soils to rice. Commun Soil Sci Plant Anal 2010; 41(7): 820-31.
[http://dx.doi.org/10.1080/00103621003592341]

[45] Kumar Yadav K, Gupta N, Kumar A, *et al.* Mechanistic understanding and holistic approach of phytoremediation: A review on application and future prospects. Ecol Eng 2018; 120: 274-98.
[http://dx.doi.org/10.1016/j.ecoleng.2018.05.039]

[46] Gupta N, Yadav KK, Kumar V, Kumar S, Chadd RP, Kumar A. Trace elements in soil-vegetables interface: Translocation, bioaccumulation, toxicity and amelioration - A review. Sci Total Environ 2019; 651(Pt 2): 2927-42.
[http://dx.doi.org/10.1016/j.scitotenv.2018.10.047] [PMID: 30463144]

[47] Kurniawan SB, Ramli NN, Said NSM, *et al.* Practical limitations of bioaugmentation in treating heavy metal contaminated soil and role of plant growth promoting bacteria in phytoremediation as a promising alternative approach. Heliyon 2022; 8(4): e08995.
[http://dx.doi.org/10.1016/j.heliyon.2022.e08995] [PMID: 35399376]

[48] Oyugi AM, Kibet JK, Adongo JO. A review of the health implications of heavy metals and pesticide

residues on khat users. Bull Natl Res Cent 2021; 45(1): 158.
[http://dx.doi.org/10.1186/s42269-021-00613-y]

[49]   Żołnowski A, Busse M, Zając P. Response of maize (*Zea mays* L.) to soil contamination with copper depending on applied contamination neutralizing substances. J Elem 2013; 18(3): 447-68.

[50]   Ayuso R, Foley N, Robinson G, Wandless G, Dillingham J. Lead isotopic compositions of common arsenical pesticides used in New England. US Geol Surv Open-File Rep 2004; 1342: 14.

[51]   Codling EE, Chaney RL, Green CE. Accumulation of lead and arsenic by carrots grown on lead-arsenate contaminated orchard soils. J Plant Nutr 2015; 38(4): 509-25.
[http://dx.doi.org/10.1080/01904167.2014.934477]

[52]   Hamilton AJ, Burry K, Mok HF, Barker SF, Grove JR, Williamson VG. Give peas a chance? Urban agriculture in developing countries. A review. Agron Sustain Dev 2014; 34(1): 45-73.
[http://dx.doi.org/10.1007/s13593-013-0155-8]

[53]   Nagajyoti PC, Lee KD, Sreekanth TVM. Heavy metals, occurrence and toxicity for plants: a review. Environ Chem Lett 2010; 8(3): 199-216.
[http://dx.doi.org/10.1007/s10311-010-0297-8]

[54]   Marrugo-Negrete J, Pinedo-Hernández J, Díez S. Assessment of heavy metal pollution, spatial distribution and origin in agricultural soils along the Sinú River Basin, Colombia. Environ Res 2017; 154: 380-8.
[http://dx.doi.org/10.1016/j.envres.2017.01.021] [PMID: 28189028]

[55]   Ayaz SÇ, Akça L. Treatment of wastewater by natural systems. Environ Int 2001; 26(3): 189-95.
[http://dx.doi.org/10.1016/S0160-4120(00)00099-4] [PMID: 11341705]

[56]   Ungureanu N, Vlăduţ V, Voicu G. Water scarcity and wastewater reuse in crop irrigation. Sustainability (Basel) 2020; 12(21): 9055.
[http://dx.doi.org/10.3390/su12219055]

[57]   Bjuhr J. Trace metals in soils irrigated with waste water in a periurban area downstream Hanoi City. Vietnam 2007.

[58]   Bougnom BP, Piddock LJV. Wastewater for urban agriculture: A significant factor in dissemination of antibiotic resistance. Environ Sci Technol 2017; 51(11): 5863-4.
[http://dx.doi.org/10.1021/acs.est.7b01852] [PMID: 28489349]

[59]   Singh RP, Agrawal M. Potential benefits and risks of land application of sewage sludge. Waste Manag 2008; 28(2): 347-58.
[http://dx.doi.org/10.1016/j.wasman.2006.12.010] [PMID: 17320368]

[60]   Singh J, Rawat KS, Kumar A, Singh A. Effect of sewage sludge and biofertilizers on physicochemical properties of alluvial soil. Biochem Cell Arch 2013; 13(2): 319-22.

[61]   Alvarenga P, Mourinha C, Farto M, *et al.* Sewage sludge, compost and other representative organic wastes as agricultural soil amendments: Benefits versus limiting factors. Waste Manag 2015; 40: 44-52.
[http://dx.doi.org/10.1016/j.wasman.2015.01.027] [PMID: 25708406]

[62]   Islam MA, Romić D, Akber MA, Romić M. Trace metals accumulation in soil irrigated with polluted water and assessment of human health risk from vegetable consumption in Bangladesh. Environ Geochem Health 2018; 40(1): 59-85.
[http://dx.doi.org/10.1007/s10653-017-9907-8] [PMID: 28101717]

[63]   Sharma B, Sarkar A, Singh P, Singh RP. Agricultural utilization of biosolids: A review on potential effects on soil and plant grown. Waste Manag 2017; 64: 117-32.
[http://dx.doi.org/10.1016/j.wasman.2017.03.002] [PMID: 28336334]

[64]   Abii TA, Okorie DO. Assessment of the level of heavy metals [Cu, Pb, Cd and Cr] contamination in four popular vegetables sold in urban and rural markets of Abia State Nigeria. Continental Water Air

and Soil Pollut 2011; 2(1): 42-7.

[65] Singh S, Kumar M. Heavy metal load of soil, water and vegetables in peri-urban Delhi. Environ Monit Assess 2006; 120(1-3): 79-91.
[http://dx.doi.org/10.1007/s10661-005-9050-3] [PMID: 16897527]

[66] Brar B, Sonu N, Rawat J. Wastewater reuse for irrigation of a vegetable crops and its impacts. Pharma Innov J 2022; (7): 111-7.

[67] Smith LA. Remedial options for metals-contaminated sites. Boca Raton, FL, US: Lewis Publishers 1995.

[68] Shen H, Luo Z, Xiong R, *et al.* A critical review of pollutant emission factors from fuel combustion in home stoves. Environ Int 2021; 157: 106841.
[http://dx.doi.org/10.1016/j.envint.2021.106841] [PMID: 34438232]

[69] Girardet H. The metabolism of cities. In: The Living City 2019; pp. 170–80.
[http://dx.doi.org/10.4324/9780429197307-11]

[70] Sądej W, Żołnowski AC, Ciećko Z, Grzybowski Ł, Szostek R. Evaluation of the impact of soil contamination with mercury and application of soil amendments on the yield and chemical composition of *Avena sativa* L. J Environ Sci Health Part A Tox Hazard Subst Environ Eng 2020; 55(1): 82-96.
[http://dx.doi.org/10.1080/10934529.2019.1667671] [PMID: 31549913]

[71] Aryal R, Beecham S, Sarkar B, *et al.* Readily wash-off road dust and associated heavy metals on motorways. Water Air Soil Pollut 2017; 228(1): 1.
[http://dx.doi.org/10.1007/s11270-016-3178-3]

[72] Dubey B, Pal AK, Singh G. Airborne particulate matter: source scenario and their impact on human health and environment. Climate Change and Environmental Concerns: Breakthroughs in Research and Practice. IGI Global 2018; pp. 447-68.
[http://dx.doi.org/10.4018/978-1-5225-5487-5.ch023]

[73] Rolka E, Zolnowski AC, Kozlowska KA. Assessment of the content of trace elements in soils and roadside vegetation in the vicinity of some gasoline stations in Olsztyn (Poland). J Elem 2020; 25(2): 549-63.

[74] Tiwari S, Tripathi IP. Lead pollution-an overview. Int Res J Environ Sci 2012; 1(5): 84-6.

[75] Majumder AK, Al Nayeem A, Islam M, Akter MM, Carter WS. Critical review of lead pollution in Bangladesh. J Health Pollut 2021; 11(31): 210902.
[http://dx.doi.org/10.5696/2156-9614-11.31.210902] [PMID: 34434594]

[76] Chen B, Stein AF, Maldonado PG, *et al.* Size distribution and concentrations of heavy metals in atmospheric aerosols originating from industrial emissions as predicted by the HYSPLIT model. Atmos Environ 1994; 2013(71): 234-44.

[77] Hwang HM, Fiala MJ, Park D, Wade TL. Review of pollutants in urban road dust and stormwater runoff: part 1. Heavy metals released from vehicles. International Journal of Urban Sciences 2016; 20(3): 334-60.
[http://dx.doi.org/10.1080/12265934.2016.1193041]

[78] Gope M, Masto RE, George J, Hoque RR, Balachandran S. Bioavailability and health risk of some potentially toxic elements (Cd, Cu, Pb and Zn) in street dust of Asansol, India. Ecotoxicol Environ Saf 2017; 138: 231-41.
[http://dx.doi.org/10.1016/j.ecoenv.2017.01.008] [PMID: 28068580]

[79] Scragg AH. Environmental Biotechnology. 2nd ed., London, England: Oxford University Press 2006.

[80] Havlin JL. Soil: Fertility and nutrient management. Landscape and land capacity. CRC Press 2020; pp. 251-65.
[http://dx.doi.org/10.1201/9780429445552-34]

[81]   Lasat MM. Phytoextraction of metals from contaminated soil: A review of plant/soil/metal interaction and assessment of pertinent agronomic issues. Journal of Hazardous Substance Research 1999; 2(1).
[http://dx.doi.org/10.4148/1090-7025.1015]

[82]   Kumar Yadav K, Gupta N, Kumar A, *et al.* Mechanistic understanding and holistic approach of phytoremediation: A review on application and future prospects. Ecol Eng 2018; 120: 274-98.
[http://dx.doi.org/10.1016/j.ecoleng.2018.05.039]

[83]   Das S, Kim GW, Hwang HY, Verma PP, Kim PJ. Cropping with slag to address soil, environment, and food security. Front Microbiol 2019; 10: 1320.
[http://dx.doi.org/10.3389/fmicb.2019.01320] [PMID: 31275262]

[84]   Savci S. An agricultural pollutant: Chemical fertilizer. Int J Environ Sci Dev 2012; •••: 73-80.
[http://dx.doi.org/10.7763/IJESD.2012.V3.191]

[85]   Chen XX, Liu YM, Zhao QY, Cao WQ, Chen XP, Zou CQ. Health risk assessment associated with heavy metal accumulation in wheat after long-term phosphorus fertilizer application. Environ Pollut 2020; 262(114348): 114348.
[http://dx.doi.org/10.1016/j.envpol.2020.114348] [PMID: 32182536]

[86]   Bekele Bahiru D, Yegrem L. Levels of heavy metal in vegetable, fruits and cereals crops in Ethiopia: A review. Int J Environ Monit Anal 2021; 9(4): 96.
[http://dx.doi.org/10.11648/j.ijema.20210904.11]

[87]   Tóth G, Hermann T, Da Silva MR, Montanarella L. Heavy metals in agricultural soils of the European Union with implications for food safety. Environ Int 2016; 88: 299-309.
[http://dx.doi.org/10.1016/j.envint.2015.12.017] [PMID: 26851498]

[88]   Gill M. Heavy metal stress in plants: a review. Int J Adv Res (Indore) 2014; 2(6): 1043-55.

[89]   Sumner ME. Beneficial use of effluents, wastes, and biosolids. Commun Soil Sci Plant Anal 2000; 31(11-14): 1701-15.
[http://dx.doi.org/10.1080/00103620009370532]

[90]   Lu Q, He ZL, Stoffella PJ. Land application of biosolids in the USA: A review. Appl Environ Soil Sci 2012; 2012: 1-11.
[http://dx.doi.org/10.1155/2012/201462]

[91]   McLaughlin MJ, Hamon RE, McLaren RG, Speir TW, Rogers SL. Review: A bioavailability-based rationale for controlling metal and metalloid contamination of agricultural land in Australia and New Zealand. Soil Res 2000; 38(6): 1037.
[http://dx.doi.org/10.1071/SR99128]

[92]   Silveira MLA, Alleoni LRF, Guilherme LRG. Biosolids and heavy metals in soils. Sci Agric 2003; 60(4): 793-806.
[http://dx.doi.org/10.1590/S0103-90162003000400029]

[93]   Shi Y, Ge Y, Chang J, Shao H, Tang Y. Garden waste biomass for renewable and sustainable energy production in China: Potential, challenges and development. Renew Sustain Energy Rev 2013; 22: 432-7.
[http://dx.doi.org/10.1016/j.rser.2013.02.003]

[94]   Keller C, McGrath SP, Dunham SJ. Trace metal leaching through a soil-grassland system after sewage sludge application. J Environ Qual 2002; 31(5): 1550-60.
[http://dx.doi.org/10.2134/jeq2002.1550a] [PMID: 12371172]

[95]   McLaren RG, Clucas LM, Taylor MD. Leaching of macronutrients and metals from undisturbed soils treated with metal-spiked sewage sludge. 3. Distribution of residual metals. Soil Res 2005; 43(2): 159.
[http://dx.doi.org/10.1071/SR04109]

[96]   Ippolito JA, Ducey TF, Diaz K, Barbarick KA. Long-term biosolids land application influences soil health. Sci Total Environ 2021; 791(148344): 148344.

[http://dx.doi.org/10.1016/j.scitotenv.2021.148344] [PMID: 34412404]

[97]   Bolan NS, Choppala G, Kunhikrishnan A, Park J, Naidu R. Microbial transformation of trace elements in soils in relation to bioavailability and remediation. Rev Environ Contam Toxicol 2013; 225: 1-56. [PMID: 23494555]

[98]   Rai V, Vajpayee P, Singh SN, Mehrotra S. Effect of chromium accumulation on photosynthetic pigments, oxidative stress defense system, nitrate reduction, proline level and eugenol content of *Ocimum tenuiflorum* L. Plant Sci 2004; 167(5): 1159-69. [http://dx.doi.org/10.1016/j.plantsci.2004.06.016]

[99]   Niassy S, Diarra K. Effects of organic inputs in Urban agriculture and their optimization for poverty alleviation in Senegal, West-Africa. Organic Fertilizers. Types, Production and Environmental Impact 2012; pp. 1-22.

[100]   Yang X, Li Q, Tang Z, *et al.* Heavy metal concentrations and arsenic speciation in animal manure composts in China. Waste Manag 2017; 64: 333-9. [http://dx.doi.org/10.1016/j.wasman.2017.03.015] [PMID: 28320622]

[101]   Panwar NR, Saha JK, Adhikari T, Kundu S. Soil and water pollution in India: some case studies. IISS Technical Bulletin 2010; 1–4.

[102]   DeVolder PS, Brown SL, Hesterberg D, Pandya K. Metal bioavailability and speciation in a wetland tailings repository amended with biosolids compost, wood ash, and sulfate. J Environ Qual 2003; 32(3): 851-64. [http://dx.doi.org/10.2134/jeq2003.8510] [PMID: 12809286]

[103]   Kumar SS, Kumar A, Singh S, *et al.* Industrial wastes: Fly ash, steel slag and phosphogypsum-potential candidates to mitigate greenhouse gas emissions from paddy fields. Chemosphere 2020; 241: 124824. [http://dx.doi.org/10.1016/j.chemosphere.2019.124824] [PMID: 31590026]

[104]   Pavilonis B, Grassman J, Johnson G, Diaz Y, Caravanos J. Characterization and risk of exposure to elements from artisanal gold mining operations in the Bolivian Andes. Environ Res 2017; 154: 1-9. [http://dx.doi.org/10.1016/j.envres.2016.12.010] [PMID: 27992737]

[105]   Naz A, Chowdhury A, Chandra R, Mishra BK. Potential human health hazard due to bioavailable heavy metal exposure *via* consumption of plants with ethnobotanical usage at the largest chromite mine of India. Environ Geochem Health 2020; 42(12): 4213-31. [http://dx.doi.org/10.1007/s10653-020-00603-5] [PMID: 32495026]

[106]   Sharma GK, Jena RK, Hota S, *et al.* Recent development in bioremediation of soil pollutants through biochar for environmental sustainability. Biochar applications in agriculture and environment management. Cham: Springer 2020; pp. 123-40. [http://dx.doi.org/10.1007/978-3-030-40997-5_6]

[107]   Sun Z, Xie X, Wang P, Hu Y, Cheng H. Heavy metal pollution caused by small-scale metal ore mining activities: A case study from a polymetallic mine in South China. Sci Total Environ 2018; 639: 217-27. [http://dx.doi.org/10.1016/j.scitotenv.2018.05.176] [PMID: 29787905]

[108]   Chen G, Shah KJ, Shi L, Chiang P-C, You Z. Red soil amelioration and heavy metal immobilization by a multi-element mineral amendment: Performance and mechanisms. Environ Pollut 2019; 254(Pt A):112964.

[109]   Hanfi MY, Mostafa MYA, Zhukovsky MV. Heavy metal contamination in urban surface sediments: sources, distribution, contamination control, and remediation. Environ Monit Assess 2020; 192(1): 32. [http://dx.doi.org/10.1007/s10661-019-7947-5] [PMID: 31823021]

[110]   Manisalidis I, Stavropoulou E, Stavropoulos A, Bezirtzoglou E. Environmental and health impacts of air pollution: A review. Front Public Health 2020; 8: 14. [http://dx.doi.org/10.3389/fpubh.2020.00014] [PMID: 32154200]

[111] Alengebawy A, Abdelkhalek ST, Qureshi SR, Wang MQ. Heavy metals and pesticides toxicity in agricultural soil and plants: Ecological risks and human health implications. Toxics 2021; 9(3): 42.
[http://dx.doi.org/10.3390/toxics9030042] [PMID: 33668829]

[112] Chibuike GU, Obiora SC. Heavy metal polluted soils: Effect on plants and bioremediation methods. Appl Environ Soil Sci 2014; 2014: 1-12.
[http://dx.doi.org/10.1155/2014/752708]

[113] Tran TA, Popova LP. Functions and toxicity of cadmium in plants: recent advances and future prospects. Turk J Bot 2013; 37(1): 1-13.
[http://dx.doi.org/10.3906/bot-1112-16]

[114] Kumar B, Smita K, Cumbal Flores L. Plant mediated detoxification of mercury and lead. Arab J Chem 2017; 10: S2335-42.
[http://dx.doi.org/10.1016/j.arabjc.2013.08.010]

[115] Gjorgieva Ackova D. Heavy metals and their general toxicity for plants. Plant Sci Today 2018; 5(1): 14-8.
[http://dx.doi.org/10.14719/pst.2018.5.1.355]

[116] Cao YY, Qi CD, Li S, *et al.* Melatonin alleviates copper toxicity *via* improving copper sequestration and ROS scavenging in cucumber. Plant Cell Physiol 2019; 60(3): 562-74.
[http://dx.doi.org/10.1093/pcp/pcy226] [PMID: 30496548]

[117] Jain D, Kour R, Bhojiya AA, *et al.* Zinc tolerant plant growth promoting bacteria alleviates phytotoxic effects of zinc on maize through zinc immobilization. Sci Rep 2020; 10(1): 13865.
[http://dx.doi.org/10.1038/s41598-020-70846-w] [PMID: 32807871]

[118] Page Z, Tokpah DP, Drame KN, Luther Z, Voor VM, King CF. Morphological variation of iron toxicity tolerance in lowland rice (*Oryza sativa* L.) varieties. Na J Adv Res Rev 2022; 13(1): 38-46.

[119] Karnwal A. Use of Bio-chemical surfactant producing endophytic bacteria isolated from rice root for heavy metal bioremediation. Pertanika, J Trop Agric Sci 2018; 41(2): 699-14.

[120] Fatima A, Kataria S, Prajapati R, *et al.* Magnetopriming effects on arsenic stress-induced morphological and physiological variations in soybean involving synchrotron imaging. Physiol Plant 2021; 173(1): 88-99.
[PMID: 32915504]

[121] Bano K, Kumar B, Alyemeni MN, Ahmad P. Protective mechanisms of sulfur against arsenic phytotoxicity in *Brassica napus* by regulating thiol biosynthesis, sulfur-assimilation, photosynthesis, and antioxidant response. Plant Physiol Biochem 2022; 188: 1-11.
[http://dx.doi.org/10.1016/j.plaphy.2022.07.026] [PMID: 35963049]

[122] Kiany T, Pishkar L, Sartipnia N, Iranbakhsh A, Barzin G. Effects of silicon and titanium dioxide nanoparticles on arsenic accumulation, phytochelatin metabolism, and antioxidant system by rice under arsenic toxicity. Environ Sci Pollut Res Int 2022; 29(23): 34725-37.
[http://dx.doi.org/10.1007/s11356-021-17927-z] [PMID: 35041168]

[123] Saikia T. Rice cultivars (*Oryza sativa* L.) susceptible to iron toxicity have poor grain nutritional quality. Inter J Agri Sci and Res 2017; 7: 453-64.

[124] Reddy MA, Francies RM, Abida PS, Kumar PS. Correlation among morphological, biochemical and physiological responses under iron toxic conditions in rice. Int J Curr Microbiol Appl Sci 2019; 8(1): 37-44.
[http://dx.doi.org/10.20546/ijcmas.2019.801.005]

[125] Onyango DA, Entila F, Dida MM, Ismail AM, Drame KN. Mechanistic understanding of iron toxicity tolerance in contrasting rice varieties from Africa: 1. Morpho-physiological and biochemical responses. Funct Plant Biol 2019; 46(1): 93-105.
[http://dx.doi.org/10.1071/FP18129] [PMID: 30939261]

[126] Szopiński M, Sitko K, Gieroń Ż, *et al.* Toxic effects of Cd and Zn on the photosynthetic apparatus of the *Arabidopsis halleri* and *Arabidopsis arenosa* pseudo-metallophytes. Front Plant Sci 2019; 10: 748. [http://dx.doi.org/10.3389/fpls.2019.00748] [PMID: 31244873]

[127] Goodarzi A, Namdjoyan S, Soorki AA. Effects of exogenous melatonin and glutathione on zinc toxicity in safflower (*Carthamus tinctorius* L.) seedlings. Ecotoxicol Environ Saf 2020; 201(110853): 110853. [http://dx.doi.org/10.1016/j.ecoenv.2020.110853] [PMID: 32563160]

[128] Wei C, Jiao Q, Agathokleous E, *et al.* Hormetic effects of zinc on growth and antioxidant defense system of wheat plants. Sci Total Environ 2022; 807(Pt 2): 150992. [http://dx.doi.org/10.1016/j.scitotenv.2021.150992] [PMID: 34662623]

[129] El-Beltagi HS, Sofy MR, Aldaej MI, Mohamed HI. Silicon alleviates copper toxicity in flax plants by up-regulating antioxidant defense and secondary metabolites and decreasing oxidative damage. Sustainability (Basel) 2020; 12(11): 4732. [http://dx.doi.org/10.3390/su12114732]

[130] Saleem MH, Ali S, Kamran M, *et al.* Ethylenediaminetetraacetic acid (EDTA) mitigates the toxic effect of excessive copper concentrations on growth, gaseous exchange and chloroplast ultrastructure of *Corchorus capsularis* L. and improves copper accumulation capabilities. Plants 2020; 9(6): 756. [http://dx.doi.org/10.3390/plants9060756] [PMID: 32560128]

[131] Zhang J, Chen XF, Huang WT, *et al.* Mechanisms for increased pH-mediated amelioration of copper toxicity in *Citrus sinensis* leaves using physiology, transcriptomics and metabolomics. Environ Exp Bot 2022; 196(104812): 104812. [http://dx.doi.org/10.1016/j.envexpbot.2022.104812]

[132] Zafar-ul-Hye M, Naeem M, Danish S, *et al.* Effect of cadmium-tolerant rhizobacteria on growth attributes and chlorophyll contents of bitter gourd under cadmium toxicity. Plants 2020; 9(10): 1386. [http://dx.doi.org/10.3390/plants9101386] [PMID: 33080896]

[133] Jawad Hassan M, Ali Raza M, Ur Rehman S, *et al.* Effect of cadmium toxicity on growth, oxidative damage, antioxidant defense system and cadmium accumulation in two sorghum cultivars. Plants 2020; 9(11): 1575. [http://dx.doi.org/10.3390/plants9111575] [PMID: 33203059]

[134] Huang H, Li M, Rizwan M, *et al.* Synergistic effect of silicon and selenium on the alleviation of cadmium toxicity in rice plants. J Hazard Mater 2021; 401(123393): 123393. [http://dx.doi.org/10.1016/j.jhazmat.2020.123393] [PMID: 32763692]

[135] Šírová K, Vaculík M. Toxic effects of cadmium on growth of Aloe ferox Mill. S Afr J Bot 2022; 147: 1181-7. [http://dx.doi.org/10.1016/j.sajb.2020.12.026]

[136] Handa N, Kohli SK, Sharma A, Thukral AK, Bhardwaj R. Selenium modulates dynamics of antioxidative s defence expression, photosynthetic attributes and secondary metabolites to mitigate chromium toxicity in *Brassica juncea* L. plants. Environ Exp Bot 2019; 161: 180-92. [http://dx.doi.org/10.1016/j.envexpbot.2018.11.009]

[137] Zaheer IE, Ali S, Saleem MH, *et al.* Role of iron–lysine on morpho-physiological traits and combating chromium toxicity in rapeseed (*Brassica napus* L.) plants irrigated with different levels of tannery wastewater. Plant Physiol Biochem 2020; 155: 70-84. [http://dx.doi.org/10.1016/j.plaphy.2020.07.034] [PMID: 32745932]

[138] Yang S, Ulhassan Z, Shah AM, *et al.* Salicylic acid underpins silicon in ameliorating chromium toxicity in rice by modulating antioxidant defense, ion homeostasis and cellular ultrastructure. Plant Physiol Biochem 2021; 166: 1001-13. [http://dx.doi.org/10.1016/j.plaphy.2021.07.013] [PMID: 34271533]

[139] Ahmad S, Mfarrej MFB, El-Esawi MA, *et al.* Chromium-resistant *Staphylococcus aureus* alleviates

chromium toxicity by developing synergistic relationships with zinc oxide nanoparticles in wheat. Ecotoxicol Environ Saf 2022; 230(113142): 113142.
[http://dx.doi.org/10.1016/j.ecoenv.2021.113142] [PMID: 34990991]

[140] Shabaan M, Asghar HN, Akhtar MJ, Ali Q, Ejaz M. Role of plant growth promoting rhizobacteria in the alleviation of lead toxicity to *Pisum sativum* L. Int J Phytoremediation 2021; 23(8): 837-45.
[http://dx.doi.org/10.1080/15226514.2020.1859988] [PMID: 33372547]

[141] Xiong Z, Yang J, Zhang K. Effects of lead pollution on germination and seedling growth of turfgrass, *Cynodon dactylon.* Pak J Bot 2021; 53(6): 2003-9.
[http://dx.doi.org/10.30848/PJB2021-6(6)]

[142] Fatemi H, Esmaiel Pour B, Rizwan M. Foliar application of silicon nanoparticles affected the growth, vitamin C, flavonoid, and antioxidant enzyme activities of coriander (*Coriandrum sativum* L.) plants grown in lead (Pb)-spiked soil. Environ Sci Pollut Res Int 2021; 28(2): 1417-25.
[http://dx.doi.org/10.1007/s11356-020-10549-x] [PMID: 32839908]

[143] Chen F, Aqeel M, Maqsood MF, *et al.* Mitigation of lead toxicity in *Vigna radiata* genotypes by silver nanoparticles. Environ Pollut 2022; 308(119606): 119606.
[http://dx.doi.org/10.1016/j.envpol.2022.119606] [PMID: 35716894]

[144] Yuan H, Liu Q, Guo Z, *et al.* Sulfur nanoparticles improved plant growth and reduced mercury toxicity *via* mitigating the oxidative stress in *Brassica napus* L. J Clean Prod 2021; 318(128589): 128589.
[http://dx.doi.org/10.1016/j.jclepro.2021.128589]

[145] Ahmad P, Alyemeni MN, Wijaya L, *et al.* Nitric oxide donor, sodium nitroprusside, mitigates mercury toxicity in different cultivars of soybean. J Hazard Mater 2021; 408(124852): 124852.
[http://dx.doi.org/10.1016/j.jhazmat.2020.124852] [PMID: 33383453]

[146] Padmavathi S, Sujatha B. Assessment of mercury toxicity on seed germination, shoot and root growth of *Cajanus cajan* (L.) Millsp. Int J Creat Res Thoughts 2021; 9(1): 428-36.

[147] Escobar-Vargas S, Vargas Aguirre CF, Páez R. Arbuscular mycorrhizal fungi prevent mercury toxicity in *Lactuca sativa* (L.) seed germination. Pollution 2022; 8: 1014-25.

[148] Kamran MA, Eqani SAMAS, Bibi S, *et al.* Bioaccumulation of nickel by E. sativa and role of plant growth promoting rhizobacteria (PGPRs) under nickel stress. Ecotoxicol Environ Saf 2016; 126: 256-63.
[http://dx.doi.org/10.1016/j.ecoenv.2016.01.002] [PMID: 26773835]

[149] Kotapati KV, Palaka BK, Ampasala DR. Alleviation of nickel toxicity in finger millet (*Eleusine coracana* L.) germinating seedlings by exogenous application of salicylic acid and nitric oxide. Crop J 2017; 5(3): 240-50.
[http://dx.doi.org/10.1016/j.cj.2016.09.002]

[150] Hasanuzzaman M, Alam MM, Nahar K, *et al.* Silicon-induced antioxidant defense and methylglyoxal detoxification works coordinately in alleviating nickel toxicity in *Oryza sativa* L. Ecotoxicology 2019; 28(3): 261-76.
[http://dx.doi.org/10.1007/s10646-019-02019-z] [PMID: 30761430]

[151] Naveed M, Bukhari SS, Mustafa A, *et al.* Mitigation of nickel toxicity and growth promotion in sesame through the application of a bacterial endophyte and zeolite in nickel contaminated soil. Int J Environ Res Public Health 2020; 17(23): 8859.
[http://dx.doi.org/10.3390/ijerph17238859] [PMID: 33260516]

[152] Mostofa MG, Hossain MA, Siddiqui MN, Fujita M, Tran LSP. Phenotypical, physiological and biochemical analyses provide insight into selenium-induced phytotoxicity in rice plants. Chemosphere 2017; 178: 212-23.
[http://dx.doi.org/10.1016/j.chemosphere.2017.03.046] [PMID: 28324842]

[153] Cabral Gouveia GC, Galindo FS, Dantas Bereta Lanza MG, *et al.* Selenium toxicity stress-induced

phenotypical, biochemical and physiological responses in rice plants: Characterization of symptoms and plant metabolic adjustment. Ecotoxicol Environ Saf 2020; 202: 110916.
[http://dx.doi.org/10.1016/j.ecoenv.2020.110916] [PMID: 32800251]

[154]   Mostofa MG, Rahman MM, Siddiqui MN, Fujita M, Tran LSP. Salicylic acid antagonizes selenium phytotoxicity in rice: selenium homeostasis, oxidative stress metabolism and methylglyoxal detoxification. J Hazard Mater 2020; 394(122572): 122572.
[http://dx.doi.org/10.1016/j.jhazmat.2020.122572] [PMID: 32283381]

[155]   Saleem M, Fariduddin Q. Novel mechanistic insights of selenium induced microscopic, histochemical and physio-biochemical changes in tomato (*Solanum lycopersicum* L.) plant. An account of beneficiality or toxicity. J Hazard Mater 2022; 434(128830): 128830.
[http://dx.doi.org/10.1016/j.jhazmat.2022.128830] [PMID: 35429754]

[156]   Babalola OO. Beneficial bacteria of agricultural importance. Biotechnol Lett 2010; 32(11): 1559-70.
[http://dx.doi.org/10.1007/s10529-010-0347-0] [PMID: 20635120]

[157]   Carvalhais LC, Dennis PG, Badri DV, Kidd BN, Vivanco JM, Schenk PM. Linking jasmonic acid signaling, root exudates, and rhizosphere microbiomes. Mol Plant Microbe Interact 2015; 28(9): 1049-58.
[http://dx.doi.org/10.1094/MPMI-01-15-0016-R] [PMID: 26035128]

[158]   Lynch JM. Some consequences of microbial rhizosphere competence for plant and soil. In: Lynch JM, Ed. The rhizosphere. Wiley Interscience New York 1990; pp. 1-10.

[159]   Santoyo G, Moreno-Hagelsieb G, del Carmen Orozco-Mosqueda M, Glick BR. Plant growth-promoting bacterial endophytes. Microbiol Res 2016; 183: 92-9.
[http://dx.doi.org/10.1016/j.micres.2015.11.008] [PMID: 26805622]

[160]   Reed MLE, Glick BR. Applications of plant growth-promoting bacteria for plant and soil systems. In: Gupta VK, Schmoll M, Maki M, Tuohy M, Mazutti MA, Eds. Applications of Microbial Engineering. Enfield, CT, USA: Taylor and Francis 2013; pp. 181-229.

[161]   Calvo P, Nelson L, Kloepper JW. Agricultural uses of plant biostimulants. Plant Soil 2014; 383(1-2): 3-41.
[http://dx.doi.org/10.1007/s11104-014-2131-8]

[162]   Amarger N. Genetically modified bacteria in agriculture. Biochimie 2002; 84(11): 1061-72.
[http://dx.doi.org/10.1016/S0300-9084(02)00035-4] [PMID: 12595134]

[163]   Barriuso J, Solano BR, Lucas JA, Lobo AP, Villaraco AG, Manero F. 2008.

[164]   Prashar P, Kapoor N, Sachdeva S. Rhizosphere: its structure, bacterial diversity and significance. Rev Environ Sci Biotechnol 2014; 13(1): 63-77.
[http://dx.doi.org/10.1007/s11157-013-9317-z]

[165]   Ahemad M, Kibret M. Mechanisms and applications of plant growth promoting rhizobacteria: Current perspective. J King Saud Univ Sci 2014; 26(1): 1-20.
[http://dx.doi.org/10.1016/j.jksus.2013.05.001]

[166]   Khoshru B, Mitra D, Khoshmanzar E, *et al.* Current scenario and future prospects of plant growth-promoting rhizobacteria: an economic valuable resource for the agriculture revival under stressful conditions. J Plant Nutr 2020; 43(20): 3062-92.
[http://dx.doi.org/10.1080/01904167.2020.1799004]

[167]   Parray JA, Jan S, Kamili AN, Qadri RA, Egamberdieva D, Ahmad P. Current perspectives on plant growth-promoting rhizobacteria. J Plant Growth Regul 2016; 35(3): 877-902.
[http://dx.doi.org/10.1007/s00344-016-9583-4]

[168]   Backer R, Rokem JS, Ilangumaran G, *et al.* Plant growth-promoting rhizobacteria: Context, mechanisms of action, and roadmap to commercialization of biostimulants for sustainable agriculture. Front Plant Sci 2018; 9: 1473.
[http://dx.doi.org/10.3389/fpls.2018.01473] [PMID: 30405652]

[169] Bhattacharyya PN, Jha DK. Plant growth-promoting rhizobacteria (PGPR): emergence in agriculture. World J Microbiol Biotechnol 2012; 28(4): 1327-50.
[http://dx.doi.org/10.1007/s11274-011-0979-9] [PMID: 22805914]

[170] Vaikuntapu PR, Dutta S, Samudrala RB, Rao VRVN, Kalam S, Podile AR. Preferential promotion of *Lycopersicon esculentum* (tomato) growth by plant growth promoting bacteria associated with tomato. Indian J Microbiol 2014; 54(4): 403-12.
[http://dx.doi.org/10.1007/s12088-014-0470-z] [PMID: 25320438]

[171] Jain D, Kour R, Bhojiya AA, Meena RH, Singh A, Mohanty SR, et al. Zinc tolerant plant growth promoting bacteria alleviates phytotoxic effects of zinc on maize through zinc immobilization. Sci Rep 2020;10(1):13865.

[172] Vejan P, Abdullah R, Khadiran T, Ismail S, Nasrulhaq Boyce A. Role of plant growth promoting rhizobacteria in agricultural sustainability - A review. Molecules 2016; 21(5): 573.
[http://dx.doi.org/10.3390/molecules21050573] [PMID: 27136521]

[173] Gyongyver M, Becze A, Vincze E-B, Varga H-M. Effect of plant growth promoting rhizobacteria on *Zea mays* development and growth under heavy metal and salt stress condition. Environ Eng Manag J 2021; 20(4): 547-57. [EEMJ].
[http://dx.doi.org/10.30638/eemj.2021.053]

[174] Swarnalakshmi K, Yadav V, Tyagi D, Dhar DW, Kannepalli A, Kumar S. Significance of plant growth promoting rhizobacteria in grain legumes: Growth promotion and crop production. Plants 2020; 9(11): 1596.
[http://dx.doi.org/10.3390/plants9111596] [PMID: 33213067]

[175] Yadav A, Yadav K. Exploring the potential of endophytes in agriculture: A mini review. Adv Plants Agric Res 2017; 6(4): 102-6.
[http://dx.doi.org/10.15406/apar.2017.06.00221]

[176] Doty S. Symbiotic plant-bacterial endospheric interactions. Microorganisms 2018; 6(2): 28.
[http://dx.doi.org/10.3390/microorganisms6020028] [PMID: 29565291]

[177] Afzal I, Shinwari ZK, Sikandar S, Shahzad S. Plant beneficial endophytic bacteria: Mechanisms, diversity, host range and genetic determinants. Microbiol Res 2019; 221: 36-49.
[http://dx.doi.org/10.1016/j.micres.2019.02.001] [PMID: 30825940]

[178] Cassán F, Coniglio A, López G, *et al.* Everything you must know about *Azospirillum* and its impact on agriculture and beyond. Biol Fertil Soils 2020; 56(4): 461-79.
[http://dx.doi.org/10.1007/s00374-020-01463-y]

[179] Shahzad R, Bilal S, Imran M, *et al.* Amelioration of heavy metal stress by endophytic Bacillus amyloliquefaciens RWL-1 in rice by regulating metabolic changes: potential for bacterial bioremediation. Biochem J 2019; 476(21): 3385-400.

[180] Rana KL, Kour D, Kaur T, Devi R, Yadav A, Yadav AN. Bioprospecting of endophytic bacteria from the Indian Himalayas and their role in plant growth promotion of maize (*Zea mays* L.). J Appl Biol Biotechnol 2021; 9(3): 41-50.

[181] Dubey A, Saiyam D, Kumar A, Hashem A, Abd Allah EF, Khan ML. Bacterial root endophytes: Characterization of their competence and plant growth promotion in soybean (*Glycine max* (L.) Merr.) under drought stress. Int J Environ Res Public Health 2021; 18(3): 931.
[http://dx.doi.org/10.3390/ijerph18030931] [PMID: 33494513]

[182] Lata R, Chowdhury S, Gond SK, White JF Jr. Induction of abiotic stress tolerance in plants by endophytic microbes. Lett Appl Microbiol 2018; 66(4): 268-76.
[http://dx.doi.org/10.1111/lam.12855] [PMID: 29359344]

[183] Okungbowa FI, Shittu HO, Obiazikwor HO. Endophytic bacteria: Hidden protective associates of plants against biotic and abiotic stresses. Not Sci Biol 2019; 11(2): 167-74.
[http://dx.doi.org/10.15835/nsb11210423]

[184] Kandel S, Joubert P, Doty S. Bacterial endophyte colonization and distribution within plants. Microorganisms 2017; 5(4): 77.
[http://dx.doi.org/10.3390/microorganisms5040077] [PMID: 29186821]

[185] Vanbleu E, Vanderleyden J. Molecular genetics of rhizosphere and plant-root colonization. Associative and endophytic nitrogen-fixing bacteria and cyanobacterial associations. Dordrecht: Springer 2007; pp. 85-112.
[http://dx.doi.org/10.1007/1-4020-3546-2_5]

[186] Eid AM, Fouda A, Abdel-Rahman MA, *et al.* Harnessing bacterial endophytes for promotion of plant growth and biotechnological applications: An overview. Plants 2021; 10(5): 935.
[http://dx.doi.org/10.3390/plants10050935] [PMID: 34067154]

[187] Taulé C, Vaz-Jauri P, Battistoni F. Insights into the early stages of plant–endophytic bacteria interaction. World J Microbiol Biotechnol 2021; 37(1): 13.
[http://dx.doi.org/10.1007/s11274-020-02966-4] [PMID: 33392741]

[188] Compant S, Reiter B, Sessitsch A, Nowak J, Clément C, Ait Barka E. Endophytic colonization of *Vitis vinifera* L. by plant growth-promoting bacterium *Burkholderia* sp. strain PsJN. Appl Environ Microbiol 2005; 71(4): 1685-93.
[http://dx.doi.org/10.1128/AEM.71.4.1685-1693.2005] [PMID: 15811990]

[189] Glick BR. Plant growth-promoting bacteria: mechanisms and applications. Scientifica (Cairo) 2012; 2012: 1-15.
[http://dx.doi.org/10.6064/2012/963401] [PMID: 24278762]

[190] Thomas J, Kim HR, Rahmatallah Y, *et al.* RNA-seq reveals differentially expressed genes in rice (*Oryza sativa*) roots during interactions with plant-growth promoting bacteria, *Azospirillum brasilense*. PLoS One 2019; 14(5): e0217309.
[http://dx.doi.org/10.1371/journal.pone.0217309] [PMID: 31120967]

[191] Schillaci M, Gupta S, Walker R, Roessner U. The role of plant growth-promoting bacteria in the growth of cereals under abiotic stresses. In: Ohyama T, Ed. Root Biology - Growth, Physiology, and Functions London: IntechOpen. 2019.
[http://dx.doi.org/10.5772/intechopen.87083]

[192] Yahaghi Z, Shirvani M, Nourbakhsh F, Pueyo JJ. Uptake and effects of lead and zinc on alfalfa (*Medicago sativa* L.) seed germination and seedling growth: Role of plant growth promoting bacteria. S Afr J Bot 2019; 124: 573-82.
[http://dx.doi.org/10.1016/j.sajb.2019.01.006]

[193] Pramanik P, Goswami AJ, Ghosh S, Kalita C. An indigenous strain of potassium-solubilizing bacteria *Bacillus pseudomycoides* enhanced potassium uptake in tea plants by increasing potassium availability in the mica waste-treated soil of North-east India. J Appl Microbiol 2019; 126(1): 215-22.
[http://dx.doi.org/10.1111/jam.14130] [PMID: 30326179]

[194] Hossain MT, Khan A, Harun-Or-Rashid M, Chung YR. A volatile producing endophytic *Bacillus siamensis* YC7012 promotes root development independent on auxin or ethylene/jasmonic acid pathway. Plant Soil 2019; 439(1-2): 309-24.
[http://dx.doi.org/10.1007/s11104-019-04015-y]

[195] Liu C, Mou L, Yi J, Wang J, Liu A, Yu J. The eno gene of *Burkholderia cenocepacia* strain 71-2 is involved in phosphate solubilization. Curr Microbiol 2019; 76(4): 495-502.
[http://dx.doi.org/10.1007/s00284-019-01642-7] [PMID: 30798378]

[196] Dinnage R, Simonsen AK, Barrett LG, *et al.* Larger plants promote a greater diversity of symbiotic nitrogen-fixing soil bacteria associated with an Australian endemic legume. J Ecol 2019; 107(2): 977-91.
[http://dx.doi.org/10.1111/1365-2745.13083]

[197] Abd-Alla MH, Nafady NA, Bashandy SR, Hassan AA. Mitigation of effect of salt stress on the

nodulation, nitrogen fixation and growth of chickpea (*Cicer arietinum* L.) by triple microbial inoculation. Rhizosphere 2019; 10(100148): 100148.
[http://dx.doi.org/10.1016/j.rhisph.2019.100148]

[198] Mehmood U, Inam-Ul-Haq M, Saeed M, Altaf A, Azam F, Hayat S. A brief review on plant growth promoting rhizobacteria (PGPR): a key role in plant growth promotion. Plant Prot 2018; 2(2): 77-82.

[199] Volpiano CG, Lisboa BB, São José JFB, *et al.* Rhizobium strains in the biological control of the phytopathogenic fungi *Sclerotium* (*Athelia*) *rolfsii* on the common bean. Plant Soil 2018; 432(1-2): 229-43.
[http://dx.doi.org/10.1007/s11104-018-3799-y]

[200] Akbar S, Sultan S. Soil bacteria showing a potential of chlorpyrifos degradation and plant growth enhancement. Braz J Microbiol 2016; 47(3): 563-70.
[http://dx.doi.org/10.1016/j.bjm.2016.04.009] [PMID: 27266625]

[201] Shahzad R, Waqas M, Khan AL, *et al.* Seed-borne endophytic *Bacillus amyloliquefaciens* RWL-1 produces gibberellins and regulates endogenous phytohormones of *Oryza sativa.* Plant Physiol Biochem 2016; 106: 236-43.
[http://dx.doi.org/10.1016/j.plaphy.2016.05.006] [PMID: 27182958]

[202] Mehta P, Walia A, Kulshrestha S, Chauhan A, Shirkot CK. Efficiency of plant growth-promoting P☐solubilizing *Bacillus circulans* CB7 for enhancement of tomato growth under net house conditions. J Basic Microbiol 2015; 55(1): 33-44.
[http://dx.doi.org/10.1002/jobm.201300562] [PMID: 24464353]

[203] Lucy M, Reed E, Glick BR. Applications of free living plant growth-promoting rhizobacteria. Antonie van Leeuwenhoek 2004; 86(1): 1-25.
[http://dx.doi.org/10.1023/B:ANTO.0000024903.10757.6e] [PMID: 15103234]

[204] Bai Y, D'Aoust F, Smith DL, Driscoll BT. Isolation of plant-growth-promoting *Bacillus* strains from soybean root nodules. Can J Microbiol 2002; 48(3): 230-8.
[http://dx.doi.org/10.1139/w02-014] [PMID: 11989767]

[205] Kaneko T, Nakamura Y, Sato S, *et al.* Complete genomic sequence of nitrogen-fixing symbiotic bacterium *Bradyrhizobium japonicum* USDA110. DNA Res 2002; 9(6): 189-97.
[http://dx.doi.org/10.1093/dnares/9.6.189] [PMID: 12597275]

[206] Verma A, Kukreja K, Pathak D, Suneja S, Narula N. *In vitro* production of plant growth regulators (PGRs). Indian J Microbiol 2001; 41: 305-7.

[207] Rana KL, Kour D, Kaur T, *et al.* Endophytic microbes from diverse wheat genotypes and their potential biotechnological applications in plant growth promotion and nutrient uptake. Proc Natl Acad Sci, India, Sect B Biol Sci 2020; 90(5): 969-79.
[http://dx.doi.org/10.1007/s40011-020-01168-0]

[208] Kumawat KC, Sharma P, Sirari A, *et al.* Synergism of *Pseudomonas aeruginosa* (LSE-2) nodule endophyte with *Bradyrhizobium* sp. (LSBR-3) for improving plant growth, nutrient acquisition and soil health in soybean. World J Microbiol Biotechnol 2019; 35(3): 47.
[http://dx.doi.org/10.1007/s11274-019-2622-0] [PMID: 30834977]

[209] Zhao L, Xu Y, Lai X. Antagonistic endophytic bacteria associated with nodules of soybean (*Glycine max* L.) and plant growth-promoting properties. Braz J Microbiol 2018; 49(2): 269-78.
[http://dx.doi.org/10.1016/j.bjm.2017.06.007] [PMID: 29117917]

[210] Sandhya V, Shrivastava M, Ali SZ, Sai Shiva Krishna Prasad V. Endophytes from maize with plant growth promotion and biocontrol activity under drought stress. Russ Agric Sci 2017; 43(1): 22-34.
[http://dx.doi.org/10.3103/S1068367417010165]

[211] Singh D, Rajawat MVS, Kaushik R, Prasanna R, Saxena AK. Beneficial role of endophytes in biofortification of Zn in wheat genotypes varying in nutrient use efficiency grown in soils sufficient and deficient in Zn. Plant Soil 2017; 416(1-2): 107-16.

[http://dx.doi.org/10.1007/s11104-017-3189-x]

[212] Kumar P, Dubey RC, Maheshwari DK, Bajpai V. ACC deaminase producing *Rhizobium leguminosarum* RPN5 isolated from root nodules of *Phaseolus vulgaris* L. Bangladesh J Bot 2016; 45: 477-84.

[213] Puri A, Padda KP, Chanway CP. Can a diazotrophic endophyte originally isolated from lodgepole pine colonize an agricultural crop (corn) and promote its growth? Soil Biol Biochem 2015; 89: 210-6.
[http://dx.doi.org/10.1016/j.soilbio.2015.07.012]

[214] Verma P, Yadav AN, Khannam KS, *et al.* Assessment of genetic diversity and plant growth promoting attributes of psychrotolerant bacteria allied with wheat (*Triticum aestivum*) from the northern hills zone of India. Ann Microbiol 2015; 65(4): 1885-99.
[http://dx.doi.org/10.1007/s13213-014-1027-4]

[215] Subramanian P, Kim K, Krishnamoorthy R, Sundaram S, Sa T. Endophytic bacteria improve nodule function and plant nitrogen in soybean on co-inoculation with *Bradyrhizobium japonicum* MN110. Plant Growth Regul 2015; 76(3): 327-32.
[http://dx.doi.org/10.1007/s10725-014-9993-x]

[216] Mbai F, Magiri E, Matiru V, Nganga J, Nyambati V. Isolation and characterization of bacterial root endophytes with potential to enhance plant growth from Kenyan basmati rice. Am Int J Contemp Res 2013; 3: 25-40.

[217] López-López A, Rogel MA, Ormeño-Orrillo E, Martínez-Romero J, Martínez-Romero E. *Phaseolus vulgaris* seed-borne endophytic community with novel bacterial species such as *Rhizobium endophyticum* sp. nov. Syst Appl Microbiol 2010; 33(6): 322-7.
[http://dx.doi.org/10.1016/j.syapm.2010.07.005] [PMID: 20822874]

[218] Govindarajan M, Balandreau J, Kwon SW, Weon HY, Lakshminarasimhan C. Effects of the inoculation of *Burkholderia vietnamensis* and related endophytic diazotrophic bacteria on grain yield of rice. Microb Ecol 2008; 55(1): 21-37.
[http://dx.doi.org/10.1007/s00248-007-9247-9] [PMID: 17406771]

[219] Chelius MK, Triplett EW. Immunolocalization of dinitrogenase reductase produced by *Klebsiella pneumoniae* in association with *Zea mays* L. Appl Environ Microbiol 2000; 66(2): 783-7.
[http://dx.doi.org/10.1128/AEM.66.2.783-787.2000] [PMID: 10653751]

[220] Jing X, Cui Q, Li X, *et al.* Engineering *Pseudomonas protegens* Pf-5 to improve its antifungal activity and nitrogen fixation. Microb Biotechnol 2020; 13(1): 118-33.
[http://dx.doi.org/10.1111/1751-7915.13335] [PMID: 30461205]

[221] Delgadillo J, Lafuente A, Doukkali B, *et al.* Improving legume nodulation and Cu rhizostabilization using a genetically modified rhizobia. Environ Technol 2015; 36(10): 1237-45.
[http://dx.doi.org/10.1080/09593330.2014.983990] [PMID: 25377353]

[222] Galleguillos C, Aguirre C, Miguel Barea J, Azcón R. Growth promoting effect of two *Sinorhizobium meliloti* strains (a wild type and its genetically modified derivative) on a non-legume plant species in specific interaction with two arbuscular mycorrhizal fungi. Plant Sci 2000; 159(1): 57-63.
[http://dx.doi.org/10.1016/S0168-9452(00)00321-6] [PMID: 11011093]

[223] Scupham AJ, Bosworth AH, Ellis WR, Wacek TJ, Albrecht KA, Triplett EW. Inoculation with *Sinorhizobium meliloti* RMBPC-2 increases alfalfa yield compared with inoculation with a nonengineered wild-type strain. Appl Environ Microbiol 1996; 62(11): 4260-2.
[http://dx.doi.org/10.1128/aem.62.11.4260-4262.1996] [PMID: 16535451]

[224] Moore LW, Warren G. *Agrobacterium radiobacter* strain 84 and biological control of crown gall. Annu Rev Phytopathol 1979; 17(1): 163-79.
[http://dx.doi.org/10.1146/annurev.py.17.090179.001115]

[225] Cheng S, Chen T, Xu W, Huang J, Jiang S, Yan B. Application research of biochar for the remediation of soil heavy metals contamination: A review. Molecules 2020; 25(14): 3167.

[http://dx.doi.org/10.3390/molecules25143167] [PMID: 32664440]

[226]   Liu L, Li W, Song W, Guo M. Remediation techniques for heavy metal-contaminated soils: Principles and applicability. Sci Total Environ 2018; 633: 206-19.
[http://dx.doi.org/10.1016/j.scitotenv.2018.03.161] [PMID: 29573687]

[227]   Li C, Zhou K, Qin W, *et al.* A review on heavy metals contamination in soil: effects, sources, and remediation techniques. Soil Sediment Contam 2019; 28(4): 380-94.
[http://dx.doi.org/10.1080/15320383.2019.1592108]

[228]   Dhaliwal SS, Singh J, Taneja PK, Mandal A. Remediation techniques for removal of heavy metals from the soil contaminated through different sources: a review. Environ Sci Pollut Res Int 2020; 27(2): 1319-33.
[http://dx.doi.org/10.1007/s11356-019-06967-1] [PMID: 31808078]

[229]   Rajendran S, Priya TAK, Khoo KS, *et al.* A critical review on various remediation approaches for heavy metal contaminants removal from contaminated soils. Chemosphere 2022; 287(Pt 4): 132369.
[http://dx.doi.org/10.1016/j.chemosphere.2021.132369] [PMID: 34582930]

[230]   Sharma S, Tiwari S, Hasan A, Saxena V, Pandey LM. Recent advances in conventional and contemporary methods for remediation of heavy metal-contaminated soils. 3 Biotech 2018; 8(4):216.

[231]   Ahemad M. Remediation of metalliferous soils through the heavy metal resistant plant growth promoting bacteria: Paradigms and prospects. Arab J Chem 2019; 12(7): 1365-77.
[http://dx.doi.org/10.1016/j.arabjc.2014.11.020]

[232]   Khanna K, Ohri P, Bhardwaj R, Ahmad P. Unsnarling plausible role of plant growth-promoting rhizobacteria for mitigating Cd-toxicity from plants: An environmental safety aspect. J Plant Growth Regul 2022; 41(6): 2514-42.
[http://dx.doi.org/10.1007/s00344-021-10445-9]

[233]   Sharma P, Kumar S. Bioremediation of heavy metals from industrial effluents by endophytes and their metabolic activity: Recent advances. Bioresour Technol 2021; 339(125589): 125589.
[http://dx.doi.org/10.1016/j.biortech.2021.125589] [PMID: 34304098]

[234]   Bramhachari PV, Nagaraju GP. Extracellular polysaccharide production by bacteria as a mechanism of toxic heavy metal biosorption and biosequestration in the marine environment. In: Naik M, Dubey S, Eds. Marine Pollution and Microbial Remediation. Singapore: Springer 2017; pp. 67-85.
[http://dx.doi.org/10.1007/978-981-10-1044-6_5]

[235]   Wang C, Tan H, Li H, *et al.* Mechanism study of Chromium influenced soil remediated by an uptake-detoxification system using hyperaccumulator, resistant microbe consortium, and nano iron complex. Environ Pollut 2020; 257(113558): 113558.
[http://dx.doi.org/10.1016/j.envpol.2019.113558] [PMID: 31708284]

[236]   Mathivanan K, Chandirika JU, Vinothkanna A, Yin H, Liu X, Meng D. Bacterial adaptive strategies to cope with metal toxicity in the contaminated environment – A review. Ecotoxicol Environ Saf 2021; 226: 112863.
[http://dx.doi.org/10.1016/j.ecoenv.2021.112863] [PMID: 34619478]

[237]   Valls M, Delorenzo V. Exploiting the genetic and biochemical capacities of bacteria for the remediation of heavy metal pollution. FEMS Microbiol Rev 2002; 26(4): 327-38.
[http://dx.doi.org/10.1016/S0168-6445(02)00114-6] [PMID: 12413663]

[238]   Yin K, Wang Q, Lv M, Chen L. Microorganism remediation strategies towards heavy metals. Chem Eng J 2019; 360: 1553-63.
[http://dx.doi.org/10.1016/j.cej.2018.10.226]

[239]   Vijayaraghavan K, Yun YS. Bacterial biosorbents and biosorption. Biotechnol Adv 2008; 26(3): 266-91.
[http://dx.doi.org/10.1016/j.biotechadv.2008.02.002] [PMID: 18353595]

[240]   Shen N, Birungi ZS, Chirwa E. Selective biosorption of precious metals by cell-surface engineered

microalgae. Chem Eng Trans 2017; 61: 25-30.

[241] Priya AK, Gnanasekaran L, Dutta K, Rajendran S, Balakrishnan D, Soto-Moscoso M. Biosorption of heavy metals by microorganisms: Evaluation of different underlying mechanisms. Chemosphere 2022; 307(Pt 4): 135957.
[http://dx.doi.org/10.1016/j.chemosphere.2022.135957] [PMID: 35985378]

[242] Samuel Sehar J, More AE, More VS, Gudibanda Ramesh A, More SS. Biosorption of heavy metals and metal-complexed dyes under the influence of various physicochemical parameters. In: Hussain CM, Kadeppagari RK, Eds. Biotechnology for Zero Waste: Emerging Waste Management Techniques. 2022; pp. 189-205.
[http://dx.doi.org/10.1002/9783527832064.ch13]

[243] Olubode TP, Amusat AI, Olawale BR, Adekola FF. Biosorption of heavy metals using bacterial isolates from e-waste soil. Afr J Microbiol Res 2022; 16(7): 268-72.

[244] Rahman Z, Thomas L, Singh VP. Biosorption of heavy metals by a lead (Pb) resistant bacterium, *Staphylococcus hominis* strain AMB 2. J Basic Microbiol 2019; 59(5): 477-86.
[http://dx.doi.org/10.1002/jobm.201900024] [PMID: 30900761]

[245] Aryal M. A comprehensive study on the bacterial biosorption of heavy metals: materials, performances, mechanisms, and mathematical modellings. Rev Chem Eng 2021; 37(6): 715-54.
[http://dx.doi.org/10.1515/revce-2019-0016]

[246] Mohapatra RK, Parhi PK, Pandey S, Bindhani BK, Thatoi H, Panda CR. Active and passive biosorption of Pb(II)using live and dead biomass of marine bacterium *Bacillus xiamenensis* PbRPSD202: Kinetics and isotherm studies. J Environ Manage 2019; 247: 121-34.
[http://dx.doi.org/10.1016/j.jenvman.2019.06.073] [PMID: 31238200]

[247] Peng H, Li D, Ye J, *et al.* Biosorption behavior of the *Ochrobactrum* MT180101 on ionic copper and chelate copper. J Environ Manage 2019; 235: 224-30.
[http://dx.doi.org/10.1016/j.jenvman.2019.01.060] [PMID: 30682675]

[248] Rizvi A, Ahmed B, Zaidi A, Khan MS. Biosorption of heavy metals by dry biomass of metal tolerant bacterial biosorbents: an efficient metal clean-up strategy. Environ Monit Assess 2020; 192(12): 801.
[http://dx.doi.org/10.1007/s10661-020-08758-5] [PMID: 33263175]

[249] Kashyap S, Chandra R, Kumar B, Verma P. Biosorption efficiency of nickel by various endophytic bacterial strains for removal of nickel from electroplating industry effluents: an operational study. Ecotoxicology 2022; 31(4): 565-80.
[http://dx.doi.org/10.1007/s10646-021-02445-y] [PMID: 34184169]

[250] Velásquez L, Dussan J. Biosorption and bioaccumulation of heavy metals on dead and living biomass of *Bacillus sphaericus*. J Hazard Mater 2009; 167(1-3): 713-6.
[http://dx.doi.org/10.1016/j.jhazmat.2009.01.044] [PMID: 19201532]

[251] Diep P, Mahadevan R, Yakunin AF. Heavy metal removal by bioaccumulation using genetically engineered microorganisms. Front Bioeng Biotechnol 2018; 6: 157.
[http://dx.doi.org/10.3389/fbioe.2018.00157] [PMID: 30420950]

[252] Ahemad M, Malik A. Bioaccumulation of heavy metals by zinc resistant bacteria isolated from agricultural soils irrigated with wastewater. Bacteriology Journal 2011; 2(1): 12-21.
[http://dx.doi.org/10.3923/bj.2012.12.21]

[253] Akhter K, Ghous T, Andleeb S, *et al.* Bioaccumulation of heavy metals by metal-resistant bacteria isolated from *Tagetes minuta* rhizosphere, growing in soil adjoining automobile workshops. Pak J Zool 2017; 49(5): 1841-6.
[http://dx.doi.org/10.17582/journal.pjz/2017.49.5.1841.1846]

[254] Mello IS, Targanski S, Pietro-Souza W, Frutuoso Stachack FF, Terezo AJ, Soares MA. Endophytic bacteria stimulate mercury phytoremediation by modulating its bioaccumulation and volatilization. Ecotoxicol Environ Saf 2020; 202: 110818.

[http://dx.doi.org/10.1016/j.ecoenv.2020.110818] [PMID: 32590206]

[255]   Ustiatik R, Nuraini Y, Suharjono S, Jeyakumar P, Anderson CWN, Handayanto E. Endophytic bacteria promote biomass production and mercury-bioaccumulation of Bermuda grass and Indian goosegrass. Int J Phytoremediation 2022; 24(11): 1184-92.
[http://dx.doi.org/10.1080/15226514.2021.2023461] [PMID: 34986046]

[256]   Zhu N, Zhang B, Yu Q. Genetic engineering-facilitated co assembly of synthetic bacterial cells and magnetic nanoparticles for efficient heavy metal removal. ACS Appl Mater Interfaces 2020; 12(20): 22948-57.
[http://dx.doi.org/10.1021/acsami.0c04512] [PMID: 32338492]

[257]   Roy JJ, Cao B, Madhavi S. A review on the recycling of spent lithium-ion batteries (LIBs) by the bioleaching approach. Chemosphere 2021; 282(130944): 130944.
[http://dx.doi.org/10.1016/j.chemosphere.2021.130944] [PMID: 34087562]

[258]   Gu T, Rastegar SO, Mousavi SM, Li M, Zhou M. Advances in bioleaching for recovery of metals and bioremediation of fuel ash and sewage sludge. Bioresour Technol 2018; 261: 428-40.
[http://dx.doi.org/10.1016/j.biortech.2018.04.033] [PMID: 29703427]

[259]   Li D, Li R, Ding Z, *et al.* Discovery of a novel native bacterium of *Providencia* sp. with high biosorption and oxidation ability of manganese for bioleaching of heavy metal contaminated soils. Chemosphere 2020; 241: 125039.
[http://dx.doi.org/10.1016/j.chemosphere.2019.125039] [PMID: 31606568]

[260]   Pirsaheb M, Zadsar S, Rastegar SO, Gu T, Hossini H. Bioleaching and ecological toxicity assessment of carbide slag waste using *Acidithiobacillus* bacteria. Environ Technol Innov 2021; 22: 101480.
[http://dx.doi.org/10.1016/j.eti.2021.101480]

[261]   Wang Z, Ni G, Xia J, *et al.* Bioleaching of toxic metals from anaerobically digested sludge without external chemical addition. Water Res 2021; 200(117211): 117211.
[http://dx.doi.org/10.1016/j.watres.2021.117211] [PMID: 34022632]

[262]   Abraham J, Chatterjee A, Sharma J. Isolation and characterization of *Bacillus licheniformis* strain for bioleaching of heavy metals. J App Biotechnol Rep 2020; 7(3): 139-44.

[263]   Liu S, Zhang F, Chen J, Sun G. Arsenic removal from contaminated soil *via* biovolatilization by genetically engineered bacteria under laboratory conditions. J Environ Sci (China) 2011; 23(9): 1544-50.
[http://dx.doi.org/10.1016/S1001-0742(10)60570-0] [PMID: 22432292]

[264]   Huang K, Chen C, Zhang J, *et al.* Efficient arsenic methylation and volatilization mediated by a novel bacterium from an arsenic-contaminated paddy soil. Environ Sci Technol 2016; 50(12): 6389-96.
[http://dx.doi.org/10.1021/acs.est.6b01974] [PMID: 27258163]

[265]   Marwa N, Singh N, Srivastava S, Saxena G, Pandey V, Singh N. Characterizing the hypertolerance potential of two indigenous bacterial strains (*Bacillus flexus* and *Acinetobacter junii*) and their efficacy in arsenic bioremediation. J Appl Microbiol 2019; 126(4): 1117-27.
[http://dx.doi.org/10.1111/jam.14179] [PMID: 30556924]

[266]   Sahinkaya E, Uçar D, Kaksonen AH. Bioprecipitation of metals and metalloids. In: Rene E, Sahinkaya E, Lewis A, Lens P, Eds. Sustainable heavy metal remediation. Cham: Springer 2017; pp. 199-231.
[http://dx.doi.org/10.1007/978-3-319-58622-9_7]

[267]   Choudhury S, Chatterjee A. Microbial application in remediation of heavy metals: an overview. Arch Microbiol 2022; 204(5): 268.
[http://dx.doi.org/10.1007/s00203-022-02874-1] [PMID: 35438381]

[268]   Han H, Cai H, Wang X, Hu X, Chen Z, Yao L. Heavy metal-immobilizing bacteria increase the biomass and reduce the Cd and Pb uptake by pakchoi (*Brassica chinensis* L.) in heavy metal-contaminated soil. Ecotoxicol Environ Saf 2020; 195(110375): 110375.
[http://dx.doi.org/10.1016/j.ecoenv.2020.110375] [PMID: 32200142]

[269] Li F, Zheng Y, Tian J, *et al.* Cupriavidus sp. strain Cd02-mediated pH increase favoring bioprecipitation of Cd^{2+} in medium and reduction of cadmium bioavailability in paddy soil. Ecotoxicol Environ Saf 2019; 184: 109655.
[http://dx.doi.org/10.1016/j.ecoenv.2019.109655] [PMID: 31525561]

[270] Zhu X, Lv B, Shang X, Wang J, Li M, Yu X. The immobilization effects on Pb, Cd and Cu by the inoculation of organic phosphorus-degrading bacteria (OPDB) with rapeseed dregs in acidic soil. Geoderma 2019; 350: 1-10.
[http://dx.doi.org/10.1016/j.geoderma.2019.04.015]

[271] Jalilvand N, Akhgar A, Alikhani HA, Rahmani HA, Rejali F. Removal of heavy metals zinc, lead, and cadmium by biomineralization of urease-producing bacteria isolated from Iranian mine calcareous soils. J Soil Sci Plant Nutr 2020; 20(1): 206-19.
[http://dx.doi.org/10.1007/s42729-019-00121-z]

[272] Iravani S, Varma RS. Bacteria in heavy metal remediation and nanoparticle biosynthesis. ACS Sustain Chem& Eng 2020; 8(14): 5395-409.
[http://dx.doi.org/10.1021/acssuschemeng.0c00292]

[273] Verma S, Bhatt P, Verma A, Mudila H, Prasher P, Rene ER. Microbial technologies for heavy metal remediation: effect of process conditions and current practices. Clean Technol Environ Policy 2021; 25(38): 1-23.

[274] Ali J, Ali N, Wang L, Waseem H, Pan G. Revisiting the mechanistic pathways for bacterial mediated synthesis of noble metal nanoparticles. J Microbiol Methods 2019; 159: 18-25.
[http://dx.doi.org/10.1016/j.mimet.2019.02.010] [PMID: 30797020]

[275] Wu M, Li Y, Li J, Wang Y, Xu H, Zhao Y. Bioreduction of hexavalent chromium using a novel strain CRB-7 immobilized on multiple materials. J Hazard Mater 2019; 368: 412-20.
[http://dx.doi.org/10.1016/j.jhazmat.2019.01.059] [PMID: 30703702]

[276] Frankenberger WT Jr, Arshad M. Bioremediation of selenium-contaminated sediments and water. Biofactors 2001; 14(1-4): 241-54.
[http://dx.doi.org/10.1002/biof.5520140130] [PMID: 11568461]

[277] Tan Y, Wang Y, Wang Y, *et al.* Novel mechanisms of selenate and selenite reduction in the obligate aerobic bacterium *Comamonas testosteroni* S44. J Hazard Mater 2018; 359: 129-38.
[http://dx.doi.org/10.1016/j.jhazmat.2018.07.014] [PMID: 30014908]

[278] He Y, Gong Y, Su Y, Zhang Y, Zhou X. Bioremediation of Cr (VI) contaminated groundwater by *Geobacter sulfurreducens*: Environmental factors and electron transfer flow studies. Chemosphere 2019; 221: 793-801.
[http://dx.doi.org/10.1016/j.chemosphere.2019.01.039] [PMID: 30684777]

[279] Sarubbo LA, Rocha RB Jr, Luna JM, Rufino RD, Santos VA, Banat IM. Some aspects of heavy metals contamination remediation and role of biosurfactants. Chem Ecol 2015; 31(8): 707-23.
[http://dx.doi.org/10.1080/02757540.2015.1095293]

[280] Das AJ, Lal S, Kumar R, Verma C. Bacterial biosurfactants can be an ecofriendly and advanced technology for remediation of heavy metals and co-contaminated soil. Int J Environ Sci Technol 2017; 14(6): 1343-54.
[http://dx.doi.org/10.1007/s13762-016-1183-0]

[281] da Rocha Junior RB, Meira HM, Almeida DG, *et al.* Application of a low-cost biosurfactant in heavy metal remediation processes. Biodegradation 2019; 30(4): 215-33.
[http://dx.doi.org/10.1007/s10532-018-9833-1] [PMID: 29725781]

[282] Wang S, Mulligan CN. Rhamnolipid biosurfactant-enhanced soil flushing for the removal of arsenic and heavy metals from mine tailings. Process Biochem 2009; 44(3): 296-301.
[http://dx.doi.org/10.1016/j.procbio.2008.11.006]

[283] Yang Z, Zhang Z, Chai L, Wang Y, Liu Y, Xiao R. Bioleaching remediation of heavy metal-

contaminated soils using *Burkholderia* sp. Z-90. J Hazard Mater 2016; 301: 145-52.
[http://dx.doi.org/10.1016/j.jhazmat.2015.08.047] [PMID: 26348147]

[284]   Jimoh AA, Lin J. Biosurfactant: A new frontier for greener technology and environmental sustainability. Ecotoxicol Environ Saf 2019; 184: 109607.
[http://dx.doi.org/10.1016/j.ecoenv.2019.109607] [PMID: 31505408]

[285]   Orozco-Mosqueda MC, Flores A, Rojas-Sánchez B, *et al.* Plant growth-promoting bacteria as bioinoculants: Attributes and challenges for sustainable crop improvement. Agronomy (Basel) 2021; 11(6): 1167.
[http://dx.doi.org/10.3390/agronomy11061167]

[286]   Oleńska E, Małek W, Wójcik M, Swiecicka I, Thijs S, Vangronsveld J. Beneficial features of plant growth-promoting rhizobacteria for improving plant growth and health in challenging conditions: A methodical review. Sci Total Environ 2020; 743: 140682.
[http://dx.doi.org/10.1016/j.scitotenv.2020.140682] [PMID: 32758827]

[287]   dos Santos RM, Diaz PAE, Lobo LLB, Rigobelo EC. Use of plant growth-promoting rhizobacteria in maize and sugarcane: Characteristics and applications. Front Sustain Food Syst 2020; 4: 136.
[http://dx.doi.org/10.3389/fsufs.2020.00136]

[288]   Chandran H, Meena M, Swapnil P. Plant growth-promoting rhizobacteria as a green alternative for sustainable agriculture. Sustainability (Basel) 2021; 13(19): 10986.
[http://dx.doi.org/10.3390/su131910986]

[289]   Rizvi A, Khan MS. Heavy metal induced oxidative damage and root morphology alterations of maize (*Zea mays* L.) plants and stress mitigation by metal tolerant nitrogen fixing *Azotobacter chroococcum.* Ecotoxicol Environ Saf 2018; 157: 9-20.
[http://dx.doi.org/10.1016/j.ecoenv.2018.03.063] [PMID: 29605647]

[290]   Ullah A, Heng S, Munis MFH, Fahad S, Yang X. Phytoremediation of heavy metals assisted by plant growth promoting (PGP) bacteria: A review. Environ Exp Bot 2015; 117: 28-40.
[http://dx.doi.org/10.1016/j.envexpbot.2015.05.001]

[291]   Jinal HN, Gopi K, Prittesh P, Kartik VP, Amaresan N. Phytoextraction of iron from contaminated soils by inoculation of iron-tolerant plant growth-promoting bacteria in *Brassica juncea* L. Czern. Environ Sci Pollut Res Int 2019; 26(32): 32815-23.
[http://dx.doi.org/10.1007/s11356-019-06394-2] [PMID: 31502049]

[292]   Billah M, Khan M, Bano A, Hassan TU, Munir A, Gurmani AR. Phosphorus and phosphate solubilizing bacteria: Keys for sustainable agriculture. Geomicrobiol J 2019; 36(10): 904-16.
[http://dx.doi.org/10.1080/01490451.2019.1654043]

[293]   Sayantan D, Das SS. Phosphorus: A boon or curse for the environment? In: Churchill DG, Sikirić MD, Čolović B, Milhofer HF, Eds. Contemporary Topics about Phosphorus in Biology and Materials. London: IntechOpen 2020.
[http://dx.doi.org/10.5772/intechopen.91250]

[294]   Bargaz A, Elhaissoufi W, Khourchi S, Benmrid B, Borden KA, Rchiad Z. Benefits of phosphate solubilizing bacteria on belowground crop performance for improved crop acquisition of phosphorus. Microbiol Res 2021; 252(126842): 126842.
[http://dx.doi.org/10.1016/j.micres.2021.126842] [PMID: 34438221]

[295]   Efe D. Potential plant growth-promoting bacteria with heavy metal resistance. Curr Microbiol 2020; 77(12): 3861-8.
[http://dx.doi.org/10.1007/s00284-020-02208-8] [PMID: 32960302]

[296]   Kour R, Jain D, Bhojiya AA, *et al.* Zinc biosorption, biochemical and molecular characterization of plant growth-promoting zinc-tolerant bacteria. 3 Biotech 2019; 9(11): 421.
[http://dx.doi.org/10.1007/s13205-019-1959-2] [PMID: 31696026]

[297]   Sun F, Ou Q, Wang N, *et al.* Isolation and identification of potassium-solubilizing bacteria from

Mikania micrantha rhizospheric soil and their effect on *M. micrantha* plants. Glob Ecol Conserv 2020; 23: e01141.
[http://dx.doi.org/10.1016/j.gecco.2020.e01141]

[298] Rehman FU, Kalsoom M, Adnan M, Toor M, Zulfiqar A. Plant growth promoting rhizobacteria and their mechanisms involved in agricultural crop production: A review. SunText Rev Biotechnol 2020; 1(2): 1-6.

[299] Korver RA, Koevoets IT, Testerink C. Out of shape during stress: A key role for auxin. Trends Plant Sci 2018; 23(9): 783-93.
[http://dx.doi.org/10.1016/j.tplants.2018.05.011] [PMID: 29914722]

[300] Rolón-Cárdenas GA, Arvizu-Gómez JL, Soria-Guerra RE, Pacheco-Aguilar JR, Alatorre-Cobos F, Hernández-Morales A. The role of auxins and auxin-producing bacteria in the tolerance and accumulation of cadmium by plants. Environ Geochem Health 2022; 44(11): 3743-64.
[http://dx.doi.org/10.1007/s10653-021-01179-4] [PMID: 35022877]

[301] Kang SM, Asaf S, Khan AL, *et al.* Complete genome sequence of *Pseudomonas psychrotolerans* CS51, a plant growth-promoting bacterium, under heavy metal stress conditions. Microorganisms 2020; 8(3): 382.
[http://dx.doi.org/10.3390/microorganisms8030382] [PMID: 32182882]

[302] Rajkumar M, Ae N, Freitas H. Endophytic bacteria and their potential to enhance heavy metal phytoextraction. Chemosphere 2009; 77(2): 153-60.
[http://dx.doi.org/10.1016/j.chemosphere.2009.06.047] [PMID: 19647283]

[303] Egamberdieva D, Wirth SJ, Alqarawi AA, Abd Allah EF, Hashem A. Phytohormones and beneficial microbes: Essential components for plants to balance stress and fitness. Front Microbiol 2017; 8: 2104.
[http://dx.doi.org/10.3389/fmicb.2017.02104] [PMID: 29163398]

[304] Singh SK, Singh PP, Gupta A, Singh AK, Keshri J. 2019.

[305] Zhang Y, He L, Chen Z, Wang Q, Qian M, Sheng X. Characterization of ACC deaminase-producing endophytic bacteria isolated from copper-tolerant plants and their potential in promoting the growth and copper accumulation of *Brassica napus*. Chemosphere 2011; 83(1): 57-62.
[http://dx.doi.org/10.1016/j.chemosphere.2011.01.041] [PMID: 21315404]

[306] Zainab N, Amna , Din BU, *et al.* Deciphering metal toxicity responses of flax (*Linum usitatissimum* L.) with exopolysaccharide and ACC-deaminase producing bacteria in industrially contaminated soils. Plant Physiol Biochem 2020; 152: 90-9.
[http://dx.doi.org/10.1016/j.plaphy.2020.04.039] [PMID: 32408178]

[307] Rosier A, Medeiros FHV, Bais HP. Defining plant growth promoting rhizobacteria molecular and biochemical networks in beneficial plant-microbe interactions. Plant Soil 2018; 428(1-2): 35-55.
[http://dx.doi.org/10.1007/s11104-018-3679-5] [PMID: 30078912]

[308] Beneduzi A, Ambrosini A, Passaglia LMP. Plant growth-promoting rhizobacteria (PGPR): their potential as antagonists and biocontrol agents. Genet Mol Biol 2012; 35(4 suppl 1) (Suppl. 1): 1044-51.
[http://dx.doi.org/10.1590/S1415-47572012000600020] [PMID: 23411488]

[309] Ulloa-Ogaz AL, Muñoz-Castellanos LN, Nevárez-Moorillón GV. Biocontrol of phytopathogens: antibiotic production as mechanism of control. In: Méndez-Vilas A, Ed. The Battle against microbial pathogens: Basic Science. Technological Advances and Educational Programs 2015; pp. 305-9.

[310] Di Cesare A, Eckert E, Corno G. Co-selection of antibiotic and heavy metal resistance in freshwater bacteria. J Limnol 2016; 75(s2.)
[http://dx.doi.org/10.4081/jlimnol.2016.1198]

[311] Rajkumar M, Ae N, Prasad MNV, Freitas H. Potential of siderophore-producing bacteria for improving heavy metal phytoextraction. Trends Biotechnol 2010; 28(3): 142-9.
[http://dx.doi.org/10.1016/j.tibtech.2009.12.002] [PMID: 20044160]

[312] Schalk IJ, Hannauer M, Braud A. New roles for bacterial siderophores in metal transport and tolerance. Environ Microbiol 2011; 13(11): 2844-54.
[http://dx.doi.org/10.1111/j.1462-2920.2011.02556.x] [PMID: 21883800]

[313] Aznar A, Dellagi A. New insights into the role of siderophores as triggers of plant immunity: what can we learn from animals? J Exp Bot 2015; 66(11): 3001-10.
[http://dx.doi.org/10.1093/jxb/erv155] [PMID: 25934986]

[314] Tripathi M, Munot HP, Shouche Y, Meyer JM, Goel R. Isolation and functional characterization of siderophore-producing lead- and cadmium-resistant *Pseudomonas putida* KNP9. Curr Microbiol 2005; 50(5): 233-7.
[http://dx.doi.org/10.1007/s00284-004-4459-4] [PMID: 15886913]

[315] Babu AG, Kim JD, Oh BT. Enhancement of heavy metal phytoremediation by *Alnus firma* with endophytic *Bacillus thuringiensis* GDB-1. J Hazard Mater 2013; 250-251: 477-83.
[http://dx.doi.org/10.1016/j.jhazmat.2013.02.014] [PMID: 23500429]

[316] Ndeddy Aka RJ, Babalola OO. Effect of bacterial inoculation of strains of *pseudomonas aeruginosa, alcaligenes feacalis* and *bacillus subtilis* on germination, growth and heavy metal (cd, cr, and ni) uptake of *brassica juncea*. Int J Phytoremediation 2016; 18(2): 200-9.
[http://dx.doi.org/10.1080/15226514.2015.1073671] [PMID: 26503637]

[317] Singh JS, Abhilash PC, Singh HB, Singh RP, Singh DP. Genetically engineered bacteria: An emerging tool for environmental remediation and future research perspectives. Gene 2011; 480(1-2): 1-9.
[http://dx.doi.org/10.1016/j.gene.2011.03.001] [PMID: 21402131]

[318] Darham S, Syed-Muhaimin SN, Subramaniam K, *et al.* Optimisation of various physicochemical variables affecting molybdenum bioremediation using Antarctic bacterium, Arthrobacter sp. Strain AQ5-05. Water 2021; 13(17): 2367.
[http://dx.doi.org/10.3390/w13172367]

[319] Arias D, Cisternas L, Rivas M. Biomineralization mediated by ureolytic bacteria applied to water treatment: A review. Crystals (Basel) 2017; 7(11): 345.
[http://dx.doi.org/10.3390/cryst7110345]

[320] Das S, Dash HR, Chakraborty J. Genetic basis and importance of metal resistant genes in bacteria for bioremediation of contaminated environments with toxic metal pollutants. Appl Microbiol Biotechnol 2016; 100(7): 2967-84.
[http://dx.doi.org/10.1007/s00253-016-7364-4] [PMID: 26860944]

[321] Agency for Toxic Substances and Disease Registry [ATSDR]. Priority list of hazardous substances. (2015). Available online at: http://www.atsdr.cdc.gov/spl/index.html#modalIdString_myTable2015

[322] Goswami D, Thakker JN, Dhandhukia PC. Portraying mechanics of plant growth promoting rhizobacteria (PGPR): A review. Cogent Food Agric 2016; 2(1): 1127500.
[http://dx.doi.org/10.1080/23311932.2015.1127500]

[323] El-Meihy RM, Abou-Aly HE, Youssef AM, Tewfike TA, El-Alkshar EA. Efficiency of heavy metals-tolerant plant growth promoting bacteria for alleviating heavy metals toxicity on sorghum. Environ Exp Bot 2019; 162: 295-301.
[http://dx.doi.org/10.1016/j.envexpbot.2019.03.005]

[324] Joshi S, Gangola S, Jaggi V, Sahgal M. Functional characterization and molecular fingerprinting of potential phosphate solubilizing bacterial candidates from Shisham rhizosphere. Sci Rep 2023; 13(1): 7003.
[http://dx.doi.org/10.1038/s41598-023-33217-9] [PMID: 37117212]

[325] Gupta R, Khan F, Alqahtani FM, Hashem M, Ahmad F. Plant Growth–Promoting Rhizobacteria (PGPR) Assisted Bioremediation of Heavy Metal Toxicity. Appl Biochem Biotechnol 2024; 196(5): 2928-56.
[http://dx.doi.org/10.1007/s12010-023-04545-3] [PMID: 37097400]

[326] Patil A, Chakraborty S, Yadav Y, Sharma B, Singh S, Arya M. Bioremediation strategies and mechanisms of bacteria for resistance against heavy metals: a review. Bioremediat J 2024; 1-33.
[http://dx.doi.org/10.1080/10889868.2024.2375204]

[327] Azhar U, Ahmad H, Shafqat H, *et al.* Remediation techniques for elimination of heavy metal pollutants from soil: A review. Environ Res 2022; 214(Pt 4): 113918.
[http://dx.doi.org/10.1016/j.envres.2022.113918] [PMID: 35926577]

[328] Perez-Vazquez A, Barciela P, Prieto MA. In Situ and Ex Situ Bioremediation of Different Persistent Soil Pollutants as Agroecology Tool. Processes (Basel) 2024; 12(10): 2223.
[http://dx.doi.org/10.3390/pr12102223]

[329] Naseem M, Syab S, Akhtar S, *et al.* Ex-Situ and In-Situ bioremediation strategies and their limitations for Solid Waste Management: A Mini-Review. JQAAS 2023; 3(1): 28-31.
[http://dx.doi.org/10.38211/jqaas.2023.3.39]

[330] Tamás M, Sharma S, Ibstedt S, Jacobson T, Christen P. Heavy metals and metalloids as a cause for protein misfolding and aggregation. Biomolecules 2014; 4(1): 252-67.
[http://dx.doi.org/10.3390/biom4010252] [PMID: 24970215]

[331] D'Alessandro A, Taamalli M, Gevi F, Timperio AM, Zolla L, Ghnaya T. Cadmium stress responses in *Brassica juncea*: hints from proteomics and metabolomics. J Proteome Res 2013; 12(11): 4979-97.
[http://dx.doi.org/10.1021/pr400793e] [PMID: 24074147]

[332] Chrestensen CA, Starke DW, Mieyal JJ. Acute cadmium exposure inactivates thioltransferase (Glutaredoxin), inhibits intracellular reduction of protein-glutathionyl-mixed disulfides, and initiates apoptosis. J Biol Chem 2000; 275(34): 26556-65.
[http://dx.doi.org/10.1074/jbc.M004097200] [PMID: 10854441]

[333] Farid M, Ali S, Rizwan M, *et al.* Microwave irradiation and citric acid assisted seed germination and phytoextraction of nickel (Ni) by *Brassica napus* L.: morpho-physiological and biochemical alterations under Ni stress. Environ Sci Pollut Res Int 2017; 24(26): 21050-64.
[http://dx.doi.org/10.1007/s11356-017-9751-5] [PMID: 28726228]

[334] Alaraidh IA, Alsahli AA, Abdel Razik ES. Alteration of antioxidant gene expression in response to heavy metal stress in *Trigonella foenum-graecum* L. S Afr J Bot 2018; 115: 90-3.
[http://dx.doi.org/10.1016/j.sajb.2018.01.012]

[335] Thomine S, Wang R, Ward JM, Crawford NM, Schroeder JI. Cadmium and iron transport by members of a plant metal transporter family in *Arabidopsis* with homology to *Nramp* genes. Proc Natl Acad Sci USA 2000; 97(9): 4991-6.
[http://dx.doi.org/10.1073/pnas.97.9.4991] [PMID: 10781110]

[336] Singh S, Parihar P, Singh R, Singh VP, Prasad SM. Heavy metal tolerance in plants: role of transcriptomics, proteomics, metabolomics, and ionomics. Front Plant Sci 2016; 6: 1143.
[http://dx.doi.org/10.3389/fpls.2015.01143] [PMID: 26904030]

[337] Colangelo EP, Guerinot ML. Put the metal to the petal: metal uptake and transport throughout plants. Curr Opin Plant Biol 2006; 9(3): 322-30.
[http://dx.doi.org/10.1016/j.pbi.2006.03.015] [PMID: 16616607]

[338] Ahmad A, Mushtaq Z, Nazir A, *et al.* Growth response of cowpea (*Vigna unguiculata* L.) exposed to *Pseudomonas fluorescens, Pseudomonas stutzeri*, and *Pseudomonas gessardii* in lead contaminated soil. Plant Stress 2023; 10: 100259.
[http://dx.doi.org/10.1016/j.stress.2023.100259]

[339] Shi G, Yan Y, Yu Z, Zhang L, Cheng Y, Shi W. Modification-bioremediation of copper, lead, and cadmium-contaminated soil by combined ryegrass (*Lolium multiflorum* Lam.) and *Pseudomonas aeruginosa* treatment. Environ Sci Pollut Res Int 2020; 27(30): 37668-76.
[http://dx.doi.org/10.1007/s11356-020-09846-2] [PMID: 32608000]

[340] Khakimova L, Chubukova O, Vershinina Z, Maslennikova D. Effects of *Pseudomonas* sp. OBA 2.4.1

on Growth and Tolerance to Cadmium Stress in *Pisum sativum* L. BioTech (Basel) 2023; 12(1): 5.
[http://dx.doi.org/10.3390/biotech12010005] [PMID: 36648831]

[341] Bashir S, Javed S, Al-Anazi KM, Farah MA, Ali S. Bioremediation of cadmium toxicity in wheat (*Triticum aestivum* L.) plants primed with L-Proline, *Bacillus subtillis* and *Aspergillus niger*. Int J Environ Res Public Health 2022; 19(19): 12683.
[http://dx.doi.org/10.3390/ijerph191912683] [PMID: 36231984]

[342] Asadullah AB, Javed H. PGPR assisted bioremediation of heavy metals and nutrient accumulation in *Zea mays* under saline sodic soil. Pak J Bot 2021; 53(1): 31-8.

[343] Zainab N, Amna , Khan AA, *et al.* PGPR-mediated plant growth attributes and metal extraction ability of *Sesbania sesban* L. in industrially contaminated soils. Agronomy (Basel) 2021; 11(9): 1820.
[http://dx.doi.org/10.3390/agronomy11091820]

CHAPTER 7

Phytohormones and Heavy Metal Detoxification: Current Status and Future Perspectives

Harpreet Kaur[1,*], Madhu Chandel[1], Gurvarinder Kaur[2], Isha Madaan[2,4], Geetika Sirhindi[2] and Renu Bhardwaj[3]

[1] *P.G. Department of Botany, Khalsa College Amritsar, Amritsar, Punjab, India*

[2] *Department of Botany, Punjabi University, Patiala, Punjab, India*

[3] *Department of Botanical and Environmental Sciences, Guru Nanak Dev University, Amritsar, Punjab, India*

[4] *Government College of Education, Jalandhar, Punjab, India*

Abstract: Escalating application of heavy metals (HMTs) due to diverse anthropogenesis is a staid problem. In contaminated soils plants are severely affected by HMTs contamination. Under such conditions, endurance of plants becomes difficult and overall vigor is highly affected under HMTs stress. Detoxification of HMTs in metal tainted soil is done by various biochemical and chemical methods. Plant hormones are signal molecules that are synthesized by plants and the main group of phytohormones includes auxins, gibberellins cytokinins, abscisic acid, brassinosteroids and jasmonic acid. It is interesting to note that using plant hormones to lessen the harmful effects of HMTs has become increasingly significant in recent years. Plant hormones are signal molecules that are synthesized in different parts of plants, which have a major impact on plant metabolism and play vital role in abiotic stress mitigation. Though, phytohormone level is altered under abiotic stress conditions, which cause plant growth and developmental inhibitions. Metal stress has also been reported to hamper the concentrations of phytohormones in plant tissues. Phytohormones act as signaling entities for mitigation of HMTs stress in plants, thus let them to regain their original growth and plasticity. The present chapter mainly focuses on the potential roles of phytohormones in recreating the mitigation of heavy metal toxicity in the environment. We will focus on the study of accumulation, uptake and translocation of heavy metals in plants and give different pathways of how different phytohormones interact with plant protection systems under heavy metal toxicity. The function of numerous genes, enzymes, proteins, and signaling compounds associated with HMTs tolerance by phytohormones and their regulation will be explored. Thus, this this chapter will cover the different aspects of HMTs toxicity tolerance in plants with the use of plant growth regulators for sustainable cultivation in the future prospects.

* **Corresponding author Harpreet Kaur:** P.G. Department of Botany, Khalsa College Amritsar, Amritsar, Punjab, India; E-mails: sekhon.harpreet123@gmail.com, harpreetskaur@khalsacollege.edu.in

Keywords: Abiotic stress, Environment, Heavy metals (HMTs), Phytohormones, Toxicity.

INTRODUCTION

Heavy metal stress has substantial influence on metabolism, plant viability, and sometimes on the ability of the plants to resist various effects of pathogens such as low crop productivity. To increase productivity extensive use of chemical fertilizers and artificial irrigation have been employed extensively, which is a common practice all over the world. Sometimes, such practices cause accumulation of metal in soil to such an extent that the land cannot support growth of any agricultural crop any more. The need of the hour is to find some other solution thatis technically simple and correct so that crop productivity can be increased without harming the environment. In this way use of plant growth regulators is simple and technically right way to make plants tolerant of heavy metal stress without harming ecology and environment. Phytohormones or plant hormones are naturally occurring organic substances that influence physiological functions in plants at very low concentrations [1]. In other words, phytohormones are growth regulators that regulate various cellular activities of plants [2]. Then [3] reported that seven categories of phytohormones, that is, Auxins, Cytokinins (CK), Gibberellins (GA), Abscisic acid (ABA), Ethylene (ETH), Brassinosteroids (BR) and Jasmonates (JA) have been identified so far. The first five phytohormones are sometimes referred to as the "classical" phytohormones, while the latter two are more additions to the growing phytohormonal family [2]. Among several heavy metals, four heavy metals Arsenic (As), Lead (Pb), Cadmium (Cd), and Mercury (Hg) are recognized as the major lethal metals based on their toxicity, occurrence, and most significantly, their revelation to plants and animals. Minute concentration of phytohormones play significant role in modulating the growth and development of plants [4]. Also, their role has been acknowledged in rendering plant HMTs stress tolerant therefore, this eco-friendly approach of using phytohormones against HMTs has been gaining greater importance [5].

The study [6] avowed that in response to abiotic stresses, it triggers signal transduction pathways and exogenous application of phytohormones is ensured to upregulate stress resistance in plants facing HMTs. The advancements in molecular studies such as mutant screening, expression profiling, microarray, genomics and proteomics have helped in looking deeper into mechanisms and pathways that phytohormones follow to shield plants against HMTs stress [7]. Still very little information is available in the stress management roles of phytohormones towards HMTs signaling pathways. In the present scenario, our chapter mainly focuses on the role of phytohormones auxins, cytokinins, gibberellins, ethylene, abscisic acid, brassinosteroids, and jasmonates in HMTs stress mitigation in plants (Fig. **1**).

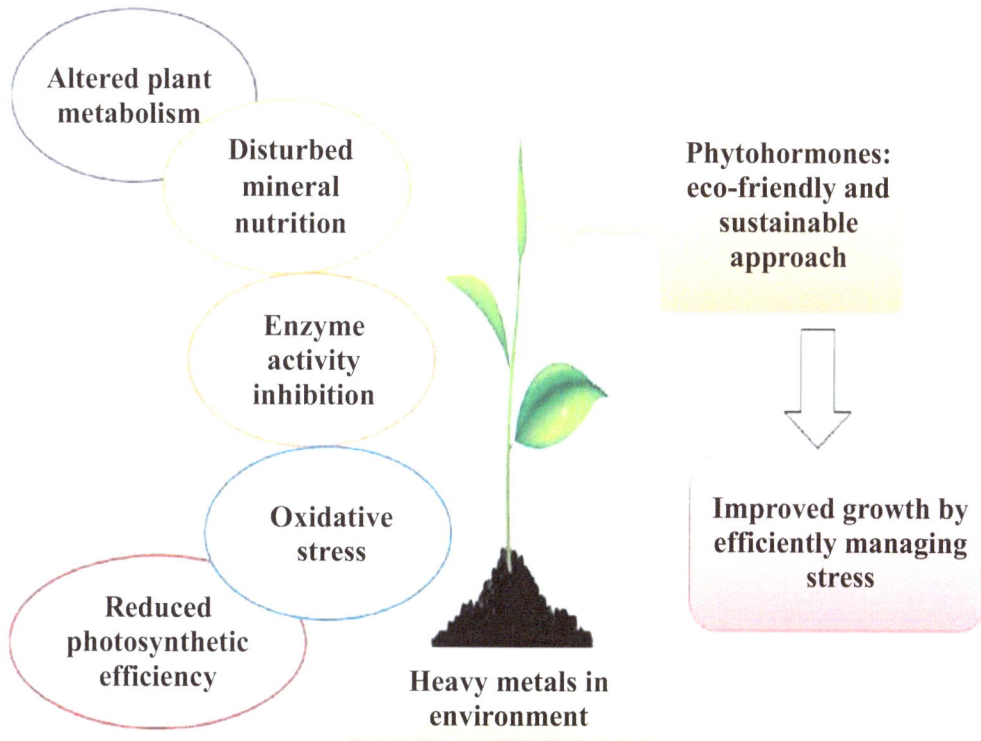

Fig. (1). Generalized mechanism of Heavy metals detoxification in plants by phytohormones.

Heavy Metals (HMTs) and their Toxicity in Plants and Environment

Arsenic (As) is a probably occurring metal which create staid healthiness to peoples across the world [8]. The contagion of As in underground water affects crop productivity and also accumulates in diverse plant tissues, grains and also contaminates food chain [9]. Different investigations have been carried out to explore the physiological and molecular means of As toxicity, accretion, detoxification, and tolerance in numerous plants including lettuce, rice, carrot and spinach [10]. **Lead(Pb)** is an abundantly distributed trace metal that exists in various forms in natural sources. It can affect plants and animals, by contaminations from leaded dust, fuels, old lead plumbing pipes, various industrial sites, or even old orchard sites in production where lead arsenate is used [11]. Pb^{2+} is non-biodegradable and its exposure is toxic to both flora and fauna [12]. **Cadmium (Cd)** is known as a heavy metal with phytotoxic nature, It is absorbed by the plants as it is soluble in water that is the main pathway for entry into food chain causing serious human health peril [13]. Cd is flown into the soil from manufacturing processes and agricultural practices [14]. Even at small concentrations Cd can alter enzyme activities including those involved in the C3 cycle, metabolism of phosphorus and carbohydrates [15, 16] eventually resultant in growth inhibition, loss of chlorophyll, leaf epinasty, alterations in chloroplast ultrastructure, inhibition of viability and pollen tube growth, lipid peroxidation induction, disruption of antioxidant machinery [16]. **Mercury (Hg)** is a natural component of Earth crust that gathered in water and land due to diverse human activities such as mining and industrial activities [17, 18]. Then further [19] avowed that increased level of Hg^{2+} can accumulate in higher plants. At very minute concentrations Hg^{2+} may not significantly affect plant growth but higher doses of mercury highly toxic to plant cells and can cause morphological and physiological abnormalities [20]. Binding of mercury to water channel proteins leads closure of stomata [20]. Lead (Pb) damage a variety of biological processes in plants including seed germination, shoot length, root length, transpiration, translation, chlorophyll metabolism, and cell division [21, 22]. It also surpasses the cell membrane permeability by reaction with different metabolic enzymes and phosphates such as ADP or ATP [22]. Pb toxicity leads to increase in MDA content and DNA denaturation by over production of reactive oxygen species (Table **1**).

Table 1. Effect of Heavy metal Stress (HMTs) on different plant species.

Plant Species	Heavy Metal Stress (HMTs)	Parameters Studied	References
Arabidopsis thaliana	Arsenite, As(III)	Auxin transporter mutants aux1, pin1 and pin2 were significantly more sensitive to As(III) indole-3- acetic acid improved As(III) tolerance of aux1	[23]
Rice	Cd	Mitogen-activated protein kinase gene and MBP kinases	[24]
Vigna radiate	Ni	Growth, nodulation, antioxidants and osmolytes	[25]
Rice	Zn	QTL mapping and gene analysis	[26]
Medicago sativa L.	Hg	Lipid peroxidation	[27]
Camellia sinensis L.	Pb	Growth parameters Root, Stem and Leaves	[28]
Vitis vinifera L.	Cu, Pb, Zn	Vegetation uptake, accumulation by roots	[29]
Maize	Cu	Leaf membrane structure and root activity	[30]
Wheat cultivars	As, Cd, and Pb	root > leaf > stem > grain	[31]

(Table 1) cont.....

Plant Species	Heavy Metal Stress (HMTs)	Parameters Studied	References
Pea Plants	Cd	Photosynthesis rate, chlorophyll content of leaves, catalase, SOD, guaiacol peroxidase activities, glutathione reductase Activity	[32]
Citrus	Cd	Leaf damage, gas exchange parameters, and hormonal Contents	[33]
Wheat	Cd	nitric oxide production	[34]
Aeluropus littoralis	Cu, Ni, Zn	superoxide dismutase (SOD), guaiacol peroxidase (GPx) and catalase (CAT)), accumulation of proline etc	[35]
Essential Food Crops	Cu	growth and yield of essential food crops	[36]
Wheat and safflower	Cd, Cu, Pb and Zn	Metal mobility and leaching in soil	[37]
Rice	Cd	seed germination, seedling growth and seed amylase activities	[38]
Medicago sativa L.	Cd^{-2}, Cr^{-6}, Cu^{-2}, Ni^{+2}, and Zn^{-2}	Seed germination and Plant Growth	[39]
Chlorella vulgaris	Cd, Cu and Pb	chlorophyll, monosaccharides, and protein content, Phytochelatins content	[40]

Auxin and Heavy Metal Detoxification

Auxin protects the plant from oxidative stress and increases the production of antioxidative enzymes. Auxin concentrations in roots are increased by various rhizospheric microorganisms and mitigate plants from HMTs. Exogenous application of auxins helped in mitigating As stress in *Brassica juncea* by modulating expression of microRNAs like miR167, miR319, and miR854 which were altered during As stress and led to toxicity symptoms [23] Similarly [41], demonstrated that exogenous application of IAA protected metabolic processes, such as nitrogen metabolism, in pea seedlings against Cr(VI) and Mn toxicity by regulating oxidative stress and the antioxidant defense system. Studies suggested that mutants of auxin transporters viz. *aux1*, *pin1*, and *pin2* were prone to toxic effects of As(III) when compared with wild type (WT), also this sensitivity increased when auxin inhibitors were used which was due to enhanced levels of oxidative stress markers like H_2O_2 [42] However, when auxin was applied exogenously in mutants *aux1*, it helped in mitigating oxidative stress by operating ROs signaling cascades thereby suggesting positive and ameliorative potential of auxin transport in HMT stress tolerance.

Exogenous application of auxin has known to impart HMT stress tolerance and this approach is also helpful in implementation of phytoremediation. Auxin, when applied exogenously helps in phytoextration of HMTs along with protecting them against their noxious effects on plant growth and development [43]. The role of improved phytoextraction of Pb in maize by IAA has also been reported [44]. An active auxins pool has also been shown to be involved in reducing the toxicity of heavy metals. In *Populus*, Glycoside hydrolase (GH) family 3 (GH3) enzymes conjugates and removes auxin from its active pool and a report suggested when *Populus* was exposed to Cd (50 µM), it enhanced the activities of GH3 which led to the deleterious impacts on photosynthesis as it triggered the activities of GH3 [45]. Nevertheless, the studies suggest significance of auxin transport and its active pool in imparting HMT stress tolerance in plants.

Gibberellins (GAs) and Heavy Metal Detoxification

Gibberellins (GAs) constitute family of tetracyclic diterpenoid plant hormones that modulate functions of growth including cell division, seed germination, hypocotyls elongation, leaf expansion, stem elongation and growth of reproductive structures, seed and bud dormancy along with controlling activities of essential hydrolytic enzymes in cereal grains [46]. Rich sources of GAs are young fruits, unripe seeds, apical root cells and young meristematic cauline tissue. GAs greatly impact phloem loading and also impacts activity of sucrose phosphate synthase and fructose-1,6- biphosphatase [47].

The role of GAs in ameliorating stress has greatly been acknowledged [48]. Transcriptome studies have revealed that GA application combats stressors like those revealed in *Arabidopsis* where GA regulates levels of ROS by activating *GAST1* gene which is a GA stimulated transcript [49]. demonstrated that GA imparts abiotic stress resistance to Arabidopsis by up-regulating the GAST1 gene (identified as a GA-stimulated transcript) which in turn modulates ROS accumulation. *OsAOP* gene in rice has been reported to express TIP (tonoplast intrinsic protein) as a response to GA application under stress conditions [50]. In *Arabidopsis thaliana*, Cd toxicity resulted in altered growth which was retrieved by exogenous application of GA (5 µM) by reducing lipid peroxidation [48]. Also, GA application down regulated *IRT1* gene, which is involved in Cd absorption. GA application helped in enhancing antioxidant defense system in pea seedlings under Cr stress and also enhanced nitrogen assimilation and growth capacity [51].

GA also ameliorated the toxic impacts of Ni (50 mM) on wheat seedlings. Ni stress led to decrease in growth by decreasing photosynthetic efficiency, reduced photosynthetic pigments and activity of enzymes like carbonic anhydrase, however, these noxious effects were overcome by GA application [52]. Another study in wheat was reported by [53] where GA application helped in ameliorating salt stress effects by shielding photosynthetic machinery and maintaining ionic homeostasis. GA (50 µM) helped in combating Cd stress in alleviated *Brassica napus* L. by scavenging free radicals as Cd stess led to oxidative stress and reduced germination [54]. It has been observed that Pb and Zn affected seed germination in *Cicer arietinum* cv. Aziziye-94, Pb and Zn led to hormonal imbalances that affected the seed germination, however, these noxious effects were reversed by GA application [55]. The same action of GA was reported by [56] in broad bean and lupin plants where GA protected the plants against Cd and Pb stress by activating antioxidant defense system (Table **2**).

Cytokinins and Heavy Metal Detoxification

Cytokinins, N6-substituted adenine derivatives, include kinetin and other substances displaying kinetin like activities and regulate wide array of physiological activities like cell division, flowering, nodulation, apical dominance, circadian rhythms and nutrient metabolism [81]. The role of cytokinins has also been reported under stressful environmental conditions. Under Mn toxicity conditions, 10 µM of kinetin helped to improve antioxidant capacity and ammonium assimilation in *Pisum sativum* seedlings [82]. Cd stress led to decrease in the germination percentage and coleoptiles elongation in barley. However, kinetin helped in decreasing the toxic symptoms posed by Cd [83]. INCYDE (2-chloro-6-(3-methoxyphenyl)aminopurine) is an inhibitor of degradation of cytokinin and INCYDE when applied to *Bulbine natalensis*, a medicinal plant, protected plan against Cd stress suggesting role of cytokinin in Cd stress tolerance [84]. Lower lipid peroxidation and enhanced expression of metallothionein gene (MT-*L2*) was suggested in cytokinin-synthesizing gene (*Ipt*) tobacco plants exposed to Cu stress [85]. The lipid peroxidation was low despite the Cu was accumulated in plants dur to enhanced levels of metallothioneins. However, several known components like cytokinin oxidase/dehydrogenases (CKXs), histidine kinase receptors and ATP/ADP-isopentenyltransferase (IPT) that are key regulators in cytokinin signaling are yet to be studied extensively in order to gain insights into cytokinin mediated HMT tolerance. Phytohormones mediate HMT tolerance in plants but still there lacks some points in interconnecting the signaling pathways of phytohormones and their crosstalk with metal binding complexes [48, 86].

Table 2. Effect of Different Phytohormones on various parameters under the influence of different HMTs stresses

Phytohormone	HeavyMetals (HMTs)	Parameters Analyzed	Reference
Auxin	Cd	Root elongation, shoot weight, hypocotyl length and Chlorophyll content in wild-type (WT) *Arabidopsis* seedlings.	[57]
	Pb and Zn	Shoot and root dry weight, root length, root volume and root surface area.	[43]
	As	Down-regulation of catalase-3 increased H2O2 levels that triggered various defense related pathways.	[42]
	Cd	Genes GH3.3, GH3.17, YDK1, MES7, MES17, CYP71A13, NIT2, AUX/IAA, Modulation of auxin biosynthesis through up-regulation of IAOX biosynthetic pathway genes while, down-regulation of genes involved in auxin conjugation, methylation and repression of auxin signaling alleviated Cd toxicity	[58]
	Cd	In Tobacco, YUCCA6, YUCCA8, YUCCA9, PIN1a, PIN1c, PIN4 up-regulation of auxin biosynthesis and transport genes by selenium resulted in increased auxin concentration and alleviation of HM stress	[59]
	Cd and Arsenate	In Rice, ASA2, YUCCA2, AUX1, PIN5b, Altered expression of auxin biosynthesis and transport genes negatively affected lateral root formation	[60]
	Cd	*Arabidopsis*, YUCCA6, Increased expression of YUCCA6 gene in response to Cd stress resulted in accumulation of auxin thus enhancing lateral and adventitious root density	[61]
Gibberellins	Cd and Pb	Growth, metal biosorption	[62]
	Cd and Mo	soluble sugar content, activities of enzymes, α- and β-amylase, amino acid content	[63]
	As	GA2ox3, GA2ox9, Up-regulated in response to stress that probably enhanced the GA levels	[64]
	Cd and Mo	α- and β-amylase, acid and alkaline phosphatase, Increased activity of different hydrolytic enzymes resulted in better seed germination	[63]
	Cr	In *Pisum sativum*, Nitrate reductase (NR), glutamine synthetase (GS), and glutamine 2-oxoglutarate aminotransferase (GOGAT) Decreased activity of enzymes involved in nitrogen assimilation altered the nitrogen levels	[5]

(Table 2) cont.....

Phytohormone	HeavyMetals (HMTs)	Parameters Analyzed	Reference
Cytokinin	Cd	Growth rate, chlorophyll (Chl) content, net photosynthesis (PN), content of soluble sugars and free amino acids	[65]
	Arsenate	*Arabidopsis*, Tobacco, Transgenic CK-deficient plants, Improved stress tolerance due to arsenic sequestration mediated by thiol compounds including phytochelatins and glutathione	[66]
INCYDE (Inhibitor of cytokinin degradation)	Zn	In, Tobacco, IPT, IPT gene transformed tobacco plants showed better photosynthesis and transpiration due to increased levels of cytokinins	[67]
Abscisic Acid	ABA	Over-expression of the NF-YC9 gene confers ABA hypersensitivity in both the early seedling growth and stomatal response, while down-regulation of NF-YC9 does not affect ABA response	[68]
Ethylene	Zn	In *Arabidopsis*, Ethylene signaling defective mutants ein2–1 and etr1–3 Enhanced activities of antioxidant enzymes such as SOD, CAT, APX, and POD	[69]
	Cd	In *Arabidopsis*, Ethylene biosynthesis double knockout acs2–1/6–1 and ethylene signaling mutant ein2–1 and ein 2–5 Controls ROS level upon Cd exposure by down-regulating RBOH expression in the mutants	[70]
	Hg	In *Arabidopsis*, Ethylene insensitive ein2–5 mutants Reduced ROS production and expression of ethylene-related genes required for HM stress amelioration	[18]
28-Homobrassinolide	Zn	Growth and antioxidant enzymes	[71]
	Cu	Antioxidant enzymes, lipid peroxidation	[72]
	Zn	Growth, lipid peroxidation & antioxidant enzymes	[73]
	Pb	Growth, carbohydrates and nucleic acids	[74]
	Cd and Hg	Glutathione-S-transferases, PPO activity	[75]
Jasmonic Acid	Cd	Osmolytes and antioxidants in faba bean	[76]

(Table 2) cont.....

Phytohormone	HeavyMetals (HMTs)	Parameters Analyzed	Reference
Salicylic Acid	Zn^{2+}, Cu^{2+}, Mn^{2+}, Cd^{2+}, Hg^{2+}, and Pb^{2+}	catalase and superoxide dismutase activity	[77]
	Pb	Growth parameters in Brassica napus	[78]
	Hg	Lipid peroxidation	[20]
	Pb	Plant height, number of tendrils, leaf length and leaf width.	[79]
	Cd	Seedling growth, antioxidant enzyme activities	[80]

Abscisic Acid (ABA) and Heavy Metal Detoxification

Abscisic acid (ABA) is an essential hormone that regulates functioning of plants not only under normal conditions but also under stressful environment [87]. It has been observed that on exposure to HMT condition the concentration of ABA increases to many folds suggesting role of ABA in HMT stress tolerance [88, 89]. ABA concentration increased when exposed to Cd stress conditions as observed in roots of *Typha latifolia* [90], in *Solanum tuberosum* [91] and also in *Oryza sativa* [92]. ABA levels were also increased during other conditions of HMT stress like Cu, Hg and Cd stress in *Triticum aestivum* during [93], in *Cicer arietinum* under Pb stress and Cu and Zn stress in cucumbers [94] and the author also identified involvement of two SnRK2 genes, PYL/PYR/RCAR and three PP2C which are essential components of signaling of ABA. Also, ABA biosynthesis genes like *OsNCED2* and *OsNCED3* were upregulated under As stress [64]. ABA mediates oxidative stress in potato tubers [95].

Ethylene and Heavy Metal Detoxification

Heavy metal toxicity can stimulate ethylene accumulation, which deliberate a potential mechanism between auxin and ethylene signaling under HMTs [96]. reported that *Arabidopsis,* biosynthesis of ethylene and ACC was induced under the stress conditions of Cd toxicity. The authors [97] reported that Pb increased the transcript levels of EIN2 genes in *Arabidopsis*. Ethylene insensitive etr1-1 and ein3-3 mutants were less sensitive to HMTs stress in *Arabidopsis* [98, 99]. Pb toxicity leads to enhanced transcript levels of EIN2 in *Arabidopsis* which is a significant component of ethylene signaling responsible for generating stress related responses. Also, it was suggested that EIN2 further cross talks with *AtPDR12* which is a membrane transporter that helps in excluding Pb which is an essential tolerance mechanism.

Brassinosteroids (BRs) and Heavy Metal Detoxification

Ever since the discovery of the plant steroids, Brassinosteroids (BRs) have gained enormous attention in helping plants to combat stress conditions. BRs regulate a wide array of functions in plant's development phases like seed germination, flowering, stomata formation, photomorphogenesis, cell expansion, male fertility, vascular development and differentiation and many more [100]. Out of the several discovered and known forms of BRs, the most active forms of BRs showing biostimulatory potential on exogenous application are Brassinolide (BL), 28-homobrassinolide (28-HomoBL) and 24-epibrassinolide (24-EpiBL) [101]. Along with developing physiological functioning under normal conditions, BRs also regulate the plant's growth and development under stress conditions [102].

BRs mediate the expression of several stress related genes which help in inculcating stress tolerance in plants [103]. Ni- stress led to lipid peroxidation in case of *B. juncea* and deteriorated growth, however the supplementation of 24-epiBL helped the plant to fight against oxidative stress by uplifting the antioxidants thereby combating Ni-stress [98]. Yusuf [104] reported alleviation of Ni- stress in *Triticum aestivum* where exogenous supplementation of 28- homoBL helped in enhancing the efficiency of enzymatic antioxidants. Similar results were reported in *Vigna radiata* where 24-epiBL supplementation proved to be beneficial for plants against Ni-Stress [105]. Mitigation of Ni stress has also been reported [98] in *B. juncea* where antioxidants were reported to be enhanced. Foliar application of 28-homoBL ameliorated Ni stress caused by variable concentrations of Ni (50 and 100 µM) in five wheat cultivars [104]. BR treated and Ni stressed *Vigna radiata* L. Wilczek plants reported enhanced nitrogen metabolism and antioxidant capacity [106].

Cadmium (Cd) is another HMT that poses deleterious effects on overall health of plants [107]. The application of different forms of BRs helped the plants to fight against Cd stress. 28- homoBL/24-epiBL application in *Solanum lycopersicum* helped the plant to mitigate Cd toxicity [108]. Also, HomoBL foliar spray mitigated Cd stress in *Brassica juncea* [109] and another report suggested similar results in *Cicer arietinum* [110]. Another study on tomatoes suggested improved fruit yield and antioxidants during Cd stress [111]. BR application enhances levels of osmolytes like proline in order to maintain osmotic potential in plant cells in *B. juncea* under Cd toxicity [109] and similar results were reported in *Phaseolus vulgaris* [112].

Copper (Cu) is an essential HMT that is required by plants at optimum levels and above which it poses deleterious impacts. BR supplementation in *Helianthus annus* exposed to Cu stress helped in protecting the reaction centers and maintained electron transport chains and O_2 evolving complexes in order to diminish effects of ROS [113]. Also BR treatment enhanced the antioxidants and diminished excess oxidative stress in *Raphanus sativus* and *Brassica juncea* as reported [114, 115]. Amelioration of NaCl+Cu stress has also been reported in *Cucumis sativus* [116]. BRs have also proved their efficiency under Hg stress in case of radish where ROS were effectively managed by BR treatment and the

enhancement in PPO and GST was observed [117]. Also, 24-epiBL ameliorated noxious effects of Hg stress in radish as reported by the study [114].

Lead (Pb) is another major contaminant that poses deleterious effects on plants. However, Br supplementation has been beneficial in mitigating harmful impacts of Pb stress. In Pb stressed *Trigonella foenum-graecum* L. plants, 28-homoBL helped in enhancing plant height, leaf area, root height and plant biomass [74]. Another report investigated effects of 28-homoBL in fenugreek exposed to Pb stress. Pb stress led to reduction in growth of fenugreek due to reduced cell division and cell elongation. However, when supplemented with 28-homoBL, fenugreek displayed better growth traits [118]. 28-homobrassinolide regulates antioxidant enzyme activities and gene expression in response to salt and temperature induced oxidative stress in *Brassica juncea* [119 - 122].

Jasmonic Acid and Heavy Metals Detoxification

Jasmonic Acid (JA) was first isolated as a plant growth inhibitor from the culture filterate of the fungus *Lasiodiplodia theobromae* by [123]. Changes in some photosynthetic and photorespiratory properties in barley leaves after treatment with Jasmonic acid were studied by [124]. Sembdner [125] studied the biochemistry and physiological and molecular actions of Jasmonates in plants. Jasmonate is essential for insect defense in *Arabidopsis* was studied by [126]. Overmyer [127], studied role of ozone sensitive *Arabidopsis* red1 mutant for ethylene and jasmonate signaling pathways in regulating superoxide-dependent cell death. Rao [128] studied Jasmonic acid signaling modulates ozone-induced hypersensitive cell death. Reymond [129] studied differential gene expression in response to mechanical wounding and insect feeding in *Arabidopsis*. Methyl Jasmonate alters polyamine metabolism and induces systematic protection against powdery mildew infection in barley seedlings studied by [130]. [131] studied inhibitory effect of Me-JA on flowering and elongation growth in *Pharbitis nil*. The level of JA in *Arabidopsis thaliana* and *Phaseolus coccineus* plants under heavy metal stress was studied by [132]. Researchers studied the toxic effect of low concentrations of Cu on nodulation of cowpea (*Vigna reticulata*) [133]. Assessment of antioxidation potential of selected plants with antisickling property was studied by [134]. Glycemic profile of *C. cajan* leaves in experimental rats was studied by [135]. Me-JA enhanced antioxidant activity and flavonoid content in blackberries (*Rubus* sp.) and promotes anti-proliferation of human cancer cells was studied by [136]. Ethanol modifies supercritical fluid extraction and antioxidant activity of pigeon pea (*C. cajan* (L) leaves was studied by [137]. Antioxidant activities of extracts and main components of pigeon pea (*C. cajan* L. Mill sp.) leaves was studied by [138]. Cobalt stress affects nitrogen metabolism, photosynthesis and antioxidant system in Chickpea (*Cicer arietinum* L.) was reported by [139]. The authors studied the effect of jasmonic acid on some biochemical and physiological parameters in salt stressed *Brassica napus* seedlings [140].

CONCLUSION AND FUTURE PROSPECTS

Heavy metal stress shave become a great concern for the researchers and agriculturists across the globe as they lead to deleterious impacts not only on plants but on all the biological organisms. It has become a threat to the plants and the economic and nutritional benefits that humans derive from them. Therefore, greater concern has aroused to protect plants with sustainable approach. Implication of phytohormones have shown a promising potential in protect plants against heavy metal stress without penalizing their growth. Exogenous application of phytohormones has helped plants to manage oxidative stress by modulating antioxidant defense machinery and modulating activity of several essential enzymes. They activate several cross-talk mechanisms and trigger signaling cascades to improve the growth of plants under cues of heavy metal stress. However, researchers still face lacunae in understanding several signaling cascades that trigger expression of stress responsive genes which can be a future prospect of research. Also, there is need to promote and explore eco-friendly approach of HMT stress alleviation using phytohormones to step towards sustainability.

REFERENCES

[1]   Davies PJ. The Plant Hormones: Their Nature, Occurrence, and Functions. In: Davies, P.J. (eds) Plant Hormones. Springer, Dordrecht. 2010: p. 1-15..
[http://dx.doi.org/10.1007/978-1-4020-2686-7_1]

[2]   Fleet C, Williams M. Gibberellins. Teaching tools in plant biology: lecture notes. Plant Cell 2011; 110. www.plantcell.org/cgi/doi/101105/tpc

[3]   Gomez-Roldan V, Fermas S, Brewer PB, *et al.* Strigolactone inhibition of shoot branching. Nature 2008; 455(7210): 189-94.
[http://dx.doi.org/10.1038/nature07271] [PMID: 18690209]

[4]   Jiang K, Asami T. Chemical regulators of plant hormones and their applications in basic research and agriculture. Biosci Biotechnol Biochem 2018; 82(8): 1265-300.
[http://dx.doi.org/10.1080/09168451.2018.1462693] [PMID: 29678122]

[5]   Piotrowska-Niczyporuk A, Bajguz A, Zambrzycka E, Godlewska-Żyłkiewicz B. Phytohormones as regulators of heavy metal biosorption and toxicity in green alga *Chlorella vulgaris* (Chlorophyceae).

Plant Physiol Biochem 2012; 52: 52-65.
[http://dx.doi.org/10.1016/j.plaphy.2011.11.009] [PMID: 22305067]

[6]   Rahman K, Rahman M, Ahmed N, *et al.* Morphophysiological changes and reactive oxygen species metabolism in *Corchorus olitorius* L. under different abiotic stresses. Open Agric 2021; 6(1): 549-62.
[http://dx.doi.org/10.1515/opag-2021-0040]

[7]   Singh S, Prasad SM. Growth, photosynthesis and oxidative responses of *Solanum melongena* L. seedlings to cadmium stress: Mechanism of toxicity amelioration by kinetin. Sci Hortic (Amsterdam) 2014; 176: 1-10.
[http://dx.doi.org/10.1016/j.scienta.2014.06.022]

[8]   Kumari B, Kumar V, Sinha AK, *et al.* Toxicology of arsenic in fish and aquatic systems. Environ Chem Lett 2017; 15(1): 43-64.
[http://dx.doi.org/10.1007/s10311-016-0588-9]

[9]   Mahajan L, Verma PK, Raina R, Sood S. Toxic effects of imidacloprid combined with arsenic: Oxidative stress in rat liver. Toxicol Ind Health 2018; 34(10): 726-35.
[http://dx.doi.org/10.1177/0748233718778993] [PMID: 30033815]

[10]   Singh N, Kumar D, Sahu AP. Arsenic in the environment: effects on human health and possible prevention. J Environ Biol 2007; 28(2) (Suppl.): 359-65.
[PMID: 17929751]

[11]   Tangahu BV, Abdullah SRS, Basri H, Idris M, Anuar N, Mukhlisin M. Phytotoxicity of wastewater containing lead (Pb) effects *scirpus grossus*. Int J Phytoremediation 2013; 15(8): 814-26.
[http://dx.doi.org/10.1080/15226514.2012.736437] [PMID: 23819277]

[12]   Bradham KD, Nelson CM, Kelly J, *et al.* Relationship between total and bioaccessible lead on children's blood lead levels in urban residential philadelphia soils. Environ Sci Technol 2017; 51(17): 10005-11.
[http://dx.doi.org/10.1021/acs.est.7b02058] [PMID: 28787152]

[13]   Buchet JP, Staessen J, Roels H, Lauwerys R, Fagard R. Geographical and temporal differences in the urinary excretion of inorganic arsenic: a Belgian population study. Occup Environ Med 1996; 53(5): 320-7.
[http://dx.doi.org/10.1136/oem.53.5.320] [PMID: 8673179]

[14]   Wagner GJ. Accumulation of cadmium in crop plants and its consequences to human health. Adv Agron 1993; 51: 173-212.
[http://dx.doi.org/10.1016/S0065-2113(08)60593-3]

[15]   Sharma P, Dubey RS. Involvement of oxidative stress and role of antioxidative defense system in growing rice seedlings exposed to toxic concentrations of aluminum. Plant Cell Rep 2007; 26(11): 2027-38.
[http://dx.doi.org/10.1007/s00299-007-0416-6] [PMID: 17653721]

[16]   Gill SS, Tuteja N. Reactive oxygen species and antioxidant machinery in abiotic stress tolerance in crop plants. Plant Physiol Biochem 2010; 48(12): 909-30.
[http://dx.doi.org/10.1016/j.plaphy.2010.08.016] [PMID: 20870416]

[17]   Järup L. Hazards of heavy metal contamination. Br Med Bull 2003; 68(1): 167-82.
[http://dx.doi.org/10.1093/bmb/ldg032] [PMID: 14757716]

[18]   Montero-Palmero MB, Martín-Barranco A, Escobar C, Hernández LE. Early transcriptional responses to mercury: a role for ethylene in mercury-induced stress. New Phytol 2014; 201(1): 116-30.
[http://dx.doi.org/10.1111/nph.12486] [PMID: 24033367]

[19]   Yadav SK. Cold stress tolerance mechanisms in plants. A review. Agron Sustain Dev 2010; 30(3): 515-27.
[http://dx.doi.org/10.1051/agro/2009050]

[20]   Zhou ZS, Huang SQ, Guo K, Mehta SK, Zhang PC, Yang ZM. Metabolic adaptations to mercury-

induced oxidative stress in roots of *Medicago sativa* L. J Inorg Biochem 2007; 101(1): 1-9.
[http://dx.doi.org/10.1016/j.jinorgbio.2006.05.011] [PMID: 17084899]

[21]   Pourrut B, Shahid M, Dumat C, Winterton P, Pinelli E. Lead uptake, toxicity, and detoxification in plants. Rev Environ Contam Toxicol 2011; 213: 113-36.
[http://dx.doi.org/10.1007/978-1-4419-9860-6_4] [PMID: 21541849]

[22]   Kumar A, Kumar A, M M S CP, *et al.* Lead Toxicity: Health Hazards, Influence on Food Chain, and Sustainable Remediation Approaches. Int J Environ Res Public Health 2020; 17(7): 2179.
[http://dx.doi.org/10.3390/ijerph17072179] [PMID: 32218253]

[23]   Srivastava S, Srivastava AK, Suprasanna P, D'Souza SF. Identification and profiling of arsenic stress-induced microRNAs in *Brassica juncea*. J Exp Bot 2013; 64(1): 303-15.
[http://dx.doi.org/10.1093/jxb/ers333] [PMID: 23162117]

[24]   Yeh CM, Hsiao LJ, Huang HJ. Cadmium activates a mitogen-activated protein kinase gene and MBP kinases in rice. Plant Cell Physiol 2004; 45(9): 1306-12.
[http://dx.doi.org/10.1093/pcp/pch135] [PMID: 15509854]

[25]   Yusuf M, Fariduddin Q, Ahmad A. 24-Epibrassinolide modulates growth, nodulation, antioxidant system, and osmolyte in tolerant and sensitive varieties of *Vigna radiata* under different levels of nickel: A shotgun approach. Plant Physiol Biochem 2012; 57: 143-53.
[http://dx.doi.org/10.1016/j.plaphy.2012.05.004] [PMID: 22705589]

[26]   Zhang J, Chen K, Pang Y, *et al.* QTL mapping and candidate gene analysis of ferrous iron and zinc toxicity tolerance at seedling stage in rice by genome-wide association study. BMC Genomics 2017; 18(1): 828.
[http://dx.doi.org/10.1186/s12864-017-4221-5] [PMID: 29078746]

[27]   Zhou Y, Lam HM, Zhang J. Inhibition of photosynthesis and energy dissipation induced by water and high light stresses in rice. J Exp Bot 2007; 58(5): 1207-17.
[http://dx.doi.org/10.1093/jxb/erl291] [PMID: 17283375]

[28]   Chen Y, Xu J, Yu M, Chen X, Shi J. Lead contamination in different varieties of tea plant (*Camellia sinensis* L.) and factors affecting lead bioavailability. J Sci Food Agric 2010; 90(9): 1501-7.
[http://dx.doi.org/10.1002/jsfa.3974] [PMID: 20549803]

[29]   Chopin EIB, Marin B, Mkoungafoko R, *et al.* Factors affecting distribution and mobility of trace elements (Cu, Pb, Zn) in a perennial grapevine (*Vitis vinifera* L.) in the Champagne region of France. Environ Pollut 2008; 156(3): 1092-8.
[http://dx.doi.org/10.1016/j.envpol.2008.04.015] [PMID: 18550238]

[30]   Liu JJ, Wei Z, Li JH. Effects of copper on leaf membrane structure and root activity of maize seedling. Bot Stud (Taipei, Taiwan) 2014; 55(1): 47.
[http://dx.doi.org/10.1186/s40529-014-0047-5] [PMID: 28510936]

[31]   Guo G, Lei M, Wang Y, Song B, Yang J. Accumulation of As, Cd, and Pb in sixteen wheat cultivars grown in contaminated soils and associated health risk assessment. Int J Environ Res Public Health 2018; 15(11): 2601.
[http://dx.doi.org/10.3390/ijerph15112601] [PMID: 30469364]

[32]   Sandalio LM, Dalurzo HC, Gómez M, Romero-Puertas MC, del Río LA. Cadmium-induced changes in the growth and oxidative metabolism of pea plants. J Exp Bot 2001; 52(364): 2115-26.
[http://dx.doi.org/10.1093/jexbot/52.364.2115] [PMID: 11604450]

[33]   López-Climent MF, Arbona V, Pérez-Clemente RM, Gómez-Cadenas A. Effects of cadmium on gas exchange and phytohormone contents in citrus. Biol Plant 2011; 55(1): 187-90.
[http://dx.doi.org/10.1007/s10535-011-0028-4]

[34]   Mahmood T, Gupta KJ, Kaiser WM. Cd stress stimulates nitric oxide production by wheat roots. Pak J Bot 2009; 41: 1285-90.

[35]   Alemzadeh AL, Rastgoo A, Tale S. Tazangi and Eslamzadeh T. Effects of copper, nickel and zinc on

biochemical parameters and metal accumulation in gouan, *Aeluropus littoralis*. Plant Knowledge J 2014; 3: 31-8.

[36]   Adrees M, Ali S, Rizwan M, *et al.* Mechanisms of silicon-mediated alleviation of heavy metal toxicity in plants: A review. Ecotoxicol Environ Saf 2015; 119: 186-97.
[http://dx.doi.org/10.1016/j.ecoenv.2015.05.011] [PMID: 26004359]

[37]   Sayyad G, Afyuni M, Mousavi S-F, Abbaspour KC, Richards BK, Schulin R. Transport of Cd, Cu, Pb and Zn in a calcareous soil under wheat and safflower cultivation— A column study. Geoderma 2010; 154(3-4): 311-20.
[http://dx.doi.org/10.1016/j.geoderma.2009.10.019]

[38]   He J, Ren Y, Zhu C, Jiang D. Effects of cadmium stress on seed germination, seedling growth and seed amylase activities in rice (*Oryza sativa*). Rice Sci 2008; 15(4): 319-25.
[http://dx.doi.org/10.1016/S1672-6308(09)60010-X]

[39]   The effects of heavy metals on seed germination and plant growth on alfalfa plant (*Medicago sativa*). Bulg J Agric Sci 2009; 15: 347-50.

[40]   Bajguz A. Suppression of *Chlorella vulgaris* growth by cadmium, lead, and copper stress and its restoration by endogenous brassinolide. Arch Environ Contam Toxicol 2011; 60(3): 406-16.
[http://dx.doi.org/10.1007/s00244-010-9551-0] [PMID: 20523975]

[41]   Gangwar S, Singh VP, Prasad SM, Maurya JN. Differential responses of pea seedlings to indole acetic acid under manganese toxicity. Acta Physiol Plant 2011; 33(2): 451-62.
[http://dx.doi.org/10.1007/s11738-010-0565-z]

[42]   Krishnamurthy A, Rathinasabapathi B. Auxin and its transport play a role in plant tolerance to arsenite☐induced oxidative stress in *A rabidopsis thaliana*. Plant Cell Environ 2013; 36(10): 1838-49.
[http://dx.doi.org/10.1111/pce.12093] [PMID: 23489261]

[43]   Fassler E, Evangelou MW, Robinson BH, Schulin R. Effects of indole-3-acetic acid (IAA) on sunflower growth and heavy metal uptake in combination with ethylene diamine disuccinic acid (EDDS). Chemosphere 2010; 80(8): 901-07, ISSN 0045-6535.

[44]   Hadi F, Bano A, Fuller MP. The improved phytoextraction of lead (Pb) and the growth of maize (*Zea mays* L.): the role of plant growth regulators (GA3 and IAA) and EDTA alone and in combinations. Chemosphere 2010; 80(4): 457-62.
[http://dx.doi.org/10.1016/j.chemosphere.2010.04.020] [PMID: 20435330]

[45]   Elobeid M, Göbel C, Feussner I, Polle A. Cadmium interferes with auxin physiology and lignification in poplar. J Exp Bot 2012; 63(3): 1413-21.
[http://dx.doi.org/10.1093/jxb/err384] [PMID: 22140243]

[46]   Matsuoka M. Gibberellin signaling: how do plant cells respond to GA signals? J Plant Growth Regul 2003; 22(2): 123-5.
[http://dx.doi.org/10.1007/s00344-003-0039-2]

[47]   Iqbal N, Nazar R, Khan MIR, Masood A, Khan NA. Role of gibberellins in regulation of source sink relations under optimal and limiting environmental conditions. Curr Sci 2011; 100: 998-1007.

[48]   Zhu XF, Jiang T, Wang ZW, *et al.* Gibberellic acid alleviates cadmium toxicity by reducing nitric oxide accumulation and expression of IRT1 in *Arabidopsis thaliana*. J Hazard Mater 2012; 239-240: 302-7.
[http://dx.doi.org/10.1016/j.jhazmat.2012.08.077] [PMID: 23021314]

[49]   Sun S, Wang H, Yu H, *et al.* GASA14 regulates leaf expansion and abiotic stress resistance by modulating reactive oxygen species accumulation. J Exp Bot 2013; 64(6): 1637-47.
[http://dx.doi.org/10.1093/jxb/ert021] [PMID: 23378382]

[50]   Liang W, Li L, Zhang F, *et al.* Effects of abiotic stress, light, phytochromes and phytohormones on the expression of OsAQP, a rice aquaporin gene. Plant Growth Regul 2013; 69(1): 21-7.
[http://dx.doi.org/10.1007/s10725-012-9743-x]

[51]   Gangwar S, Singh VP, Srivastava PK, Maurya JN. Modification of chromium (VI) phytotoxicity by exogenous gibberellic acid application in *Pisum sativum* (L.) seedlings. Acta Physiol Plant 2011; 33(4): 1385-97.
[http://dx.doi.org/10.1007/s11738-010-0672-x]

[52]   Siddiqui MH, Al-Whaibi MH, Basalah MO. Interactive effect of calcium and gibberellin on nickel tolerance in relation to antioxidant systems in *Triticum aestivum* L. Protoplasma 2011; 248(3): 503-11.
[http://dx.doi.org/10.1007/s00709-010-0197-6] [PMID: 20730631]

[53]   Iqbal M, Ashraf M. Gibberellic acid mediated induction of salt tolerance in wheat plants: Growth, ionic partitioning, photosynthesis, yield and hormonal homeostasis. Environ Exp Bot 2013; 86: 76-85.
[http://dx.doi.org/10.1016/j.envexpbot.2010.06.002]

[54]   Meng H, Hua S, Shamsi IH, Jilani G, Li Y, Jiang L. Cadmium-induced stress on the seed germination and seedling growth of *Brassica napus* L., and its alleviation through exogenous plant growth regulators. Plant Growth Regul 2009; 58(1): 47-59.
[http://dx.doi.org/10.1007/s10725-008-9351-y]

[55]   Atici Ö, Ağar G, Battal P. Changes in phytohormone contents in chickpea seeds germinating under lead or zinc stress. Biol Plant 2005; 49(2): 215-22.
[http://dx.doi.org/10.1007/s10535-005-5222-9]

[56]   El-Monem A, Sharaf AE-MM, Farghal II, Sofy MR. Role of gibberellic acid in abolishing the detrimental effects of Cd and Pb on broad bean and lupin plants. Res J Agric Biol Sci 2009; 5: 6-13.

[57]   Hu YF, Zhou G, Na XF, *et al.* Cadmium interferes with maintenance of auxin homeostasis in *Arabidopsis* seedlings. J Plant Physiol 2013; 170(11): 965-75.
[http://dx.doi.org/10.1016/j.jplph.2013.02.008] [PMID: 23683587]

[58]   Pacenza M, Muto A, Chiappetta A, *et al.* In *Arabidopsis thaliana* Cd differentially impacts on hormone genetic pathways in the methylation defective ddc mutant compared to wild type. Sci Rep 2021; 11(1): 10965.
[http://dx.doi.org/10.1038/s41598-021-90528-5] [PMID: 34040101]

[59]   Luo Y, Wei Y, Sun S, *et al.* Selenium modulates the level of auxin to alleviate the toxicity of cadmium in Tobacco. Int J Mol Sci 2019; 20(15): 3772.
[http://dx.doi.org/10.3390/ijms20153772] [PMID: 31374993]

[60]   Ronzan M, Piacentini D, Fattorini L, *et al.* Cadmium and arsenic affect root development in *Oryza sativa* L. negatively interacting with auxin. Environ Exp Bot 2018; 151: 64-75.
[http://dx.doi.org/10.1016/j.envexpbot.2018.04.008]

[61]   Fattorini L, Ronzan M, Piacentini D, *et al.* Cadmium and arsenic affect quiescent centre formation and maintenance in *Arabidopsis thaliana* post-embryonic roots disrupting auxin biosynthesis and transport. Environ Exp Bot 2017; 144: 37-48.
[http://dx.doi.org/10.1016/j.envexpbot.2017.10.005]

[62]   Falkowska M, Pietrvczuk A, Piotrowska A, Bajguz A, Grvgoruk A, Czerpak R. The effect of gibberellic acid (GA3) on growth, metal biosorption and metabolism of the green algae *Chlorella vulgaris* (Chlorophyceae) Beijerinck exposed to cadmium and lead stress. Pol J Environ Stud 2011; 20: 53-9.

[63]   Amri B, Khamassi K, Ali MB, Teixeira da Silva JA, Bettaieb Ben Kaab L, Kaab LBB. Effects of gibberellic acid on the process of organic reserve mobilization in barley grains germinated in the presence of cadmium and molybdenum. S Afr J Bot 2016; 106: 35-40.
[http://dx.doi.org/10.1016/j.sajb.2016.05.007]

[64]   Huang TL, Nguyen QTT, Fu SF, Lin CY, Chen YC, Huang HJ. Transcriptomic changes and signalling pathways induced by arsenic stress in rice roots. Plant Mol Biol 2012; 80(6): 587-608.
[http://dx.doi.org/10.1007/s11103-012-9969-z] [PMID: 22987115]

[65]   Al-Hakimi AMA. Modification of cadmium toxicity in pea seedlings by kinetin. Plant Soil Environ

2007; 53(3): 129-35.
[http://dx.doi.org/10.17221/2228-PSE]

[66]   Mohan TC, Castrillo G, Navarro C, *et al.* Cytokinin determines thiol-mediated arsenic tolerance and accumulation in Arabidopsis thaliana. Plant Physiol 2016; 171(2): pp.00372.2016.
[http://dx.doi.org/10.1104/pp.16.00372] [PMID: 27208271]

[67]   Pavlíková D, Pavlík M, Procházková D, Zemanová V, Hnilička F, Wilhelmová N. Nitrogen metabolism and gas exchange parameters associated with zinc stress in tobacco expressing an ipt gene for cytokinin synthesis. J Plant Physiol 2014; 171(7): 559-64.
[http://dx.doi.org/10.1016/j.jplph.2013.11.016] [PMID: 24655392]

[68]   Bi C, Ma Y, Wang XF, Zhang DP. Overexpression of the transcription factor NF-YC9 confers abscisic acid hypersensitivity in *Arabidopsis*. Plant Mol Biol 2017; 95(4-5): 425-39.
[http://dx.doi.org/10.1007/s11103-017-0661-1] [PMID: 28924726]

[69]   Khan AR, Wakeel A, Muhammad N, *et al.* Involvement of ethylene signaling in zinc oxide nanoparticle-mediated biochemical changes in *Arabidopsis thaliana* leaves. Environ Sci Nano 2019; 6(1): 341-55.
[http://dx.doi.org/10.1039/C8EN00971F]

[70]   Keunen E, Schellingen K, Vangronsveld J, Cuypers A. Ethylene and Metal Stress: Small Molecule, Big Impact. Front Plant Sci 2016; 7: 23.
[http://dx.doi.org/10.3389/fpls.2016.00023] [PMID: 26870052]

[71]   Sharma P, Bhardwaj R. Effect of 24-Epibrassinolide on seed germination, seedling growth and heavy metals uptake in *Brassica juncea* L. Gen Appl Plant Physiol 2007; 32: 1-2.

[72]   Arora M, Kiran B, Rani S, Rani A, Kaur B, Mittal N. Heavy metal accumulation in vegetables irrigated with water from different sources. Food Chem 2008; 111(4): 811-5.
[http://dx.doi.org/10.1016/j.foodchem.2008.04.049]

[73]   Arora P, Bhardwaj R, Kumar Kanwar M. 24-epibrassinolide induced antioxidative defense system of Brassica juncea L. under Zn metal stress. Physiol Mol Biol Plants 2010; 16(3): 285-93.
[http://dx.doi.org/10.1007/s12298-010-0031-9] [PMID: 23572978]

[74]   Swamy KN, Vardhini BV, Ramakrishna B, Anuradha S, Siddulu N, Rao SSR. Role of 28-homobrassinolide on growth biochemical parameters of *Trigonella foenugraecum* L. plants subjected to lead toxicity. Int J Multidiscip Curr Res 2014; 2: 2321-4.

[75]   Bhardwaj R, Hundal G, Sharma N, Sharma I. 28-Homobrassinolide alters protein content and activities of glutathione-S-transferase and polyphenol oxidase in *Raphanus sativus* L. plants under heavy metal stress. Toxicol Int 2014; 21(1): 45.
[http://dx.doi.org/10.4103/0971-6580.128792] [PMID: 24748734]

[76]   Ahmad P, Alyemeni MN, Wijaya L, Alam P, Ahanger MA, Alamri SA. Jasmonic acid alleviates negative impacts of cadmium stress by modifying osmolytes and antioxidants in faba bean (*Vicia faba* L.). Arch Agron Soil Sci 2017; 63(13): 1889-99.
[http://dx.doi.org/10.1080/03650340.2017.1313406]

[77]   Song Y, Dong Y, Kong J, Tian X, Bai X, Xu L. Effects of root addition and foliar application of nitric oxide and salicylic acid in alleviating iron deficiency induced chlorosis of peanut seedlings. J Plant Nutr 2017; 40(1): 63-81.
[http://dx.doi.org/10.1080/01904167.2016.1201491]

[78]   Boroumand JShH, Lari Y, Ranjbar M. Effect of salicylic acid on some plant growth parameters under lead stress in *Brassica napus* var. Okapi. Iran J Plant Physiol 2011; 3: 177-85.

[79]   Khan I GA, Khan I, Ahmed I, Mustafa I. Amelioration of lead toxicity in *Pisum sativum* (l.) by foliar application of salicylic acid. J Environ Anal Toxicol 2015; 5(4): 292.
[http://dx.doi.org/10.4172/2161-0525.1000292]

[80]   Gondor OK, Pál M, Darkó É, Janda T, Szalai G. Salicylic acid and sodium salicylate alleviate

cadmium toxicity to different extents in Maize (*Zea mays* L.). PLoS One 2016; 11(8): e0160157.
[http://dx.doi.org/10.1371/journal.pone.0160157] [PMID: 27490102]

[81]   Perilli S, Moubayidin L, Sabatini S. The molecular basis of cytokinin function. Curr Opin Plant Biol 2010; 13(1): 21-6.
[http://dx.doi.org/10.1016/j.pbi.2009.09.018] [PMID: 19850510]

[82]   Gangwar S, Singh VP, Prasad SM, Maurya JN. Modulation of manganese toxicity in *Pisum sativum* L. seedlings by kinetin. Sci Hortic (Amsterdam) 2010; 126(4): 467-74.
[http://dx.doi.org/10.1016/j.scienta.2010.08.013]

[83]   Munzuroglu O, Zengin FK. Effect of cadmium on germination, coleoptile and root growth of barley seeds in the presence of gibberellic acid and kinetin. J Environ Biol 2006; 27(4): 671-7.
[PMID: 17405329]

[84]   Gemrotová M, Kulkarni MG, Stirk WA, Strnad M, Van Staden J, Spíchal L. Seedlings of medicinal plants treated with either a cytokinin antagonist (PI-55) or an inhibitor of cytokinin degradation (INCYDE) are protected against the negative effects of cadmium. Plant Growth Regul 2013; 71(2): 137-45.
[http://dx.doi.org/10.1007/s10725-013-9813-8]

[85]   Thomas JC, Perron M, LaRosa PC, Smigocki AC. Cytokinin and the regulation of a tobacco metallothionein□like gene during copper stress. Physiol Plant 2005; 123(3): 262-71.
[http://dx.doi.org/10.1111/j.1399-3054.2005.00440.x]

[86]   Masood A, Khan MIR, Fatma M, Asgher M, Per TS, Khan NA. Involvement of ethylene in gibberellic acid-induced sulfur assimilation, photosynthetic responses, and alleviation of cadmium stress in mustard. Plant Physiol Biochem 2016; 104: 1-10.
[http://dx.doi.org/10.1016/j.plaphy.2016.03.017] [PMID: 26998941]

[87]   Finkelstein R. Abscisic Acid synthesis and response. Arabidopsis Book 2013; 11: e0166.
[http://dx.doi.org/10.1199/tab.0166] [PMID: 24273463]

[88]   Rauser WE, Dumbroff EB. Effects of excess cobalt, nickel and zinc on the water relations of *Phaseolus vulgaris*. Environ Exp Bot 1981; 21(2): 249-55.
[http://dx.doi.org/10.1016/0098-8472(81)90032-0]

[89]   Hollenbach B, Schreiber L, Hartung W, Dietz KJ. Cadmium leads to stimulated expression of the lipid transfer protein genes in barley: implications for the involvement of lipid transfer proteins in wax assembly. Planta 1997; 203(1): 9-19.
[http://dx.doi.org/10.1007/s00050159] [PMID: 9299788]

[90]   Fediuc E, Lips SH, Erdei L. O-acetylserine (thiol) lyase activity in Phragmites and Typha plants under cadmium and NaCl stress conditions and the involvement of ABA in the stress response. J Plant Physiol 2005; 162(8): 865-72.
[http://dx.doi.org/10.1016/j.jplph.2004.11.015] [PMID: 16146312]

[91]   Stroiński A, Chadzinikolau T, Giżewska K, Zielezińska M. ABA or cadmium induced phytochelatin synthesis in potato tubers. Biol Plant 2010; 54(1): 117-20.
[http://dx.doi.org/10.1007/s10535-010-0017-z]

[92]   Kim YH, Khan AL, Kim DH, *et al.* Silicon mitigates heavy metal stress by regulating P-type heavy metal ATPases, *Oryza sativa*low silicon genes, and endogenous phytohormones. BMC Plant Biol 2014; 14(1): 13.
[http://dx.doi.org/10.1186/1471-2229-14-13] [PMID: 24405887]

[93]   Munzuro O, Fikriye KZ, Yahyagil Z. The abscisic acid levels of wheat (*Triticum aestivum* L. cv. Çakmak 79) seeds that were germinated under heavy metal (Hg^{++}, Cd^{++}, Cu^{++}) stress. G U J Sci 2008; 21: 1-7.

[94]   Wang Y, Wang Y, Kai W, *et al.* Transcriptional regulation of abscisic acid signal core components during cucumber seed germination and under Cu2+, Zn2+, NaCl and simulated acid rain stresses.

Plant Physiol Biochem 2014; 76: 67-76.
[http://dx.doi.org/10.1016/j.plaphy.2014.01.003] [PMID: 24486581]

[95] Stroiński A, Kozłowska M. Cadmium-induced oxidative stress in potato tuber. Acta Soc Bot Pol 2014; 66(2): 189-95.
[http://dx.doi.org/10.5586/asbp.1997.024]

[96] Trinh NN, Huang TL, Chi WC, Fu SF, Chen CC, Huang HJ. Chromium stress response effect on signal transduction and expression of signaling genes in rice. Physiol Plant 2014; 150(2): 205-24.
[http://dx.doi.org/10.1111/ppl.12088] [PMID: 24033343]

[97] Schellingen K, Van Der Straeten D, Vandenbussche F, *et al.* Cadmium-induced ethylene production and responses in *Arabidopsis thaliana* rely on ACS2 and ACS6 gene expression. BMC Plant Biol 2014; 14(1): 214.
[http://dx.doi.org/10.1186/s12870-014-0214-6] [PMID: 25082369]

[98] Kanwar MK, Bhardwaj R, Chowdhary SP, Arora P. Priyanka Sharma and Kumar S. Isolation and characterization of 24-Epibrassinolide from *Brassica juncea* L. and its effects on growth, Ni ion uptake, antioxidant defence of *Brassica* plants and in vitro cytotoxicity. Acta Physiol Plant 2013; 35: 1351-62.
[http://dx.doi.org/10.1007/s11738-012-1175-8]

[99] Roy M, McDonald LM. Metal uptake in plants and health risk assessments in metal-contaminated smelter soils. Land Degrad Dev 2015; 26(8): 785-92.
[http://dx.doi.org/10.1002/ldr.2237]

[100] Mandava NB. Plant growth-promoting brassinosteroids. Annu Rev Plant Physiol Plant Mol Biol 1988; 39(1): 23-52.
[http://dx.doi.org/10.1146/annurev.pp.39.060188.000323]

[101] Vardhini BV. Brassinosteroids role for amino acids, peptides and amines modulation in stressed plants - A review. In: Anjum NA, Gill SS, Gill R, Eds. Plant Adaptation to Environmental Change: Significance of Amino Acids and their Derivatives. Wallingford: CAB International 2014; pp. 300-16.
[http://dx.doi.org/10.1079/9781780642734.0300]

[102] Mahesh K, Balaraju P, Ramakrishna B, Ram Rao SS, Rao R. Effect of brassinosteroids on germination and seedling growth of radish (*Raphanus sativus* L.) under PEG-6000 induced water stress. Am J Plant Sci 2013; 4(12): 2305-13.
[http://dx.doi.org/10.4236/ajps.2013.412285]

[103] Xia XJ, Wang YJ, Zhou YH, *et al.* Reactive oxygen species are involved in brassinosteroid-induced stress tolerance in cucumber. Plant Physiol 2009; 150(2): 801-14.
[http://dx.doi.org/10.1104/pp.109.138230] [PMID: 19386805]

[104] Yusuf M, Fariduddin Q, Hayat S, Hasan SA, Ahmad A. Protective response of 28-homobrassinolide in cultivars of *Triticum aestivum* with different levels of nickel. Arch Environ Contam Toxicol 2011; 60(1): 68-76.
[http://dx.doi.org/10.1007/s00244-010-9535-0] [PMID: 20464550]

[105] Yusuf M, Fariduddin Q, Ahmad A. 24-Epibrassinolide modulates growth, nodulation, antioxidant system, and osmolyte in tolerant and sensitive varieties of *Vigna radiata* under different levels of nickel: A shotgun approach. Plant Physiol Biochem 2012; 57: 143-53.
[http://dx.doi.org/10.1016/j.plaphy.2012.05.004] [PMID: 22705589]

[106] Yusuf M, Fariduddin Q, Ahmad I, Ahmad A. Brassinosteroid-mediated evaluation of antioxidant system and nitrogen metabolism in two contrasting cultivars of *Vigna radiata* under different levels of nickel. Physiol Mol Biol Plants 2014; 20(4): 449-60.
[http://dx.doi.org/10.1007/s12298-014-0259-x] [PMID: 25320468]

[107] Vázquez MN, Guerrero YR, González LM, Noval WT. Brassinosteroids and plant responses to heavy metal stress. An overview. Open J Met 2013; 3(2): 34-41.
[http://dx.doi.org/10.4236/ojmetal.2013.32A1005]

[108] Hasan SA, Hayat S, Ahmad A. Brassinosteroids protect photosynthetic machinery against the cadmium induced oxidative stress in two tomato cultivars. Chemosphere 2011; 84(10): 1446-51.
[http://dx.doi.org/10.1016/j.chemosphere.2011.04.047] [PMID: 21565386]

[109] Hayat S, Ali B, Aiman Hasan S, Ahmad A. Brassinosteroid enhanced the level of antioxidants under cadmium stress in *Brassica juncea*. Environ Exp Bot 2007; 60(1): 33-41.
[http://dx.doi.org/10.1016/j.envexpbot.2006.06.002]

[110] Hasan SA, Hayat S, Ali B, Ahmad A. 28-Homobrassinolide protects chickpea (*Cicer arietinum*) from cadmium toxicity by stimulating antioxidants. Environ Pollut 2008; 151(1): 60-6.
[http://dx.doi.org/10.1016/j.envpol.2007.03.006] [PMID: 17481788]

[111] Hayat S, Alyemeni MN, Hasan SA. Foliar spray of brassinosteroid enhances yield and quality of *Solanum lycopersicum* under cadmium stress. Saudi J Biol Sci 2012; 19(3): 325-35.
[http://dx.doi.org/10.1016/j.sjbs.2012.03.005] [PMID: 23961193]

[112] Rady MM. Effect of 24-epibrassinolide on growth, yield, antioxidant system and cadmium content of bean (*Phaseolus vulgaris* L.) plants under salinity and cadmium stress. Sci Hortic (Amsterdam) 2011; 129(2): 232-7.
[http://dx.doi.org/10.1016/j.scienta.2011.03.035]

[113] Filová A, Sytar O, Krivosudská E. Effects of brassinosteroid on the induction of physiological changes in Helianthus annuus L. under copper stress. Acta Univ Agric Silvic Mendel Brun 2013; 61(3): 623-9.
[http://dx.doi.org/10.11118/actaun201361030623]

[114] Kapoor D, Rattan A, Gautam V, *et al.* 24-Epibrassinolide mediated changes in photosynthetic pigments and antioxidative defence system of radish seedlings under cadmium and mercury stress. Physiol Biochem 2014; 10: 110-21.

[115] Fariduddin Q, Yusuf M, Hayat S, Ahmad A. Effect of 28-homobrassinolide on antioxidant capacity and photosynthesis in *Brassica juncea* plants exposed to different levels of copper. Environ Exp Bot 2009; 66(3): 418-24.
[http://dx.doi.org/10.1016/j.envexpbot.2009.05.001]

[116] Fariduddin Q, Khalil RRAE, Mir BA, Yusuf M, Ahmad A. 24-Epibrassinolide regulates photosynthesis, antioxidant enzyme activities and proline content of Cucumis sativus under salt and/or copper stress. Environ Monit Assess 2013; 185(9): 7845-56.
[http://dx.doi.org/10.1007/s10661-013-3139-x] [PMID: 23443638]

[117] Sharma N, Hundal GS, Sharma I, Bhardwaj R. Effect of 24-epibrassinolide on protein content and activies of glutathione S-transferase and polyphenol oxidase in Raphanus sativus L. plants under cadmium and mercury metal stress. Terre Aquat Environ Toxicol 2012; 6(1): 1-7. Global Science Books.

[118] Bajguz A. An enhancing effect of exogenous brassinolide on the growth and antioxidant activity in *Chlorella vulgaris* cultures under heavy metals stress. Environ Exp Bot 2010; 68(2): 175-9.
[http://dx.doi.org/10.1016/j.envexpbot.2009.11.003]

[119] Sirhindi G, Kaur H, Bhardwaj R, Sharma P, Mushtaq R. 28-Homobrassinolide potential for oxidative interface in *Brassica juncea* under temperature stress. Acta Physiol Plant 2017; 39(10): 228.
[http://dx.doi.org/10.1007/s11738-017-2524-4]

[120] Kaur H, Sirhindi G, Bhardwaj R, Alyemeni MN, Siddique KHM, Ahmad P. 28-homobrassinolide regulates antioxidant enzyme activities and gene expression in response to salt- and temperature-induced oxidative stress in *Brassica juncea*. Sci Rep 2018; 8(1): 8735.
[http://dx.doi.org/10.1038/s41598-018-27032-w] [PMID: 29880861]

[121] Kaur H, Singh J, & Kaur K. Impact of temperature stress on the functional efficiency of Brassica napus seedlings. Asian J Plant Soil Sci 2021; 5(1): 18-23. Available from: https://www.ikprress.org/index.php/AJOPSS/article/view/6045

[122] Kaur H, Kaur M, Rafiq R. Regulation of morpho-physiological and biochemical traits of garden pea

(*pisum sativum* l.) through different levels of indole butyric acid under salt stress conditions. J Adv Food Sci Technol 2021; 7(1): 38-46. https://www.ikprress.org/index.php/JAFSAT/article/view/6013

[123] Aldridge DC, Galt S, Giles D, Turner WB. Metabolites of *Lasiodiplodia theobromae*. J Chem Soc C 1971; 1623-27: 1623.
[http://dx.doi.org/10.1039/j39710001623]

[124] Popova LP, Tsonev TD, Vaklinova SG. Changes in some photosynthetic and photorespiratory properties in barley leaves after treatment with jasmonic acid. J Plant Physiol 1988; 132(3): 257-61.
[http://dx.doi.org/10.1016/S0176-1617(88)80101-9]

[125] Sembdner G, Atzorn R, Schneider G. Plant hormone conjugation. Plant Mol Biol 1994; 26(5): 1459-81.
[http://dx.doi.org/10.1007/BF00016485] [PMID: 7858200]

[126] McConn M, Browse J. The critical requirement for linolenic acid is pollen development, not photosynthesis, in an *Arabidopsis* mutant. Plant Cell 1996; 8(3): 403-16.
[http://dx.doi.org/10.2307/3870321] [PMID: 12239389]

[127] Overmyer K, Tuominen H, Kettunen R, *et al*. Ozone-sensitive *arabidopsis rcd1* mutant reveals opposite roles for ethylene and jasmonate signaling pathways in regulating superoxide-dependent cell death. Plant Cell 2000; 12(10): 1849-62.
[http://dx.doi.org/10.1105/tpc.12.10.1849] [PMID: 11041881]

[128] Rao MV, Lee H, Creelman RA, Mullet JE, Davis KR. Jasmonic acid signaling modulates ozone-induced hypersensitive cell death. Plant Cell 2000; 12(9): 1633-46.
[http://dx.doi.org/10.1105/tpc.12.9.1633] [PMID: 11006337]

[129] Reymond P, Weber H, Damond M, Farmer EE. Differential gene expression in response to mechanical wounding and insect feeding in Arabidopsis. Plant Cell 2000; 12(5): 707-19.
[http://dx.doi.org/10.1105/tpc.12.5.707] [PMID: 10810145]

[130] Walters D, Cowley T, Mitchell A. Methyl jasmonate alters polyamine metabolism and induces systemic protection against powdery mildew infection in barley seedlings. J Exp Bot 2002; 53(369): 747-56.
[http://dx.doi.org/10.1093/jexbot/53.369.747] [PMID: 11886895]

[131] Maciejewska B, Kopcewicz J. Inhibitory effect of methyl jasmonate on flowering and elongation growth on *Pharbitis nil.* J Plant Growth Regul 2002; 21(3): 216-23.
[http://dx.doi.org/10.1007/s003440010061]

[132] Maksymiec W, Wianowska D, Dawidowicz AL, Radkiewicz S, Mardarowicz M, Krupa Z. The level of jasmonic acid in *Arabidopsis thaliana* and *Phaseolus coccineus* plants under heavy metal stress. J Plant Physiol 2005; 162(12): 1338-46.
[http://dx.doi.org/10.1016/j.jplph.2005.01.013] [PMID: 16425452]

[133] Kopittke PM, Dart PJ, Menzies NW. Toxic effects of low concentrations of Cu on nodulation of cowpea (Vigna unguiculata). Environ Pollut 2007; 145(1): 309-15.
[http://dx.doi.org/10.1016/j.envpol.2006.03.007] [PMID: 16678321]

[134] Ngozi OA, Imaga SO, Adenekan GA, Yussuph TI, Nwoyimi OO. Balogun and Eguntola TA. Assessment of antioxidation potential of selected plants with antisickling property. J Med Plants Res 2010; 4(21): 2217-21.
[http://dx.doi.org/10.5897/JMPR10.175]

[135] Jaiswal D, Rai PK, Kumar A, Watal G. Study of glycemic profile of *Cajanus cajan* leaves in experimental rats. Indian J Clin Biochem 2008; 23(2): 167-70.
[http://dx.doi.org/10.1007/s12291-008-0037-z] [PMID: 23105745]

[136] Wang SY. Antioxidant capacity and phenolic content of berry fruits as affected by genotype, preharvest conditions, maturity, and handling. In: Zhao Y, Ed. Berry Fruit: Value added Products for Health Promotion. Boca Raton: Taylor and Francis Group 2007; pp. 147-86.

[http://dx.doi.org/10.1201/9781420006148.ch5]

[137]   Kong Y, Fu YJ, Zu YG, *et al.* Cajanu slactone a new coumarin with anti bacterial activity from pigeon pea leaves. Food Chem 2010; 121: 1150-5.
[http://dx.doi.org/10.1016/j.foodchem.2010.01.062]

[138]   Wu N, Fu K, Fu YJ, *et al.* Antioxidant activities of extracts and main components of Pigeonpea [*Cajanus cajan* (L.) Millsp.] leaves. Molecules 2009; 14(3): 1032-43.
[http://dx.doi.org/10.3390/molecules14031032] [PMID: 19305357]

[139]   Ali B, Hayat S, Hayat Q, Ahmad A. Ali B, Hayat S, Hayat Q & Ahmad A. Cobalt stress affects nitrogen metabolism, photosynthesis and antioxidant system in chickpea (*Cicer arietinum* L.). Journal of Plant Interac 2010; 5:3: 223-31
[http://dx.doi.org/10.1080/17429140903370584]

[140]   Kaur H, Sirhindi G, Sharma P. Effect of jasmonic acid on some biochemical and physiological parameters in salt-stressed *Brassica napus* seedlings. Int J Plant Physiol Biochem 2017; 9(4): 36-42.
[http://dx.doi.org/10.5897/IJPPB2016.0245]

CHAPTER 8

Microalgae-based Remediation of Heavy Metals Polluted Environment

Arun Dev Singh[1], Shalini Dhiman[1], Jaspreet Kour[1], Tamanna Bhardwaj[1], Raman Tikoria[2], Mohd Ali[2], Parkirti[2], Roohi Sharma[2], Kanika Khanna[1], Puja Ohri[2] and Renu Bhardwaj[1,*]

[1] *Department of Botanical and Environmental Sciences, Guru Nanak Dev University, Amritsar, Punjab, India*

[2] *Department of Zoology, Guru Nanak Dev University, Amritsar, Punjab, India*

Abstract: Population explosion and rapid industrialization surge are posing a serious threat to plants, the human population, and the world's environment. These vigorous developmental prospects lead to the production of serious pollutants and higher concentrations of toxic heavy metals (HMs) in the environment. These toxic HMs are severely compromising the global environment, induce toxicity to the living systems, and cause the deterioration of water and land ecosystems globally. However, to minimize these toxic pollutants, certain remediation methods have been adopted to bring these pollutants to a minimum threatening level. Thus, remediation mechanisms like biological and non-biological methods are brought into consideration. Among these methods, biological methods like novel, phytoremediation techniques by employing "microalgae" are considered to be the most effective, inexpensive, easy to implement, and eco-friendly among all the recommended methods. Phytoremediation requires sunlight as energy input whereas it undergoes environmental reclamation such as nitrate, phosphate, and HMs removal from the wastewater sources as well as mediates HMs elimination from the soil. Microalgae act as bio-accumulators and further lead to HMs precipitation and fixation inside the algal tissues. This chapter reviews the application of different microalgae strains, their bioremediation strategies, and mechanisms adopted under HMs stress environments.

Keywords: Bio-accumulators, Biofilm, Biomagnification, Biosorption, Carcinogenic, Food chain, Heavy metals, Microalgae, Photo-bioreactors.

INTRODUCTION

Heavy metals (HMs) are among the natural constituents that are known to exist in the earth's crust and soil. However, HMs are defined by their density, or generally

---

* **Corresponding author Kanika Khanna:** Department of Botanical and Environmental Sciences, Guru Nanak Dev University, Amritsar, Punjab, India; E-mail: kanika.27590@gmail.com

as an element with metallic properties and having an atomic weight ranging between 63.5 to 200.6, such as Copper (Cu), Lead (Pb), Silver (Ag), Iron (Fe), *etc* [1, 2].

Whereas, from an ecological point of view, HMs are those elements that are posing toxic impacts on the environment and are non-biodegradable, and highly toxic at low concentrations [3]. Among these HMs, some are considered to be micronutrients that are potentially required for the growth and development of plants such as Zinc (Zn), Cobalt (Co), Copper (Cu), Nickel (Ni), Manganese (Mn), *etc* [4]. However, extensive industrialization, urbanization, and false agricultural practices have globally increased the concentrations of these HMs in aquatic as well as terrestrial ecosystems [5, 6]. These rapid developments and technological innovations have further increased human contact with these toxic heavy metals through agriculture, and wastewater globally. These heavy metal residues exist in different chemical states in the environment and pose serious impacts on human health, plants, and the associated ecosystems [7]. Plants grown under HM-contaminated soils are found to have visible symptoms like chlorosis, stunted growth, root browning, and even death. Also, extensive concentrations are found to show hindrance in the normal functioning of the cellular, metabolic, and genetic potential of the affected plants [8]. In humans, these HMs can lead to multiple ill impacts from mutagenicity, bioaccumulating amplification, and carcinogenic and teratogenic effects. Also, after their entry into the human system, these HMs lead to arthralgia and headache, mental disorders, abnormal kidney, and liver functioning [9, 10]. In aquatic ecosystems, these HMs will flow through the wastewater and accumulate in oceans too. Their concentrations will lead to multiple detrimental impacts on aquatic organisms like fish. These aquatic organisms need good quality water for their survival and their contamination through different pollutants has existential pressure on these aquatic organisms. HMs also get their entry into humans through the aquatic food chains and accumulate along the different food chains [11]. Moreover, other health issues associated with the HMs accumulated in drinking water may lead to muscle atrophy and loss of appetite. Whereas, Cd toxicity may lead to bone mineral deficiency as well as kidney poisoning. Hexavalent Cd is well known to cause birth defects in newborn babies and other issues like diarrhoea and vomiting even at low doses [12 - 14]. However, in soil ecosystems, these HMs accumulate through multiple human activities like energy production, fuel production, mining, electroplating, agriculture, and wastewater sludge treatment. HMs like iron, arsenic, mercury, nickel, chromium, lead, cadmium, and zinc after reaching certain amounts get infiltrated into the groundwater and subsequently enter different food chains through crop plants, leading to the disruption of biological processes [15]. Furthermore, the elimination and treatment of HMs from polluted ecosystems is a challenge and thus requires an economical and efficient method to treat these pollutants. These methods may include physical, chemical, and biological methods to treat these ecosystems. Among these methods, biological methods have gained importance due to their eco-friendly and feasible nature with no secondary pollutants and thus gained importance in treating HMs [16, 17]. Microalgae have been introduced in order to eliminate these HMs from aquatic as well as terrestrial ecosystems. "Microalgae" are microscopic plants that are available in aquatic ecosystems and have been extensively exploited to treat sewage treatment plants since 1950 [15, 18]. However, microalgae have emerged as a potential bioremediation agent with promising solutions to clean up the HMs from the environment [19]. These microalgae cells are acclimatized to accumulate HMs, as well as act tolerant through ROS detoxification mechanisms under HMs toxicity and further undergo bioabsorption, chelation, and bioaccumulation of these HMs [20 - 22]. Also, the microalgal populations are efficient as they have a high capability of reproducing themself. High nutrient loads in the wastewater are the priority requirements for algal growth like nitrogen and phosphorous *etc*, whereas microalgal photosynthesis is triggered by CO_2 eliminated through wastes [23]. These microalgal populations are typically used as biosorbents to eliminate HMs, which undergo multiple steps from the biomass selection process, pre-treatment, and immobilization [24]. However, most of the time, a natural biofilm or mat is synthesized as a complex matrix that is typically made up of several microorganisms, including microalgae, bacteria, and fungal populations [25]. Microorganisms and microalgae upon colonizing the surface undergo secretions of multiple compounds such as polysaccharides, nucleic acids, and phospholipids and help to adhere to micronutrients [26]; thus, are efficient in the removal of organic pollutants, nutrients, pathogens, and HMs from the wastewater to get treated [27]. Therefore, microalgae-based heavy metals bioremediation in wastewater has been thoroughly discussed in this study and is considered to be a potential answer to address current environmental issues related to heavy metal contamination in aquatic ecosystems.

GENERAL OVERVIEW OF DIFFERENT HMS AND THEIR SOURCES

Heavy metals (As, Pb, Cd, Hg, Ni, Co, Cr, Zn, La, *etc*.) are generally considered non-essential, and toxic because they pose a health risk to the life of all living beings. These heavy metals get into the ecosystem through both natural and man-made processes. According to the findings of a large number of studies, natural sources of heavy metals in the environment are often less frequent as compared to anthropogenic activities [18]. 95% of Earth's crust is igneous rocks and 5% is sedimentary [28]. In general, basaltic ingenious rocks are abundant in a variety of heavy metals. These heavy metals in rocks can get into the soil environment through a variety of natural processes like erosion, leaching, surface winds, rain, weathering, biological, geological, terrestrial, and volcanic processes [29]. The slow geochemical cycle of heavy metals in nature is disrupted by human activities. This causes one or more heavy metals to build up in the soil [30]. Recent progress in the field of agriculture, urbanization, rural-urban transformation, industrialization, and many other advanced technology-based modern sectors all made a big difference in the amount of heavy metals in the soil. In addition to these factors, further humanistic activities, use of phosphate fertilizers, pesticides, insecticides, herbicides, fungicides, Cattle feed additives, industrial wastes, industrial effluents, sewage sludge, metalliferous ore, smelting, mining, refining, electroplating, wood preservatives, biosolids, paints, tanneries, kitchen appliances, semiconductors, medical equipment, steel alloys, energy plants, car batteries, municipal solid waste, landfills, fossil fuels, coal, peat, wood, and forest fires contribute to increasing the concentrations of heavy metals in soil, water, and air (Fig. **1**). All these factors finally lead to an increased risk of mounting the heavy metals in the ecosystem that further have a deleterious impact on the primary, secondary and territory consumers.

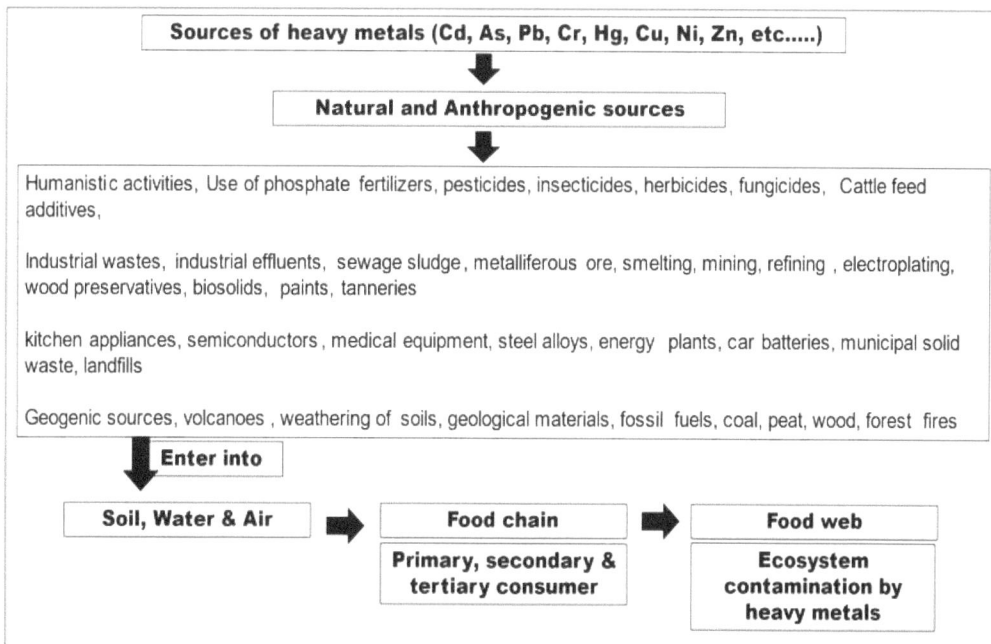

Fig. (1). Heavy metal sources and their impact on the ecosystem.

TOXIC EFFECTS OF HEAVY METALS ON PLANTS AND THE HUMAN POPULATION

The concentration of HMs in soil is rapidly rising as a result of anthropogenic and natural processes. These HMs are extremely hazardous to humans, even at low concentrations. When present in higher concentrations than permitted, certain metals can be hazardous to plants. The chief sources of HM pollution are considered to be the burning of fossil fuels, the smelting and mining of metallic resources, fertilisers, municipal garbage, pesticides, and sewage [31]. According to Upadhyay *et al.* [32], arsenic (As) is a key hazardous element that contaminates water in many nations, including India. In the same way, cadmium (Cd) is regarded as a poisonous invasive heavy metal that is toxic to both plants and people [33]. Another HM chromium (Cr) is utilised in a number of industrial processes and is available in Cr III and Cr VI, among other oxidation states. According to Pradhan *et al.* [34], Cr VI is the most unstable and hazardous form of Cr and affects both the normal physiology of plants and people. In addition to these HMs, a number of additional metals that are hazardous to both people and plants are also covered in detail in (Table **1**).

Table 1. Effect of different HMs on plants and humans.

S. No.	Heavy Metal	Effect on Plants	Effect on Humans	References
1.	As	Impairment to DNA, cell organelle, protein degradation, and lipid peroxidation.	A low concentration of $0.1\,\mu g/ml$ causes cancer in the lung, bladder, and kidneys. At higher concentrations causes arsenical dermatitis, arsenicosis, diabetes, cardiovascular disease, infant morbidity, impairment of the central nervous system, immune alteration, hyperkeratosis, and hepatic damage.	[32, 35]
2.	Cd	Reduces the growth of cells and chlorophyll content. Excess production of superoxide dismutase, phytochelatins, peroxidase, and catalase.	Damage reproductive organs, lungs, liver, and kidneys. Cause cancer, Alzheimer's, and Parkinson's Disease. Responsible for osteoporosis, kidney stone, gastrointestinal disorder, respiratory insufficiency, peripheral neuropathy, and hypertension.	[36, 37]
3.	Cr	Affects seed germination, and slows down the growth of roots and shoots. Due to Cr accumulation chlorophyll content and photosynthesis are affected. This leads to excess production of ROS, which results in oxidative damage.	Its ingestion is responsible for skin, lung, and stomach cancer. Chronic bronchitis, liver damage, epigastric pain, tissue neurosis, kidney problem ulcers, internal haemorrhage, and DNA impairment are also caused due to Cr intake.	[38 - 40]

(Table 1) cont.....

S. No.	Heavy Metal	Effect on Plants	Effect on Humans	References
4.	Hg	Affects plant growth, nutrient uptake, yield production, and homeostasis. It induces oxidative stress even at very low concentrations. Hg results in increasing the antioxidative activity. When bound with DNA, it is responsible for damaging the chromosomes.	Causes antibiotic resistance, reproductive retardation, and mental retardation and shows biomagnification if entered into the food chain.	[41, 42]
5.	Pb	Disturbs plant growth by affecting seed germination, transpiration, lamellar organization, chlorophyll production, seedling development, and root elongation. Due to overproduction of ROS, it inhibits lipid peroxidation, and ATP production and causes DNA damage.	Anaemia, central nervous system damage, reproductive abnormality, brain disease, kidney malfunction, dementia and even death.	[43, 44]

HMS TOXICITY IN AQUATIC ECOSYSTEMS

The amount of heavy metals (HMs) rises to dangerously harmful levels in the water as a result of growing industrialization and urbanization, demonstrating that anthropogenic activities are closely connected to the abundance of heavy metals in many ecosystems. Among various HMs, Pb, Cr, Cd, Cu, Hg, and Zn are some of the common pollutants found in industrial effluent, freshwater bodies, and marine ecosystems and they contribute to a global ecological problem [45]. These elements may exist in the aquatic system in a variety of chemical compositions (species), dispersed between the soil and the solution. Their recalcitrance and subsequent persistence in water bodies suggest that, even if they are present in minute, undetectable amounts, concentrations may rise to the point where they begin to display hazardous properties due to natural processes like biomagnifications [46]. According to Khan *et al.* [47] under this situation, exposure to HMs in aquatic environments over the allowable limits directly harms humans and other living forms. HMs may be found in the sediment as dissolved inorganic complexes, suspended particulates, or in combination with organic colloidal particles or might exist as free metal ions, as well as in the form of organic and inorganic compounds, in solutions [48]. HMs are extremely difficult to remove from wastewater compared to other contaminants since they cannot be reduced chemically or biologically. The ability of an aquatic system to self-purify is destroyed when HMs are fatally introduced into the aquatic environment through waste [47]. Monitoring the levels of these pollutants in marine creatures is crucial due to heavy metal toxicity in aqueous organisms, lengthy residence times within food chains, and the possibility of human exposure. The balance between all of these metal species is replaceable and is influenced by environmental variables including temperature, pH, and alkalinity as well as by the aquatic biota [49]. The chemical form of a trace metal affects its biological availability as a necessary nutrient (such as iron and zinc) or as a toxicant (such as cadmium and lead). Since the most poisonous species are often the free metal ions, which are the bioavailable forms any procedure that expedites their conversion into bound forms causes a decrease in toxicity to the biota [50]. Since water is one of the main entry points for these harmful substances into the body, the EPA [51], established maximum contaminant levels for specific metals (*i.e.*, concentrations of Hg, Pb, Cr, Cu, Cd, Zn, and Ni in drinking water should not exceed 0.002, 0.015, 0.1, 1.3, 0.005, and 0.04 mg/L, respectively).

If HMs are not properly eliminated from the aquatic environment, they will eventually accumulate throughout the food chain due to their long-term persistence or bio-transformations [49]. Numerous physical, chemical, and biological processes may occur as a result of the introduction of heavy metals into aquatic systems [52]. Physical alterations in the receiving water habitats may also result from HM emission. These take into account variations in the pH of the water, the substrate's organic content, and the size of the water particle. As a result of these annoyances, the number, variety, and species composition of aquatic plants decrease. Further, HMs accumulation in aquatic plants and animals affects their physiology by modulating various enzymes (antioxidative and digestive enzymes) and hormone levels (auxin, cytokinin, gibberellins, ethylene) leading to the death of particular life forms [53 - 55]. Depicting the effects of HM pollution different from those physical effects that are activated indirectly might thus be problematic [56]. The impacts of HMs on the environment and the effects of the environment on HMs may be separated into two broad categories [57]. The first categorization is dependent on the environment; diversity, density, species makeup of the population, and community organisation may all change. The second classification emphasises that changes in the speciation and harmfulness of HMs may be caused by circumstances in receiving waters. These factors include the relative contributions of anthropogenic and geochemical material, the type of industrial effluents, the concentration of chelators, and suspended solids. The impact of HMs on aquatic plants varies with species and habitat ecology [58]. Exposure to HM pollutants increases the risk of cancer and cancer-related disorders by causing cytotoxicity, carcinogenicity, teratogenicity and mutagenicity [59, 60]. Because of the underlying toxicity of HMs pollution, the aquatic life exposed to it may undergo a variety of metabolic changes. These changes may limit growth through (i) conformational changes in biological molecules that compromise the integrity of whole cells and/or their membranes;

(ii) disruptions of the osmotic balance between the cytoplasm and the medium surrounding the cells; and (iii) disruptions of enzyme functions caused by the blocking of specific functional groups and competition with essential nutrients [61]. Accidental HMs' ingestion by humans *via* the aquatic system can result in a number of health issues. High quantities of HMs in the body cause renal issues, anaemia, sclerosis, and diseases of the neurological system [62]. Wilson's illness is brought on by the accumulation of Cu, whereas metal fume fever is brought on by Zn and can limit bone formation and result in nephritis [63]. Infertility can also result from HMs exposure in people. HMs' contamination poses a serious hazard to fish and other aquatic resources. Humans will be directly impacted by the accumulation of HMs in aquatic products like fish. In aquaculture, it is essential to make sure the water is free of HM contamination in advance. *Percocypris ping* freshwater fish were found to be 50% less likely to survive after being exposed to mercury at a concentration of 0.441 mg/L and cadmium at a concentration of 2.551 mg/L within 24 hours. Additionally, within 24 hours, a concentration of 59.56 mg/L aluminium resulted in the death of *Catla catla*, *Labeorohita*, and *Cirrhina mrigala* [64].

HEAVY METAL TOXICITY IN SOIL

Metals and metalloids with a density of more than 5 g/cm³ are together referred to as heavy metals. Some of these are crucial micronutrients, while others have no beneficial effects on living things. Heavy metal ion emissions from human activity are rising, and this poses a major threat to ecosystems in both water and land. Heavy metals are mostly found in rocks and are discharged into the environment by both natural and artificial processes. The weathering of rocks and volcanic activity are two examples of natural sources of heavy metals [65]. Heavy metal soil pollution is a major cause for concern due to its negative impact on the living biota. Heavy metal's ability to endure and lack of biodegradability facilitates their accumulation in the environment. The number of contaminants reaching agricultural soils from diverse sources is enormous. Due to their negative impact on the living biota, heavy metal soil pollution is a major cause for concern. Heavy metal accumulation in agricultural soils is a severe issue because it poses a potential risk to human health and food safety as well as has a negative impact on the soil's ecosystem [66]. Heavy metal concentrations in the soil are related to geochemical and biological cycles. Human-made activities like transportation, waste disposal, industrialisation, and social, and agricultural activities have an impact on environmental contamination and the health of the planet's ecology. These processes have a detrimental impact on human health as well as the health of all living things. Since the start of the industrial revolution, there has been a dramatic rise in the amount of harmful metal pollution in the environment. It is a problem of concern when heavy metals, such as cadmium, lead, chromium, and copper, pollute the soil [67]. Arsenic is one of the heavy metals that are widely distributed throughout soils and aquatic bodies. Arsenic, the 20th most common element on Earth, can be found as an oxide, sulphide, or metal salt. In almost every habitat, it is frequently found in extremely low amounts. Natural processes such as rock weathering, volcanic eruptions, biological activity, and geochemical reactions release arsenic into the environment [68]. Heavy metal contamination of the environment is a result of melting operations and sludge dumping, although sewage sludge contains a lot of organic matter, it also has a lot of hazardous heavy metals, such as cadmium, nickel, chromium, mercury, copper, lead, and zinc [69]. Furthermore, the high levels of heavy metals in sewage sludge may lead to soil contamination, which would harm living things. Hazardous metals released into the soil drastically lower plant quality and yield, which poses a serious threat to people and animals due to their bio-magnification in food chains. This is because soil health is essential for the development of food crops [70].

Sources of Heavy Metal Toxicity in the Soil

Both anthropogenic activities and natural processes contribute heavy metals to the agroecosystem (Fig. **2**). Numerous studies have shown that natural sources of heavy metals in the environment are often of low importance when compared to anthropogenic activity. Some soils take these metals from parent material that has a high background of naturally occurring heavy metals. As the parent material typically has high concentrations of these metals, this inheritance may be harmful to plants and other species [71].

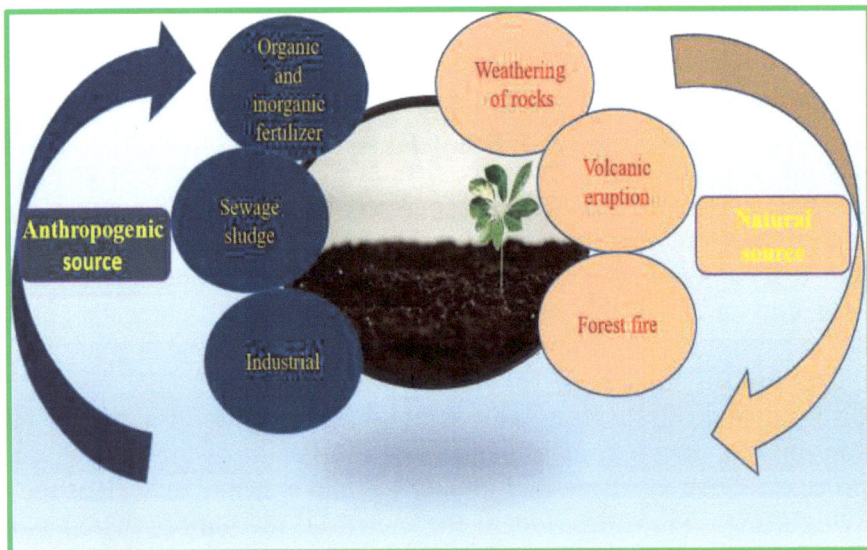

Fig. (2). Sources of heavy metal toxicity in soil.

The parent component from which they were formed is the main source of heavy metals in soils. 95% of the crust of the Earth is composed of innovative rocks, while only 5% is made up of sedimentary rocks. In general, heavy metals like Cu, Cd, Ni, and cobalt (Co) are abundant in basaltic intelligent rocks, whereas Pb, Cu, Zn, manganese (Mn), and Cd are abundant in shales. Through natural processes such as meteoric, biological, terrestrial, and volcanic processes, as well as erosion, leaching, and surface winds, heavy metals contained in rocks can infiltrate the soil environment [72]. The primary sources of Cu, Zn, and Cd in agricultural soils are also thought to be phosphate fertilizers, herbicides, and fungicides [73]. Consequently, pinpointing the source of heavy metals is crucial for pollution reduction and prevention. To assess the variability and potential sources of heavy metals in soils, multivariate statistical methods are typically utilized in conjunction with geostatistical techniques [74].

MICROALGAE IN THE BIOREMEDIATION OF HMS IN AQUATIC ECOSYSTEMS

Heavy metals are high-density metallic chemical compounds that occur naturally on earth's crust *e.g* Arsenic (As), Lead (Pb), Zinc (Zn), Copper (Cu), Palladium (Pd), Uranium (U), Radium (Ra), *etc.* These heavy metals are essential for living beings to some extent but at higher concentrations, they may act as toxic substances [75]. Moreover, after reacting with other organic compounds, they form a metal-organic complex, which is a very toxic pollutant [76]. Accumulation of these heavy metals in water bodies at toxic concentrations can cause disability to an aquatic ecosystem by entering into the food chain thus causing the death of fish. Economic growth depending upon urbanization and industrialization are the main reasons behind ever-increasing heavy metal pollution in water bodies [77]. An increase in the concentration of heavy metals in aquatic ecosystems causes bioaccumulation of heavy metals in the body of aquatic organisms, affects normal embryonic development, induces immunotoxicity, damages the endocrine system, destroys the cell membrane structure, brings genetic mutations, and causes carcinogenesis and abnormal functioning of cellular organelles in aquatic organisms while it inhibits photosynthesis, growth, and respiration and reduces pigments in plants. It induces reactive oxygen species by generating free radicals, disturbs metabolic pathways, and can damage cells of microorganisms in aquatic ecosystems causing their death [76]. As heavy metals are non-biodegradable in nature, appropriate methods should be advanced to remove these toxicants from water bodies. To date, various methods are available that are effective in the removal of heavy metals from the water like the ion exchange method, floatation method, membrane filtration method, chemical precipitation method, *etc.*, but technology is advancing toward the development of some biological methods to overcome this issue, as chemical methods can alter the physiochemical properties of water. Moreover, these methods are costlier and need high maintenance to establish [78, 79]. Biological methods are far better than chemical methods. These biological methods include the use of microorganisms like fungi, bacteria, and algae to remove heavy metals from water [80]. Of all microorganisms, bioremediation of heavy metals using microalgae is more in demand due to some of its advantageous characteristics over other microorganisms like rapid growth, abundancy, shorter life span, high carbon utilization efficiency, excellent metal uptake efficacy, value as both living cells or non-living biomass, high photosynthetic rate, *etc*. They grow well even under harsh conditions, and this microalgae has more surface area, abundant metal binding sites, high binding affinity and is non-toxic, cost-efficient, eco-friendly and provides valuable byproducts [81, 82, 33]. Microalgae used for bioremediation can be reused to obtain products like fertilizers, bioplastics, feed, microfibers, *etc*. It also has the ability to reduce other organic and inorganic pollutants [79]. Various microalgal strains are found effective in the removal of heavy metals from the water like *Scenedesmus, Chlorella, Penium, Desmodesmus, Neochloris, Botryococcus, Nannochloropsis, Chlamydomonas, Chaetoceros, Porphyridiun, Spirulina, Picochlorum, etc* [83, 84]. Various studies have been conducted on different strains of microalgae against different metals as given below (Table **2**).

Table 2. Interaction of different microalgae with a variety of heavy metals.

S. No	Microalgae	Heavy metal	Reference
1.	*Chlorella vulgaris* and *Spirulina maxima*	Copper and Zinc	[85]
2.	*Scenedesmus spinosus* and *Chlorella vulgaris*	Copper and Molybdenum	[21]
3.	*Arthrospira platensis* (Spirulina) and *Chlorella vulgaris*	Nickel, Cadmium, Copper and Zinc	[86]
4.	*Chlorella vulgaris*	Lead, Cadmium, Copper and Chromium	[87]
5.	*Spirulina platensis, Chlorella vulgaris* and *Scenedesmus quadricuda*	Nickel, Cadmium and Lead	[88]
6.	*Heterochlorella sp.* MAS3 and *Desmodesmus sp.* MAS1	Zinc and Manganese	[33]
7.	*Chlorella kessleri*	Lead, Copper, Cobalt, Cadmium and Chromium	[89]
8.	*Chlorella sorokiniana*	Cadmium	[90]
9.	*Arthrospira platensis* (*A. platensis*)	Cadmium, Copper, Nickel, Lead, Chromium and Cobalt	[91]
10.	*Desmodesmus abundans* M3456	Lead, Manganese, Copper, Cadmium and Zinc	[92]

(Table 2) cont.....

S. No	Microalgae	Heavy metal	Reference
11.	*Oscillatoria limosa, Nostoc commune* and *Chlorella vulgaris*	Zinc, Cadmium, Iron, Copper, Nickel and Lead	[93]
12.	*Chlorella sorokiniana*	Chromium	[94]
13.	*Scenedesmus abundances* and *Chlorella vulgaris*	Aluminium, Iron and Copper	[95]
14.	*Chlorella vulgaris*	Nickel, Cadmium and Zinc	[96]
15.	*Scenedesmus acutus* and *Chlorella pyrenoidosa*	Cadmium	[97]
16.	*Dunaliella salina, Navicula salinicola* and *Amphora coffaeiformis*	Lead, Cadmium and Chromium	[98]
17.	*Arthrospira platensis*	Zinc, Lead, Copper and Iron	[99]
18.	*Chlorella sorokiniana*	Lead	[100]
19.	*Monoraphidium pusillum* and *Desmodesmus communis*	Zinc and Copper	[101]
20.	Scenedesmus almeriensis and *Chlorella vulgaris*	Copper, Arsenic, Manganese and Zinc	[102]
21.	*Tetraselmis sp.*	Nickel, Cobalt and Chromium	[103]
22.	*Parachlorella kessleri*	Cadmium and Chromium	[104]
23.	*Chaetoceros calcitrans*	Cadmium and Copper	[105]
24.	*Polyedriopsis spinulosa* and *Chlorosarcinopsis bastropiensis*	Lead and Cadmium	[100]
25.	*Chlorella vulgaris*	Lead	[106]
26.	*Arthrospira maxima*	Iron	[107]
27.	*Phacus* strain	Nickel, Aluminium and Lead	[108]
28.	*Tetraselmis marina* AC16-MESO	Iron, Copper and Manganese	[109]
29.	*Botryococcus sp.*	Arsenic	[110]
30.	*Nannochloropsis oculata*	Cadmium	[111]

Each microalgal strains show different response towards different metal ions at different rates from one another which is clearly depicted by a study where three microalgae *Chlorella, Porphyridium,* and *Spirulina* highly prefer to reduce Pb, Cu, and Cd, respectively among other metals in the given media at different metal reduction percentages [112]. The mechanism of heavy metal removal by microalgae involves either extracellular and intracellular strategies or both (Fig. **3**). Extracellular strategy is a rapid and passive process that includes binding of metals on cell wall surface by biosorption. It takes place in living as well as in non-living biomass of microalgae [113]. Cell wall composition and structure on which functional groups like hydroxyl, carboxyl, amino, phosphate, sulphydryl, *etc.*, and extracellular polymeric material like lipids, proteins, sugar, nucleic acids *etc.* are attached, play vital roles in metal binding *via* any covalent bond, ionic bond, Vander Waals force or electrostatic interaction in case of complex formation, ion exchange mechanism, microprecipitation or physical adsorption on cell surface [114 - 116]. While the intracellular strategy takes place only in living cell biomass, it involves bioaccumulation of heavy metals in cells. It mainly consists of transporting heavy metals through the cell membrane. It is a biphasic process in which the first step is the rapid passive absorption of heavy metals on the cell wall similar to biosorption while the second unlike biosorption is a slow and active process that includes the transport of metal ions across the cell membrane through the transporter protein that helps in the accumulation of metal ions in the cells. It can maintain its intracellular ion concentration using various strategies like chelation, compartmentalization, and biotransformation of heavy metals [116, 114]. Accumulation of heavy metals in higher concentrations in microalgae induces free radicals resulting in the production of reactive oxygen species. The microalgae have certain antioxidant mechanisms including both enzymatic as well as non-enzymatic mechanisms that can combat this oxidative stress and prevent cell death, the accumulated metal is finally detoxified by forming complexes by chelation with the cellular components or by transporting them to other cell compartments [117]. There are various factors that influence the bioremediation of heavy metals like concentration of biomass, pH of the system, temperature, initial metal ion concentration, electronegative interactions, interference of multi-metal ions, time of contact between metal, concentration of heavy metals, *etc.* [118, 59]. Studies show that microalgae like dried *Synechocystis*, *Scenedesmus,* and *Chlorella* sp. increase the uptake of metal ions like copper, chromium, or nickel at the initial concentration of metal ions to some extent due to abundantly available metal binding sites [119]. While adsorption process has a certain limit due to the overall occupancy of adsorption active sites on the cell wall. This was observed in a study on microalgae *Nannochloropsis oculata* against copper metal whose average adsorption limit was observed at $10.70 \pm 1.92\%$ approximately [120]. The greater the contact time between the metal and the wall, the more the biosorption, as proper time is required to pass the metal into the cell, which is initiated by the passive process followed by an active process [118]. Another study shows that the efficiency of *C. vulgaris* in removing heavy metals from the water was highly influenced by the pH of the media and temperature. High temperature can be a reason for weakening bonds between metals and functional groups present on the cell wall thus reducing adsorption capacity [121, 122]. A study shows that in a mixture of metals, the microalgae was more effective in removing lead from media. The reason can be the high electronegativity of lead among other metals present in that media. The respective electronegativity of different metals and their removal rate was in the order of lead followed by cobalt, copper, cadmium, and chromium, respectively [89].

Another study shows an increase in the efficacy of microalgae *Chlorella kessleri* of heavy metal biosorption by increasing biomass concentration from 0.5 to 1.5 grams per litre [89]. While the efficiency of microalgal bioremediation decreases with an increase in metal concentration. This is shown by a study in which an increase in the concentration of zinc and copper leads to a decrease in cell growth and photosynthetic activity of the algae *Scenedesmus obliquus* and *Chlorella pyrenoidosa* [84, 123]. The biomass obtained after the bioremediation of heavy metals can be reused in various ways, it can be used as biofuel, and precious metal ions like gold, silver, platinum, *etc.* can also be recovered from it, producing various bioactive products like vitamins, proteins, pigments, *etc.* Moreover, they can also act as biofertilizers and biochar [82, 84, 124].

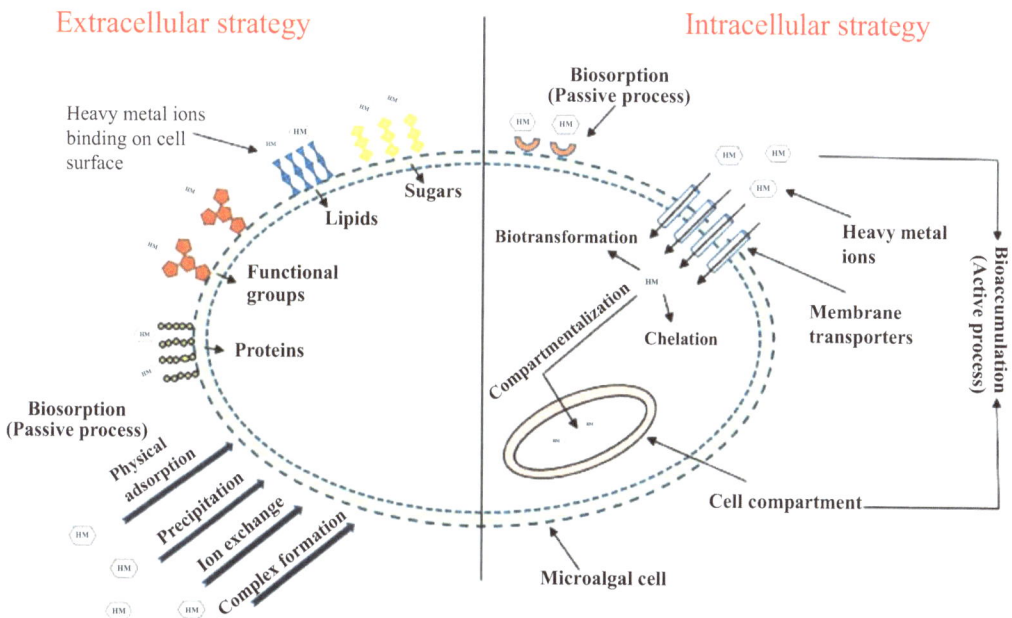

Fig. (3). Extracellular and intracellular strategies involved in the mechanism for the bioremediation of heavy metals by microalgae.

Potential Mechanisms Adopted by Microalgae for Wastewater Treatments

Researchers have been investigating the utilisation of algae for wastewater reclamation, a topic that is currently trending since the 1960s [125]. In light of their capacity to use both organic and inorganic carbon, nitrogen, and phosphorus, as well as their capacity to accumulate biomass and reduce COD, N and P in the wastewater, microalgae have been used in the treatment of wastewater [26, 126, 127]. Although microalgae are not currently widely used in the cleanup of wastewater, important commercial systems do emerge [128]. Microalgae have the capacity to absorb micronutrients from waste water and reduce carbon dioxide emissions. The floating photo-bioreactor (PBR) from the US-based enterprise Algae Technologies LLC is constructed for use with ambient light and CO_2 to flush out nutrients downstream from their origin [129, 130]. A controlled PBR setup linked with an anaerobic digester to produce biogas (methane) is the basis of Algal Enterprises' (Australian) approach to an array of wastewaters [131, 132]. Microbial Engineering's (United States) RNEW technology employs wide raceway ponds enhanced with CO_2 for eliminating N and P from wastewater from municipalities and creates biomass for biofuels [128, 133]. However, a plethora of other elements influence nutrient removal from wastewater and the establishment of microalgae [134]. Availability and accessibility of nitrogen or phosphorus have distinct effects on microalgal metabolism. Under nitrogen starvation, many microalgae continue to generate biomass, but they will generate more lipids and/or carbohydrates and less protein [135, 136]. The pH value is also very essential for nutrients [137]. For instance, it influences the soluble state of ammonium or phosphate as well as the development of precipitates. Calcium phosphate or magnesium hydroxides, which are inaccessible to microalgae, can be formed at a high pH of 9 or above [138, 139]. The values of the pH also control the electrical charge of functional groups at the exterior of microalgae, which affects their capacity for binding substances like heavy metal ions in wastewater [140, 141]. Due to its impact on the elimination of nutrients and microalgal development, wastewater colour plays a significant role in wastewater cleansing [142, 143]. Dark brown-greyish and thick wastewater, such as that from livestock, pulp, and paper exhibit substantial light absorbances, which restricts the development and assimilation of nutrients by microalgae by preventing them from performing photosynthetic absorption [144, 145]. Using the processes of biosorption, bioaccumulation, biofilm-forming, and biodegradation, microalgae eliminate several kinds of poisons [79]. Toxins from agricultural runoffs, and wastewater from the textile and pharmaceutical industries, are a few examples. Certain microalgae can even use biosorption to remove heavy metals from industrial wastes also [81, 146].

Biosorption

A biological substance that may attach to and absorb contaminants from water is the sorbent in the passive biosorption process. In simple terms, biosorption is a method of mass transfer where a substance is removed from the liquid phase and anchored to the surface of a solid [147]. Precipitation, ion exchange, surface complexation, electrostatic contact, absorption, and adsorption are just a few of the various mechanisms that promote the physicochemical and metabolic-independent processes that makeup biosorption [148]. The microbes, including their parts, could be either living or deceased when it comes to the biological material involved. The attraction of sorbate species is caused by the biosorbent's strong affinity for its intended sorbate, and the overall capacity is what determines how many sorbate molecules are produced [149, 150]. Until a balance is attained involving the amount of the material still present in the liquid and how much has been taken by the biosorbent, the procedure is repeated. Furthermore, the split of a particular sorbate between the solid and liquid phases depends on the degree of biosorbent affinity for that sorbate [116].

When the radioactive substances and HMs emitted from a nuclear reactor were collected in microalgae in the early 1970s, biosorption was initially noted in a number of microalgae [79]. Microalgae's cell wall, which is solely accountable for biosorption, performs a crucial role in the procedure and dictates the mechanism by which this occurrence occurs [151, 152]. The intercellular spaces found on the cell wall, paired with the fibril matrix that generally gives microalgae's cell walls their considerable rigidity and the amorphous fraction that gives them elasticity, may all help to improve the method of biosorption [153]. Compact cell sizes provide a better contact area per biomass unit, improving the adsorbent surface area [154]. However, greater biomass does not always imply more biosorption properties, as metal cation recorded in an aqueous medium may be minimized by active site repulsion within microalgae surfaces, which lowers the removal efficiency. The number of biosorbent particles is directly proportional to the number of active sites for biosorption [34]. Some heavy metals are generally very toxic to microalgae. However, owing to the hormesis phenomenon, heavy metals promote microalgae development and metabolic processes in the midst of minute amounts of hazardous metal ions [155]. A two-step method is used by microalgae to biosorb HMs ions. A metabolism-independent phase confronts quick and reversible adsorption of adsorbate onto active sites on the surface of microalgae following a slow phase that relates to affirmative intracellular diffusion and predominantly entails the metabolizing processes of microalgae [156]. Extracellular polymeric substances (EPS) created from algae can hasten the biosorption of metals and their efficiency is determined by the EPS framework, solution characteristics, metal species, and operational circumstances [157]. For instance, the ability of *Nostoc linckia*'s EPS to biosorb Co (II) and Cr (IV) at a low pH level was believed to be due to interactions involving the negatively charged functional components of EP and metal ions [158]. In contrast to ion exchange, Mota *et al.* [159] discovered that *Cyanothece* sps. EPS can remove HMs from contaminated water because of organic functional groups like carboxyl and hydroxyl. Similarly, the two green algae species, *Chlorella* sp. and *Micractinium* sp. were utilized in wastewater treatment to absorb nutrients [160]. It is possible to estimate a biosorbent's selectivity and capability for particular HMs ions using detailed and accurate information about the chemical structure of the material. Scientists can characterise the specific functional groups responsible for complexity with HMs and pinpoint the precise root causes by using nuclear magnetic resonance spectroscopy and Fourier-transform infrared spectroscopy (FTIR) techniques [161, 162].

The best pH for biosorption depends on the type of microalgae and the metal being absorbed. The pH ranges from 3.0 to 6.5 typically. Metals can precipitate in their hydroxide versions at values over 6.5, leaving biosorbents useless and causing the buildup of highly toxic waste whereas values less than 3.0 cause dissociated H^+ ions to compete for active sites on the surface of cells [163]. The adsorption of Uranium was achieved by *Nostoc* and *Scenedesmus* species at pH 4.5 and 5.6 respectively [164]. On the other hand, *Nostoc linckia* and *Nannochloris oculate* were successful in adsorbing Cr(IV) at pH 2.0 [158, 165]. Also, metal abundance and competition are two key elements for biomass saturation since there are only so many active sites on the surface of cells [166]. Numerous studies have also demonstrated that ambient temperature has an impact on absorption, which may be advantageous or disadvantageous for the application in wastewater treatment [163, 167]. However, in other investigations, the temperature had a little effect on metal sorption [168]. Therefore, there are numerous factors that affect biosorption, such as temperature, pH, biomass parameters, and the presence of other cations [116].

Bioaccumulation

The concept of utilizing microalgae for the bioaccumulation of heavy metals was initially put forth in 1957 by Oswald and Gootas [169]. Additionally, during that time it was also discovered that, in the aquatic environment, when microalgae settle out during algal blooms, the amount of heavy metals concentration in the water decreased by 20 to 75 percent [170]. Heavy metal accumulation is also aided by algal polysaccharides (carrageenan). Furthermore, Green algae have high tolerances for lead (Pb), copper (Cu), and zinc (Zn) better than blue-green algae and diatoms [15]. Moreover, *Spirogyra hatillensis*, a freshwater filamentous alga, has been shown to be capable of performing heavy metal biosorption from aqueous solutions [15]. In addition to these reports, it was also investigated that marine microalgae are used as cheap adsorbents to remove mercury (Hg), lead (Pb), and cadmium (Cd) from water. Recovery of metals from wastewater can be accomplished by the use of microalgae. In a microalgae pond, a fixed-bed anaerobic reactor can be used to remediate wastewater pollutants. The immobilization of algal cells is a promising way to clean up wastewater because it does not require harvesting, which is the hardest part of this process. The algal cells cannot move around freely because of a gel matrix. More so, their higher cell density results in faster reaction times. Immobilized algal cells are chosen since they do not demonstrate any cell washout when tested. *Chlorella vulgaris* immobilized on sodium alginate beads matrix removes more nutrient pollutants from raw sewage [15].

Presently, microalgae adopt a two-step process to remove toxic heavy metals. The first step is adsorption (*i.e.* fast passive adsorption) outside the cell. The second step is bioaccumulation (*i.e.* slower positive diffusion and HMs buildup) inside the cell [82]. Bioaccumulation only happens in living cells. It happens when metal ions pass through the cell membrane and bind to cytoplasmic proteins, polysaccharides, or certain cellular compartments like vacuoles, *etc.* In addition to this, the term "biosorption" refers to the two processes mentioned above that involve both adsorption and bioaccumulation by alive or dead microorganisms [118].

Research has shown that algae have significant sorption capacities and a high selectivity for the sorption of metal ions [118, 171]. Some microalgae and cyanobacteria can handle large amounts of metal stress and have a high tolerance for metal toxicity, because of their large surface area, and a high affinity for binding [172]. Therefore, microalgae are used as an effective bioremediation tool for the removal of toxic compounds. Moreover, it is also considered a far more effective bioremediation technique as compared to other biological agents because microalgae possess great characteristics like cell wall structure, surface area, and presence of tiny peptides Phytochelatins (PCs) and metallothioneins (MTs) that enhance adaptability in absorbing harmful components from both the soil as well as from water [173].

Biofilm Based Wastewater Treatments

The idea of a microalgal biofilm may have first been inspired by the microbial mats found in the natural world. Most of the time, a natural biofilm or mat is a complex matrix made up of several microorganisms, including bacteria, fungi, and microalgae [25]. For instance, it has been observed that biofilm that had developed on the white sea coast and found that *Haematococcus lacustris* cells were detected in the biofilm's upper layer (the photoautotrophic layer), while heterotrophic microorganisms were found in the matrix [174]. The trapping of the cells in hydrogel polymer matrices and the production of biofilms, either naturally or artificially are used to immobilize the microalgae. The ability of microalgae to produce biofilms is exploited by immobilizing the cells. The surface of a bedding material becomes coated with organic and inorganic substances, favouring the growth of microorganisms [128]. Upon colonizing the surface, microalgae and bacteria start to secrete extracellular materials comprised of nucleic acids, proteins, polysaccharides, and phospholipids that help the organisms adhere to the bedding material. These substances not only help the microalgae and bacteria adhere better to the bedding material but also trap and concentrate the nutrients required for cell growth [26]. Applications of immobilized microalgae in wastewater treatment offer a number of benefits over the conventional suspended method such as higher cell density caused the photobioreactor design to be more flexible, which in turn enhanced reaction and absorption rates, improved operational stability preventing cell washouts, simpler replacement of the algae through simple cultivation, harvesting, and management of the biomass production [175].

Production of Biomass

It has been found that growing microalgae based on recovering nutrients from wastewater is a more affordable method of producing microbial biomass [176]. In a study, it was observed that the biomass production of microalgae on biofilm was 1.474 g/m^2/day, 47 g/m^2/day, and 7 g/m^2/day in sludge thickening supernatant [177]. In addition, it was reported that a typical mixotrophic artificial medium (TAP media) was found to have a suspended microalgae biomass productivity of roughly 0.367 g/L/ day with a maximum biomass production of 1.1 g/L on the third day of cultivation [178]. According to Quan *et al.* [179], in experiments with algal biofilm in landfill leachate, algal biomass was produced at various rates by biofilms in wastewater from various sources. This phenomenon is mostly caused by the diverse nutritional profiles of wastewater. The highest biofilm density of 28.0 g/m^2 was attained when the molar ratio of nitrogen to phosphorus (N/P) was 16:1. Similarly, the system factors may play a role in the algal biofilm's biomass productivity. For instance, biomass productivity (footprint) reached 4.5 and 6.5 g/m^2/day when the vertical heights of algal biofilms were established at 0.9 and 1.8 m, respectively [180]. Therefore, wastewater nutrient and algal biofilm parameters need to be tuned in order to maximize biomass productivity.

Removal of Nutrients by Algal Biofilm

In addition to affecting the biomass yield of an algal biofilm, nutrient removal affects the water quality of treated wastewater. Few algal biofilms were effective at treating both natural and manufactured wastewater (Fig. **4**). First, in synthetic wastewater made from the primary settling sewage, a vertical algal biofilm-enhanced raceway pond eliminated 86.37% of the total nitrogen (TN), 91.20%

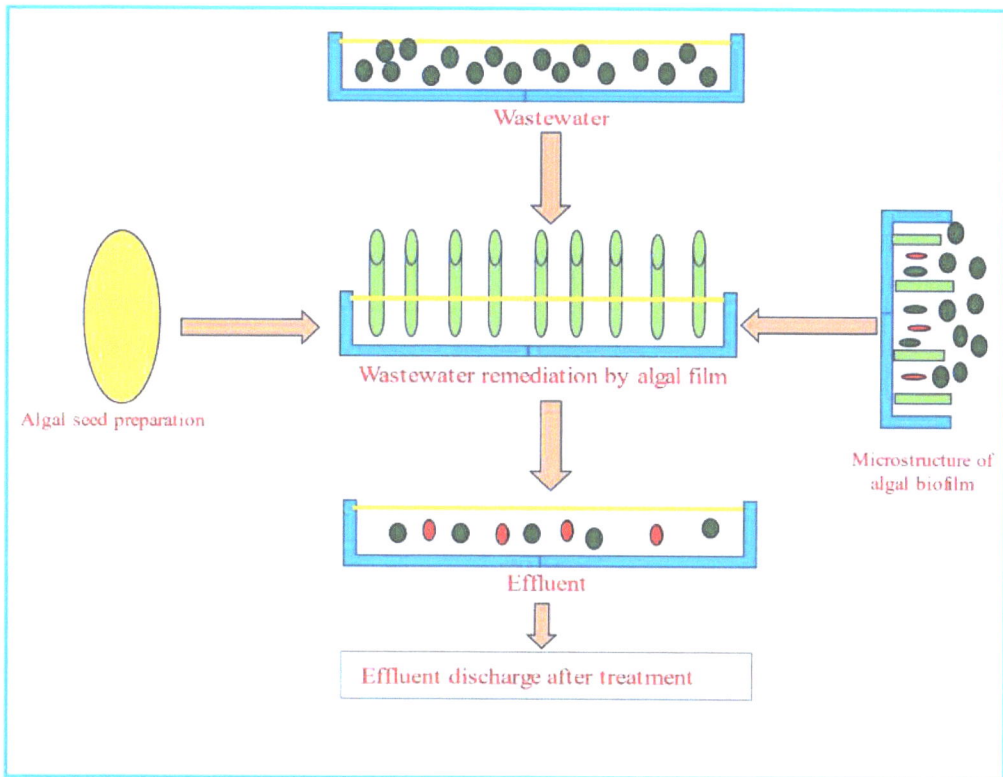

Fig. (4). Algal biofilm for waste remediation.

of the chemical oxygen demand (COD),and 95.19% of the total phosphorus (TP) [181]. According to Tenore *et al.*, light is shown to be the most important component in the ecology of phototrophic-diverse biofilms [182]. In addition, algal biofilms functioned under red light and blue light to remove various amounts of nutrients from wastewater [177]. Similarly, the amount of algal biomass may also be influenced by light intensity and the day-dark cycle, which further affects how well biofilms remove nutrients [183]. The biofilm-based wastewater treatment may be impacted by algal-bacterial interactions in addition to physical variables [184]. According to Katam *et al.* [185], the removal of carbon and nutrients by algal biofilm in domestic wastewater benefits from algal-bacterial collaboration.

Mechanism of Heavy Metal Bioaccumulation in Microalgae

In addition to the biosorption process, bioaccumulation is a dependent metabolic pathway. It describes how HMs accumulate intracellularly through living microalgae cell membranes *via* passive and/or active transport channels. It is differentiated by two continuous processes, first is the quick, passive, and non-specific uptake of metal ions on the cell wall. After that, there is either active or passive transport to the cytoplasm *via* the cell wall and plasma membrane (Filote *et al.*, 2021). In fact, *Tetraselmis suecica* assisted Cd (II)-uptake was reported by Pérez-Rama *et al.* (2002) as a two-stage process that involved an energy-dependent accumulation in the cytosol after the first phase was aided by adsorption to proteins or polysaccharides. Additionally, cations can be carried by the negatively charged groups of the cell surface to the intracellular compartment *via* active transport across the plasma membrane after binding to thiol molecules, mainly cysteine, when the concentration of metal in the extracellular environment is much higher than the inside cells concentration. Other amino acids, like proline, glutamate, and histidine, can also be very important for detoxification and metal chelation [59]. Due to the hydrophilia of the majority of HMs, a particular protein known as a metal transporter is primarily responsible for their trafficking across the plasma membrane. In intracellular compartments, many detoxification mechanisms can then occur [82, 157].

MICROALGAE-BASED REMEDIATION OF HMS FROM SOIL ECOSYSTEM

Microalgae consume HMs such as boron, copper, molybdenum, and zinc as part of their nutrition and cell metabolism. However, other HMs, such as arsenic (As), lead (Pb), cadmium (Cd), mercury (Hg), and chromium (Cr) are highly lethal for microalgal growth. Microalgal species display various strategies to protect themselves against various HMs. They facilitate HMs immobilization, gene regulation, exclusion, chelation and production of antioxidants to reduce HMs. Also, microalgae also synthesize phytochelatins (PCs) that reduce the impact of HM stress [186]. In order to counter free radicals, microalgae are reported to form antioxidant enzymes (superoxide dismutase (SOD), ascorbate peroxidase, catalase, glutathione reductase, and peroxidase) and non-enzymatic antioxidants (carotenoids, cysteine, ascorbic acid (ASC), glutathione (GSH) and proline) [32]. The removal of HMs by microalgae can be explained by a two-stage mechanism. The first stage is defined as rapid extracellular passive adsorption (biosorption) while the second is slow intracellular diffusion and accumulation (bioaccumulation) [187]. The interaction between various HMs and microalgae is explained in the following subsection.

Arsenic (As)

Arsenic (As) is categorised as a class A and category 1 carcinogen as per the reports by the United States Environmental Protection Agency (USEPA) [188]. The pollution of As results from excessive anthropogenic activities, such as mining, the use of fertilizers, pesticides, smelting, and pigment manufacturing [189]. It has been reported by Upadhyay *et al.* [32] that As results in the impairment of cell organelles, DNA, lipid peroxidation, and protein degradation. Microalgae diminishes As toxicities by As (III) oxidation, complex formation with phytochelatins, and its biotransformation into methylated arsenic forms or arsenolipids [190]. According to Arora *et al.* [188], oxidation of As (III) occurs outside microalgal cells after its interaction with functional groups such as -OH, -NH, -CH of aldehyde and aliphatic as well as -CN of amino groups. It was reported by Baker and Wallschläger [191], that microalgae excrete As in two forms, methylated arsenic species and reduced As (III). Due to the similarity of outer electronic configuration with arsenate As(V), the process of bioaccumulation and detoxification of As occurs in microalgae. Mainly modification of proteins by phosphorylation plays an important role in the uptake and detoxification of As [192]. In corroboration to this, it was observed that the limitation of phosphate induces the synthesis of As transporters, which further facilitates As uptake [193]. Also, it was reported by Wang and his coworkers that phosphate and As had a synergistic interaction in the growth promotion of *Dunaliella salina*. The interaction of symbiotic bacteria and phosphate limitation encourages the reduction of As (V) and excretion of As (III) by *D. salina* [193]. It was observed by Jiang *et al.* [194] in 2011 that *Chlorella vulgaris* stimulated the removal of As (V) up to 200 mg/L. Huang *et al.* [195] reported that intracellular As bioaccumulation occurs in microalgae *M. aeruginosa*. Many other microalgal species are reported for their As accumulation potential, such as *Diatoms, Hydrodictiyon reticulatom, Pithophora sp., Phormidium sp.* and *Oscillatoria* sp [189]. Another microalgal form, *Nannochloropsis* sp. showed decreased carotenoids and lipid contents. Upon treatment with 100 μm As (III) it has the highest lipid productivity and meets the standards of a biofuel [32].

Cadmium (Cd)

Microalgae responds variably to Cd stress. Microalgae such as *Chlamydomonas moewusii* and *Monoraphidium* sp. produce phytochelatins and antioxidative enzymes (catalase, peroxidases, superoxidases) as defense mechanisms against Cd stress [196]. Also, it has been reported that there is enhanced lipid content and productivity in comparison to control forms. Due to typical cell wall components, enhanced phytohormone levels, and better antioxidation capacity, microalgae such as *C. vulgaris* JSC-7 have higher Cd^{2+} tolerance (21.5 mg/g) in comparison to non-flocculating species (6.5 mg/g) [197]. In *S. obliquus* AS-6-1, the presence of functional groups, phosphate, and carboxyl groups is responsible for Cd adsorption. Cadmium chloride failed to hamper the flocculating properties of S. obliquus AS-6-1 [198]. However, strategies such as genetic engineering, bio-immobilization, microalgae immobilization, and bio-pellets have been studied to improve cadmium biosorption. With the assistance of genetic manipulation, the overexpressed metal tolerance protein showed 3.06 folds Cd^{2+} tolerance in comparison to the wild form (2.29) of *C. reinhardtii* [199]. Still, the natural strains, such as *Parachlorella hussii, P. kessleri,* and *Chlorella luteoviridis* performed better in heavy metal tolerance and accumulation. Also, the reducing biomolecules (NADPH, glutathione) have played a crucial role in the synthesis of CdSe nanoparticles [200]. With time, more advancements are coming our way, as bioremediation of Cd by both water hyacinth and biochar complex attained 214.7 mg/g higher Cd than *Chlorella* sp. and biochar alone [201]. Application of microalgae (*C. vulgaris*) and fungi (*Aspergillus niger*) showed better Cd removal (up to 56%) in comparison to microalgae alone (40%) [202]. On the other hand, the focus is shifting to the application of aerogel beads comprised of Nostoc commune and *Chlamydomonas angulosa* for Cd^{2+} removal up to 63.1% adsorption capacity at pH 6 [203].

Chromium (Cr)

There are multiple factors on which the removal of Cr (VI) by macroalgae depends, such as functional groups and nature of prevailing conditions. Majorly, biosorption of Cr (VI) on extracellular polymeric substances has been reported as a major mechanism for chromium bioremediation by *Phaeodactylum tricornutum* and *Navicula pelliculosa* [204]. Apart from extracellular and intracellular accumulation of Cr (VI), enzymatic chromium reductase is also in charge of the removal of heavy metal ions [205]. Additionally, Cr (VI) also interacts with reducing agents such as hydroxyl group and alcohol groups in order to transform into Cr (III) [206]. Also, a few of the Cr (III) are released into the medium due to complex formation with negatively charged functional groups [207]. As per the reports of Husien *et al.* [94], *Chlorella sorokiniana* can endure Cr up to 100 ppm levels for three days. Similarly, Cherifi *et al.* [208] reported that microalgae can remove 98% heavy metal in a culture medium containing 20mg/L Cr (VI). Nath *et al.* [209] stated that microalgae have Cr (VI) removal efficiency in decreasing order *Lyngbya* sp. > *Chlorella* sp.> *Scenedesmus dimorphus*> *Oscillatoria* sp. Balaji *et al.* [210] reported that functional groups such as carboxylate, eater, and hydroxyl are responsible for Cr (VI) removal. Cr (VI) removal by microbial biomass was improved by pre-treatment of NaOH and SDS [209].

Lead (Pb)

There are many reports that are cited in the literature for the use of microalgae (*Spirulina platensis, Rhizoclonium hookeri, Chaetoceros, Chlorella* sp.) as an efficient biosorbent of Lead [82]. The presence of functional groups, such as acyl-amino, hydroxyl, phenols, amide, amine, carbonyl, carboxyl, and phosphate aided in the biosorption of Pb^{2+}. The adsorption of Pb^{2+} decreases at lower pH as there is an increase in positive charge on the binding sites and higher competition with H^+ at sorption sites. It has been stated that the adsorption of Pb^{2+} by microalgae is highly endothermic and spontaneous. It followed a second-order kinetic model [211]. Hence, the reaction of ion exchange and electron sharing between Pb^{2+} and biomass is the rate-limiting step. Interestingly, Batsalova *et al.* [212] reported that microalgae, *Scenedesmus incrassatulus* (peripheral cell type "incrassatulus") form is employed as an effective indicator of Pb pollution. Correspondingly, microalgae *Nitzchia closterium* has become a prime indicator of Pb toxicity by employing chlorophyll fluorescence technology [213].

Mercury (Hg)

Microalgal forms have the capability to transform reducible form (Hg^{2+}) to elemental form (Hg°) [214]. Post enzymatic reduction by mercuric reductase, (Hg^{2+} to Hg°) the volatile Hg° is removed by volatilization. The rest of the remaining unreduced Hg^{2+} form is converted into β-HgS. Volatilization of Hg is brought about by microalgal forms, such as *Selenastrum minutum, Chlorella fusca var. fusca, diatom Navicula pellicosa,* and thermophilic alga *Galdiera sulphuraria*. Devars *et al.* [215] reported that biological volatilization was dependent on metal concentration and cell density. Huang *et al.* [216] reported that overexpression of mercuric reductase (Mer A) from *Bacillus megaterium* B1 improved Hg removal ability in *Chlorella* sp. coupled with reduced oxidative stress. Mangal *et al.* [217] stated that Hg-binding ligands are more homologous and aromatic in nature than the ones in darker growth conditions. Bioreactor systems for wastewater treatment employing the use of activated sludge and membrane filtration have gained a lot of importance due to their efficient Hg cleansing ability. Hence, *C. vulgaris* powder was interestingly used in the removal of Hg^{2+} up to 300-800 ppb concentrations [218]. Hg toxicology assessment in aquatic habitats employs the use of frustule morphology, SOD gene expression levels, and Chlorophyll content in *Halamphora veneta*. It aids in the assessment of Hg toxicities in aquatic habitats [219].

CONCLUSION

Expanding population, extensive industrialization, and agricultural practices have contributed to the high concentrations of toxic heavy metals in the soil and aquatic ecosystems. However, to deal with these toxic HMs concentrations, and their elimination from different ecosystems, certain methods that are low-cost, eco-friendly, and with no secondary pollutants are selected. Among these methods, biological methods have immense efficacy in dealing with these toxic HMs concentrations. Thus, microalgae have been introduced to ameliorate these HMs from aquatic and terrestrial ecosystems. These microalgae possess high binding affinity, are easy to culture, and with no toxic sludge available but despite all these characteristic features, their application has met with limited commercial success. Furthermore, the work also focuses on the mechanism of adsorption, accumulation, and algal biofilm formation, to treat these toxic heavy metal concentrations in soil and aquatic ecosystems.

REFERENCES

[1]   Jing Y, He Z, Yang X. Role of soil rhizobacteria in phytoremediation of heavy metal contaminated soils. J Zhejiang Univ Sci B 2007; 8(3): 192-207.
[http://dx.doi.org/10.1631/jzus.2007.B0192] [PMID: 17323432]

[2]   Carolin CF, Kumar PS, Saravanan A, Joshiba GJ, Naushad M. Efficient techniques for the removal of toxic heavy metals from aquatic environment: A review. J Environ Chem Eng 2017; 5(3): 2782-99.
[http://dx.doi.org/10.1016/j.jece.2017.05.029]

[3]   Kong Q, Shi X, Ma W, *et al.* Strategies to improve the adsorption properties of graphene-based adsorbent towards heavy metal ions and their compound pollutants: A review. J Hazard Mater 2021; 415: 125690.
[http://dx.doi.org/10.1016/j.jhazmat.2021.125690] [PMID: 33773257]

[4]   Gaur A, Adholeya A. Prospects of arbuscular mycorrhizal fungi in phytoremediation of heavy metal contaminated soils. Curr Sci 2004; 528-34.

[5]   Wu X, Xue Y, Wang S. Application of the microorganism in remediation of the heavy metal pollution in water resources. Basic Clin Pharmacol Toxicol 2019; 125: 42-2.

[6]   Herrera-Estrella LR, Guevara-García AA, Lo'pez-Bucio J. Heavy metal adaptation. London, UK: Encyclopedia of Life Science Macmillian Publishers 1999; pp. 1-5.

[7]   Li X, Shi X, Wang A, Li Y. Heavy metals contamination and assessment in gas station dust of Xi'an, a mega-city of China. Environ Earth Sci 2017; 76(7): 288.
[http://dx.doi.org/10.1007/s12665-017-6582-0]

[8]   Riyazuddin R, Nisha N, Ejaz B, *et al.* A comprehensive review on the heavy metal toxicity and sequestration in plants. Biomolecules 2021; 12(1): 43.
[http://dx.doi.org/10.3390/biom12010043] [PMID: 35053191]

[9]   Long X, Liu F, Zhou X, *et al.* Estimation of spatial distribution and health risk by arsenic and heavy metals in shallow groundwater around Dongting Lake plain using GIS mapping. Chemosphere 2021; 269: 128698.
[http://dx.doi.org/10.1016/j.chemosphere.2020.128698] [PMID: 33121802]

[10]   Kong Q, Shi X, Ma W, *et al.* Strategies to improve the adsorption properties of graphene-based adsorbent towards heavy metal ions and their compound pollutants: A review. J Hazard Mater 2021; 415: 125690.
[http://dx.doi.org/10.1016/j.jhazmat.2021.125690] [PMID: 33773257]

[11]   Asiandu AP, Wahyudi A. Phycoremediation: heavy metals green-removal by microalgae and its application in biofuel production. J Environ Treat Tech 2021; 9(3): 647-56.

[12]   Saravanan A, Kumar PS, Renita AA. Hybrid synthesis of novel material through acid modification followed ultrasonication to improve adsorption capacity for zinc removal. J Clean Prod 2018; 172: 92-105.
[http://dx.doi.org/10.1016/j.jclepro.2017.10.109]

[13]   GracePavithra K, Jaikumar V, Kumar PS, SundarRajan PS. A review on cleaner strategies for chromium industrial wastewater: Present research and future perspective. J Clean Prod 2019; 228: 580-93.

[http://dx.doi.org/10.1016/j.jclepro.2019.04.117]

[14]   Imran M, Haq Khan ZU, Iqbal J, *et al.* Potential of siltstone and its composites with biochar and magnetite nanoparticles for the removal of cadmium from contaminated aqueous solutions: Batch and column scale studies. Environ Pollut 2020; 259: 113938.
[http://dx.doi.org/10.1016/j.envpol.2020.113938] [PMID: 31952099]

[15]   Priya M, Gurung N, Mukherjee K, Bose S. Microalgae in removal of heavy metal and organic pollutants from soil. Microb Biodegrad Bioremediat 2014; 519-37.
[http://dx.doi.org/10.1016/B978-0-12-800021-2.00023-6]

[16]   Sarojini G, Venkateshbabu S, Rajasimman M. Facile synthesis and characterization of polypyrrole - iron oxide – seaweed (PPy-Fe$_3$O$_4$-SW) nanocomposite and its exploration for adsorptive removal of Pb(II) from heavy metal bearing water. Chemosphere 2021; 278: 130400.
[http://dx.doi.org/10.1016/j.chemosphere.2021.130400] [PMID: 33819882]

[17]   Wu Q, Li H, Hu X, *et al.* Full-scale evaluation of reversed A2/O process for removal of multiple pollutants in sewage. Chin Chem Lett 2020; 31(10): 2825-30.
[http://dx.doi.org/10.1016/j.cclet.2020.06.029]

[18]   Li K, Liu Q, Fang F, *et al.* Microalgae-based wastewater treatment for nutrients recovery: A review. Bioresour Technol 2019; 291: 121934.
[http://dx.doi.org/10.1016/j.biortech.2019.121934] [PMID: 31395401]

[19]   Qin H, Hu T, Zhai Y, Lu N, Aliyeva J. The improved methods of heavy metals removal by biosorbents: A review. Environ Pollut 2020; 258: 113777.
[http://dx.doi.org/10.1016/j.envpol.2019.113777] [PMID: 31864928]

[20]   Moenne A, González A, Sáez CA. Mechanisms of metal tolerance in marine macroalgae, with emphasis on copper tolerance in Chlorophyta and Rhodophyta. Aquat Toxicol 2016; 176: 30-7.
[http://dx.doi.org/10.1016/j.aquatox.2016.04.015] [PMID: 27107242]

[21]   Urrutia C, Yañez-Mansilla E, Jeison D. Bioremoval of heavy metals from metal mine tailings water using microalgae biomass. Algal Res 2019; 43: 101659.
[http://dx.doi.org/10.1016/j.algal.2019.101659]

[22]   Khatiwada B, Hasan MT, Sun A, *et al.* Proteomic response of Euglena gracilis to heavy metal exposure – Identification of key proteins involved in heavy metal tolerance and accumulation. Algal Res 2020; 45: 101764.
[http://dx.doi.org/10.1016/j.algal.2019.101764]

[23]   Rath B. Microalgal bioremediation: current practices and perspectives. J Biochem Technol 2012; 3(3): 299-304.

[24]   Abdelwahab O, Amin NK, El-Ashtoukhy E-SZ. Removal of zinc ions from aqueous solution using a cation exchange resin. Chem Eng Res Des 2013; 91(1): 165-73.
[http://dx.doi.org/10.1016/j.cherd.2012.07.005]

[25]   Nikitin DA, Marfenina OE, Kudinova AG, *et al.* Microbial biomass and biological activity of soils and soil-like bodies in coastal oases of Antarctica. Eurasian Soil Sci 2017; 50(9): 1086-97.
[http://dx.doi.org/10.1134/S1064229317070079]

[26]   Mohsenpour SF, Hennige S, Willoughby N, Adeloye A, Gutierrez T. Integrating micro-algae into wastewater treatment: A review. Sci Total Environ 2021; 752: 142168.
[http://dx.doi.org/10.1016/j.scitotenv.2020.142168] [PMID: 33207512]

[27]   Khan MI, Shin JH, Kim JD. The promising future of microalgae: current status, challenges, and optimization of a sustainable and renewable industry for biofuels, feed, and other products. Microb Cell Fact 2018; 17(1): 36.
[http://dx.doi.org/10.1186/s12934-018-0879-x] [PMID: 29506528]

[28]   Sarwar N, Imran M, Shaheen MR, *et al.* Phytoremediation strategies for soils contaminated with heavy metals: Modifications and future perspectives. Chemosphere 2017; 171: 710-21.

[http://dx.doi.org/10.1016/j.chemosphere.2016.12.116] [PMID: 28061428]

[29]   Li C, Zhou K, Qin W, *et al.* A review on heavy metals contamination in soil: effects, sources, and remediation techniques. Soil Sediment Contam 2019; 28(4): 380-94.
[http://dx.doi.org/10.1080/15320383.2019.1592108]

[30]   Dixit R, Wasiullah X, Malaviya D, *et al.* Bioremediation of heavy metals from soil and aquatic environment: an overview of principles and criteria of fundamental processes. Sustainability (Basel) 2015; 7(2): 2189-212.
[http://dx.doi.org/10.3390/su7022189]

[31]   Kumar B, Smita K, Cumbal Flores L. Plant mediated detoxification of mercury and lead. Arab J Chem 2017; 10: S2335-42.
[http://dx.doi.org/10.1016/j.arabjc.2013.08.010]

[32]   Upadhyay AK, Mandotra SK, Kumar N, Singh NK, Singh L, Rai UN. Augmentation of arsenic enhances lipid yield and defense responses in alga Nannochloropsis sp. Bioresour Technol 2016; 221: 430-7.
[http://dx.doi.org/10.1016/j.biortech.2016.09.061] [PMID: 27665531]

[33]   Abinandan S, Subashchandrabose SR, Panneerselvan L, Venkateswarlu K, Megharaj M. Potential of acid-tolerant microalgae, Desmodesmus sp. MAS1 and Heterochlorella sp. MAS3, in heavy metal removal and biodiesel production at acidic pH. Bioresour Technol 2019; 278: 9-16.
[http://dx.doi.org/10.1016/j.biortech.2019.01.053] [PMID: 30669030]

[34]   Pradhan D, Sukla LB, Mishra BB, Devi N. Biosorption for removal of hexavalent chromium using microalgae Scenedesmus sp. J Clean Prod 2019; 209: 617-29.
[http://dx.doi.org/10.1016/j.jclepro.2018.10.288]

[35]   Li B, Zhang T, Yang Z. Immobilizing unicellular microalga on pellet-forming filamentous fungus: Can this provide new insights into the remediation of arsenic from contaminated water?. Bioresour Technol 2019; 284: 231-9.
[http://dx.doi.org/10.1016/j.biortech.2019.03.128] [PMID: 30947137]

[36]   Zhao Y, Song X, Yu L, Han B, Li T, Yu X. Influence of cadmium stress on the lipid production and cadmium bioresorption by Monoraphidium sp. QLY-1. Energy Convers Manage 2019; 188: 76-85.
[http://dx.doi.org/10.1016/j.enconman.2019.03.041]

[37]   Zhang Z, Yan K, Zhang L, *et al.* A novel cadmium-containing wastewater treatment method: Bio-immobilization by microalgae cell and their mechanism. J Hazard Mater 2019; 374: 420-7.
[http://dx.doi.org/10.1016/j.jhazmat.2019.04.072] [PMID: 31035092]

[38]   Daneshvar E, Zarrinmehr MJ, Kousha M, *et al.* Hexavalent chromium removal from water by microalgal-based materials: Adsorption, desorption and recovery studies. Bioresour Technol 2019; 293: 122064.
[http://dx.doi.org/10.1016/j.biortech.2019.122064] [PMID: 31491650]

[39]   Dotaniya ML, Thakur JK, Meena VD, Jajoria DK, Rathor G. Chromium pollution: A threat to environment-A review. Agric Rev (Karnal) 2014; 35(2): 153-7.
[http://dx.doi.org/10.5958/0976-0741.2014.00094.4]

[40]   Patra DK, Pradhan C, Patra HK. Chromium bioaccumulation, oxidative stress metabolism and oil content in lemon grass *Cymbopogon flexuosus* (Nees ex Steud.) W. Watson grown in chromium rich over burden soil of Sukinda chromite mine, India. Chemosphere 2019; 218: 1082-8.
[http://dx.doi.org/10.1016/j.chemosphere.2018.11.211] [PMID: 30609487]

[41]   Huang CC, Chen MW, Hsieh JL, Lin WH, Chen PC, Chien LF. Expression of mercuric reductase from Bacillus megaterium MB1 in eukaryotic microalga Chlorella sp. DT: an approach for mercury phytoremediation. Appl Microbiol Biotechnol 2006; 72(1): 197-205.
[http://dx.doi.org/10.1007/s00253-005-0250-0] [PMID: 16547702]

[42]   Cenkci S, Yıldız M, Ciğerci İH, Konuk M, Bozdağ A. Toxic chemicals-induced genotoxity detected

by random amplified polymorphic DNA (RAPD) in bean (*Phaseolus vulgaris* L.) seedlings. Chemosphere 2009; 76(7): 900-6.
[http://dx.doi.org/10.1016/j.chemosphere.2009.05.001] [PMID: 19477479]

[43]   Das S, Dash HR, Chakraborty J. Genetic basis and importance of metal resistant genes in bacteria for bioremediation of contaminated environments with toxic metal pollutants. Appl Microbiol Biotechnol 2016; 100(7): 2967-84.
[http://dx.doi.org/10.1007/s00253-016-7364-4] [PMID: 26860944]

[44]   Shiyab S, Chen J, Han FX, *et al.* Mercury-induced oxidative stress in Indian mustard (*Brassica juncea* L.). Environ Toxicol 2009; 24(5): 462-71.
[http://dx.doi.org/10.1002/tox.20450] [PMID: 19003913]

[45]   Hong KS, Lee HM, Bae JS, *et al.* Removal of heavy metal ions by using calcium carbonate extracted from starfish treated by protease and amylase. J Anal Sci Technol 2011; 2(2): 75-82.
[http://dx.doi.org/10.5355/JAST.2011.75]

[46]   Street RA. Heavy metals in medicinal plant products — An African perspective. S Afr J Bot 2012; 82: 67-74.
[http://dx.doi.org/10.1016/j.sajb.2012.07.013]

[47]   Khan MA, Rao RA, Ajmal M. Heavy metal pollution and its control through nonconventional adsorbents (1998-2007): a review. J Int Environ Appl Sci 2008; 3(2): 101-41.

[48]   Ummalyma SB, Pandey A, Sukumaran RK, Sahoo D. Bioremediation by microalgae: current and emerging trends for effluents treatments for value addition of waste streams. Biosynt Technol Environ Challeng 2018; 355-75.
[http://dx.doi.org/10.1007/978-981-10-7434-9_19]

[49]   Suresh Kumar K, Dahms HU, Won EJ, Lee JS, Shin KH. Microalgae – A promising tool for heavy metal remediation. Ecotoxicol Environ Saf 2015; 113: 329-52.
[http://dx.doi.org/10.1016/j.ecoenv.2014.12.019] [PMID: 25528489]

[50]   Kaplan D. Absorption and adsorption of heavy metals by microalgae. Handb Microalgal Cult 2013; 602-11.
[http://dx.doi.org/10.1002/9781118567166.ch32]

[51]   EPA U. National primary drinking water regulations, EPA 816-F09–004.

[52]   Guo Q, Li N, Bing Y, *et al.* Denitrifier communities impacted by heavy metal contamination in freshwater sediment. Environ Pollut 2018; 242(Pt A): 426-32.
[http://dx.doi.org/10.1016/j.envpol.2018.07.020] [PMID: 30005255]

[53]   Dhiman S, Ibrahim M, Devi K, *et al.* Biochar assisted remediation of toxic metals and metalloids. Handb Assist Amend: Enhanc Sustain Remediat Technol 2021; 131-62.
[http://dx.doi.org/10.1002/9781119670391.ch7]

[54]   Kaur R, Sharma N, Tikoria R, *et al.* Insights into biosynthesis and signaling of cytokinins during plant growth, development and stress tolerance. Auxins, Cytokinins Gibberellins Signal Plants 2022; 153-87.
[http://dx.doi.org/10.1007/978-3-031-05427-3_7]

[55]   Tikoria R, Kaur A, Ohri P. Modulation of various phytoconstituents in tomato seedling growth and Meloidogyne incognita–induced stress alleviation by vermicompost application. Front Environ Sci 2022; 10: 891195.
[http://dx.doi.org/10.3389/fenvs.2022.891195]

[56]   Martínez-Cortijo J, Ruiz-Canales A. Effect of heavy metals on rice irrigated fields with waste water in high pH Mediterranean soils: The particular case of the Valencia area in Spain. Agric Water Manage 2018; 210: 108-23.
[http://dx.doi.org/10.1016/j.agwat.2018.07.037]

[57]   Torres-Cruz TJ, Hesse C, Kuske CR, Porras-Alfaro A. Presence and distribution of heavy metal

tolerant fungi in surface soils of a temperate pine forest. Appl Soil Ecol 2018; 131: 66-74.
[http://dx.doi.org/10.1016/j.apsoil.2018.08.001]

[58]   Huang J, Yuan F, Zeng G, *et al.* Influence of pH on heavy metal speciation and removal from wastewater using micellar-enhanced ultrafiltration. Chemosphere 2017; 173: 199-206.
[http://dx.doi.org/10.1016/j.chemosphere.2016.12.137] [PMID: 28110009]

[59]   Zeraatkar AK, Ahmadzadeh H, Talebi AF, Moheimani NR, McHenry MP. Potential use of algae for heavy metal bioremediation, a critical review. J Environ Manage 2016; 181: 817-31.
[http://dx.doi.org/10.1016/j.jenvman.2016.06.059] [PMID: 27397844]

[60]   Afonne OJ, Ifediba EC. Heavy metals risks in plant foods – need to step up precautionary measures. Curr Opin Toxicol 2020; 22: 1-6.
[http://dx.doi.org/10.1016/j.cotox.2019.12.006]

[61]   Monteiro CM, Castro PML, Malcata FX. Metal uptake by microalgae: Underlying mechanisms and practical applications. Biotechnol Prog 2012; 28(2): 299-311.
[http://dx.doi.org/10.1002/btpr.1504] [PMID: 22228490]

[62]   Asiandu AP, Wahyudi A. Phycoremediation: heavy metals green-removal by microalgae and its application in biofuel production. J Environ Treat Tech 2021; 9(3): 647-56.

[63]   Nimisha P, Joseph S. Assessment of Phycoremediation Efficiency of Spirogyra Maxima by using Heavy Metals Manganese and Lead. International Journal of Environment. Agric Biotechnol 2020; 5(3)

[64]   Yuan D, Huang L, Zeng L, *et al.* Acute toxicity of mercury chloride (hgcl2) and cadmium chloride (cdcl2) on the behavior of freshwater fish, percocyprispingi. Int J Aquac Fish Sci 2017; 3(3): 66-70.

[65]   Nowicka B. Heavy metal–induced stress in eukaryotic algae—mechanisms of heavy metal toxicity and tolerance with particular emphasis on oxidative stress in exposed cells and the role of antioxidant response. Environ Sci Pollut Res Int 2022; 29(12): 16860-911.
[http://dx.doi.org/10.1007/s11356-021-18419-w] [PMID: 35006558]

[66]   Islam MS, Ahmed MK, Habibullah-Al-Mamun M. Apportionment of heavy metals in soil and vegetables and associated health risks assessment. Stochastic Environ Res Risk Assess 2016; 30(1): 365-77.
[http://dx.doi.org/10.1007/s00477-015-1126-1]

[67]   Proshad R, Kormoker T, Mursheed N, *et al.* Heavy metal toxicity in agricultural soil due to rapid industrialization in Bangladesh: a review. Int J Adv Geosci 2018; 6(1): 83-8.
[http://dx.doi.org/10.14419/ijag.v6i1.9174]

[68]   Ahmed SF, Kumar PS, Rozbu MR, *et al.* Heavy metal toxicity, sources, and remediation techniques for contaminated water and soil. Environ Technol Innov 2022; 25: 102114.
[http://dx.doi.org/10.1016/j.eti.2021.102114]

[69]   Tytła M, Widziewicz K, Zielewicz E. Heavy metals and its chemical speciation in sewage sludge at different stages of processing. Environ Technol 2016; 37(7): 899-908.
[http://dx.doi.org/10.1080/09593330.2015.1090482] [PMID: 26419833]

[70]   Alsherif EA, Al-Shaikh TM, Almaghrabi O, AbdElgawad H. AbdElgawad H. High redox status as the basis for heavy metal tolerance of *Sesuvium portulacastrum* L. inhabiting contaminated soil in Jeddah, Saudi Arabia. Antioxidants 2021; 11(1): 19.
[http://dx.doi.org/10.3390/antiox11010019] [PMID: 35052523]

[71]   Sun L, Guo D, Liu K, *et al.* Levels, sources, and spatial distribution of heavy metals in soils from a typical coal industrial city of Tangshan, China. Catena 2019; 175: 101-9.
[http://dx.doi.org/10.1016/j.catena.2018.12.014]

[72]   Sun L, Liao X, Yan X, Zhu G, Ma D. Evaluation of heavy metal and polycyclic aromatic hydrocarbons accumulation in plants from typical industrial sites: potential candidate in phytoremediation for co-contamination. Environ Sci Pollut Res Int 2014; 21(21): 12494-504.

[http://dx.doi.org/10.1007/s11356-014-3171-6] [PMID: 24946706]

[73]   Pan L, Ma J, Wang X, Hou H. Heavy metals in soils from a typical county in Shanxi Province, China: Levels, sources and spatial distribution. Chemosphere 2016; 148: 248-54.
[http://dx.doi.org/10.1016/j.chemosphere.2015.12.049] [PMID: 26807946]

[74]   Chai Y, Guo J, Chai S, Cai J, Xue L, Zhang Q. Source identification of eight heavy metals in grassland soils by multivariate analysis from the Baicheng–Songyuan area, Jilin Province, Northeast China. Chemosphere 2015; 134: 67-75.
[http://dx.doi.org/10.1016/j.chemosphere.2015.04.008] [PMID: 25911049]

[75]   Koller M, Saleh HM. Introductory chapter: introducing heavy metals. Heavy Metals 2018; 1: 3-11.

[76]   Hong Y, Liao W, Yan Z, *et al.* Progress in the research of the toxicity effect mechanisms of heavy metals on freshwater organisms and their water quality criteria in China. J Chem 2020; 2020: 1-12.
[http://dx.doi.org/10.1155/2020/9010348]

[77]   Fatima S, Muzammal M, Rehman A, *et al.* Water pollution on heavy metals and its effects on fishes. Int J Fish Aquat Stud 2020; 8(3): 6-14.

[78]   Renu NA, Agarwal M, Singh K. Methodologies for removal of heavy metal ions from wastewater: an overview. Interdiscip Environ Rev 2017; 18(2): 124-42.
[http://dx.doi.org/10.1504/IER.2017.087915]

[79]   Abdelfattah A, Ali SS, Ramadan H, *et al.* Microalgae-based wastewater treatment: mechanisms, challenges, recent advances, and future prospects. Environ Sci Ecotechnol 2022; 100205.

[80]   Liu J, Luo X, Sun Y, *et al.* Thallium pollution in China and removal technologies for waters: A review. Environ Int 2019; 126: 771-90.
[http://dx.doi.org/10.1016/j.envint.2019.01.076] [PMID: 30884277]

[81]   Priya AK, Jalil AA, Vadivel S, *et al.* Heavy metal remediation from wastewater using microalgae: Recent advances and future trends. Chemosphere 2022; 305: 135375.
[http://dx.doi.org/10.1016/j.chemosphere.2022.135375] [PMID: 35738200]

[82]   Leong YK, Chang JS. Bioremediation of heavy metals using microalgae: Recent advances and mechanisms. Bioresour Technol 2020; 303: 122886.
[http://dx.doi.org/10.1016/j.biortech.2020.122886] [PMID: 32046940]

[83]   Spain O, Plöhn M, Funk C. The cell wall of green microalgae and its role in heavy metal removal. Physiol Plant 2021; 173(2): 526-35.
[http://dx.doi.org/10.1111/ppl.13405] [PMID: 33764544]

[84]   Goswami RK, Agrawal K, Shah MP, Verma P. Bioremediation of heavy metals from wastewater: a current perspective on microalgae-based future. Lett Appl Microbiol 2022; 75(4): 701-17.
[http://dx.doi.org/10.1111/lam.13564] [PMID: 34562022]

[85]   Chan A, Salsali H, McBean E. Heavy metal removal (copper and zinc) in secondary effluent from wastewater treatment plants by microalgae. ACS Sustain Chem& Eng 2014; 2(2): 130-7.
[http://dx.doi.org/10.1021/sc400289z]

[86]   Piccini M, Raikova S, Allen MJ, Chuck CJ. A synergistic use of microalgae and macroalgae for heavy metal bioremediation and bioenergy production through hydrothermal liquefaction. Sustain Energy Fuels 2019; 3(1): 292-301.
[http://dx.doi.org/10.1039/C8SE00408K]

[87]   Mubashar M, Naveed M, Mustafa A, *et al.* Experimental investigation of *Chlorella vulgaris* and Enterobacter sp. MN17 for decolorization and removal of heavy metals from textile wastewater. Water 2020; 12(11): 3034.
[http://dx.doi.org/10.3390/w12113034]

[88]   Abdel-Razek MA, Abozeid AM, Eltholth MM, *et al.* Bioremediation of a pesticide and selected heavy metals in wastewater from various sources using a consortium of microalgae and cyanobacteria. Slov

Vet Res 2019; 56 (Suppl. 22): 61-73.
[http://dx.doi.org/10.26873/SVR-744-2019]

[89]   Sultana N, Hossain SMZ, Mohammed ME, *et al.* Experimental study and parameters optimization of microalgae based heavy metals removal process using a hybrid response surface methodology-crow search algorithm. Sci Rep 2020; 10(1): 15068.
[http://dx.doi.org/10.1038/s41598-020-72236-8] [PMID: 32934284]

[90]   León-Vaz A, León R, Giráldez I, Vega JM, Vigara J. Impact of heavy metals in the microalga *Chlorella sorokiniana* and assessment of its potential use in cadmium bioremediation. Aquat Toxicol 2021; 239: 105941.
[http://dx.doi.org/10.1016/j.aquatox.2021.105941] [PMID: 34469852]

[91]   Kumar N, Hans S, Verma R, Srivastava A. Acclimatization of microalgae *Arthrospira platensis* for treatment of heavy metals in Yamuna River. Water Sci Eng 2020; 13(3): 214-22.
[http://dx.doi.org/10.1016/j.wse.2020.09.005]

[92]   Gevorgyan G, Mamyan A, Boshyan T, Vardanyan T, Vaseashta A. Heavy metal contamination in an industrially affected river catchment basin: assessment, effects, and mitigation. Int J Environ Res Public Health 2021; 18(6): 2881.
[http://dx.doi.org/10.3390/ijerph18062881] [PMID: 33799803]

[93]   Atoku DI, Ojekunle OZ, Taiwo AM, Shittu OB. Evaluating the efficiency of Nostoc commune, *Oscillatoria limosa* and *Chlorella vulgaris* in a phycoremediation of heavy metals contaminated industrial wastewater. Sci Am 2021; 12: e00817.

[94]   Husien S, Labena A, El-Belely EF, Mahmoud HM, Hamouda AS. Absorption of hexavalent chromium by green micro algae *Chlorella sorokiniana*: live planktonic cells. Water Pract Technol 2019; 14(3): 515-29.
[http://dx.doi.org/10.2166/wpt.2019.034]

[95]   Adhoni SA, Raikar SM, Shivasharana CT. Bioremediation of industrial effluents with heavy metals using immobilised microalgae. Int J Appl Nat Sci 2018; 7(5): 67.

[96]   Santos FM, Mazur LP, Mayer DA, Vilar VJP, Pires JCM. Inhibition effect of zinc, cadmium, and nickel ions in microalgal growth and nutrient uptake from water: An experimental approach. Chem Eng J 2019; 366: 358-67.
[http://dx.doi.org/10.1016/j.cej.2019.02.080]

[97]   P S C, Sanyal D, Dasgupta S, Banik A. Cadmium biosorption and biomass production by two freshwater microalgae *Scenedesmus acutus* and *Chlorella pyrenoidosa*: An integrated approach. Chemosphere 2021; 269: 128755.
[http://dx.doi.org/10.1016/j.chemosphere.2020.128755] [PMID: 33143896]

[98]   Elleuch J, Hmani R, Drira M, Michaud P, Fendri I, Abdelkafi S. Potential of three local marine microalgae from Tunisian coasts for cadmium, lead and chromium removals. Sci Total Environ 2021; 799: 149464.
[http://dx.doi.org/10.1016/j.scitotenv.2021.149464] [PMID: 34388883]

[99]   Elleuch J, Hmani R, Drira M, Michaud P, Fendri I, Abdelkafi S. Potential of three local marine microalgae from Tunisian coasts for cadmium, lead and chromium removals. Sci Total Environ 2021; 799: 149464.
[http://dx.doi.org/10.1016/j.scitotenv.2021.149464] [PMID: 34388883]

[100] Nanda M, Jaiswal KK, Kumar V, *et al.* Bio-remediation capacity for Cd(II) and Pb(II) from the aqueous medium by two novel strains of microalgae and their effect on lipidomics and metabolomics. J Water Process Eng 2021; 44: 102404.
[http://dx.doi.org/10.1016/j.jwpe.2021.102404]

[101] Novák Z, Harangi S, Baranyai E, Gonda S, B-Béres V, Bácsi I. Effects of metal quantity and quality to the removal of zinc and copper by two common green microalgae (Chlorophyceae) species. Phycol Res 2020; 68(3): 227-35.

[http://dx.doi.org/10.1111/pre.12422]

[102] Saavedra R, Muñoz R, Taboada ME, Bolado S. Influence of organic matter and CO_2 supply on bioremediation of heavy metals by *Chlorella vulgaris* and *Scenedesmus almeriensis* in a multimetallic matrix. Ecotoxicol Environ Saf 2019; 182: 109393.
[http://dx.doi.org/10.1016/j.ecoenv.2019.109393] [PMID: 31299473]

[103] Dammak M, Ben Hlima H, Tounsi L, Michaud P, Fendri I, Abdelkafi S. Effect of heavy metals mixture on the growth and physiology of Tetraselmis sp.: Applications to lipid production and bioremediation. Bioresour Technol 2022; 360: 127584.
[http://dx.doi.org/10.1016/j.biortech.2022.127584] [PMID: 35798164]

[104] Bauenova MO, Sadvakasova AK, Mustapayeva ZO, *et al.* Potential of microalgae *Parachlorella kessleri* Bh-2 as bioremediation agent of heavy metals cadmium and chromium. Algal Res 2021; 59: 102463.
[http://dx.doi.org/10.1016/j.algal.2021.102463]

[105] Pratiwi DC, Pratiwi N, Yona D, Sasmita RD. Cadmium and copper removal using microalgae *Chaetoceros calcitrans* for bioremediation potential test. Ecology. Environ Conserv J 2020; 26(1): 314-7.

[106] Dewi ER, Nuravivah R. Potential of microalgae *Chlorella vulgaris* as bioremediation agents of heavy metal pb (lead) on culture media. E3S Web Conf 2018; 31: 05010.

[107] Blanco-Vieites M, Suárez-Montes D, Delgado F, Álvarez-Gil M, Battez AH, Rodríguez E. Removal of heavy metals and hydrocarbons by microalgae from wastewater in the steel industry. Algal Res 2022; 64: 102700.
[http://dx.doi.org/10.1016/j.algal.2022.102700]

[108] Ahmad N, Mounsef JR, Abou Tayeh J, Lteif R. Bioremediation of Ni, Al and Pb by the living cells of a resistant strain of microalga. Water Sci Technol 2020; 82(5): 851-60.
[http://dx.doi.org/10.2166/wst.2020.381] [PMID: 33031065]

[109] Cameron H, Mata MT, Riquelme C. The effect of heavy metals on the viability of *Tetraselmis marina* AC16-MESO and an evaluation of the potential use of this microalga in bioremediation. PeerJ 2018; 6: e5295.
[http://dx.doi.org/10.7717/peerj.5295] [PMID: 30065883]

[110] Hubadillah SK, Othman MHD, Gani P, *et al.* Integrated green membrane distillation-microalgae bioremediation for arsenic removal from Pengorak River Kuantan, Malaysia. Chem Eng Process 2020; 153: 107996.
[http://dx.doi.org/10.1016/j.cep.2020.107996]

[111] Rizal A, Permana R, Apriliani IM. The effect of phosphate addition with different concentration on the capability of *Nannochloropsis oculata* as a bioremediation agent of medium heavy metal (Cd2+). World Sci News 2020; (145): 286-97.

[112] Soeprobowati TR, Hariyati R. The potential used of microalgae for heavy metals remediation. Proc 2nd Int Sem New Paradigm Innov Nat Sci Its Appl. 3: 72-87.

[113] Chugh M, Kumar L, Shah MP, Bharadvaja N. Algal bioremediation of heavy metals: an insight into removal mechanisms, recovery of by-products, challenges, and future opportunities. Energy Nexus 2022; 100129.

[114] Danouche M, El Ghachtouli N, El Arroussi H. Phycoremediation mechanisms of heavy metals using living green microalgae: physicochemical and molecular approaches for enhancing selectivity and removal capacity. Heliyon 2021; 7(7): e07609.
[http://dx.doi.org/10.1016/j.heliyon.2021.e07609] [PMID: 34355100]

[115] Singh DV, Bhat RA, Upadhyay AK, Singh R, Singh DP. Microalgae in aquatic environs: A sustainable approach for remediation of heavy metals and emerging contaminants. Environ Technol Innov 2021; 21: 101340.

[http://dx.doi.org/10.1016/j.eti.2020.101340]

[116] Mantzorou A, Navakoudis E, Paschalidis K, Ververidis F. Microalgae: a potential tool for remediating aquatic environments from toxic metals. Int J Environ Sci Technol 2018; 15(8): 1815-30.
[http://dx.doi.org/10.1007/s13762-018-1783-y]

[117] Xiao X, Li W, Jin M, Zhang L, Qin L, Geng W. Responses and tolerance mechanisms of microalgae to heavy metal stress: A review. Mar Environ Res 2023; 183: 105805.
[http://dx.doi.org/10.1016/j.marenvres.2022.105805] [PMID: 36375224]

[118] Sibi G. Factors influencing heavy metal removal by microalgae—A review. J Crit Rev 2019; 6: 29-32.

[119] Çetinkaya Dönmez G, Aksu Z, Öztürk A, Kutsal T. A comparative study on heavy metal biosorption characteristics of some algae. Process Biochem 1999; 34(9): 885-92.
[http://dx.doi.org/10.1016/S0032-9592(99)00005-9]

[120] Martínez-Macias MR, Correa-Murrieta MA, Villegas-Peralta Y, *et al.* Uptake of copper from acid mine drainage by the microalgae Nannochloropsis oculata. Environ Sci Pollut Res Int 2019; 26(7): 6311-8.
[http://dx.doi.org/10.1007/s11356-018-3963-1] [PMID: 30617876]

[121] Manzoor F, Karbassi A, Golzary A. Removal of heavy metal contaminants from wastewater by using chlorella vulgaris beijerinck: a review. Curr Environ Manag 2019; 6(3): 174-87.

[122] Znad H, Awual MR, Martini S. The utilization of algae and seaweed biomass for bioremediation of heavy metal-contaminated wastewater. Molecules 2022; 27(4): 1275.
[http://dx.doi.org/10.3390/molecules27041275] [PMID: 35209061]

[123] Zhou GJ, Peng FQ, Zhang LJ, Ying GG. Biosorption of zinc and copper from aqueous solutions by two freshwater green microalgae *Chlorella pyrenoidosa* and *Scenedesmus obliquus*. Environ Sci Pollut Res Int 2012; 19(7): 2918-29.
[http://dx.doi.org/10.1007/s11356-012-0800-9] [PMID: 22327643]

[124] Ranjbar S, Malcata FX. Is genetic engineering a route to enhance microalgae-mediated bioremediation of heavy metal-containing effluents?. Molecules 2022; 27(5): 1473.
[http://dx.doi.org/10.3390/molecules27051473] [PMID: 35268582]

[125] Hwang JH, Church J, Lee SJ, Park J, Lee WH. Use of microalgae for advanced wastewater treatment and sustainable bioenergy generation. Environ Eng Sci 2016; 33(11): 882-97.
[http://dx.doi.org/10.1089/ees.2016.0132]

[126] Ferro L, Gentili FG, Funk C. Isolation and characterization of microalgal strains for biomass production and wastewater reclamation in Northern Sweden. Algal Res 2018; 32: 44-53.
[http://dx.doi.org/10.1016/j.algal.2018.03.006]

[127] Salama ES, Kurade MB, Abou-Shanab RAI, *et al.* Recent progress in microalgal biomass production coupled with wastewater treatment for biofuel generation. Renew Sustain Energy Rev 2017; 79: 1189-211.
[http://dx.doi.org/10.1016/j.rser.2017.05.091]

[128] Plöhn M, Spain O, Sirin S, *et al.* Wastewater treatment by microalgae. Physiol Plant 2021; 173(2): 568-78.
[http://dx.doi.org/10.1111/ppl.13427] [PMID: 33860948]

[129] Novoveská L, Zapata AKM, Zabolotney JB, Atwood MC, Sundstrom ER. Optimizing microalgae cultivation and wastewater treatment in large-scale offshore photobioreactors. Algal Res 2016; 18: 86-94.
[http://dx.doi.org/10.1016/j.algal.2016.05.033]

[130] Han SF, Jin WB, Tu RJ, Wu WM. Biofuel production from microalgae as feedstock: current status and potential. Crit Rev Biotechnol 2015; 35(2): 255-68.
[http://dx.doi.org/10.3109/07388551.2013.835301] [PMID: 24641484]

[131] Arutselvan C, Seenivasan H, Lewis Oscar F, *et al.* Review on wastewater treatment by microalgae in different cultivation systems and its importance in biodiesel production. Fuel 2022; 324: 124623.
[http://dx.doi.org/10.1016/j.fuel.2022.124623]

[132] Montingelli ME, Tedesco S, Olabi AG. Biogas production from algal biomass: A review. Renew Sustain Energy Rev 2015; 43: 961-72.
[http://dx.doi.org/10.1016/j.rser.2014.11.052]

[133] Craggs RJ, Heubeck S, Lundquist TJ, Benemann JR. Algal biofuels from wastewater treatment high rate algal ponds. Water Sci Technol 2011; 63(4): 660-5.
[http://dx.doi.org/10.2166/wst.2011.100] [PMID: 21330711]

[134] Ajeng AA, Rosli NSM, Abdullah R, Yaacob JS, Qi NC, Loke SP. Resource recovery from hydroponic wastewaters using microalgae-based biorefineries: A circular bioeconomy perspective. J Biotechnol 2022; 360: 11-22.
[http://dx.doi.org/10.1016/j.jbiotec.2022.10.011] [PMID: 36272573]

[135] Gojkovic Z, Lu Y, Ferro L, Toffolo A, Funk C. Modeling biomass production during progressive nitrogen starvation by North Swedish green microalgae. Algal Res 2020; 47: 101835.
[http://dx.doi.org/10.1016/j.algal.2020.101835]

[136] Beuckels A, Smolders E, Muylaert K. Nitrogen availability influences phosphorus removal in microalgae-based wastewater treatment. Water Res 2015; 77: 98-106.
[http://dx.doi.org/10.1016/j.watres.2015.03.018] [PMID: 25863319]

[137] Larsdotter K. Wastewater treatment with microalgae-a literature review. Vatten 2006; 62(1): 31.

[138] Ahmed SF, Mofijur M, Parisa TA, *et al.* Progress and challenges of contaminate removal from wastewater using microalgae biomass. Chemosphere 2022; 286(Pt 1): 131656.
[http://dx.doi.org/10.1016/j.chemosphere.2021.131656] [PMID: 34325255]

[139] Vandamme D, Foubert I, Muylaert K. Flocculation as a low-cost method for harvesting microalgae for bulk biomass production. Trends Biotechnol 2013; 31(4): 233-9.
[http://dx.doi.org/10.1016/j.tibtech.2012.12.005] [PMID: 23336995]

[140] Qasem NAA, Mohammed RH, Lawal DU. Removal of heavy metal ions from wastewater: a comprehensive and critical review. NPJ Clean Water 2021; 4(1): 36.
[http://dx.doi.org/10.1038/s41545-021-00127-0]

[141] Abdi O, Kazemi M. A review study of biosorption of heavy metals and comparison between different biosorbents. J Mater Environ Sci 2015; 6(5): 1386-99.

[142] Chai WS, Tan WG, Halimatul Munawaroh HS, Gupta VK, Ho SH, Show PL. Multifaceted roles of microalgae in the application of wastewater biotreatment: A review. Environ Pollut 2021; 269: 116236.
[http://dx.doi.org/10.1016/j.envpol.2020.116236] [PMID: 33333449]

[143] Pathak VV, Singh DP, Kothari R, Chopra AK. Phycoremediation of textile wastewater by unicellular microalga *Chlorella pyrenoidosa*. Cell Mol Biol 2014; 60(5): 35-40.
[PMID: 25535710]

[144] Gupta S, Pawar SB, Pandey RA. Current practices and challenges in using microalgae for treatment of nutrient rich wastewater from agro-based industries. Sci Total Environ 2019; 687: 1107-26.
[http://dx.doi.org/10.1016/j.scitotenv.2019.06.115] [PMID: 31412448]

[145] Lee H, Shoda M. Removal of COD and color from livestock wastewater by the Fenton method. J Hazard Mater 2008; 153(3): 1314-9.
[http://dx.doi.org/10.1016/j.jhazmat.2007.09.097] [PMID: 17988792]

[146] Chu WL, Phang SM. Biosorption of heavy metals and dyes from industrial effluents by microalgae. Microalgae Biotechnol Dev Biofuel Wastew Treat 2019; 599-634.
[http://dx.doi.org/10.1007/978-981-13-2264-8_23]

[147] Chia WY, Ying Tang DY, Khoo KS, Kay Lup AN, Chew KW. Nature's fight against plastic pollution: Algae for plastic biodegradation and bioplastics production. Environ Sci Ecotechnol 2020; 4: 100065.
[http://dx.doi.org/10.1016/j.ese.2020.100065] [PMID: 36157709]

[148] Acheampong MA. Sustainable gold mining wastewater treatment by sorption using low-cost materials. Wageningen University and Research 2013.

[149] Gan PP, Li SF. Biosorption of elements Elements Recovery and Sustainability. Cambridge, UK: Royal Society of Chemistry 2013; pp. 80-113.
[http://dx.doi.org/10.1039/9781849737340-00080]

[150] Lesmana SO, Febriana N, Soetaredjo FE, Sunarso J, Ismadji S. Studies on potential applications of biomass for the separation of heavy metals from water and wastewater. Biochem Eng J 2009; 44(1): 19-41.
[http://dx.doi.org/10.1016/j.bej.2008.12.009]

[151] Ghaffar I, Hussain A, Hasan A, Deepanraj B. Microalgal-induced remediation of wastewaters loaded with organic and inorganic pollutants: An overview. Chemosphere 2023; 320: 137921.
[http://dx.doi.org/10.1016/j.chemosphere.2023.137921] [PMID: 36682632]

[152] Pathak B, Gupta S, Verma R. Biosorption and biodegradation of polycyclic aromatic hydrocarbons (pahs) by microalgae. Green Adsorbents Pollut Removal: Fundam Des 2018; 215-47.

[153] Kumar V, Jaiswal KK, Verma M, *et al.* Algae-based sustainable approach for simultaneous removal of micropollutants, and bacteria from urban wastewater and its real-time reuse for aquaculture. Sci Total Environ 2021; 774: 145556.
[http://dx.doi.org/10.1016/j.scitotenv.2021.145556]

[154] Li X, Xiao W, He G, Zheng W, Yu N, Tan M. Pore size and surface area control of MgO nanostructures using a surfactant-templated hydrothermal process: High adsorption capability to azo dyes. Colloids Surf A Physicochem Eng Asp 2012; 408: 79-86.
[http://dx.doi.org/10.1016/j.colsurfa.2012.05.034]

[155] Manikandan A, Suresh Babu P, Shyamalagowri S, Kamaraj M, Muthukumaran P, Aravind J. Emerging role of microalgae in heavy metal bioremediation. J Basic Microbiol 2022; 62(3-4): 330-47.
[http://dx.doi.org/10.1002/jobm.202100363] [PMID: 34724223]

[156] Hadiyanto H, Khoironi A, Dianratri I, Suherman S, Muhammad F, Vaidyanathan S. Interactions between polyethylene and polypropylene microplastics and *Spirulina* sp. microalgae in aquatic systems. Heliyon 2021; 7(8): e07676.
[http://dx.doi.org/10.1016/j.heliyon.2021.e07676] [PMID: 34401570]

[157] Chen Z, Osman AI, Rooney DW, Oh WD, Yap PS. Remediation of heavy metals in polluted water by immobilized algae: current applications and future perspectives. Sustainability (Basel) 2023; 15(6): 5128.
[http://dx.doi.org/10.3390/su15065128]

[158] Mona S, Kaushik A. Chromium and cobalt sequestration using exopolysaccharides produced by freshwater cyanobacterium Nostoc linckia. Ecol Eng 2015; 82: 121-5.
[http://dx.doi.org/10.1016/j.ecoleng.2015.04.037]

[159] Mota R, Rossi F, Andrenelli L, Pereira SB, De Philippis R, Tamagnini P. Released polysaccharides (RPS) from Cyanothece sp. CCY 0110 as biosorbent for heavy metals bioremediation: interactions between metals and RPS binding sites. Appl Microbiol Biotechnol 2016; 100(17): 7765-75.
[http://dx.doi.org/10.1007/s00253-016-7602-9] [PMID: 27188779]

[160] Wang M, Kuo-Dahab WC, Dolan S, Park C. Kinetics of nutrient removal and expression of extracellular polymeric substances of the microalgae, Chlorella sp. and Micractinium sp., in wastewater treatment. Bioresour Technol 2014; 154: 131-7.
[http://dx.doi.org/10.1016/j.biortech.2013.12.047] [PMID: 24384320]

[161] Golnaraghi Ghomi A, Asasian-Kolur N, Sharifian S, Golnaraghi A. Biosorpion for sustainable

recovery of precious metals from wastewater. J Environ Chem Eng 2020; 8(4): 103996.
[http://dx.doi.org/10.1016/j.jece.2020.103996]

[162] Liu R, Li S, Tu Y, Hao X. Capabilities and mechanisms of microalgae on removing micropollutants from wastewater: A review. J Environ Manage 2021; 285: 112149.
[http://dx.doi.org/10.1016/j.jenvman.2021.112149] [PMID: 33607565]

[163] Furuhashi Y, Honda R, Noguchi M, *et al.* Optimum conditions of pH, temperature and preculture for biosorption of europium by microalgae Acutodesmus acuminatus. Biochem Eng J 2019; 143: 58-64.
[http://dx.doi.org/10.1016/j.bej.2018.12.007]

[164] Ismaiel MMS, El-Ayouty YM, Abdelaal SA, Fathey HA. Biosorption of uranium by immobilized Nostoc sp. and Scenedesmus sp.: kinetic and equilibrium modeling. Environ Sci Pollut Res Int 2022; 29(55): 83860-77.
[http://dx.doi.org/10.1007/s11356-022-21641-9] [PMID: 35771321]

[165] Kim EJ, Park S, Hong HJ, Choi YE, Yang JW. Biosorption of chromium (Cr(III)/Cr(VI)) on the residual microalga *Nannochloris oculata* after lipid extraction for biodiesel production. Bioresour Technol 2011; 102(24): 11155-60.
[http://dx.doi.org/10.1016/j.biortech.2011.09.107] [PMID: 22014703]

[166] Kumar D, Pandey LK, Gaur JP. Metal sorption by algal biomass: From batch to continuous system. Algal Res 2016; 18: 95-109.
[http://dx.doi.org/10.1016/j.algal.2016.05.026]

[167] Lee H, Shim E, Yun HS, *et al.* Biosorption of Cu(II) by immobilized microalgae using silica: kinetic, equilibrium, and thermodynamic study. Environ Sci Pollut Res Int 2016; 23(2): 1025-34.
[http://dx.doi.org/10.1007/s11356-015-4609-1] [PMID: 25953610]

[168] Das N, Das D. Recovery of rare earth metals through biosorption: An overview. J Rare Earths 2013; 31(10): 933-43.
[http://dx.doi.org/10.1016/S1002-0721(13)60009-5]

[169] Oswald WJ. Micro-algal and waste-water treatment. Micro-algal Biotechnol 1988; 305-28.

[170] Luoma SN, van Geen A, Lee BG, Cloern JE. Metal uptake by phytoplankton during a bloom in South San Francisco Bay: Implications for metal cycling in estuaries. Limnol Oceanogr 1998; 43(5): 1007-16.
[http://dx.doi.org/10.4319/lo.1998.43.5.1007]

[171] Markou G, Mitrogiannis D, Çelekli A, Bozkurt H, Georgakakis D, Chrysikopoulos CV. Biosorption of Cu2+ and Ni2+ by Arthrospira platensis with different biochemical compositions. Chem Eng J 2015; 259: 806-13.
[http://dx.doi.org/10.1016/j.cej.2014.08.037]

[172] Bux F, Chisti Y, Eds. Algae biotechnology: products and processes. Springer 2016.
[http://dx.doi.org/10.1007/978-3-319-12334-9]

[173] Mondal M, Halder G, Oinam G, Indrama T, Tiwari ON. Bioremediation of organic and inorganic pollutants using microalgae. New Future Dev Microb Biotechnol Bioeng 2019; 223-35.
[http://dx.doi.org/10.1016/B978-0-444-63504-4.00017-7]

[174] Kublanovskaya A, Baulina O, Chekanov K, Lobakova E. The microalga *Haematococcus lacustris* (Chlorophyceae) forms natural biofilms in supralittoral White Sea coastal rock ponds. Planta 2020; 252(3): 37.
[http://dx.doi.org/10.1007/s00425-020-03438-7] [PMID: 32778946]

[175] Escudero-Oñate C, Ferrando-Climent L. Microalgae for biodiesel production and pharmaceutical removal from water. Nanoscience Biotechnol Environ Appl 2019; 1-28.
[http://dx.doi.org/10.1007/978-3-319-97922-9_1]

[176] Lu Q, Li J, Wang J, *et al.* Exploration of a mechanism for the production of highly unsaturated fatty acids in *Scenedesmus* sp. at low temperature grown on oil crop residue based medium. Bioresour

Technol 2017; 244(Pt 1): 542-51.
[http://dx.doi.org/10.1016/j.biortech.2017.08.005] [PMID: 28803104]

[177]   Chaiwong C, Koottatep T, Polprasert C. Comparative study on attached-growth photobioreactors under blue and red lights for treatment of septic tank effluent. J Environ Manage 2020; 260: 110134.
[http://dx.doi.org/10.1016/j.jenvman.2020.110134] [PMID: 32090830]

[178]   Lu Q, Zhou W, Min M, *et al.* Growing Chlorella sp. on meat processing wastewater for nutrient removal and biomass production. Bioresour Technol 2015; 198: 189-97.
[http://dx.doi.org/10.1016/j.biortech.2015.08.133] [PMID: 26386422]

[179]   Quan X, Hu R, Chang H, *et al.* Enhancing microalgae growth and landfill leachate treatment through ozonization. J Clean Prod 2020; 248: 119182.
[http://dx.doi.org/10.1016/j.jclepro.2019.119182]

[180]   Zhao X, Kumar K, Gross MA, Kunetz TE, Wen Z. Evaluation of revolving algae biofilm reactors for nutrients and metals removal from sludge thickening supernatant in a municipal wastewater treatment facility. Water Res 2018; 143: 467-78.
[http://dx.doi.org/10.1016/j.watres.2018.07.001] [PMID: 29986255]

[181]   Zhang Q, Yu Z, Zhu L, *et al.* Vertical-algal-biofilm enhanced raceway pond for cost-effective wastewater treatment and value-added products production. Water Res 2018; 139: 144-57.
[http://dx.doi.org/10.1016/j.watres.2018.03.076] [PMID: 29635151]

[182]   Tenore A, Mattei MR, Frunzo L. Modelling the ecology of phototrophic-heterotrophic biofilms. Commun Nonlinear Sci Numer Simul 2021; 94: 105577.
[http://dx.doi.org/10.1016/j.cnsns.2020.105577]

[183]   Huang J, Chu R, Chang T, *et al.* Modeling and improving arrayed microalgal biofilm attached culture system. Bioresour Technol 2021; 331: 124931.
[http://dx.doi.org/10.1016/j.biortech.2021.124931] [PMID: 33812139]

[184]   Avendaño-Herrera RE, Riquelme CE. Production of a diatom-bacteria biofilm in a photobioreactor for aquaculture applications. Aquacult Eng 2007; 36(2): 97-104.
[http://dx.doi.org/10.1016/j.aquaeng.2006.08.001]

[185]   Katam K, Tiwari Y, Shimizu T, Soda S, Bhattacharyya D. Start-up of a trickling photobioreactor for the treatment of domestic wastewater. Water Environ Res 2021; 93(9): 1690-9.
[http://dx.doi.org/10.1002/wer.1554] [PMID: 33715232]

[186]   Gómez-Jacinto V, García-Barrera T, Gómez-Ariza JL, Garbayo-Nores I, Vílchez-Lobato C. Elucidation of the defence mechanism in microalgae *Chlorella sorokiniana* under mercury exposure. Identification of Hg–phytochelatins. Chem Biol Interact 2015; 238: 82-90.
[http://dx.doi.org/10.1016/j.cbi.2015.06.013] [PMID: 26079052]

[187]   Priatni S, Ratnaningrum D, Warya S, Audina E. Phycobiliproteins production and heavy metals reduction ability of porphyridium sp. IOP Conf Ser Earth Environ Sci. 160: 012006.
[http://dx.doi.org/10.1088/1755-1315/160/1/012006]

[188]   Arora N, Dubey D, Sharma M, *et al.* NMR-based metabolomic approach to elucidate the differential cellular responses during mitigation of arsenic (III, V) in a green microalga. ACS Omega 2018; 3(9): 11847-56.
[http://dx.doi.org/10.1021/acsomega.8b01692] [PMID: 30320279]

[189]   Singh NK, Raghubanshi AS, Upadhyay AK, Rai UN. Arsenic and other heavy metal accumulation in plants and algae growing naturally in contaminated area of West Bengal, India. Ecotoxicol Environ Saf 2016; 130: 224-33.
[http://dx.doi.org/10.1016/j.ecoenv.2016.04.024] [PMID: 27131746]

[190]   Papry RI, Ishii K, Mamun MAA, *et al.* Arsenic biotransformation potential of six marine diatom species: effect of temperature and salinity. Sci Rep 2019; 9(1): 10226.
[http://dx.doi.org/10.1038/s41598-019-46551-8] [PMID: 31308398]

[191] Baker J, Wallschläger D. The role of phosphorus in the metabolism of arsenate by a freshwater green alga, Chlorella vulgaris. J Environ Sci (China) 2016; 49: 169-78.
[http://dx.doi.org/10.1016/j.jes.2016.10.002] [PMID: 28007172]

[192] Sun J, Cheng J, Yang Z, Li K, Zhou J, Cen K. Microstructures and functional groups of Nannochloropsis sp. cells with arsenic adsorption and lipid accumulation. Bioresour Technol 2015; 194: 305-11.
[http://dx.doi.org/10.1016/j.biortech.2015.07.041] [PMID: 26210144]

[193] Wang Y, Zhang CH, Lin MM, Ge Y. A symbiotic bacterium differentially influences arsenate absorption and transformation in Dunaliella salina under different phosphate regimes. J Hazard Mater 2016; 318: 443-51.
[http://dx.doi.org/10.1016/j.jhazmat.2016.07.031] [PMID: 27450336]

[194] Jiang Y, Purchase D, Jones H, Garelick H. Effects of arsenate (AS5+) on growth and production of glutathione (GSH) and phytochelatins (PCS) in *Chlorella vulgaris*. Int J Phytoremediation 2011; 13(8): 834-44.
[http://dx.doi.org/10.1080/15226514.2010.525560] [PMID: 21972522]

[195] Huang WJ, Wu CC, Chang WC. Bioaccumulation and toxicity of arsenic in cyanobacteria cultures separated from a eutrophic reservoir. Environ Monit Assess 2014; 186(2): 805-14.
[http://dx.doi.org/10.1007/s10661-013-3418-6] [PMID: 24046239]

[196] Zhao Y, Song X, Yu L, Han B, Li T, Yu X. Influence of cadmium stress on the lipid production and cadmium bioresorption by Monoraphidium sp. QLY-1. Energy Convers Manage 2019; 188: 76-85.
[http://dx.doi.org/10.1016/j.enconman.2019.03.041]

[197] Alam MA, Vandamme D, Chun W, *et al.* Bioflocculation as an innovative harvesting strategy for microalgae. Rev Environ Sci Biotechnol 2016; 15(4): 573-83.
[http://dx.doi.org/10.1007/s11157-016-9408-8]

[198] Zhang X, Zhao X, Wan C, Chen B, Bai F. Efficient biosorption of cadmium by the self-flocculating microalga Scenedesmus obliquus AS-6-1. Algal Res 2016; 16: 427-33.
[http://dx.doi.org/10.1016/j.algal.2016.04.002]

[199] Ibuot A, Dean AP, McIntosh OA, Pittman JK. Metal bioremediation by CrMTP4 over-expressing Chlamydomonas reinhardtii in comparison to natural wastewater-tolerant microalgae strains. Algal Res 2017; 24: 89-96.
[http://dx.doi.org/10.1016/j.algal.2017.03.002]

[200] Wang D, Xia X, Wu S, Zheng S, Wang G. The essentialness of glutathione reductase GorA for biosynthesis of Se(0)-nanoparticles and GSH for CdSe quantum dot formation in *Pseudomonas stutzeri* TS44. J Hazard Mater 2019; 366: 301-10.
[http://dx.doi.org/10.1016/j.jhazmat.2018.11.092] [PMID: 30530022]

[201] Shen Y, Li H, Zhu W, *et al.* Microalgal-biochar immobilized complex: A novel efficient biosorbent for cadmium removal from aqueous solution. Bioresour Technol 2017; 244(Pt 1): 1031-8.
[http://dx.doi.org/10.1016/j.biortech.2017.08.085] [PMID: 28847109]

[202] Bodin Ö. Collaborative environmental governance: Achieving collective action in social-ecological systems. Science 2017; 357(6352): eaan1114.
[http://dx.doi.org/10.1126/science.aan1114] [PMID: 28818915]

[203] Hwang K, Kwon GJ, Yang J, *et al.* Chlamydomonas angulosa (Green Alga) and Nostoc commune (Blue-Green Alga) microalgae-cellulose composite aerogel beads: manufacture, physicochemical characterization, and Cd (II) adsorption. Materials (Basel) 2018; 11(4): 562.
[http://dx.doi.org/10.3390/ma11040562] [PMID: 29621190]

[204] Hedayatkhah A, Cretoiu MS, Emtiazi G, Stal LJ, Bolhuis H. Bioremediation of chromium contaminated water by diatoms with concomitant lipid accumulation for biofuel production. J Environ Manage 2018; 227: 313-20.

[http://dx.doi.org/10.1016/j.jenvman.2018.09.011] [PMID: 30199727]

[205] Yen HW, Chen PW, Hsu CY, Lee L. The use of autotrophic *Chlorella vulgaris* in chromium (VI) reduction under different reduction conditions. J Taiwan Inst Chem Eng 2017; 74: 1-6.
[http://dx.doi.org/10.1016/j.jtice.2016.08.017]

[206] Gagrai MK, Das C, Golder AK. Reduction of Cr(VI) into Cr(III) by Spirulina dead biomass in aqueous solution: Kinetic studies. Chemosphere 2013; 93(7): 1366-71.
[http://dx.doi.org/10.1016/j.chemosphere.2013.08.021] [PMID: 24053944]

[207] Ye H, Shen Z, Cui J, *et al.* Hypoglycemic activity and mechanism of the sulfated rhamnose polysaccharides chromium(III) complex in type 2 diabetic mice. Bioorg Chem 2019; 88: 102942.
[http://dx.doi.org/10.1016/j.bioorg.2019.102942] [PMID: 31028988]

[208] Cherifi O, Sbihi K, Bertrand M, Cherifi K. The siliceous microalga *Navicula subminuscula* (Manguin) as a biomaterial for removing metals from tannery effluents: a laboratory study. J Mater Environ Sci 2017; 8(3): 884-93.

[209] Nath A, tiwari PK, Rai AK, Sundaram S. Microalgal consortia differentially modulate progressive adsorption of hexavalent chromium. Physiol Mol Biol Plants 2017; 23(2): 269-80.
[http://dx.doi.org/10.1007/s12298-017-0415-1] [PMID: 28461716]

[210] Balaji S, Kalaivani T, Sushma B, Pillai CV, Shalini M, Rajasekaran C. Characterization of sorption sites and differential stress response of microalgae isolates against tannery effluents from ranipet industrial area—An application towards phycoremediation. Int J Phytoremediation 2016; 18(8): 747-53.
[http://dx.doi.org/10.1080/15226514.2015.1115960] [PMID: 26587690]

[211] Suganya S, Saravanan A, Senthil Kumar P, Yashwanthraj M, Sundar Rajan P, Kayalvizhi K. Sequestration of Pb(II) and Ni(II) ions from aqueous solution using microalga *Rhizoclonium hookeri*: adsorption thermodynamics, kinetics, and equilibrium studies. J Water Reuse Desalin 2017; 7(2): 214-27.
[http://dx.doi.org/10.2166/wrd.2016.200]

[212] Batsalova T, Teneva I, Belkinova D, Stoyanov P, Rusinova-Videva S, Dzhambazov B. Assessment of cad-mium, nickel and lead toxicity by us-ing green algae Scenedesmus incras-satulus and human cell lines: Potential *in vitro* test-systems for monitoring of heavy metal pollution. Toxicol Forensic Med 2017; 2(2): 63-73.
[http://dx.doi.org/10.17140/TFMOJ-2-121]

[213] Gan T, Zhao N, Yin G, *et al.* Optimal chlorophyll fluorescence parameter selection for rapid and sensitive detection of lead toxicity to marine microalgae *Nitzschia closterium* based on chlorophyll fluorescence technology. J Photochem Photobiol B 2019; 197: 111551.
[http://dx.doi.org/10.1016/j.jphotobiol.2019.111551] [PMID: 31306954]

[214] Kelly DJA, Budd K, Lefebvre DD. Biotransformation of mercury in pH-stat cultures of eukaryotic freshwater algae. Arch Microbiol 2006; 187(1): 45-53.
[http://dx.doi.org/10.1007/s00203-006-0170-0] [PMID: 17031617]

[215] Devars S, Avilés C, Cervantes C, Moreno-Sánchez R. Mercury uptake and removal by *Euglena gracilis*. Arch Microbiol 2000; 174(3): 175-80.
[http://dx.doi.org/10.1007/s002030000193] [PMID: 11041348]

[216] Huang CC, Chen MW, Hsieh JL, Lin WH, Chen PC, Chien LF. Expression of mercuric reductase from Bacillus megaterium MB1 in eukaryotic microalga Chlorella sp. DT: an approach for mercury phytoremediation. Appl Microbiol Biotechnol 2006; 72(1): 197-205.
[http://dx.doi.org/10.1007/s00253-005-0250-0] [PMID: 16547702]

[217] Mangal V, Phung T, Nguyen TQ, Guéguen C. Molecular characterization of mercury binding ligands released by freshwater algae grown at three photoperiods. Front Environ Sci 2019; 6: 155.
[http://dx.doi.org/10.3389/fenvs.2018.00155]

[218] Fard GH, Mehrnia MR. Investigation of mercury removal by Micro-Algae dynamic membrane bioreactor from simulated dental waste water. J Environ Chem Eng 2017; 5(1): 366-72.
[http://dx.doi.org/10.1016/j.jece.2016.11.031]

[219] Mu W, Jia K, Liu Y, Pan X, Fan Y. Response of the freshwater diatom *Halamphora veneta* (Kützing) Levkov to copper and mercury and its potential for bioassessment of heavy metal toxicity in aquatic habitats. Environ Sci Pollut Res Int 2017; 24(34): 26375-86.
[http://dx.doi.org/10.1007/s11356-017-0225-6] [PMID: 28944446]

CHAPTER 9

Earthworm-mediated Remediation and Mitigation of Heavy Metals Toxicity in Plants

Ashutosh Sharma[1], **Pooja Sharma**[2,3], **Sahil Dhiman**[1], **Sandeep**[1,4], **Sanjay Kumar**[5], **Sukhwinder Kaur**[5], **Arun Dev Singh**[3], **Raj Bala**[4], **Renu Bhardwaj**[3] and **Indu Sharma**[5,*]

[1] *Department of Agricultural Sciences, DAV University, Sarmastpur, Jalandhar, Punjab, India*

[2] *Department of Microbiology, DAV University, Sarmastpur, Jalandhar, Punjab, India*

[3] *Department of Botanical and Environmental Sciences, Guru Nanak Dev University, Amritsar, Punjab, India*

[4] *Department of Botany and Environment Studies, DAV University, Jalandhar, Punjab, India*

[5] *Department of Life Sciences & Allied Health Sciences, Sant Baba Bhag Singh University, Jalandhar, Punjab, India*

Abstract: Earthworms are the ecosystem engineers that convert waste into vermicompost. The vermicompost is further utilized to improve the soil's organic composition, condition, and health. Recently, earthworms have been explored as an effective, efficient, and eco-friendly remediation approach called 'vermiremediation' to mitigate the toxic elements from the soil. The soil contains different types of essential or non-essential elements. The presence of these elements above threshold levels in the soil leads to its contamination. The major soil contaminants include xenobiotic compounds, agrochemicals, and heavy metals. The plants exposed to higher amounts of heavy metal-containing soils show symptoms of metal-induced phyto-toxicities that result from the loss of soil fertility, disturbance in nutrient uptake and translocation, and interruption in the regular physiological functions of affected plants. To overcome heavy metal-induced toxicities in plants and soils, the treatments of earthworms, either alone or in combination with PGPR or other soil amendments, are being tried. The present chapter is an attempt to compile information about phytoremediation and vermiremediation, distribution of earthworms in contaminated soils, remediation and amelioration of heavy metals by earthworms, and factors affecting bioaccumulation of heavy metals in earthworms.

Keywords: Bioaccumulation, Bioremediation, Phytoremediation, Phytotoxicity, Vermi-remediation.

---

* **Corresponding author Indu Sharma:** Department of Life Sciences & Allied Health Sciences, Sant Baba Bhag Singh University, Jalandhar, Punjab, India;
E-mails:indu.gndu@yahoo.com and indu.gndu@gmail.com

INTRODUCTION

Earthworms can help in the remediation of pollutants, potential toxicants, heavy metals, and contaminants in the soil [1]. Common soil contaminants that lead to degradation include organic materials, agrochemical residues, pesticides, fertilizers, and other xenobiotic compounds, including heavy metals. The heavy metal contamination poses risks to fertile soils and the health of nearby organisms [2]. These toxic metals reduce soil fertility and hinder photosynthesis and metabolic processes in plants, ultimately lowering crop yields. For remediation of contaminated soils, nano-biochar is being used because of its high hydrodynamic dispersivity and strong rhizospheric interactions [3]. However, its impact on the rhizosphere and biogeochemical behaviour makes it unsuitable for soil remediation due to environmental risks. Thus, biosafety guidelines are needed for the application of nano-biochar. The biochar and earthworms influence soil heavy metal bioavailability, with earthworm-mediated nitrification and gut digestion facilitating remobilization [4]. Rice-husk and sludge biochar immobilize heavy metals, while earthworm gut digestion enhances bioavailability and mobility. The epigeic species of earthworm, namely *Amynthas cortices* and an endogeic species of earthworm *Amynthas robustus* (*A. robustus*), have been found to restore soil properties by enhancing the bioavailability of heavy metals through interaction with soil [5]. Moreover, earthworms effectively absorb heavy metals from polluted soils, acting as pollution bioindicators and soil decontaminators, thereby enhancing organic matter decomposition and plant availability [6]. Hence, to clean up the contaminated soil ecosystem, earthworms can be used along with conventional methods [7]. Fig. (**1**) highlights the major useful roles of earthworms in heavy metal remediation.

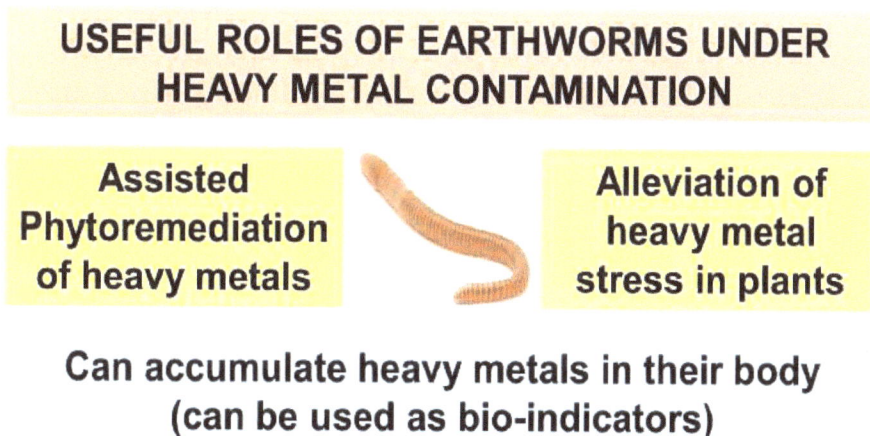

USEFUL ROLES OF EARTHWORMS UNDER HEAVY METAL CONTAMINATION

Assisted Phytoremediation of heavy metals

Alleviation of heavy metal stress in plants

Can accumulate heavy metals in their body (can be used as bio-indicators)

Fig. (1). Major useful roles of earthworms in heavy metal remediation.

As per existing physical remediation practices, polluted sites are restored through soil replacement and thermal desorption methods [8]. Chemical remediation approaches utilize chemical leaching and fixation, vitrifying technology, and electrokinetics remediation methods for soil cleanup. Also, the physicochemical approaches include techniques like soil encapsulation, immobilization techniques, electro-remediation, etc [9]. Soil properties, pH, acidity, and potential release of pollutants limit soil remediation methods. These methods are costly, time-consuming, and eco-friendly, making them unsuitable for large-scale pollution removal. Eco-friendly green technologies like plants, microbes, and earthworms may be more practical. *In situ* and *ex situ* bioremediation employs various biological organisms to eradicate soil toxicants, and it is an economically- and environmentally-sustainable remediation approach [10].

The biological remediation approach includes bioremediation, phytoremediation, plant growth regulators like brassinosteroids, microbes including Plant Growth Promoting Rhizobacteria (PGPR), integrated biological remediation methods, or the use of earthworms has emerged as a green and eco-friendly remediation method [2, 6, 7, 9, 11 - 17]. Several studies highlighted that the use of earthworms is beneficial to improve soil health and productivity [1, 2, 16, 18]. Earthworms can effectively remediate soil pollutants through a process called *vermiremediation*. This involves earthworms accumulating pollutants like heavy metals, crude oil, and non-recyclable and organic contaminants, which are then transformed and degraded by the soil [19]. This process can be used alone or in combination with other soil amendments or microorganisms [1]. The supplementation of phytoremediation, microorganisms, biochar, and surfactants helps in improving the potential of earthworms to remediate polluted soils. Recently, a study emphasized the use of *Eisenia fetida* (an earthworm) as a potential organism for the remediation of agricultural soils affected by sewage sludge [18]. Sewage sludge, used in agriculture, contains heavy metals, increasing soil concentrations. *E. fetida* can reduce these metals and improve soil fertility. Earthworm mucus and microbial symbionts release enzymes that activate biochar, reducing heavy metal levels and enhancing soil nutrients [20]. This hydrophobic interaction helps maintain soil fertility.

Earthworm-mediated activated biochar, known as *vermibiochar*, can effectively remediate heavy metals in textile mill dyeing sludge. This eco-friendly and cost-effective strategy uses *E. fetida* to convert dyeing sludge into nutrient-rich vermicompost [21]. The process enhances nutrient levels, nitrogen, phosphorus, and sodium levels. It also restores pH, Carbon-to-Nitrogen (C/N) ratio, and electrical conductivity, demonstrating that *E. fetida* has tremendous potential for waste sludge conversion into useful manure. The ability of earthworms to detoxify toxic pollutants from the soil into less harmful ones through biotransformation and bio-degradation was highlighted in a report by Sinha *et al.* [22]. Consequently, earthworms' activity improves the quality of soil and also produces valuable vermicompost. Vermicompost, which is a superior soil amendment consisting of digested compost and is rich in nutrients, is known as "organic gold" [23] or "black gold" [24]. Thus, earthworms can be used to stabilize garbage into gold, an eco-friendly and bio-oxidative process that is beneficial for long-term sustainable agriculture [25].

Keeping in view the immense potential roles of earthworms as soil decontaminators, this chapter discusses their role in soil decontamination, comparing phytoremediation and vermiremediation. It discusses how plants absorb heavy metals, their phytotoxic effects, and the health risks they face in contaminated soils. Further, the chapter also focuses on earthworm-mediated improvement of soil structure and aeration, facilitating plant growth for remediation. The chapter also explores the role of earthworms in the bioaccumulation of heavy metals and their ability to reduce soil contamination through natural processes.

PHYTOREMEDIATION AND VERMIREMEDIATION

Ecology and sustainable development face a threat from environmental pollution. The diagnosis, environmental pollution risk assessment, pollution remediation, and reclamation are crucial steps for the true management of pollution. The toxicants, contaminants, or pollutants are removed from the soil or watercourses through remediation techniques, which may be chemical, physical, physicochemical, or biological remediation. The process of biological remediation utilizes living organisms, such as plants, microorganisms, earthworms, etc., for the restoration and reclamation of polluted soils. The process that uses plants for environmental remediation is called phytoremediation, whereas the remediation process that utilizes earthworms is referred to as vermiremediation.

A study highlighted contaminants like alkanes, total petroleum hydrocarbons, and polycyclic aromatic hydrocarbons could be eradicated from the soil by using a plant species, *Panicum maximum*, for phytoremediation; an earthworm species, *Pontoscolex corethrurus*, for vermiremediation, and encapsulated bacterial consortium for bioaugmentation [26]. This experiment concluded that the employment of earthworms (vermiremediation) and plants (phytoremediation) for the alleviation of contaminants in any contaminated site would be an efficient and green alternative to existing remediation technologies. Therefore, in the following sub-sections, both phytoremediation and vermiremediation are discussed.

Phytoremediation

The cost-effective, plant-based, non-invasive alternative approach for environmental cleanup and pollutant stabilization followed by their extraction, degradation, or volatilization is referred to as phytoremediation [27]. Plants are being utilized to eradicate diverse types of pollutants, such as organic and inorganic pollutants, depending upon the species of plants. Other factors that affect phytoremediation include the availability of pollutants, the interaction of pollutants with the rhizosphere, and uptake, translocation, chelation, degradation, and volatilization of pollutants. The process of plant-mediated remediation includes (i) translocation of pollutants, (ii) accumulation of pollutants, (iii) transport, (iv) biotransformation or biodegradation of pollutants, and (v) volatilization [28]. Several other strategies, like physicochemical, agronomic, microbiological, and genetic engineering methods, when applied with phytoremediation, enhanced the efficiency of remediation of contaminated soils. However, phytoremediation has the drawback of insufficient methods for the disposal of plant biomass that has bioaccumulated higher amounts of heavy metals. The current major biomass disposal strategies include compaction, pyrolysis, incineration, composting, leaching, *etc.*, which are either costly or release secondary pollutants in the ecosystem. Oladoye *et al.* described the proper methods for the disposal of plant biomass with bioaccumulated toxicants and the methodologies to utilize the by-products of phytoremediation to enhance the applicability of phytotechnology [29].

The practical application of phytoremediation is dependent on the purification rate and factors affecting phytoremediation technology for various environments like soil, air, and water [30]. This report highlighted the purification potential of various plant species for eradicating pollutants to restore the environment and associated purification factors. Phytoremediation efficiency is dependent on stomatal conductance, light intensity, the interaction of plants with microbial species, the plant's species-specific metabolism, and heavy metal uptake. To enhance the efficiency of phytoremediation and a sustainable plant-based remediation technology and its large-scale application, integrated remediation techniques involving a combination of biological composts, plant growth regulators, biostimulants, and utilization of plant-growth-promoting rhizobacteria or microorganisms are recommended [29]. Further insights into underlying mechanisms of accumulation of pollutants by the plants might also help in understanding and improving the efficiency of a plant to remediate the contaminated soils. Pollutants like heavy metals accumulate in the plants through the following processes [31].

i. **Bioconversion and Bioavailability of Heavy Metals**: Plant root-mediated conversion of insoluble heavy metals present in soil into soluble forms to enhance their bioavailability to the hyperaccumulator plants. Plant roots interact with microorganisms in the soil and produce root exudates, which alter the pH of the rhizosphere.

ii. **Uptake of Heavy Metals by Plant**: When heavy metals are bioavailable in soil, they are sorbed on the surface of roots when coming in contact with the root surface, mobilized, and then cross the root cell membrane through diffusion or active transport. In the root cells, some heavy metals may undergo a process where they bind with metal chelators to form complexes, which get immobilized in extracellular root space or vacuoles, known as metal sequestration and cellular compartmentation.

iii. **Translocation of Heavy Metals from Root to Shoot**: Heavy metals translocate from the root cells to shoot cells*via* common symplastic pathway transport through xylem vessels by the process of xylem loading. The sequestered metals present in the vacuole can also translocate from root xylem vessels to the shoots. Both apoplastic and symplastic pathways are used for the transportation and distribution of heavy metals in leaves.

iv. **Heavy Metal Sequestration and Compartmentation**: In above-ground plant parts, metal ions are seized in the vacuole, cell walls, or extracellular spaces. But there is the least accumulation of heavy metal ions or free radicals in the cytosol.

Steps (i) to (iv), which include heavy metal mobilization, uptake, translocation, sequestration, and compartmentation, are collectively called "phytoextraction." The immobilization of heavy metals by plants in below-ground parts, ensuring their reduced bioavailability and barring their migration to above-ground parts, is called "phytostabilization." The process of "phytovolatilization" occurs when hazardous heavy metals or pollutants that are volatile are changed into less toxic and more volatile forms, which are then released from the leaves of the hyperaccumulator plants in the form of volatile gases. Another phytoremediation technique called "phytofiltration" involves the removal or filtering of heavy metals from wastewater by plant shoots (caulofiltration), roots (rhizofiltration), or seedlings (blastofiltration), which lessens the leaching of heavy metals into underground water bodies. The mobility of heavy metals and their availability to plants is limited by various factors such as adsorption, absorption, mineralization, protonation, precipitation of heavy metals, physicochemical properties of soil, and interactions with rhizosphere and rhizospheric effects [32]. Although the phytoremediation technique has several benefits, one of its main drawbacks is that remediation takes a long time and requires numerous plant cycles to complete the remediation of heavy metals.

Numerous researchers have proposed combining phytoremediation with the use of microorganisms, plant growth regulators, heavy metal accelerators, arbuscular mycorrhizal fungi, transgenic transformation, and transgenically modified plants to cut down on the time required and the need for multiple plant cycles [2, 11 - 16, 33, 34]. However, the disposal of heavy metal accumulated plant biomass is one of the major biological concerns associated with phytoremediation. Several disposal and utilization methods, such as extraction, compression landfill, thermo-treatment, microbial treatment, and nanomaterial-based disposal methods for heavy metals accumulated in plants, are explained by Liu and Tran [33]. The method for biomass disposal depends on the type of heavy metals, their characteristics, the characteristics of hyperaccumulator plants and the operation strategy of the disposal approach. Thus, the appropriate disposal technique can enhance the phytostabilization, phytoextraction, phytodegradation, phytovolatilization, and phytoremediation potential of the plants [34]. Furthermore, to better understand the metabolic processes involved in heavy metals tolerance in hyper-accumulator plants and open up new avenues for phytoremediation, more in-depth field research, the selection of the most beneficial hyperaccumulator plants, the identification of new genes, and the development of transgenic plants will be helpful.

Vermiremediation

Earthworms are reported to be tolerant to an array of toxicants or chemical contaminants, such as heavy metals and organic pollutants in soil [22]. Various species of earthworms, namely *Eisenia fetida, E. tetraedra, Lumbricus terrestris, L. rubellus,* and *Allobophora chlorotica* are reported to eliminate heavy metals like cadmium (Cd), lead (Pb), copper (Cu), mercury (Hg), *etc.*, harmful pesticides, and lipophilic organic micropollutants (*e.g.*, polycyclic aromatic hydrocarbons) from the contaminated soils. The earthworms are reported to 'absorb' the dissolved pollutants/chemicals *via* their moist 'body wall' in the interstitial water, and when the soil/dust/dirt particles enter through their mouth, they also 'ingest' pollutants/chemicals. Thereafter, the earthworms pass the pollutants/chemicals into their digestive tract or gut, where they either 'bio-transform' or 'biodegrade' the harmful pollutants/chemical contaminants into harmless products through their bodies. Earthworms can also bio-accumulate harmful contaminants or heavy metals in their tissue. The ingested material is digested and then harmlessly released into the soil as nutrient-rich vermicast. At the same time, the biological and physicochemical properties of soil are also enhanced. The resulting vermicast is a highly nutritious organic fertilizer that has ample amounts of nitrogen, phosphorous, potassium, humus, micronutrients, and beneficial soil microbes, such as actinomycetes, nitrogen-fixing bacteria, phosphate-solubilizing bacteria, and plant growth regulators like auxins, gibberellins, cytokinins, *etc.* [23].

Vermicast, which is comparable to compost and is highly valuable as an organic fertilizer, is also known as "organic gold" or "vermicompost."

An equally nutrient-rich, natural liquid extract with an enormous decomposer bacteria count and mucus called "vermiwash" is created from vermicompost [35]. Both vermiwash and vermicompost are beneficial for enhancing plant growth and development. The entire process of **conversion of raw material/raw organic matter** into useful vermicompost, vermicast, and/or vermiwash by the activity of earthworms is referred to as **vermicomposting** [24]. Through the consumption of raw nutrients and the production of beneficial metabolites or by-products, the process of vermicomposting increases the bioavailability of essential nutrients to plants. However, when the raw material/raw organic matter is a toxicant, pollutant, or contaminant, then the process of **conversion of toxicant/pollutant/contaminant** by the earthworms into valuable vermicompost or by-products is described as **vermiremediation**. The harmful polycyclic aromatic hydrocarbons from the soil were degraded by vermiremediation employing a species of earthworm, *i.e.*, utilizing *Lumbricus rubellus* and using *Pleurotus sajor-caju*, spent fungal mycelia (mycoremediation) in 30 days of incubation [36]. Therefore, the study revealed the potential of vermiremediation for the on-site clean-up of soil pollutants.

The species of earthworm, namely *E. fetida, Eudrilus eugeniae, Drawida willsi, Lampito mauritii*, and *Perionyx excavates,* were analyzed for remediation of non-recyclable waste paper amended with cow dung [37]. The earthworm *E. fetida* was further studied and found to be efficient in converting non-recyclable paper waste in feed mixtures (cow dung) into good-quality vermicompost. Allopathic sludge was obtained from the pharmaceutical industry [66]. Earthworms could not survive on 100% sludge because of the sludge's high toxicity, inappropriateness, and non-palatability for the earthworms. Therefore, sludge was mixed with different proportions of cattle dung to make the feed material for earthworms. On the cattle dung amended sludge, earthworms (*E. fetida*) were cultured. The study revealed that earthworms converted harmful sludge into vermicompost containing more N, P, K, and a less toxic and stable product. Vermiremediation may be a practical long-term biological solution for soil remediation, especially in mildly contaminated areas [38]. This methodology can be used as a backup measure to get rid of any leftover toxins and/or simultaneously employed with other physical, chemical, or biological remediation methods to clean up polluted sites. Table **1** highlights various reports related to vermiremediation done by employing different species of earthworms.

Table 1. Vermiremediationby different species of Earthworms.

Earthworm Species	Contaminant	Effect of Treatment of Earthworms	Reference
Eisenia fetida	Non-recyclable post-consumer paper waste	♦ Non-recyclable paper waste from consumable items, used paper products, and non-recyclable post-consumer paper waste were modified by adding cow dung and then converted into vermicompost by utilizing earthworm to bio-transform waste into a value-added product.	[37]
Lumbricus rubellus	Polycyclic aromatic hydrocarbons	♦ Earthworms degraded harmful polycyclic aromatic hydrocarbons, namely anthracene, phenanthrene, and benzo(α)pyrene, from the soil in the 30 days of incubation.	[36]
Eisenia fetida	Heavy metals in wastewater sludge from the paper and pulp industry	♦ Earthworms removed the heavy metals from paper mill wastewater sludge, which was amended with cow dung and also caused vermistabilization. ♦ Earthworms accumulated heavy metals in their body in the order Cd>Cr>Pb>Cu.	[39]
Metaphire posthuman	Water Treatment Plant Sludge produced from Sewage treatment	♦ Earthworms reduced total C in sludge. ♦ Enhanced levels of N. ♦ Maintained the pH from lower to neutral pH. ♦ Improved the soil quality. ♦ Reduced the accumulation of heavy metals, namely Cr and Cu, in soil. ♦ Bioprocessed the toxic water treatment plant sludge.	[40]
Eisenia andrei	Polycyclic aromatic hydrocarbons in sewage sludge	♦ Earthworms metabolized polycyclic aromatic hydrocarbons into compost and also lowered their concentration in sludge.	[41]
Eisenia andrei	Heavy metals (Cd, Cu, Zn, Ni, Cr, and Pb) in sewage sludge	♦ Earthworms were found to accumulate heavy metals in their body in the following order: Cd> Cu > Zn > Ni > Cr > Pb. ♦ Also, the levels of heavy metals were amended in sludge by the activity of earthworms.	[41]

(Table 1) cont.....

Earthworm Species	Contaminant	Effect of Treatment of Earthworms	Reference
Eisenia fetida (epigeic species) and *Amynthas robustus* (endogeic species)	Atrazine (2-chloro-4-ethylamino-6-isopropylamino-1,3,5-triazine), a triazine herbicide	♦ Earthworms significantly enhanced the degradation of atrazine. ♦ Levels of residual atrazine were reduced by both species of earthworms. ♦ Humus-fixed atrazine was reduced in the soil. ♦ Available atrazine was enhanced in bulk soil. ♦ Neutralized soil pH. ♦ Earthworms stimulated the metabolism of atrazine degraders.	[42]
Eudrilus eugeniae	Cotton textile sludge	♦ About 50-70% decrease in levels of Pb, Cd, Cr, and Zn was observed in textile sludge by the action of earthworms. ♦ Earthworms stimulated humic compound-mediated chelation of heavy metals. ♦ Conversion of more toxic Cr^{6+} to less toxic Cr^{3+} was enhanced by earthworms during vermicomposting. ♦ Concentrations of total N and bioavailability of P were increased in vermicomposting.	[43]
Alma millsoni, Eudrilus eugeniae, and *Libyodrilus violaceus*	Glyphosate-based herbicide (commercial name: Roundup® Alphée)	♦ All three species of earthworms bioaccumulated the glyphosate in their body tissues and biomagnified the glyphosate. ♦ Glyphosate was eradicated from the contaminated soil. ♦ Earthworms were found to reduce glyphosate residues in the soil.	[44]
Eisenia fetida	Soil fortified with 0, 5, 10, and 20 mg Cd kg^{-1}	♦ Earthworms reduced the bioavailability of Cd in heavy metal-contaminated soil. ♦ Enhanced the fertility of the soil. ♦ Levels of nitrogen, phosphorous, and potassium in soil were increased by earthworms.	[45]

Earthworm Species	Contaminant	Effect of Treatment of Earthworms	Reference
Eisenia fetida	Allopathic pharmaceutical industry sludge	♦ Allopathic sludge obtained from the pharmaceutical industry was amended with cattle dung, and on this feed material, earthworms were cultured. ♦ Earthworms converted harmful sludge into vermicompost containing more N, P, and K, making it a less toxic and stable product.	[25]
Eisenia fetida	0.5 mM Chromium (Cr) metal solution applied to soil in the form of potassium chromate (K_2CrO_4)	♦ Application of two PGPR, namely *Pseudomonas aeruginosa* and *Burkholderia gladioli*, along with earthworms, was found to reduce ROS levels, lipid peroxidations, upregulated the activities of antioxidant enzyme genes (*SOD, CAT, POD, APOX, GR, DHAR,* and *GST*) in *Brassica juncea* seedlings under Cr stress. ♦ Co-treatment of both PGPR and earthworm was effective in enhancing the plant biomass. ♦ Co-application of both PGPR and earthworms decreased the expression of ROS-producing *RBOH1* gene in *B. juncea* plants under Cr stress.	[16, 17]

EARTHWORMS IN CONTAMINATED SOIL

The effective degradation or removal of soil pollutants through vermiremediation depends on the capacity of earthworms to withstand high concentrations of pollutants in highly polluted soils [19]. Due to the potential of earthworms to survive in polluted soils, the remediation potential of earthworms to clean up an array of inorganic and organic contaminants like polycyclic aromatic hydrocarbons, polybrominated diphenyl ethers, polychlorinated biphenyls, crude oils, pesticides, and heavy metals is being explored. Earthworms can be applied alone or with microbes, plant-based supplements or plant growth regulators, soil amendments, etc., to restore the heavy metal-contaminated soils and to eradicate the potentially toxic elements from diversely polluted soils [1]. The earthworms are ubiquitously distributed in the soils throughout the world, except for the regions having extreme conditions, such as the driest and coldest zones. To date, there are more than 4000 species of earthworms on earth that have been identified, and earthworms from different species differ in terms of their appearance, size, and ecological behaviour.

The population of earthworms alone accounts for up to 90% of the biomass found below ground in soil [46]. In the coelenteron of earthworms, where numerous bacteria communities have been recorded, there is a vast amount of space and nutrients, which aid in the survival of microbes. Moreover, according to estimates, about 89.5 million litres of soil residue were found in the earthworm gut. Approximately 50% of soil passes through the earthworm's gut in about 10 years, whereas 90% of soil will pass through the earthworm's gut in 40 years. The density of earthworms in soils typically ranges from 5 to 150 worms per square meter. However, under ideal conditions such as plenty of food, suitable soil moisture content (60-85%), moderate temperature (15-25 °C), and neutral soil pH (6.8-7.20), the population of earthworms would surpass 3000 worms per square meter [47]. The lower pH values reduced the availability of both calcium and soil carbon, which further limited the biodiversity of earthworms. The decreased soil organic components, food supplies, genotoxicity from chemical fertilisers and pesticides, and agricultural fields have been found to lower earthworm densities [25].

The diversity of earthworms is measured by their species richness, which is a global pattern in community metrics that also includes their abundance, diversity, and biomass [47]. The number, biomass, ecological group of the species of earthworms, and distribution of above- and below-ground taxa are other factors that affect earthworm diversity. The diversity of earthworms can be predicted for many aboveground taxa by climatic or energy-related variables like primary productivity, evapotranspiration, *etc*. Additionally, it has been discovered that the diversity of aboveground species is favourably influenced by mean annual temperature, whereas the diversity of belowground microbes is negatively impacted. The distribution of earthworm communities in soils is determined by a combined influence of soil, vegetation, soil temperature, habitat cover, and precipitation rather than by a single environmental parameter [1]. Earthworms can be divided into the following three eco-physiological groups based on their living and feeding patterns [1, 48 - 51]:

i. **Epigeic Earthworms:** spend most of their lives above ground, creating only a few burrows in the upper few centimeters of topsoil. These show very fast movements and are relatively short-lived. *e.g.*, *E. fetida*, *Perionyx excavatus*, *Perionyx ceylanensis*, *Drawida sulcata*, and *Eudrilus eugeniae*.

ii. **Anecic Earthworms:** These earthworms feed by tunnelling vertically through soils and dragging organic materials from the surface into underground tunnels. These are long-lived and make large, long, and vertical permanent burrows in the soil. *E.g.*, *Aporrectodea longa*, *Aporrectodea giardia*, *Lampito mauritii*, *Lumbricus centralis*, and *Lumbricus terrestris*.

iii. **Endogeic Earthworms**: Endogeic earthworms are tenacious in soils and consume organic substances found in soil. These have intermediate longevity and make sub-horizontal, extensive, and continuous burrows in the soil. *e.g.,* *Amynthas robustu* and *Aporrectodea molleri*.

When studying how earthworms respond to soil metal contamination and the traits of metal bioaccumulation in different ecotypes, the species of earthworm is a key determining factor. In a study, *E. fetida*-mediated vermicomposting of duckweed (*Spirodela polyrhiza*) mixed with cow dung was found to be effective in altering the levels of nutrients, activities of microbial enzymes, and dynamics of earthworms [52]. Although *E. fetida* is a frequently utilized species of earthworm in laboratory experiments, it has little ecological significance and is rather uncommon in areas that have historically been poisoned [53]. Therefore, field surveys are more effective and insightful than conventional ecotoxicology studies for analyzing the impacts of soil metals on populations of wild earthworms.

UPTAKE OF HEAVY METALS BY EARTHWORMS

Earthworms make heavy metals more accessible to plants, and they also help to preserve the soil's structure and quality. The trophic transmission of metals from soils to predators and detritivores in terrestrial food webs depends on earthworms' ability to absorb trace metals, which is important both for earthworms' potential toxicity and for the transfer of metals to other organisms [54]. The prevailing consensus is that earthworms can absorb metals both dermally and by soil eating. Depending upon the species of earthworms, they have varied effects on the trophic network of soil and how soil functions, such as maintaining soil structure and cycling carbon and nutrients [49]. By restoring the pH of the soil to normal and eating organic materials, earthworm amendment was observed to predominantly accelerate the breakdown of tetracycline by promoting both direct dehydration and epimerization-dehydration pathways. The utilization of earthworms was found effective in changing the microenvironment of the soil and hastening the degradation of tetracycline, acting as a possible method for soil remediation in contaminated areas [55].

Another study recommended the use of earthworms in metal-contaminated soils to accelerate the phytoremediation process [56]. A century of industrial metallurgical activity in Wallonia, Belgium, has resulted in significant soil contamination by heavy metals, including copper (Cu), zinc (Zn), lead (Pb), and cadmium (Cd). *E. fetida* co-supplemented with two plant species, namely, *Vicia faba* and *Zea mays*, were subjected to various levels of long-term contaminated soils for 42 days. The co-application of plants and earthworms altered the metal availability in contaminated soils. It was discovered that the presence of both metal and earthworms affects how much metal accumulates in plants. Except for Cd, the amounts of metals in plants were unaffected by the presence of earthworms. When compared to *Z. mays*, which accumulated more Cd, *V. faba*accumulated larger quantities of Cu and Zn in both the presence and absence of earthworms. Moreover, the population of *E. fetida* was observed to be insensitive to the presence of heavy metals (Cu, Zn, Pb, and Cd) in the soil. Therefore, the study emphasized that *E. fetida* affect soil metal availability and uptake of heavy metals at various soil concentrations. Through the release of coelomic fluid, earthworms play a role in the dynamics and dispersion of beneficial bacteria in the soil during their peristaltic motions [51]. The earthworm-mediated metal uptake is dependent on its interaction with other microbes, the bioavailability of metals, the movement of earthworms, particularly endogenous types, through the soil matrix, and the significant population of microbes or PGPR in the drilosphere and rhizosphere. The presence of bacteria and other factors affect the uptake and bioaccumulation of heavy metals by the earthworms.

BIOACCUMULATION OF HEAVY METALS BY EARTHWORMS

Earthworms in a subtropical region of China are found to accumulate heavy metals, namely cadmium (Cd), zinc (Zn), lead (Pb), and copper (Cu) through bioaccumulation [57]. The ability of four species of earthworms, *i.e.*, *Metaphire californica, Amynthas homochaetus, Amynthas pecteniferus*, and *Amynthas heterochaetus,* to live in *in-situ* polluted environments, as well as their high level of heavy metal intake and capacity for their accumulation in tissues, confirmed their remarkable adaptability in metal contaminated soils. Cadmium (Cd) has the highest bioaccumulation factor and was ranked in the following order: Cd follows Zn, then Cu, and then Pb. The earthworm species with identical bioaccumulation factors of heavy metals are included in an ecological group. The metal uptake and accumulation of heavy metals by various earthworm species in polluted soils varies from species to species or is dependent on their ecological group. The relationship between the bioaccumulation of heavy metals in earthworms, namely *Aporrectodea caliginosa* and *Lumbricus rubellus,* and the total and accessible metal concentrations in field soils were studied by Hobbelen *et al.* [58].

Generally, the earthworm-mediated accumulation of heavy metals occurs either (i) through absorption of heavy metals through direct dermal contact (through exposure to dissolved metals in soil pore water) or (ii) through ingestion of organic matter by the earthworms and their adsorption *via* their gut tissues [22, 54, 57, 58]. In a study, metal pools accumulated by *Lumbricus rubellus* and *Aporrectodea caliginosa* in soils with high binding capacity were investigated [58]. At 15 metal-polluted field sites, the concentrations of Cd, Cu, and Zn in soil, pore water, and $CaCl_2$ extracts of soil, in earthworms, and in the leaves of the plant species namely *Urtica dioica*, were determined. The strong relationships between metal concentrations in earthworms and pore water provide evidence for the significance of pore water uptake of soluble metals. Cu and Cd concentrations were more accumulated in *L. rubellus* whereas *A. caliginosa* accumulated Cd. The variance in Cd and Cu concentrations in *L. rubellus* and *A. caliginosa*, however, can be best described by total soil concentrations. There was no correlation between the Zn concentrations in *L. rubellus* and *A. caliginosa*.

A recent study reported that sludge-treatment wetlands were found to possess higher amounts of heavy metals that limit their application as sludge-based manure or fertilizer [59]. The application of earthworm was found to uptake and accumulate Cd from the sludge. In plants, heavy metals uptake was recorded more in lower 1.0-35.9 times more in underground parts as compared to aboveground plant parts. However, the application of earthworms did not affect the uptake and accumulation of heavy metals in the plants. Overall, earthworms were found to have more accumulation factor for Cd and least for iron (Fe) in the following descending order: cadmium (Cd) > manganese (Mn) > nickel (Ni) > zinc (Zn) > lead (Pb) > chromium (Cr) > iron (Fe). Numerous research on the bioaccumulation of trace metals by earthworms demonstrated that the heavy metals accumulated in the tissues of earthworms are a valid indicator of the bioavailability of heavy metals in soils [54]. These studies revealed that while copper (Cu), arsenic (As), Zn, Cr, and Ni were deposited within limits, mercury (Hg), Pb, and Cd were accumulated over the threshold levels in earthworm tissues. Only Hg and Cd contents in earthworm tissue were found to have a strong linear correlation with soil levels. However, earthworm concentrations were strongly linked with soil concentrations of log-transformed Hg, Cd, Cu, Zn, As, and antimony (Sb), according to generalized linear mixed-effect models.

The effects of micronized tyre wear particles, microplastics, and associated heavy metals, *i.e.*, Cd, Pb, and Zn, on the earthworm (*E. fetida*) in soil were studied from an ecotoxicological perspective by Sheng *et al.* [60]. The earthworms were recorded to consume the smaller particles preferentially. The consumed tyre microplastics were found to change the surface shape, degraded to create more surface area, and allowed heavy metals to enter the organisms. Also, earthworms were reported to alleviate the activities of catalase and peroxidase enzymes and levels of lipid peroxidation. Also, the activity of glutathione S-transferase and superoxide dismutase was restored by earthworms. Further, the heavy metals bioaccumulation by earthworms also affects their longevity and correlates with their mortality [61]. Various species of earthworms have shown the following morphological signs and behavioural reactions to exposure to metals in soil and water media: inability to burrow, body curling or violent coiling, sluggish and quick movements, body lengthening or shortening, oozing or seeping coelomic fluid, preclitellar swelling, body constriction, and segmental bulging. Hence, further studies on the toxicity of heavy metals to earthworms can help in assessing the safe concentrations of heavy metals, risk assessment and addition of soil supplements to enhance earthworm-mediated heavy metal accumulation.

FACTORS AFFECTING BIOACCUMULATION OF HEAVY METALS IN EARTHWORMS

The unsustainable development, rapid urbanisation, industrialization, extreme applications of agrochemicals, and inappropriate waste disposal strategies cause an alarming pace of soil degradation, which raises the concentration of various contaminants, such as heavy metals, in the soil. The uptake and accumulation of heavy metals by earthworms are dependent on various factors, such as:

i. Characteristics of Soil: Physico-chemical properties of soil like pH, electrical conductivity, texture, nutrient availability, moisture, temperature etc., pollution index of the soil, presence of microbes and/or PGPR and microbial respiration, biostimulants, soil enzyme activity, soil-amendments, phytochelators, metallothioneins, plant growth regulators, interactions of earthworms with microbes or other factors or miscellaneous contaminants in soil, *etc*.

ii. Characteristics of Heavy Metal: Type of metal, its toxicity and availability in the rhizosphere or drilosphere, distribution, uptake, biomagnification by earthworms, the potential for compartmentalization and volatilization, *etc*.

iii. Characteristics of Earthworms: Species and type of earthworm, feeding practices of earthworms and their population in the soil, bioaccumulation factor of earthworms or the capacity of earthworms to accumulate heavy metals in their body tissues, *etc*.

iv. Miscellaneous environmental factors like rain, relative humidity, *etc*.

Through their feeding practices in the soil, earthworms promote the mineralization of soil organic matter, the production of nutrients, and the growth of plants [80]. The heavy metals, pH, and organic carbon concentrations in agricultural, non-agricultural, and industrial soils impact earthworm community structures. Industrial, non-agricultural, and agricultural soils have high, moderate, and/or low pollution levels, affecting earthworm-mediated remediation. High levels of Cd and As were found in all soil types. *Metaphire posthuman* was the most abundant earthworm in agricultural soil, with organic matter and microbes potentially aiding in heavy metal remediation through earthworms. Supplementation of biochar was found to lower the survival rate and weight of earthworms and reduce the bioaccumulation factors of the potentially toxic elements by *E. fetida* [62].

The impact of antimony on bioaccumulation and the gut bacterial community of *E. fetida* was analysed by Huang *et al.* [63]. To investigate the distribution of Sb species in earthworm guts and the way bacterial communities react to Sb pollution, *E. fetida* was cultivated in four Sb-contaminated soils. Sb was found to be accumulated in the earthworms' tissues and stomach in a dose-respons--dependent manner. There were notable differences in the types of bacteria present in soil and earthworm guts. In comparison to the soil-based network, the earthworm-gut bacterial community's network was less stable and more susceptible to Sb metal ions. The earthworm's gut bacterial community was most significantly altered by Sb (III) through the surrounding bacterial communities in soil *viz., Xanthomonadaceae, Rhodomicrobiaceae,* and *Anaerolineaceae*. The development, behaviour, oxidative reactions, gene expression, and gut microbiota of earthworms are impacted by the microplastics available in soil [64]. The interactions and combined impacts of microplastics potentially impact the earthworm-mediated heavy metal remediation because microplastics are frequently present at the same time as other pollutants. The earthworms have the potential to selectively consume microplastics of relatively smaller particle sizes, particularly under 50 meters. However, the bigger size and higher concentrations (mainly > 0.5%, w/w) of microplastics adversely affect the population of earthworms, thus negatively affecting the remediation of heavy metal contaminants.

Eisenia veneta is grown in soils supplemented with various materials like bentonite, iron oxide, rock scrap, limestone, calcium phosphate, compost, and biosolid. These amendments significantly reduced the biomass and survival rates of earthworm species [65]. Compost and bentonite caused the most significant decline in earthworm weight and survival, while rock waste and iron oxide caused the least. Biosolids were the most hazardous substance, killing all earthworms. In three wetland environments, the architecture of the earthworm communities and how they relate to the physicochemical characteristics of the soil were examined [67]. In these sites, six different species of earthworms *viz. Amynthasmorrisi, Eutyphoeus waltoni, E. nichloson, Lampito mauritii, Metaphire posthuma,* and *Octochaetona beatrix* were found. The soil properties like electrical conductivity, total dissolved solids, and levels of Cr, Ni, Zn, and Fe were reported to be inversely correlated with the population of earthworms. Whereas clay, silt, K, and moisture enhanced the earthworm population, thereby establishing a positive correlation among them. It is commonly known that vermiremediation works well at phytostabilizing soils that have been contaminated with metals. Heavy metal mobilisation may be prevented by adding soil amendments or other methods to the vermeremediation process, which may enhance the potential of earthworm-assisted heavy metal remediation [65].

REMEDIATION OF HEAVY METALS BY EARTHWORMS

The earthworms can accumulate and remediate various heavy metals in a concentration-dependent manner because of their strong gut microbioata having high metal tolerance and binding abilities that eradicate heavy metals through the absorption or adsorption of heavy metals [68]. From tiny bacteria to large crustaceans, the soil is the home to an unlimited variety of living things. The latter categories are represented by ants, termites, and earthworms. These organisms are notable for their ability to excavate the soil and have a significant impact on the soil's physical characteristics by forming organomineral structures such as macro-voids, nests, mounds, galleries, and caves [69]. The use of fungi, known as mycoremediation, is also a promising method to remove heavy metals from contaminated ecosystems. The tolerance and removal of single and multiple metals by *Trichoderma brevicompactum*, an earthworm gut fungus, was analysed by Zhang *et al.* [68]. The different minimum inhibitory concentrations of the different heavy metals, namely Cu(II), Cr(VI), Cd(II), Zn(II), and Pb(II), were observed to affect fungus and earthworms in a metal-dependent manner.

The remediation of heavy metals, such as Cd, Cu, Ni, Mn, Pb, and Zn during the composting of sewage sludge was stimulated by the synergistic interactions of earthworms and woodlice [70]. Also, the physicochemical characteristics of the compost were boosted by the combined action of woodlice and earthworms, as well as increasing its potential for application as an agricultural soil amendment. Moreover, a larger relative abundance of Proteobacteria (45.7%) and Bacteroidetes (18.8%) was found in the bacterial community of earthworm guts as compared to that of the soil [42]. Compared to the bacterial population in the soil, the bacterial community in the earthworm's gut seems to be more vulnerable to heavy metal pollution. Additionally, the heavy metals promoted the development of potential plant growth-promoting bacteria in the guts of earthworms, particularly those that produce the compounds indole-3-acetic acid and 1-aminocyclopropane-1-carboxylic acid deaminase, which may be useful for earthworm-assisted phytoremediation of heavy metals. Recently, a two-stage biodegradation strategy, *i.e.*, rotary drum composting followed by vermicomposting, was analysed for its potential role in the management of toxic lignocellulosic terrestrial weeds containing Zn and Pb [71]. The vermiculture-mediated technique was found effective and coherent in alleviating the Pb and Zn toxicities in the noxious weeds.

Wu *et al.* studied the underlying mechanisms of earthworm-mediated remediation of Cd metal-contaminated soil [72]. Earthworms absorb a significant amount of Cd in soil, which is accumulated in their midgut and hindgut. The bacterial community plays a crucial role in metabolism and Cd enrichment. Enteric bacteria, like *Pseudomonas brenneri*, stimulate Cd adsorption and digestion to reduce toxicity and enhance vermiremediation. Earthworms' hydrogen critical enzymes increase in response to Cd stress. To detoxify and manage Cd toxicity, a balance between physiological adaptation and gut bacterial population is observed. Cd can be eradicated from soils using earthworms, microbes, and plants.

EARTHWORM-MEDIATED MITIGATION OF HEAVY METALS TOXICITY IN PLANTS

Earthworms have immense potential to improve soil health by removing contaminants like heavy metals. Also, they can nullify the toxicity of noxious weed plants and, thus, enhance the use of such plants for phytoremediation. A dreaded weed named *Ipomea* was recorded to lose its toxicity after being vermicomposted by *Eisenia fetida* [73]. After that, this toxic weed turned into a safe organic fertilizer with pest-repelling properties. With the help of earthworms, the *Ipomea*-mediated phytoremediated soils can be better managed by converting dead *Ipomea* plants and debris into non-toxic fertilisers. Ganguly and Chakraborty reported that an indigenous earthworm, *Perionyx excavatus,* bioconverted the toxic heavy metals present in the paper mill wastes through the activity of aminopeptidase, glycosyl hydrolase, and esterase enzymes [74]. In vermicomposted primary sludge, the levels of heavy metals were reduced in the following manner: Zn > Pb > Cr > Cu, whereas in the vermicomposted secondary sludge reduction of heavy metals followed the following trend: Zn > Pb > Cu > Cr.

Eisenia fetida was found to affect the uptake of heavy metals by plants *Vicia faba* and *Zea mays* in metal-contaminated soils [56]. The tissues of the earthworms (anecic: *Lampito mauritii*; epigeic: *Drawida sulcata*) and their vermicasts, as well as soil samples, were collected from four industrial sites, such as the Sago industry, Chemplast industry, Dairy industry, and Dye industry [75]. The vermicast, tissue samples of earthworms, and soil samples were found to possess varied contents of heavy metals, namely cadmium (Cd), copper (Cu), chromium (Cr), lead (Pb), and zinc (Zn). In soils, the contents of heavy metals ranged from 0.01 to 326.42 mgkg^{-1} having maximum levels of Cu followed by Cr and Zn. However, the heavy metal load in the vermicasts was significantly lower. Also, the vermicasts were recorded to have enhanced humic substances and better physicochemical properties. The more the amounts of humic substances, the lesser bioaccumulation factors of heavy metals were recorded in vermicast. Also, earthworms were found to have higher levels of metallothionein protein, which was more in *Drawida sulcate* than *Lampito mauritii*. Hence earthworms' mediated production of metallothionein proteins might have neutralized the heavy metals toxicity through the management of ingestion of essential elements. In comparison, Pb was found to interact with non-metallothioneins followed by its bio-accumulation in the earthworm's internal chloragogenous tissues.

An investigation on the application of two Plant-Growth-Promoting Rhizobacteria (PGPR), namely *Pseudomonas aeruginosa* and *Burkholderia gladioli,* along with *E. fetida* applied to *Brassica juncea* seedlings under chromium (Cr) metal toxicity revealed that co-application of the earthworm and PGPR was effective in reducing heavy metal-induced phytotoxicity [16, 17]. Both earthworm and PGPR treatments could resume the activities of the antioxidant enzymes, such as superoxide dismutase, ascorbate peroxidase, guaiacol peroxidase, glutathione peroxidase, glutathione reductase, dehydroascorbate reductase. Also, the contents of superoxide anion ($O_2^{\cdot-}$), hydrogen peroxide (H_2O_2), and malondialdehyde were maintained when vermicompost and PGPR treatment were applied. The electrolyte leakage was restored, membrane and nuclear damage were alleviated, and both treatments improved cell viability. Hence, this study emphasized the co-application of PGPR to earthworms can be an effective strategy to ameliorate heavy metal-induced toxicities in plants. Moreover, co-application of earthworms was also found to improve the Cr-induced decline in photosynthetic attributes [17].

Two earthworm species (*Amynthas cortices* and *A. robustus*) based on their biota to soil accumulation factor were analyzed for their response to metal bioavailability in contaminated soils [5]. The earthworms were cultured in soil contaminated with four heavy metals *viz.*, Cu, Zn, Cd, and Pb. After 120 days, the biomass, total number, and biota to soil accumulation factor of earthworms, physicochemical properties of soil, and metals extracted by diethylenetriaminepentaacetic acid (DTPA) were determined in the soil. The activity of earthworms significantly enhanced DTPA-extracted heavy metals (Cu and Zn). Also, soil properties such as mineralized carbon (C), pH, and dissolved nitrogen (N) were amended by earthworms, which revealed a significant interaction between soil and earthworms. It also resulted in increasing the heavy metal bioavailability in the soil. *Eisenia andrei* was found to be effective in modulating the heavy metals-induced oxidative stress in *Vicia faba* plants irrigated with treated wastewater [76] by lowering the activity of glutathione--transferase and levels of malondialdehyde accumulation in bean plants. Fig. 2 explains the mechanism of earthworm-mediated accumulation and remediation of heavy metals by plants.

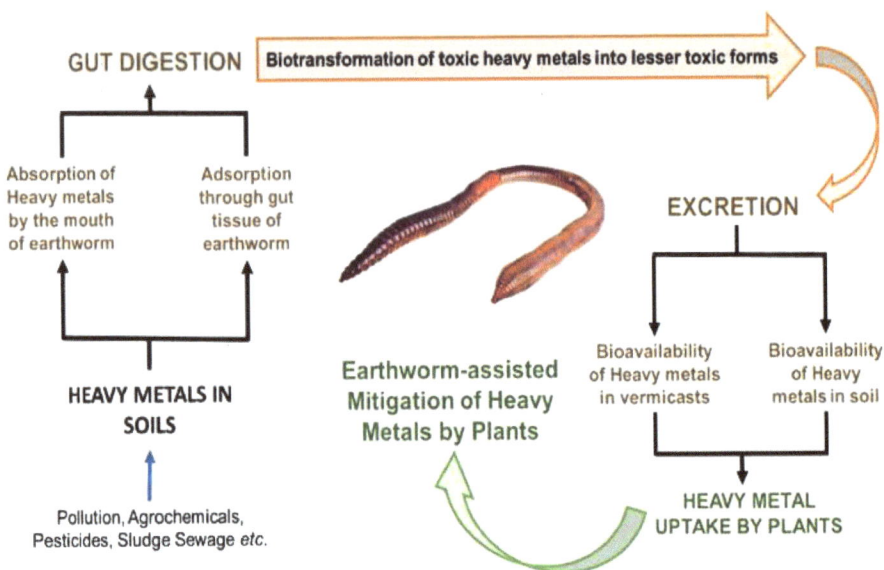

Fig. (2). Proposed mechanism of earthworm-mediated remediation of heavy metals by plants.

Fu *et al.* reported that the co-application of arbuscular mycorrhizal fungi and earthworms could enhance the biomass of *Tagetes patula* plants in Cu-contaminated soils [77]. Also, Cu extraction by *T. patula* was enhanced by 270% under the combined treatment of earthworms and fungi. It also enhanced the available phosphorus (P), potassium (K), and Cu but lowered the pH of the soil. Moreover, the abundance and biodiversity of various soil microbes such as *Actinobacteria, Bacteroides*, and *Proteobacteria,* including genera *Algoriphagus, Arthrobacter, Flavobacterium, Gaetbulibacter, Luteimonas, Pedobacter, Pseudomonas* were enhanced by both earthworm and fungi in the soils. Hence, the application of earthworms, along with alleviated Cu stress positively stimulated the structure of the soil and its fertility.

Roy *et al.* reported that recycling arsenic metal-saturated sorbing agents (particularly biochars) using vermitechnology may be a viable solution [78]. The waste sludge produced by the tannery industry has higher amounts of Cr metal, which further contaminates the nearby soil and water bodies [79]. The earthworm-mediated conversion of toxic tannery waste sludge into a sanitized, safer, eco-friendly, and nutrient-enriched product may be beneficial for agriculture. Thus, *E. fetida* was used to incubate tannery waste sludge-based vermireactors (vermibeds) for efficient vermiremediation and preparation of vermicompost. The population of *E. fetida* was found to be enhanced in these vermireactors or vermibeds, while the contents of chromium (Cr), lead (Pb), nickel (Ni), and copper(Cu) were decreased. It also enhanced the availability of nitrogen, phosphorus, and potassium, increased microbial biomass, and stimulated the activities of enzymes, such as urease, fluorescein diacetate hydrolysis, sulphatase, glucosidase, phosphatase in tannery waste sludge-vermibeds. Thus, earthworms decontaminate the metal-polluted soils by influencing the microbial communities, stimulating enzyme activities, and enhancing metal detoxification. Moreover, these soil engineers stimulate the mineralization of soil organic matter, enhance the production of nutrients, and improve plant growth [80].

Further, the earthworms also play an important in the balancing of soil ecosystem through nutrient cycling. However, their presence can also impact other micro- and macro-organism populations, particularly when involved in heavy metal detoxification processes. After ingesting heavy metal-containing soil, they can break down or sequester heavy metals, potentially reducing the toxicity of the soil [81]. However, this process might interfere with other species in the ecosystem in several ways. This earthworm-mediated detoxification of heavy metals may alter soil chemistry too. Moreover, earthworms might accumulate toxic metals in their bodies, which could enter the food chain, affecting predators and impacting biodiversity [82, 83]. Earthworms can help in mitigating heavy metal pollution, which, however, may inadvertently disrupt the balance of other biological populations by altering the habitat or transferring toxins up the food chain.

CONCLUSION AND FUTURE PROSPECTS

Earthworms are the real soil engineers that not only mitigate heavy metal toxicity but also produce valuable and nutrient-rich vermicompost to enhance soil fertility. Earthworm-mediated soil clean-up is an eco-friendly and cost-effective approach that is generally successful for mildly polluted soils because earthworms cannot survive in very harsh conditions. Therefore, to make vermiremediation a success, the polluted soils are amended with different proportions of cow dung or any other substrate, which acts as a feed mixture for the culture of earthworms. The earthworms either bio-accumulate heavy metals in their body tissues or alter the bioavailability of heavy metals in plants. Also, earthworms can convert highly toxic forms of heavy metals into less toxic forms through the process of bio-transformations. The knowledge of earthworm species remediating the type and amount of particular heavy metal, capacity of earthworm-mediated biotransformation of heavy metal, tolerance of earthworm species to particular heavy metal or more than one heavy metal, and exact requirement of catalysts/substrate for making feed mixtures might help devise more advanced practical vermiremediation strategy for eradication of heavy metals. But still, the application of earthworms alone for the eradication of heavy metal pollution has some limitations like low survival in highly contaminated sites, low remediation efficiency, relatively less knowledge of specificity for metal uptake and detoxification by earthworms, and mass culturing of earthworms for vermiremediation, *etc*. Earthworms, vital for soil ecosystem balance, can disrupt heavy metal detoxification processes, potentially affecting predators and biodiversity and disrupting habitats and food chains. To overcome these limitations, integrated strategies employing earthworms, bacteria, fungi, plants, and other green methods might be a more efficacious approach to managing and alleviating environmental pollution. Besides this, the large-scale production of the most efficient species of earthworms and their commercialization is required to reduce the dependency on chemical-based soil remediation and reclamation methods. This might prove to be a step towards achieving pollution-free soils and safe and sustainable agriculture development goals.

ACKNOWLEDGEMENTS

The author, Ashutosh Sharma acknowledges the financial grant received from the DAV University, Jalandhar (sanction letter no. DAVU/2020/DA/122). The corresponding author, Indu Sharma, is thankful to the administration of Sant Baba Bhag Singh University for providing all the necessary infrastructural research facilities and financial grant received as Seed Money for Research (sanction letter no. SBBSU/R&D/SMP/004/202/22).

REFERENCES

[1]   Xiao R, Ali A, Xu Y, *et al*. Earthworms as candidates for remediation of potentially toxic elements contaminated soils and mitigating the environmental and human health risks: A review. Environ Int 2022; 158: 106924.
[http://dx.doi.org/10.1016/j.envint.2021.106924] [PMID: 34634621]

[2]   Sharma P, Bakshi P, Kour J, Singh AD, Dhiman S, Kumar P. PGPR and earthworm-assisted phytoremediation of heavy metals. Earthworm assisted remediation of effluents and wastes. 2020: 227-45.

[3]   Zhang X, Wells M, Niazi NK, *et al*. Nanobiochar-rhizosphere interactions: Implications for the remediation of heavy-metal contaminated soils. Environ Pollut 2022; 299: 118810.
[http://dx.doi.org/10.1016/j.envpol.2022.118810] [PMID: 35007673]

[4]   Wang J, Shi L, Liu J, *et al*. Earthworm-mediated nitrification and gut digestive processes facilitate the remobilization of biochar-immobilized heavy metals. Environ Pollut 2023; 322: 121219.
[http://dx.doi.org/10.1016/j.envpol.2023.121219] [PMID: 36746291]

[5]   Xiao L, Li M, Dai J, *et al*. Assessment of earthworm activity on Cu, Cd, Pb and Zn bioavailability in contaminated soils using biota to soil accumulation factor and DTPA extraction. Ecotoxicol Environ Saf 2020; 195: 110513.
[http://dx.doi.org/10.1016/j.ecoenv.2020.110513] [PMID: 32213370]

[6]   Seribekkyzy G, Saimova RU, Saidakhmetova AK, Saidakhmetova GK, Esimov BK. Heavy metal effects on earthworms in different ecosystems. J Anim Behav Biometeorol 2022; 10(3): 2228.
[http://dx.doi.org/10.31893/jabb.22028]

[7]   Kour J, Khanna K, Sharma P, *et al*. Ecosystem Engineering by Earthworms.Earthworms and their Ecological Significance. Nova Science Publishers USA 2022; pp. 211-34.

[8]   Qayyum S, Khan I, Meng K, Zhao Y, Peng C. A review on remediation technologies for heavy metals

contaminated soil. Central Asian Journal of Environmental Science and Technology Innovation 2020; 1(1): 21-9.

[9]   Chen X, Kumari D, Cao CJ, Plaza G, Achal V. A review on remediation technologies for nickel-contaminated soil. Hum Ecol Risk Assess 2020; 26(3): 571-85.
[http://dx.doi.org/10.1080/10807039.2018.1539639]

[10]   Kumar V, Shahi SK, Singh S. Bioremediation: an eco-sustainable approach for restoration of contaminated sites. Microbial bioprospecting for sustainable development 2018; 115-36.
[http://dx.doi.org/10.1007/978-981-13-0053-0_6]

[11]   Kour J, Khanna K, Bakshi P, *et al.* Role of Beneficial Microbes in the Molecular Phytotoxicity of Heavy Metals. Cellular and Molecular Phytotoxicity of Heavy Metals 2020; pp. 227-62.
[http://dx.doi.org/10.1007/978-3-030-45975-8_13]

[12]   Sharma I, Pati PK, Bhardwaj R. Regulation of growth and antioxidant enzyme activities by 28-homobrassinolide in seedlings of *Raphanus sativus* L. under cadmium stress. Indian J Biochem Biophys 2010; 47(3): 172-7.
[PMID: 20653289]

[13]   Sharma I, Pati PK, Bhardwaj R. Effect of 28-homobrassinolide on antioxidant defence system in *Raphanus sativus* L. under chromium toxicity. Ecotoxicology 2011; 20(4): 862-74.
[http://dx.doi.org/10.1007/s10646-011-0650-0] [PMID: 21448625]

[14]   Sharma I, Sharma A, Pati P, Bhardwaj R. Brassinosteroids reciprocates heavy metals induced oxidative stress in radish by regulating the expression of key antioxidant enzyme genes. Braz Arch Biol Technol 2018; 61(0): 61.
[http://dx.doi.org/10.1590/1678-4324-2018160679]

[15]   Sharma I, Bhardwaj R, Gautam V, Kaur R, Sharma A. Brassinosteroids: occurrence, structure and stress protective activities. Frontiers in Natural Product Chemistry 2018; 4: 204-39.
[http://dx.doi.org/10.2174/9781681087252118040008]

[16]   Sharma P, Chouhan R, Bakshi P, *et al.* Amelioration of chromium-induced oxidative stress by combined treatment of selected plant-growth-promoting rhizobacteria and earthworms via modulating the expression of genes related to reactive oxygen species metabolism in *Brassica juncea.* Front Microbiol 2022; 13: 802512.
[http://dx.doi.org/10.3389/fmicb.2022.802512] [PMID: 35464947]

[17]   Sharma P, Bakshi P, Kaur R, *et al.* Inoculation of plant-growth-promoting rhizobacteria and earthworms in the rhizosphere reinstates photosynthetic attributes and secondary metabolites in *Brassica juncea* L. under chromium toxicity. Plant Soil 2023; 483(1-2): 573-87.
[http://dx.doi.org/10.1007/s11104-022-05765-y]

[18]   Žaltauskaitė J, Kniuipytė I, Praspaliauskas M. Earthworm *Eisenia fetida* potential for sewage sludge amended soil valorization by heavy metal remediation and soil quality improvement. J Hazard Mater 2022; 424(Pt A): 127316.
[http://dx.doi.org/10.1016/j.jhazmat.2021.127316] [PMID: 34583161]

[19]   Zeb A, Li S, Wu J, Lian J, Liu W, Sun Y. Insights into the mechanisms underlying the remediation potential of earthworms in contaminated soil: A critical review of research progress and prospects. Sci Total Environ 2020; 740: 140145.
[http://dx.doi.org/10.1016/j.scitotenv.2020.140145] [PMID: 32927577]

[20]   Yuvaraj A, Thangaraj R, Karmegam N, *et al.* Activation of biochar through exoenzymes prompted by earthworms for vermibiochar production: A viable resource recovery option for heavy metal contaminated soils and water. Chemosphere 2021; 278: 130458.
[http://dx.doi.org/10.1016/j.chemosphere.2021.130458] [PMID: 34126688]

[21]   Bhat SA, Singh J, Vig AP. Vermiremediation of dyeing sludge from textile mill with the help of exotic earthworm *Eisenia fetida* Savigny. Environ Sci Pollut Res Int 2013; 20(9): 5975-82.
[http://dx.doi.org/10.1007/s11356-013-1612-2] [PMID: 23508537]

[22]   Sinha RK, Bharambe G, Ryan D. Converting wasteland into wonderland by earthworms—a low-cost nature's technology for soil remediation: a case study of vermiremediation of PAHs contaminated soil. Environmentalist 2008; 28(4): 466-75.
[http://dx.doi.org/10.1007/s10669-008-9171-7]

[23]   Adhikary S. Vermicompost, the story of organic gold: A review. Agric Sci 2012; 3(7): 905-17.
[http://dx.doi.org/10.4236/as.2012.37110]

[24]   Ahmad A, Aslam Z, Bellitürk K, *et al.* Earth worms and vermicomposting: A review on the story of black gold. Journal of Innovative Sciences 2021; 7(1): 167-73.
[http://dx.doi.org/10.17582/journal.jis/2021/7.1.167.173]

[25]   Singh J, Singh S, Vig AP, *et al.* Conventional farming reduces the activity of earthworms: Assessment of genotoxicity test of soil and vermicast. Agric Nat Resour (Bangk) 2018; 52(4): 366-70.
[http://dx.doi.org/10.1016/j.anres.2018.10.012]

[26]   Rodriguez-Campos J, Perales-Garcia A, Hernandez-Carballo J, *et al.* Bioremediation of soil contaminated by hydrocarbons with the combination of three technologies: bioaugmentation, phytoremediation, and vermiremediation. J Soils Sediments 2019; 19(4): 1981-94.
[http://dx.doi.org/10.1007/s11368-018-2213-y]

[27]   Pilon-Smits E. Phytoremediation. Annu Rev Plant Biol 2005; 56(1): 15-39.
[http://dx.doi.org/10.1146/annurev.arplant.56.032604.144214] [PMID: 15862088]

[28]   Shen X, Dai M, Yang J, *et al.* A critical review on the phytoremediation of heavy metals from environment: Performance and challenges. Chemosphere 2022; 291(Pt 3): 132979.
[http://dx.doi.org/10.1016/j.chemosphere.2021.132979] [PMID: 34801572]

[29]   Oladoye PO, Olowe OM, Asemoloye MD. Phytoremediation technology and food security impacts of heavy metal contaminated soils: A review of literature. Chemosphere 2022; 288(Pt 2): 132555.
[http://dx.doi.org/10.1016/j.chemosphere.2021.132555] [PMID: 34653492]

[30]   Wei Z, Van Le Q, Peng W, *et al.* A review on phytoremediation of contaminants in air, water and soil. J Hazard Mater 2021; 403: 123658.
[http://dx.doi.org/10.1016/j.jhazmat.2020.123658] [PMID: 33264867]

[31]   Yan A, Wang Y, Tan SN, Mohd Yusof ML, Ghosh S, Chen Z. Phytoremediation: a promising approach for revegetation of heavy metal-polluted land. Front Plant Sci 2020; 11: 359.
[http://dx.doi.org/10.3389/fpls.2020.00359] [PMID: 32425957]

[32]   Shah V, Daverey A. Phytoremediation: A multidisciplinary approach to clean up heavy metal contaminated soil. Environ Technol Innov 2020; 18: 100774.
[http://dx.doi.org/10.1016/j.eti.2020.100774]

[33]   Liu Z, Tran KQ. A review on disposal and utilization of phytoremediation plants containing heavy metals. Ecotoxicol Environ Saf 2021; 226: 112821.
[http://dx.doi.org/10.1016/j.ecoenv.2021.112821] [PMID: 34571420]

[34]   Nedjimi B. Phytoremediation: a sustainable environmental technology for heavy metals decontamination. SN Appl Sci 2021; 3(3): 286.
[http://dx.doi.org/10.1007/s42452-021-04301-4]

[35]   Gudeta K, Julka JM, Kumar A, Bhagat A, Kumari A. Vermiwash: An agent of disease and pest control in soil, a review. Heliyon 2021; 7(3): e06434.
[http://dx.doi.org/10.1016/j.heliyon.2021.e06434] [PMID: 33732941]

[36]   Azizi AB, Liew KY, Noor ZM, Abdullah N. Vermiremediation and mycoremediation of polycyclic aromatic hydrocarbons in soil and sewage sludge mixture: a comparative study. Int J Environ Sci Dev 2013; 4(5): 565-8.
[http://dx.doi.org/10.7763/IJESD.2013.V4.414]

[37]   Gupta R, Garg VK. Vermiremediation and nutrient recovery of non-recyclable paper waste employing

Eisenia fetida. J Hazard Mater 2009; 162(1): 430-9.
[http://dx.doi.org/10.1016/j.jhazmat.2008.05.055] [PMID: 18573612]

[38]   Dada EO, Akinola MO, Owa SO, *et al.* Efficacy of vermiremediation to remove contaminants from soil. J Health Pollut 2021; 11(29): 210302.
[http://dx.doi.org/10.5696/2156-9614-11.29.210302] [PMID: 33815900]

[39]   Suthar S, Sajwan P, Kumar K. Vermiremediation of heavy metals in wastewater sludge from paper and pulp industry using earthworm *Eisenia fetida.* Ecotoxicol Environ Saf 2014; 109: 177-84.
[http://dx.doi.org/10.1016/j.ecoenv.2014.07.030] [PMID: 25215882]

[40]   Das S, Bora J, Goswami L, *et al.* Vermiremediation of Water Treatment Plant Sludge employing *Metaphire posthuma*: A soil quality and metal solubility prediction approach. Ecol Eng 2015; 81: 200-6.
[http://dx.doi.org/10.1016/j.ecoleng.2015.04.069]

[41]   Rorat A, Wloka D, Grobelak A, *et al.* Vermiremediation of polycyclic aromatic hydrocarbons and heavy metals in sewage sludge composting process. J Environ Manage 2017; 187: 347-53.
[http://dx.doi.org/10.1016/j.jenvman.2016.10.062] [PMID: 27836561]

[42]   Liu P, Yang Y, Li M. Responses of soil and earthworm gut bacterial communities to heavy metal contamination. Environ Pollut 2020; 265(Pt B): 114921.
[http://dx.doi.org/10.1016/j.envpol.2020.114921] [PMID: 32540597]

[43]   Paul S, Goswami L, Pegu R, Sundar Bhattacharya S. Vermiremediation of cotton textile sludge by *Eudrilus eugeniae*: Insight into metal budgeting, chromium speciation, and humic substance interactions. Bioresour Technol 2020; 314: 123753.
[http://dx.doi.org/10.1016/j.biortech.2020.123753] [PMID: 32619804]

[44]   Owagboriaye F, Dedeke G, Bamidele J, *et al.* Biochemical response and vermiremediation assessment of three earthworm species (*Alma millsoni, Eudrilus eugeniae* and *Libyodrilus violaceus*) in soil contaminated with a glyphosate-based herbicide. Ecol Indic 2020; 108: 105678.
[http://dx.doi.org/10.1016/j.ecolind.2019.105678]

[45]   Cheng Q, Lu C, Shen H, Yang Y, Chen H. The dual beneficial effects of vermiremediation: Reducing soil bioavailability of cadmium (Cd) and improving soil fertility by earthworm (*Eisenia fetida*) modified by seasonality. Sci Total Environ 2021; 755(Pt 2): 142631.
[http://dx.doi.org/10.1016/j.scitotenv.2020.142631] [PMID: 33065505]

[46]   Ding J, Zhu D, Hong B, *et al.* Long-term application of organic fertilization causes the accumulation of antibiotic resistome in earthworm gut microbiota. Environ Int 2019; 124: 145-52.
[http://dx.doi.org/10.1016/j.envint.2019.01.017] [PMID: 30641258]

[47]   Phillips HRP, Guerra CA, Bartz MLC, *et al.* Global distribution of earthworm diversity. Science 2019; 366(6464): 480-5.
[http://dx.doi.org/10.1126/science.aax4851] [PMID: 31649197]

[48]   Chaudhuri PS, Datta R. Studies on cocoons of three epigeic earthworm species *Perionyx excavatus* (Perrier), *Perionyx ceylanensis* (Michaelsen) and *Eudrilus eugeniae* (Kinberg) with SEM observations. *In Proceedings of the Zoological Society* 2020 Dec, 73: 430-440. Springer India.

[49]   Hoeffner K, Butt KR, Monard C, Frazão J, Pérès G, Cluzeau D. Two distinct ecological behaviours within anecic earthworm species in temperate climates. Eur J Soil Biol 2022; 113: 103446.
[http://dx.doi.org/10.1016/j.ejsobi.2022.103446]

[50]   Lin Z, Zhen Z, Liang Y, *et al.* Changes in atrazine speciation and the degradation pathway in red soil during the vermiremediation process. J Hazard Mater 2019; 364: 710-9.
[http://dx.doi.org/10.1016/j.jhazmat.2018.04.037] [PMID: 30412844]

[51]   Yakkou L, Houida S, Dominguez J, Raouane M, Amghar S, El Harti A. Identification and Characterization of Microbial Community in the Coelomic Fluid of Earthworm (*Aporrectodea molleri*). Microbiol Biotechnol Lett 2021; 49(3): 391-402.

[52]   Gusain R, Suthar S. Vermicomposting of duckweed (*Spirodela polyrhiza*) by employing *Eisenia fetida*: Changes in nutrient contents, microbial enzyme activities and earthworm biodynamics. Bioresour Technol 2020; 311: 123585.
        [http://dx.doi.org/10.1016/j.biortech.2020.123585] [PMID: 32492602]

[53]   Huang C, Ge Y, Yue S, Qiao Y, Liu L. Impact of soil metals on earthworm communities from the perspectives of earthworm ecotypes and metal bioaccumulation. J Hazard Mater 2021; 406: 124738.
        [http://dx.doi.org/10.1016/j.jhazmat.2020.124738] [PMID: 33316673]

[54]   Richardson JB, Görres JH, Sizmur T. Synthesis of earthworm trace metal uptake and bioaccumulation data: Role of soil concentration, earthworm ecophysiology, and experimental design. Environ Pollut 2020; 262: 114126.
        [http://dx.doi.org/10.1016/j.envpol.2020.114126] [PMID: 32120252]

[55]   Lin Z, Zhen Z, Luo S, *et al.* Effects of two ecological earthworm species on tetracycline degradation performance, pathway and bacterial community structure in laterite soil. J Hazard Mater 2021; 412: 125212.
        [http://dx.doi.org/10.1016/j.jhazmat.2021.125212] [PMID: 33524732]

[56]   Lemtiri A, Liénard A, Alabi T, *et al.* Earthworms *Eisenia fetida* affect the uptake of heavy metals by plants *Vicia faba* and *Zea mays* in metal-contaminated soils. Appl Soil Ecol 2016; 104: 67-78.
        [http://dx.doi.org/10.1016/j.apsoil.2015.11.021]

[57]   Wang K, Qiao Y, Zhang H, *et al.* Bioaccumulation of heavy metals in earthworms from field contaminated soil in a subtropical area of China. Ecotoxicol Environ Saf 2018; 148: 876-83.
        [http://dx.doi.org/10.1016/j.ecoenv.2017.11.058]

[58]   Hobbelen PHF, Koolhaas JE, van Gestel CAM. Bioaccumulation of heavy metals in the earthworms *Lumbricus rubellus* and *Aporrectodea caliginosa* in relation to total and available metal concentrations in field soils. Environ Pollut 2006; 144(2): 639-46.
        [http://dx.doi.org/10.1016/j.envpol.2006.01.019] [PMID: 16530310]

[59]   Chen Z, Hu S. Heavy metals distribution and their bioavailability in earthworm assistant sludge treatment wetland. J Hazard Mater 2019; 366: 615-23.
        [http://dx.doi.org/10.1016/j.jhazmat.2018.12.039] [PMID: 30579227]

[60]   Sheng Y, Liu Y, Wang K, Cizdziel JV, Wu Y, Zhou Y. Ecotoxicological effects of micronized car tire wear particles and their heavy metals on the earthworm (*Eisenia fetida*) in soil. Sci Total Environ 2021; 793: 148613.
        [http://dx.doi.org/10.1016/j.scitotenv.2021.148613] [PMID: 34182439]

[61]   Parihar K, Kumar R, Sankhla MS. 2019. Impact of heavy metals on survivability of earthworms. International Medico-Legal Reporter Journal, 2019; 2(3): 51-7.

[62]   Garau M, Sizmur T, Coole S, Castaldi P, Garau G. Impact of *Eisenia fetida* earthworms and biochar on potentially toxic element mobility and health of a contaminated soil. Sci Total Environ 2022; 806(Pt 3): 151255.
        [http://dx.doi.org/10.1016/j.scitotenv.2021.151255] [PMID: 34710424]

[63]   Huang B, Long J, Li J, Ai Y. Effects of antimony contamination on bioaccumulation and gut bacterial community of earthworm *Eisenia fetida*. J Hazard Mater 2021; 416: 126110.
        [http://dx.doi.org/10.1016/j.jhazmat.2021.126110] [PMID: 34492908]

[64]   Cui W, Gao P, Zhang M, Wang L, Sun H, Liu C. Adverse effects of microplastics on earthworms: A critical review. Sci Total Environ 2022; 850: 158041.
        [http://dx.doi.org/10.1016/j.scitotenv.2022.158041] [PMID: 35973535]

[65]   Ukalska-Jaruga A, Siebielec G, Siebielec S, Pecio M. The effect of soil amendments on trace elements' bioavailability and toxicity to earthworms in contaminated soils. Appl Sci (Basel) 2022; 12(12): 6280.
        [http://dx.doi.org/10.3390/app12126280]

[66]   Singh S, Singh J, Vig AP. Earthworm community structures in three wetland ecosystems with reference to soil physicochemical properties. Proceedings of the Zoological Society 2022; 75: 231–41.
[http://dx.doi.org/10.1007/s12595-022-00436-3]

[67]   IndraKumar Singh S, Singh WR, Bhat SA, *et al.* Vermiremediation of allopathic pharmaceutical industry sludge amended with cattle dung employing *Eisenia fetida.* Environ Res 2022; 214(Pt 1): 113766.
[http://dx.doi.org/10.1016/j.envres.2022.113766] [PMID: 35780853]

[68]   Zhang D, Yin C, Abbas N, Mao Z, Zhang Y. Multiple heavy metal tolerance and removal by an earthworm gut fungus *Trichoderma brevicompactum* QYCD-6. Sci Rep 2020; 10(1): 6940.
[http://dx.doi.org/10.1038/s41598-020-63813-y] [PMID: 32332813]

[69]   Biswas JK, Banerjee A, Sarkar B, *et al.* Exploration of an extracellular polymeric substance from earthworm gut bacterium (*Bacillus licheniformis*) for bioflocculation and heavy metal removal potential. Appl Sci (Basel) 2020; 10(1): 349.
[http://dx.doi.org/10.3390/app10010349]

[70]   Ahadi N, Sharifi Z, Hossaini SMT, Rostami A, Renella G. Remediation of heavy metals and enhancement of fertilizing potential of a sewage sludge by the synergistic interaction of woodlice and earthworms. J Hazard Mater 2020; 385: 121573.
[http://dx.doi.org/10.1016/j.jhazmat.2019.121573] [PMID: 31761649]

[71]   Pottipati S, Kundu A, Kalamdhad AS. Performance evaluation of a novel two-stage biodegradation technique through management of toxic lignocellulosic terrestrial weeds. Waste Manag 2022; 144: 191-202.
[http://dx.doi.org/10.1016/j.wasman.2022.03.026] [PMID: 35381446]

[72]   Wu Y, Chen C, Wang G, *et al.* Mechanism underlying earthworm on the remediation of cadmium-contaminated soil. Sci Total Environ 2020; 728: 138904.
[http://dx.doi.org/10.1016/j.scitotenv.2020.138904] [PMID: 32570329]

[73]   Hussain N, Patnaik P, Abbasi T, Khamrang C, Abbasi SA, Abbasi T. Role of different earthworm species in nullifying the toxicity of *Ipomea carnea* and enhancing its utility as a phytoremediator. Int J Phytoremediation 2022; 24(13): 1385-94.
[http://dx.doi.org/10.1080/15226514.2022.2031864] [PMID: 35166609]

[74]   Ganguly RK, Chakraborty SK. Succession of enzymes and microbial biomarkers in the process of vermicomposting: An insight towards valorization of toxic paper mill wastes using Perionyx excavatus (Oligochaeta; Perrier, 1872). J Environ Health Sci Eng 2021; 19(2): 1457-72.
[http://dx.doi.org/10.1007/s40201-021-00701-1] [PMID: 34900280]

[75]   Yuvaraj A, Govarthanan M, Karmegam N, *et al.* Metallothionein dependent-detoxification of heavy metals in the agricultural field soil of industrial area: Earthworm as field experimental model system. Chemosphere 2021; 267: 129240.
[http://dx.doi.org/10.1016/j.chemosphere.2020.129240] [PMID: 33341732]

[76]   Mkhinini M, Helaoui S, Boughattas I, Amemou C, Banni M. Earthworm *Eisenia andrei* modulates oxidative stress in bean plants *Vicia faba* irrigated with treated wastewater. Ecotoxicology 2020; 29(7): 1003-16.
[http://dx.doi.org/10.1007/s10646-020-02243-y] [PMID: 32617728]

[77]   Fu L, Zhang L, Dong P, *et al.* Remediation of copper-contaminated soils using *Tagetes patula* L., earthworms and arbuscular mycorrhizal fungi. Int J Phytoremediation 2022; 24(10): 1107-19.
[http://dx.doi.org/10.1080/15226514.2021.2002809] [PMID: 34775850]

[78]   Roy S, Sarkar D, Datta R, Bhattacharya SS, Bhattacharyya P. Assessing the arsenic-saturated biochar recycling potential of vermitechnology: Insights on nutrient recovery, metal benignity, and microbial activity. Chemosphere 2022; 286(Pt 1): 131660.
[http://dx.doi.org/10.1016/j.chemosphere.2021.131660] [PMID: 34315078]

[79]   Chakraborty P, Sarkar S, Mondal S, *et al. Eisenia fetida* mediated vermi-transformation of tannery waste sludge into value added eco-friendly product: An insight on microbial diversity, enzyme activation, and metal detoxification. J Clean Prod 2022; 348: 131368.
[http://dx.doi.org/10.1016/j.jclepro.2022.131368]

[80]   Verma F, Singh S, Singh J, *et al.* Assessment of heavy metal contamination and its effect on earthworms in different types of soils. Int J Environ Sci Technol 2021; 1-4.

[81]   Sizmur T, Hodson ME. Do earthworms impact metal mobility and availability in soil? – A review. Environ Pollut 2009; 157(7): 1981-9.
[http://dx.doi.org/10.1016/j.envpol.2009.02.029] [PMID: 19321245]

[82]   Nesterkova DV, Vorobeichik EL, Reznichenko IS. The effect of heavy metals on the soil-earthwor--European mole food chain under the conditions of environmental pollution caused by the emissions of a copper smelting plant. Contemp Probl Ecol 2014; 7(5): 587-96.
[http://dx.doi.org/10.1134/S1995425514050096]

[83]   Roodbergen M, Klok C, van der Hout A. Transfer of heavy metals in the food chain earthworm Black-tailed godwit (*Limosa limosa*): Comparison of a polluted and a reference site in The Netherlands. Sci Total Environ 2008; 406(3): 407-12.
[http://dx.doi.org/10.1016/j.scitotenv.2008.06.051] [PMID: 18752837]

CHAPTER 10

Mycorrhizae as an Effective Tool for Heavy Metal Detoxification

Kamini Devi[1], Tamanna Bhardwaj[1], Jaspreet Kour[1], Shalini Dhiman[1], Arun Dev Singh[1], Deepak Kumar[2], Roohi Sharma[2], Parkriti[2], Gurvarinder Kaur[3], Kanika Khanna[1,*], Puja Ohri[2], Geetika Sirhindi[3] and Renu Bhardwaj[1]

[1] *Department of Botanical and Environmental Sciences, Guru Nanak Dev University, Amritsar, Punjab, India*

[2] *Department of Zoology, Guru Nanak Dev University, Amritsar, Punjab, India*

[3] *Department of Botany, Punjabi University, Patiala, Punjab, India*

Abstract: Heavy metal toxicity in the ecosystems is hazardous for living beings as it enters into the food chains through water and soil. The plants absorb the excessive toxic metal ions from the environment resulting in the disruption of various metabolic and physiological pathways in plants. The removal and stabilization of toxic metals from a plant's surroundings take place by Plant Growth-Promoting Rhizobacteria (PGPR), Endophytic Bacteria, and Arbuscular Mycorrhizal Fungi (AMF). AMF is used as a biological indicator for toxic metal pollution. Mycorrhizae use various mechanisms (avoidance and tolerance) for metal detoxification and help in maintaining the nutrition, growth, ecological processes and functioning, nutrient cycles, and diversity of plants. They have vesicles that act as vacuoles and store an excessive amount of toxic metal ions in them. About 80% of plant families showed a symbiotic connection with Mycorrhizal fungi. AMF produces 20% C (Carbon) from the plant and, in return, benefits the plant with more water uptake and other essential nutrients from the soil *via* their hyphal networks. This chapter presents the importance of Mycorrhizae-assisting phytoremediation by increasing the activities of defense enzymes, expanding the area of absorption, stimulating the expression of genes, and enhancing the chelation of toxic metal ions. This document also focuses on the toxic metal uptake, their amassing, and the mechanism associated with AMF-assisted phytoremediation of metalliferous soil.

Keywords: Arbuscular mycorrhizal fungi, Heavy metal, Phytoremediation.

* **Corresponding author Kanika Khanna:** Department of Botanical and Environmental Sciences, Guru Nanak Dev University, Amritsar, Punjab, India; E-mail: kanika.27590@gmail.com

INTRODUCTION

Soil contains heavy metals in a variety of forms. They can exist as free ions, complexes, or compounds such as silicates, oxides, and hydroxides. They can also bind organic and inorganic substances. There are two ways, heavy metals can enter the roots once they are available for uptake. Heavy metal stress is an important stress that has adverse effects on plant health. High concentrations of heavy metals contradict normal plant metabolism pathways and other structural and functional processes in plants. The apoplastic pathway occurs first, followed by the symplastic pathway. Heavy metal ions present in the soil enter the plant epidermis through ion channels or as organic compounds formed after combining with chelates released by plant roots [1]. These metal-ligand complexes then enter the root epidermal layer. Heavy metals accumulate in the roots of non-hyperaccumulators because they lack the capacity to translocate heavy metals to aboveground parts [2]. The flow of metals into xylem vessels, which are involved in the transport of water and dissolved salts, causes the translocation of heavy metals from roots to shoots. The breakdown of metals is difficult; thus, when concentrations inside a plant surpass ideal levels, they have a negative direct and indirect impact on the plant [3, 4]. The replenishment of necessary nutrients at cation exchange sites in plants has an indirect, harmful impact. Furthermore, the detrimental effects of heavy metals on the development and function of soil microbes may indirectly affect plant growth [5]. For instance, a high metal content may result in fewer beneficial soil microbes, which may slow the decomposition of organic waste and reduce soil nutrients. The common mechanisms involved in HM stress alleviation by AMF include reduced uptake from the root or storage of HMS in fungal structures, improved nutrient uptake, maintenance of redox status, antioxidant system activation, reduced root-to-shoot translocation, and synthesis of organic acids, proline, and glomalin [6, 7]. Arbuscular mycorrhizal fungi (AMF) are an essential bioagent because they can significantly improve the efficiency of the terrestrial ecosystem by producing fungal structure networks such as arbuscules. Additionally, they facilitate the exchange of inorganic compounds and minerals necessary for plant growth, which in turn provides plants with a significant amount of strength. In addition, they function as biological filters for toxic HMs and contribute to the mitigation of their toxic effects. They do this by supporting the plant by increasing its water uptake, photosynthetic rate, and nutritional intake. This allows the plant to continue growing normally despite adverse conditions. Most of the time, HMs are stuck in the hyphae of fungi that live in a symbiotic relationship with plants. These fungi reduce the availability of heavy metals to plants by retaining them in the cell walls, vacuoles, or cytoplasm through chelation. This results in a reduction of the toxicity of metals to plants. When AMFs colonize the roots of higher vascular plants, the host plant can hold on to the HMs in its roots and external mycelium. Mycorrhizal plant roots have fungal vesicles that are similar to plant vacuoles, with a similar biological function, that is, sequestration or accumulation of harmful chemicals or toxic heavy metals. An increase in the number of enzymes in the soil, *e.g.*, acid phosphatase, which enhances growth and nutrient uptake in plants, was also elevated by AMF under heavy metal stress. The cell wall is the main subcellular fraction that participates in heavy metal detoxification [8, 9]. AMF can boost the generation of phytochelatins [10]. Such peptides are produced by the enzyme phytochelatin synthase (PCS), which is induced by the presence of heavy metals using glutathione (GSH) as a substrate.

Uptake and Amassing of Heavy Metal Ions

The bioavailability and uptake of metal ions through roots initiate the uptake and accumulation of heavy metal ions. Roots take up heavy metals from the soil, which then translocate to other plant parts and get amassed in different regions of plant cells (Fig. **1**). Owing to the barrier effect, heavy metals frequently remain in roots [11]. The uptake and amassing of heavy metal ions in plant tissues depend on several factors, such as plant species, nutrient availability, organic matter, pH, temperature, moisture, and the most crucial factor: the presence of metals in the soil. The apoplastic pathway enables the soluble metal portion to pass across intracellular gaps without entering cells. In contrast, the symplastic pathway uses energy to transfer non-essential metals, such as Cd, Ni, and Pb through the cytoplasm [12]. Various plant species use diverse mechanisms to absorb various heavy metals, resulting in variations in their transport pathways. H+ ATPase/pumps are one of the pathways through which heavy metals enter the roots. They aid in maintaining a negative potential across the epidermal membrane of the roots. The uptake of heavy metal ions by plants involves other transporters, such as those belonging to the heavy metal ATPase (HMA), Zrt/Irt protein (ZIP), and natural resistance-associated macrophage protein (NRAMP) family [13]. Typically, root hairs increase the root surface area for heavy metal absorption, allowing ions to move swiftly inside the apoplast. Once absorbed by the roots, heavy metal ions exhibit apoplastic migration from the root epidermis to the cortex and are directly bonded to the carboxyl groups of mucilage uronic acid or to the polysaccharides in the rhizodermis. Because Casparian strips block the apoplastic pathway in the endodermis, these ions primarily build up in the root cells [14]. After passing through the endodermis, heavy metal ions move to the aboveground parts of the plant along the symplastic pathway through vascular tissues [15]. This phrase refers to all substances that must be carried from the roots to the leaves and vice versa through xylem tissues. It is an energy-intensive process that allows the entry of metals following the symplastic pathway. This is due to the Casparian strip of the endodermis, which prevents intracellular mobility during normal transport [16]. As a result, metals traveling through the apoplast pathway become blocked, making the symplast pathway the only route for metals to reach the xylem vessels.

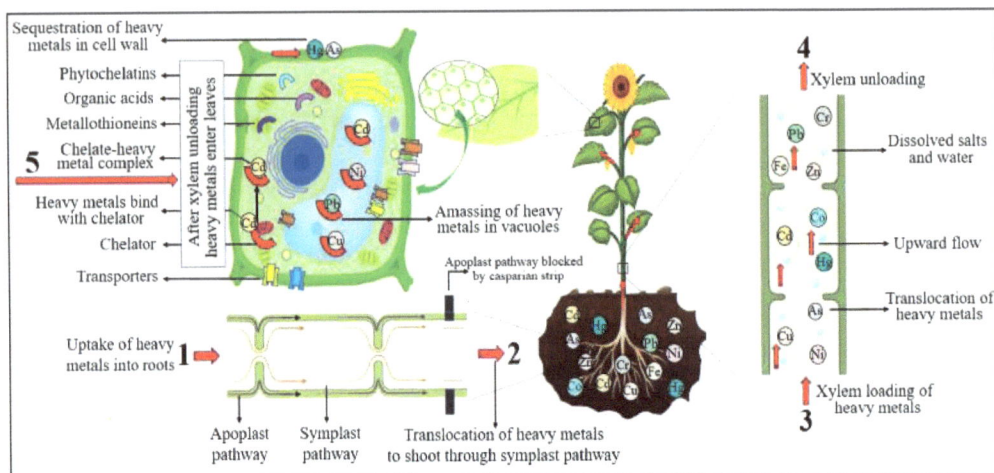

Fig. (1). Generalized mechanism of uptake and amassing of heavy metal ions.

Heavy metal detoxification and sequestration in the leaves is the next phase. Heavy metals are either intracellularly or extracellularly sequestered in the cells of the leaves when they are delivered thereafter detoxification by chelate ligands such as metallothioneins, phytochelatins, organic acids, or citrate [17]. Different hyperaccumulators store metals in different leaf regions. Vacuoles are the principal organelles involved in metal accumulation. The storage organ regulates the concentration of essential and non-essential metals in cells and prevents metals from endangering cells. Vacuole transport is mediated by metal ion transporters that specifically allow transport through the vacuoles [18]. Sun *et al.* [19] reported the storage of Hg in the cell wall and vacuoles of *Tillandsia usneoides* leaves, where hemicellulose was found to be the primary Hg binding component in the cell wall. Similarly, Cd accumulated in the vacuoles and cell walls of *Brassica napus* BN06 leaves, leading to reduced toxicity to the stomata and thylakoids. Consequently, vacuoles control the distribution of heavy metals in plants by compartmentalizing the metals.

Effects of Heavy Metal Ions on Plants

Due to its negative ecological effects, heavy metal poisoning of agricultural soils has become a vital and serious concern. In many regions of the world, it has been observed that the land used for food cultivation is contaminated by excess heavy metals, particularly Cd, Cr, As, Hg, Al, Co, Cu, Fe, Pb, and Zn [20]. Furthermore, it has been established that agricultural land with elevated heavy metal concentrations has detrimental effects on plant growth, metabolism, soil fertility, biological activity, biodiversity, and animal and human health [21]. In-depth studies have been undertaken globally to evaluate the impact of harmful heavy metals on plants [22, 23]. Some of the direct harmful consequences of heavy-metal toxicity include cytoplasmic enzyme inhibition and oxidative stress-mediated cell structural destruction. Owing to the interference of heavy metals with the actions of soil microorganisms, plant metabolism-related enzyme activities may also be hindered. The loss of plant development caused by these toxic effects (both direct and indirect) can occasionally result in plant death. Furthermore, reduced photosynthetic yield and membrane stability, nutrient and hormonal imbalance, decreased photosynthetic pigment production, inhibition of cell division, DNA replication, and gene expression are just a few biochemical and physiological effects of heavy metal toxicity in plants. The stage of plant development, uptake of heavy metals, their type, and their concentration affect the stimulation of various stress responses in plant cells. When present in excess, heavy metals cause cytotoxicity and damage to various cellular components, including disruption of proteins, nucleic acids, and membranes [24]. Furthermore, in the nucleus, heavy metals result in promutagenic damage, which includes DNA strand breaks, DNA base alterations, intra-and intermolecular cross-linking of proteins and DNA, rearrangements, and depurination. This causes abiotic stress in plants by activating genotoxic and oxidative stress responses [25, 26]. For instance, increased oxidative stress in terms of hydrogen peroxide (H_2O_2) content was reported in As- and Hg-stressed wheat plants, along with increased activity of antioxidant enzymes such as peroxidase (POD) and ascorbate peroxidase (APOX) [27]. Cadmium toxicity has also been shown to decrease the germination rate and survival of *Solanum melongena*. Similar results were obtained by Alam *et al.* [28], who observed decreased length and dry weight of tomato plants upon treatment with 300 µM Cr. Increased electrolyte leakage, malondialdehyde (MDA) content, and activity of enzymatic antioxidants, such as superoxide dismutase (SOD), catalase (CAT), glutathione reductase (GR), and APOX, were also observed after Cr application. Fatemi *et al.* reported an increase in flavonoids, MDA, and antioxidant activity, leading to a decrease in the biomass of Coriandrum sativum growing in soil contaminated with Pb in a concentration-dependent manner, where the maximum decrease was observed upon treatment with 1500 mg/kg soil. Furthermore, an increase in the activity of antioxidative enzymes to compensate for the increase in oxidative stress markers was observed in *Zea mays* growing in soil with 12 mg/kg As. Cu-mediated reductions in growth, photosynthetic pigments, transpiration rate, and photosynthetic rate were observed in *Citrus aurantium* [29]. In one study, decreased shoot length, number of leaves and leaf area, shoot weight and membrane stability, and increased electrolyte leakage were reported on foliar application of 0.5 mM Pb and/or 2.5 mM Ni on *Phaseolus vulgaris* [30]. Another study reported that the treatment of *Pisum sativum* seedlings with As resulted in a decline in plant length, photosynthetic pigments, and gas exchange properties [31]. A summary of the harmful effects of several metals on the growth, biochemistry, and physiology of plants is provided in Table 1.

Table 1. Effect of different heavy metals on plant.

Heavy metal	Concentration	Plant Species	Reported Effect	References
Cr	50, 100 and 200 mg/kg	*T. aestivum*	Decreased shoot and root length, fresh shoots, and dry weight.	[32]
Cr	50, and 100 mg/kg	*O. sativa*	Increased electrolyte leakage leads to reduced plant growth.	[33]
Co	25 and 50 mg/kg	*Phaseolus vulgaris*	Decreased protein and carbohydrate content.	[34]
Cd	35 μM	Tomato	Modulated nutrient uptake and reduced plant growth and root activity.	[35]
Ni	80 mg/kg	*Trigonella foenum-graecum*	Reduced plant growth, leaf area and activity of carbonic anhydrase and nitrate reductase.	[36]
Cr	100 μM	*Triticum aestivum*	Enhanced levels of superoxide radicals and activity of lipoxygenase.	[37]
Ni	50 μM	Tomato	Lowered growth and chlorophyll content and increased MDA, phenol, and proline content.	[38]
Cr	100 and 200 μM	*Calendula officinalis*	Induced oxidative stress in leaves by enhancing levels of oxidative stress markers.	[39]
Co	5, 25, 65, 125 and 185 mg/kg	Corn	Declined growth and nutrient concentration in aerial parts at high concentrations of Co.	[40]
Hg	100 mg/L	*Allium cepa*	Decreased germination percentage, root elongation, weight gain, mitotic index, increased chromosomal abnormalities, and micronucleus frequency.	[41]
Pb	25 and 50 μM	*Vigna radiata*	Suppression of photosynthetic activity.	[42]
Fe	0.09 and 1.8 mM	*Glycine max*	Decreased content of leaf gas exchange, increased activity of antioxidants such as APOX, CAT, and SOD, and impaired mineral accumulation.	[43]
Ni	1 and 5 ppm	*Sorghum bicolor*	Reduced plant dry weight.	[44]

(Table 1) cont.....

Heavy metal	Concentration	Plant Species	Reported Effect	References
Hg	40 μM	*Medicago sativa*	Hindered Fe acquisition and Sulfur uptake.	[45]
As	50 μM	*T. aestivum*	Decreased biomass and photosynthetic pigment content and increased oxidative stress.	[46]
Hg	1, 10, 100 and 1000 mg/kg	*Lactuca sativa*	Reduced seed germination.	[47]
Pb	500 mg/kg	*Daucus carota*	Decreased uptake of nutrients.	[48]
Mn	100, 250 and 500 μM	*Nicotiana benthamiana*	Increased MDA and H_2O_2 accumulation.	[49]
As	0.2, 0.4. 0.8, 1.6 and 3.2 mg/L	*S. lycopersicum*	Reduced growth and number of leaves and increased activity of APOX and CAT.	[50]
Mn	4 mM	*Arabidopsis thaliana*	Increased content of ROS and increased activity of antioxidants like POD, SOD, APOX, and CAT.	[51]
Cr	50 μM	*Vigna mungo* and *V. radiata*	Increased accumulation of superoxide radicals, H_2O_2, and MDA.	[52]
Hg	5, 10, and 20 μM	*T. aestivum*	Increased MDA, H_2O_2, and proline content and decreased levels of SOD.	[53]
Zn	1.95 g/kg	*O. sativa*	Induced accumulation of H_2O_2, MDA, and superoxide radicals.	[54]
Hg	0.6, 2.0 and 4.0 mg/L	*Bidens pilosa, Heliocarpus americanus* and *Taraxacum officinale*	Reduced seed germination and radical length.	[55]
Zn	2 and 5 mM	Cucumber	Increased accumulation of salicylic acid, abscisic acid, superoxide anion, and MDA.	[56]
Cd	200 μM	*Capsicum annuum*	Inhibited germination, root growth, and shoot growth.	[57]
As	50 μM	*O. sativa*	Reduced root and shoot length. Enhanced lipid peroxidation and activity of APOX, CAT, SOD, and GR.	[58]
Cu	1 and 5 mM	*Cajanus cajan*	Increased activity of SOD, CAT, and glutathione reductase. Decreased content of ascorbic acid.	[59]
As	50 and 100 μM	*O. sativa*	Reduced germination, plumule-radicle length, fresh weight, and amylase activity.	[60]
As	1.0 mM	*Phoenix dactylifera*	Increased ROS and CAT activity.	[61]

(Table 1) cont.....

Heavy metal	Concentration	Plant Species	Reported Effect	References
Cu	20 μM	Rice	Reduced expression of Cu transport proteins and increased soluble sugar and protein content.	[62]
Mn	0.1, 0.2 and 0.5 mM	Sugarcane	Increased production of nitric oxide.	[63]
Pb	200 μM	*Solanum lycopersicum*	Reduced epidermis thickness from the abaxial and adaxial leaf side.	[64]
Cu	30 or 60 mg/kg	*Brassica juncea*	Reduced stomatal size. Increased glutathione and proline content.	[65]
Ag	31.25, 62.5, 125, 250 and 500 mg/kg	*Hordeum vulgare* and *Folsomia candida*	Decreased photosynthetic pigments and dehydrogenase activity.	[66]
Ni	100, 200 and 400 μM	*Alyssum inflatum*	Increased protein, proline, and H_2O_2 content. Increased activity of antioxidant enzymes like SOD, POD, CAT, and APOX.	[67]
Cr	120 μM	*Vitis vinifera*	Reduced stem water potential, stomatal conductance, and net CO_2 assimilation rate.	[68]
Mn	1 mM	*Prunus persica*	Disrupted chloroplast ultrastructure by enhancing oxidative stress.	[69]
Ag	30 mg/L	*S. lycopersicum*	Increased relative gene expression of CAT.	[70]
Fe	300 ppm	*O. sativa*	Reduced plant height, number of leaves, and tiller.	[71]
Cu	100 mg/kg	*B. juncea*	Reduced activity of nitrate and nitrite reductase. Increased lignin deposition in root cell walls.	[72]
Ni	100 and 200 mM	Wheat	Lowered stomatal conductance, net photosynthesis, transpiration rate, and total soluble proteins and sugar.	[73]
Cu	400 μM	*C. grandis*	Increased bioaccumulation of flavonoids, alkaloids, lignin, and total phenolics. Upregulated and downregulated expression of Cu homeostasis and uptake-related genes.	[74]
Cd	500 μM	*Zea mays*	Reduced shoot length, root length, root weight, and shoot weight.	[75]
Cd	50 mg/kg	*Coriandrum sativum*	Decreased chlorophyll content, photosynthetic rate, and plant growth.	[76]
Cd	50 μM/L	*Aloe ferox*	Results in shorter and narrower leaves and decreased thickness of leaves by 50%.	[77]

(Table 1) cont.....

Heavy metal	Concentration	Plant Species	Reported Effect	References
Cd	100 μM	*Salvia sclarea*	Disrupted chloroplast ultrastructure and reduced photosystem II (PSII) photochemistry.	[78]
Zn	300 μM	Wheat	Induced membrane injuries in the leaf and root by increased oxidative stress.	[79]
Mn	0.5 Mm/L	*Saccharum officinarum*	Decreased expression of *GluTR*(encoding glutamyl-tRNA reductase).	[80]
As	50, 70 and 100 μM	*T. aestivum*	Disrupted cell membrane stability and reduced stomatal functions and water uptake.	[81]
Cu	300 μM	*Citrus sinensis*	Increased generation of superoxide anion, electrolyte leakage, and levels of MDA.	[82]
Pb	500 and 800 μM	*Citrus aurantium*	Increased lipid peroxidation and disrupted membrane integrity.	[83]
Pb	200 μM	*Oryza sativa*	Reduced stomatal and trichome density and content of photosynthetic pigments.	[84]
Zn	1.95 g/kg	*Z. mays*	Reduction in chlorophyll and carotenoid content.	[85]
Zn	300 and 700 mg/kg	*T. aestivum*	Reduced biomass.	[86]

The Mechanism Involved in the Detoxification of Heavy Metal Ions in Plants through Mycorrhizae.

There are several other approved mechanisms through which AMF affect HMs uptake, absorption, transport, and accumulation, and thus improve plant tolerance to HM stress. These approved mechanisms include toxic HMs retention in mycorrhizal roots and exterior hyphae, nutrient absorption enhancements, vacuolar sequestration of HMs, AMF-mediated binding of heavy metal ions on the cell wall of fungi, enhancement of the antioxidant response of plants, AMF-assisted chelation of heavy metal ions, and mycorrhizae-assisted phytoremediation (shown in Fig. (**2**).

Retention of Heavy Metal Ions on AMF Hyphal Network

To reduce HM toxicity in plants, AMF keep them from migrating to the plant's leaves or aboveground plant parts, mainly by sequestering them in the fungus mycelium and vesicles. The external mycelium of roots has a larger surface area than that of plant roots, which makes them better at absorbing metals than plant roots, which is why the external mycelium has the greatest effect on keeping metals from moving around in stressed soil. Additionally, the capacity of mycorrhizal plants to tolerate HM is linked to the amount of mycelial biomass produced and the rate of fungal development because the fungus takes up metals. In addition, when the concentration of HMs increases, the number of vesicles contained within the fungus also increases [87]. Investigatory reports also confirmed that after the addition of mycorrhizal fungi, such as *Rhizophagusintraradices*, *Glomus versiforme*, and *Funneliformismosseae*, dramatically decreased amounts of lead (Pb) metal ions were observed in pak choi shoots [88]. Therefore, this retention mechanism provided by AMF helps to reduce the transfer of HMs to the aerial parts of plants, which in turn contributes to phytostabilization. By retaining toxicants inside the roots, this retention mechanism also helps protect leaf tissues from damage.

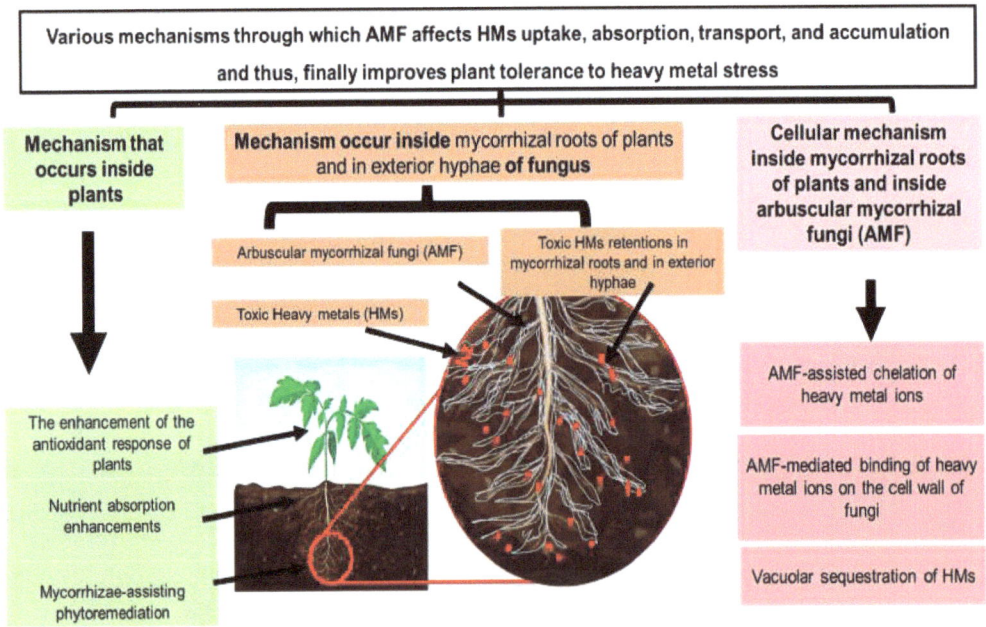

Fig. (2). Illustration of various mechanisms through which AMF affect HMs uptake, absorption, transport, and accumulation—which finally improves plant tolerance to HM stress.

Mycorrhizal Fungal-based Nutrition Absorption in Plants

Plants have several homeostatic mechanisms to mitigate stress. One of these mechanisms involves the association of plants with fungi as symbionts. AMFs are soil fungi that develop symbiotic relationships with plants and enhance host plant tolerance by providing nutrients to them. Nutrient exchange occurs in both directions; plants take mineral nutrients from fungi and, in return, provide carbohydrates for fungal growth. The entire process begins with a germinating fungal hypha from a spore that comes in contact with the host to form specialized cells or adhesion structures called appressoria. A penetration peg arises from it and pierces or enters into the parenchymal cortex for the intense proliferation of mycelium, which finally develops into differentiated intracellular haustoria, which are called arbuscules. The peri arbuscular membrane becomes an area of interface between the plant protoplasm and fungal structure, which plays an important role in the bidirectional nutrient exchange between AMF and plant cells. Nutrients exchanged between host plants and fungi include carbon, phosphorous, nitrogen, potassium, sulfur, fatty acids, and micronutrients, including copper, iron, and cobalt zinc. The hyphal length density in soil associated with plants is greater than the root length density in the soil, which aids in the enhancement of the absorption area of plant roots beyond its limit, resulting in better nutrient absorption and exploration at a greater distance of more than 25 cm from the root [89]. Various mechanisms are involved that assist the host plant in taking up nutrients from the soil in the presence of AMF under stressed conditions, such as AMF modifying the structure of soil by forming micro- and macro-aggregates by intertwining hyphae and releasing Glomalin-Related Soil Protein (GRSP), which is a special glycoprotein that improves soil structure by providing support to hyphae and by increasing organic matter content, thus providing mineral nutrients to the host. AMF also enhance the level of micronutrients in plants, which play several roles in the metabolism of lipids, carbohydrates, nucleic acids, and proteins. Phosphorus is an essential element in plant growth development and productivity, but it is insoluble. The AMF mycelium releases extracellular enzymes to convert P into a soluble form. AMF also alter the characteristics of some enzymes, increasing the activities of antioxidant enzymes, such as SOD, APX, and CAT in plants [90]. In close association of AMF mycelia with the host plant, the enhancement in the production of exocellular enzymes and soil aggregates was found to be resilient [91]. Mycorrhizal fungi also have the ability to produce phenol-oxidizing enzymes that can degrade compounds, such as phenolics and phenolic lignins, which act as barriers to mineral uptake. It also enhances the activity of photosynthetic pigment-synthesizing enzymes by increasing N and Mg^+ uptake. AMF association protects CO_2 fixing enzymes, *i.e.*, RuBisCO and Rubisco activase. To transfer mineral nutrients across the peri arbuscular membrane, various transporters (Pi transporters, mycorrhiza-inducible ammonium transporters (AMTs), and sulfate transporters) in the peri arbuscular membrane on the extra-radial part of the mycelium mediate nutrients (K, P, N, Ca, Mn, Mg, and Zn) to root cells in a controlled manner.

Sequestration of Heavy Metal Ions in the Vacuoles Through AMF and Plant Symbiosis

Heavy metal uptake and assimilation into the target sites and their sequestration mechanism undergo activation and application of polyphosphate granules in the vacuoles of fungal partners. Vacuolar sequestration is generally adopted by both plants and AMF to neutralize the harmful effects of these heavy metal ions, resulting in the compartmentalization of these toxic metal concentrations into the vacuoles. Multiple studies have reported the activities of H^+pumps for the uptake of HMs in vacuoles and subsequently advocated the roles of vacuolar pyrophosphatase (V-PPase) and vacuolar proton ATPase (V-ATPase). However, symbiosis leads to an increase in the phosphorus content of host plants, which ensures high ATP availability for metabolic processes and mediates the sequestration of heavy metals into vacuoles. This mechanism requires high energy consumption as ATP molecules, and finally hampers the translocation of heavy metals to other plant parts and restricts them inside the confined boundaries of the cell. Vacuolar sequestration stores high concentrations of heavy metals in the aboveground plant tissues and infers no phytotoxicity. Essential and non-essential HMs are sequestered in the vacuolar tissue of leaves [92]. However, cross-membrane metal transport is regulated by metal homeostasis inside the plant tissues. González-Guerrero *et al.* found that vacuolar accumulation of HMs like Zn, Cd, and Cu was observed when *R. irregularis* extra radical mycelium was grown in a culture with carrot root tissues under heavy metal concentrations [93]. Polyphosphate is considered a counter ion that is required to stabilize heavy metal ions inside fungal vacuoles and to mediate co-localization processes. However, studies have also shown that GintZnT1, a Zn transporter in *R. irregularis*, is a homolog of zrt1, a Zn transporter found in yeast vacuolar membranes, and has roles in Zn detoxification. GintZnT1 activities in yeast help to lower the cytosolic levels of Zn [93]. Another *R. irregularis* transporter, known as ABC transporter GintABC1, encodes another homolog related to the yeast Cd factor, which is employed for the transportation of complexes, such as bis-GSH-Cd, through the tonoplast membrane.

AMF-mediated Binding of Heavy Metal Ions on the Cell Wall of Fungi

The main mechanism behind fungi controlling the increasing concentrations of heavy metals in plants is the adsorption of these metal ions on the cell wall. Studies have shown that adsorbed heavy metals are mainly present on the extraradical cell walls of ectomycorrhizal fungal hyphae. AMF immobilizes heavy metals on the extra-radical or intra-radical cell walls of hyphae, forming complexes with root exudates or sequestering metal ions in vacuolar compartments. The fungal cell wall usually consists of polysaccharides, melanin, chitosan, and chitin as the main components that act as barriers in the path of heavy metals entering cells. In the case of excess concentrations of heavy metals, some metal ions can enter fungal cells, where the fungus stores these metal ions in specialized structures [94]. Immobilization of trace elements along with cell walls also occurs in various other structures like vesicles, vacuoles, and arbuscules of extraradical mycelium, all of which also store toxic ions. Other mechanisms may involve the retention of these HMs in fungal cell walls as sequestered subcellular locations and complex organic-inorganic molecules, such as phosphates. In a previous study, chitin and chitosan were reported as strong adsorbents for arsenic, cadmium, and chromium [94]. It is also effective in taking up lead metal in the cell wall, inner chamber, inner membrane, and inner chamber membrane of the vacuole of fungal hyphae [95]. Another mechanism that participates in absorbing metals to fungal cell walls is the secretion of some compounds by fungi, such as organic acids and glomalin, in the soil to seize the movement of HMs by forming precipitates of heavy metals as phosphate granules and glycol-protein metal complexes in soil.

Increased Activity of the Antioxidative Defense ystem

Overproduction of Reactive Oxygen Species (ROS) is caused by heavy metal stress in plants, which disrupts their regular functioning [96]. During exposure to heavy metals, the antioxidative defense systems of plants protect them from stress. Different enzymatic antioxidants, such as catalase (CAT), superoxide peroxidase (SOD), glutathione peroxidase (GPOX), and ascorbate peroxidase (APOX), act as ROS scavengers and protect plants from heavy-metal-induced malfunction and cell disruption [97]. The antioxidative defense system of plants has been reported to be highly active during metal stress. It has been claimed that, under stress, SOD transforms superoxide radicals into hydrogen peroxide, which is then transformed into water by APOX and CAT. AMF aid in relieving stress during stressful conditions by boosting the antioxidative defence system. According to previous studies, AMF increases the activity of antioxidants and enhances their production. According to previous reports [96], *Funneliformis mosseae* assists in reducing metal toxicity by boosting the activity of CAT and SOD. Additionally, it reduced MDA and hydrogen peroxide levels in *Zea mays* plants. Similarly, AMF have been demonstrated to decrease the toxicity of Zn, Pb, and Cd in *Cajanus cajan*. In another investigation, it was found that during Cd stress, AMF boosted the activity of CAT, GR, SOD, and APOX [98]. Numerous studies have demonstrated the effect of AMF in reducing metal stress in plants (Table **2**).

Table 2. Role of AMF in the alleviation of heavy metal stress in different plants.

S. No.	Plant	AMF	Metal	References
1.	*Robinia pseudoacacia* L.	*Glomus* *Scutellospora* *Claroideoglomus* *Paraglomus*	Cd	[99]
2.	*Bidens parviflora*	*Funneliformis mosseae*	Pb	[100]
3.	*Iris pseudacorus*	AMF	Cr	[101]
4.	*Medicago truncatula*	AMF	Cd	[102]
5.	*Zea mays*	*Rhizophagus irregularis*	Cd	[103]
6.	*Zea mays* L.	*Funneliformis mosseae and Diversispora spurcum*	P S Pb Zn Cd As	[104]
7.	*Solanum melongena* L.	AMF	Pb Cd As	[105]
8.	*Triticum aestivum* L.	*Rhizoglomus intraradices* *Glomus etunicatum*	As	[106]
9.	*Solanum nigrum*	*Funneliformis mosseae* BGC XJ02	Cd	[107]
10.	*Lonicera japonica*	*Glomus versiforme* *Rhizophagus intraradices*	Cd	[108]
11.	*Cassia italica* Mill	*Funneliformis mosseae* *Rhizophagus intraradices* *Claroideoglomus etunicatum*	Cd	[109]
12.	*Solanum photeinocarpum*	*Glomus versiforme* BGC GD01C	Cd	[110]
13.	*Robinia pseudoacacia* L.	*Funneliformis mosseae* and *Rhizophagus intraradices*	Pb	[111]

AMF-assisted Chelation of Heavy Metal Ions

The chelation of heavy metals with ligand compounds that exhibit a strong affinity for metal ions in the cytoplasm is a distinctive rectification approach used by AMF. Chelation aids in the prevention of HMs by primarily reducing their solubility, reactivity, and availability throughout the cytoplasm. Numerous chelating compounds, including metallothioneins (MTs), organic acids, phytochelatins (PCs), and nucleic acids, are found in AMF and flora [112, 113]. The entry of heavy metals inside plant cells typically occurs *via* one of the members of the metal carrier family, namely Zrt-Irt Proteins (ZIP). The heavy-metal sensitivity technique involves the creation of phytochelatins, which are thiol peptides that interact with heavy metals to form a complex. Metal-thiol binding is facilitated by phytochelatin (PC), whereas the metal chelator (HM-PC) aggregates travel through the cytosol and chief vacuole membrane on their way to the cell vacuoles, as shown in Fig. (**3**) [114]. The complexes of Phytochelatin-metal are propelled into the vacuoles, where another P1B-ATPase transporter (ABC-P1B) and ATP-binding cassette render the complex and nuanced metals unreactive.

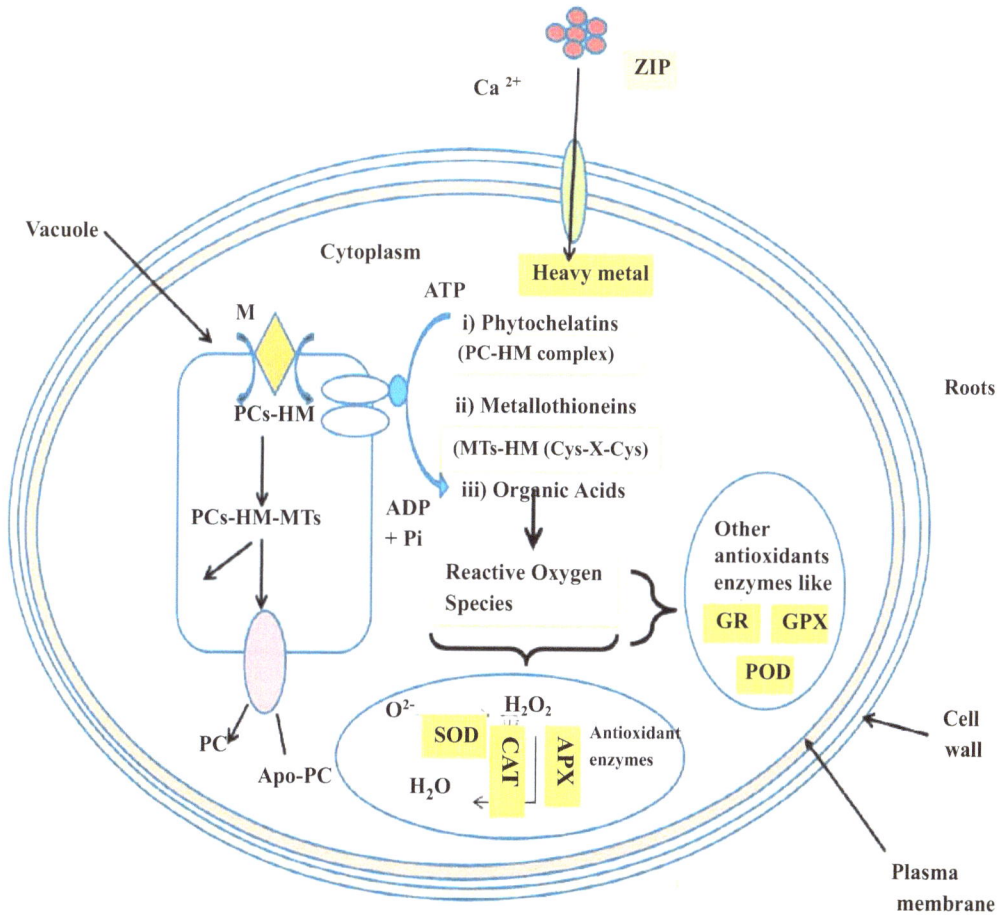

Fig. (3). Cellular mechanisms involved in the amelioration of Heavy metal stress by chelation involving the intake of heavy metal by plant roots *via* ZIP proteins, the formation of Phytochelators (PC) – Heavy metal (HM) complexes (Metallothioneins, Organic acids, stress proteins also do form complexes). Metal ions (M) enter plant roots *via* channels in the cytosol and activate the enzyme Phytochelatinsynthase (PCS) that form PCs where they bind to HMs and form HM-PCs complex, which are transferred to the vacuole *via* transporters through tonoplast of the cells.

Zea mays mycorrhizal treatment lowers Cd toxic effects and transportation by processing Cd into such a resting state which may be attributed to the higher Phytochelatins and glutathione content. Similarly, AMF *(Glomus etunicatum and Glomus mosseae)* inoculated with *Lolium perenne* showed an increase in chlorophyll synthesis, photosynthesis, and overall growth performance, where tolerance and detoxification are attributed to such mechanisms under Cd toxicity [115]. More phytochelatin-cadmium complexes are generated and transported to the plant vacuole as a by-product of AMF, increasing PC production, which ultimately reduces the adverse effects of PCs [116]. According to previous studies, culturing *Nicotiana tabacum* with the endomycorrhizal fungus *Funneliformis mosseae* increased the amount of GSH and further lowered the Cd and As concentrations in plant plants, such as leaves and roots [117, 118].

The sequestration of heavy metals is carried out similarly by a different group known as metallothioneins (MTs) [119]. These metal ligands are linked to metal ions to form metal-binding thiolates that amass and sequester heavy metals in plant vacuoles, which are responsible for regulating the endogenous metal contents between understrength and lethal levels. In *Festuca arundinacea,* AMF affected Ni movement and the expression of ABC transporters and MTs genes [120].

Mycorrhizae-assisting Phytoremediation

The mechanism involving mycorrhizal fungi-metal interactions depends on the AMF species and the plants exposed. Fungi might reduce the metal accumulation capacity of plants (phytostabilization) or enhance metal bioavailability in soils (phytoextraction) [121, 122]. The translocation of Cd from roots to shoots was reduced upon inoculation of *Glomus versiforme* and *Rhizophagus intraradices* in Lonicera japonica grown in Cd-treated soils [123].

In contrast, a study on Helianthus annuus cultivated in Cd-treated soils showed enhanced metal mobilization and immobilization upon inoculation with *Rhizophagus*and *Funneliformismosseae*, hastening both phytoextraction and phytostabilization processes.

a) Potential role of AMF in phytoextraction: A few strains such as *Glomus intraradices, G. desertico, Hebelomasinapizans, Suillus luteus,* and *Rhizopogonluteolus* have shown enhanced phytoextraction efficiency in HM polluted soils [124].

b) Potential role of AMF in phytostabilization: Numerous mycorrhizal fungi (*Glomus macrocarpum, Glomus mosseae, Glomus etunicatum*) have been reported to diminish metal translocation from underground to aerial parts of plants

[125]. Metal sequestration by polyphosphate granules forms the basis for phytostabilisation [126]. Additionally, HMs are halted by extracellular polymeric substances (*e.g.*, Glomoline) released by mycorrhizae as they are impounded inside plant vacuoles.

c) Potential role of AMF in phytotransformation: Mycorrhizae fungi have the ability to modify the metal contaminants for plant uptake. The process is termed 'phytotransformation.'

Upon uptake, the contaminant undergoes breakdown and mineralization processes involving different enzymatic reactions and metabolic processes (Fig. **4**). In addition, fungi can alter the root exudate profiles, which can ultimately change the metal oxidation state (biotransformation) and bioavailability in the rhizosphere. A list of various strains of mycorrhizae involved in phytoextraction, phytostabilization, and phytotransformation (Table **3**).

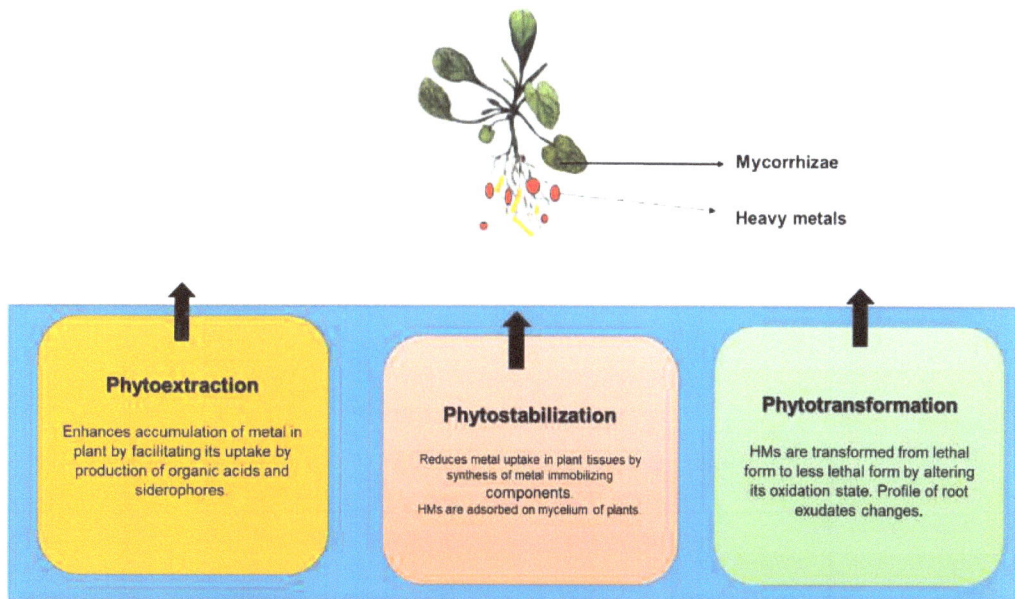

Fig. (4). Interaction of HM with AMF.

AMF INDUCED THE EXPRESSION OF GENES ON METAL ION TOXICITY

AMF display their activity in plants irrespective of sol pollution, culminating in HM uptake in plants to protect plants against HM stress, thereby improving plant growth [138]. They regulate a wide array of stress-related gene expressions to shield plants against HM stress (Fig. **4**).

A study in which Medicago sativa was exposed to toxic levels of arsenic (As) revealed upregulation of the AMF-induced metallothionein gene MsMT2 [139]. *M. truncatula,* when raised under Zn deficient and Zn toxic concentration along with colonized with *Rhizophagus irregularis,* revealed differential gene expression of ZIP (Zrt-, Irt-like protein) transporter genes [140]. *Hebelomamesophaeum* strains collected from contaminated soils expressed higher levels of genes required for metallothionein synthesis when exposed to Cadmium (Cd) toxic levels, which accounts for metal tolerance when inoculated in plants. Furthermore, LbMT1 and LbMT2 genes in *Laccaria bicolor* were upregulated during exposure to toxic levels of copper (Cu) and Cd [141]. In addition, the phytochelatin synthase gene has been reported in the genome of *Rhizophagus irregularis*, which is a metal-chelating compound required for HM detoxification [142]. When *Populus alba* colonized with *Glomus mosseae* was grown on polluted soils, the expression of MT and PCS genes was upregulated, which revealed enhanced sequestration of HMs induced by AMF. A study conducted on *Lolium perenne* exposed to nickel stress and colonized with *Funneliformis fasciculatum* revealed the positive impact of AMF on aquaporin-encoding genes. Lptip1;1 and Lptip1;2 are thought to be involved in Ni stress tolerance [143].

Table 3. Role of mycorrhizae in processes such as phytoextraction, phytostabilization, and phytotransformation.

Mechanism	Types of Mycorrhizae	Metal	Plant	Observations	References
Phytoextraction	*Claroideoglomus etunicata*	Mo	*Sorghum bicolor*	Plants inoculated with AMF picked up Mo 4 times more than non-mycorrhizal plants.	[127]
	Claroideoglomus claroideum (BEG210)	Ni	*Helianthus annuus*	*C. claroideum* improved Ni accumulation in *H. annuus* by 38%.	[129]
	Glomus mosseae	Cd, Co, & Pb	*Medicago sativa*	Inoculated plants have greater Co and Pb concentrations.	[130]
	Glomus deserticola	Cd, Zn	*Solanum melongena*	The metal content was enhanced in the tissues of plants.	[131]
	Glomus etunicatum	Pb	*Canavalia gladiata*	The Pb accumulation was encouraged in plants at medium concentrations.	[132]

(Table 3) cont.....

Mechanism	Types of Mycorrhizae	Metal	Plant	Observations	References
Phytostabilization	*Rhizophagus irregularis* (FR717169)	Cd	*Solanum nigrum*	Enhanced Cd accumulation in roots at lower concentrations of Cd.	[133]
	Glomus versiforme, Glomus mosseae, and Glomus diaphanum.	Cu, Zn, Pb, Cd.	*Oryza sativa*	Higher accumulation of metals in roots, although a slower translocation of metals from roots to shoots was observed.	[134]
	Glomus intraradices, G. mosseae	Pb, Zn	*Chrysopogon zizanioides*	Metal accumulation declined in the plant's roots.	[135]
	Glomus etunicatum	Pb	*Calopogonium mucunoides*	Amplified metal accumulation in roots with a decreased flow rate in the plant shoots.	[136]
	Glomus macrocarpum	Cd	*Apium graveolens*	Increased Cd content in roots.	[137]

Several genes associated with the antioxidant system have been found to be differentially regulated under the influence of AMF under HM toxic conditions [144, 145]. In *Glomus intraradices*, genes encoding glutathione-S-transferase were found to be upregulated along with the activation of zinc transporters when exposed to metal-toxic conditions [146].

AMF also induce the expression of genes for Metal Tolerance Proteins (MTP), as in the case of tea plants; induction of CsMTP8 promotes manganese (Mn) stress and reduces Mn accumulation under Mn-toxic conditions [147 - 149]. *P. albus* strains exhibit a Ni tolerance mechanism that can be attributed to the enhanced expression of C3578 (p-type ATPase), C5339 (ATP binding cassette), and C17235 (major facilitator superfamily). HM efflux mediated HM efflux protects plants against HMs (Fig. **5**).

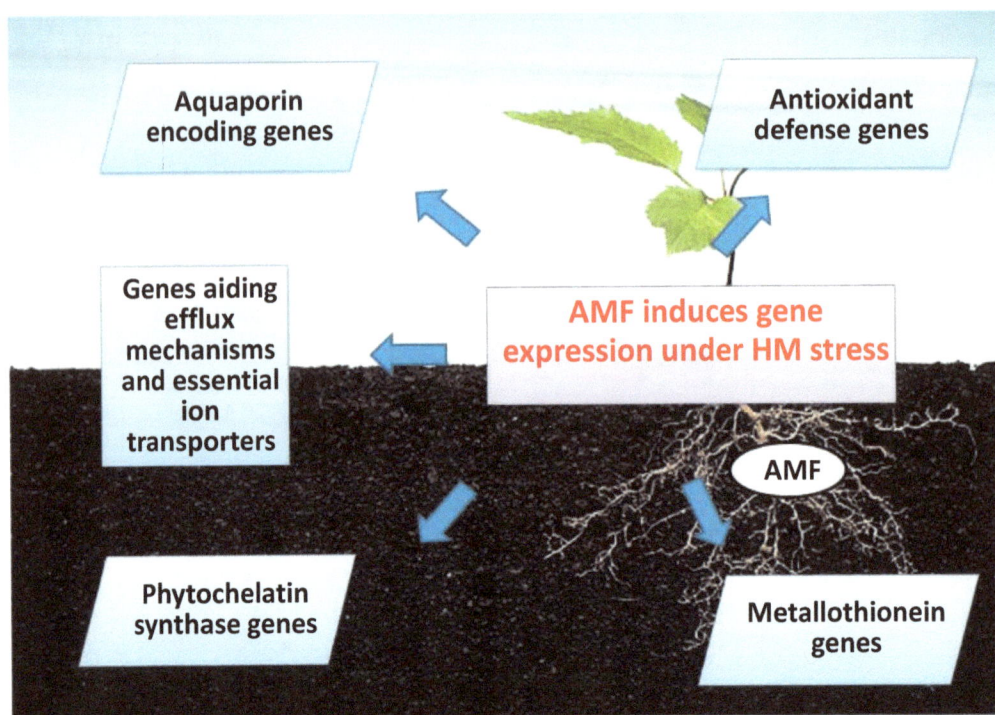

Fig. (5). Regulation of various HM stress-related gene expressions in plants by AMF.

CONCLUSION

Mycorrhizae act as lockboxes for HMs and nutrients in the rhizosphere. They concurrently enhance plant growth, development, and metal accumulation, and AMF colonization with plant roots affects the uptake of metal ions from the rhizospheric surface. Heavy metals are deposited in the cell walls of hyphae and associate with various biomolecules to form larger molecules that are trapped in the cell walls. In this way, hyphae act as barriers to various heavy metals and restrict their transport to plant root cells to help plants combat various kinds of heavy metal stress, altering gene expression of several defense-related genes and shielding plants from HM stress, thereby elucidating their potential for bioremediation without compromising plant growth. The mechanism of AMF-induced Heavy Metal (HM) tolerance in plants is being extensively reviewed and researched by the scientific community, as it not only protects plants against the toxic effects of metal ions but also opens up new approaches for phytoremediation.

REFERENCES

[1]   Mishra P, Singh A, Roy S. Plasma membrane H+-ATPase in plants. Cation transporters in plants. Academic Press 2022; pp. 357-73.
[http://dx.doi.org/10.1016/B978-0-323-85790-1.00012-9]

[2]   Azzeme AM. Plant scavenging potential to heavy metals.Augmenting Crop Productivity in Stress Environment. Singapore: Springer Nature Singapore 2022; pp. 191-203.
[http://dx.doi.org/10.1007/978-981-16-6361-1_12]

[3]   Rehman AU, Nazir S, Irshad R, *et al.* Toxicity of heavy metals in plants and animals and their uptake by magnetic iron oxide nanoparticles. J Mol Liq 2021; 321: 114455.
[http://dx.doi.org/10.1016/j.molliq.2020.114455]

[4]   Mukherjee S, Chatterjee N, Sircar A, *et al.* A comparative analysis of heavy metal effects on medicinal plants. Appl Biochem Biotechnol 2023; 195(4): 2483-518.
[http://dx.doi.org/10.1007/s12010-022-03938-0] [PMID: 35488955]

[5]   Varma S, Jangra M. Heavy metals stress and defense strategies in plants: An overview. J Pharmacogn Phytochem 2021; 10(1): 608-14.

[6]   Chandrasekaran M, Boopathi T, Manivannan P. Comprehensive assessment of ameliorative effects of AMF in alleviating abiotic stress in tomato plants. J Fungi (Basel) 2021; 7(4): 303.
[http://dx.doi.org/10.3390/jof7040303] [PMID: 33921098]

[7]   Mitra D, Djebaili R, Pellegrini M, *et al.* Arbuscular mycorrhizal symbiosis: plant growth improvement and induction of resistance under stressful conditions. J Plant Nutr 2021; 44(13): 1993-2028.
[http://dx.doi.org/10.1080/01904167.2021.1881552]

[8]   Li H, Luo N, Zhang LJ, *et al.* Do arbuscular mycorrhizal fungi affect cadmium uptake kinetics, subcellular distribution and chemical forms in rice?. Sci Total Environ 2016; 571: 1183-90.
[http://dx.doi.org/10.1016/j.scitotenv.2016.07.124] [PMID: 27450963]

[9]   Herath BM, Madushan KW, Lakmali JP, Yapa PN. Arbuscular mycorrhizal fungi as a potential tool for bioremediation of heavy metals in contaminated soil. N.a. J Adv Res Rev 2021; 10(3): 217-28.

[10]   Begum N, Qin C, Ahanger MA, *et al.* Role of arbuscular mycorrhizal fungi in plant growth regulation: implications in abiotic stress tolerance. Front Plant Sci 2019; 10: 1068.
[http://dx.doi.org/10.3389/fpls.2019.01068] [PMID: 31608075]

[11]   Podar D, Maathuis FJM. The role of roots and rhizosphere in providing tolerance to toxic metals and metalloids. Plant Cell Environ 2022; 45(3): 719-36.
[http://dx.doi.org/10.1111/pce.14188] [PMID: 34622470]

[12]   Chen YG, He XLS, Huang JH, *et al.* Impacts of heavy metals and medicinal crops on ecological systems, environmental pollution, cultivation, and production processes in China. Ecotoxicol Environ Saf 2021; 219: 112336.
[http://dx.doi.org/10.1016/j.ecoenv.2021.112336] [PMID: 34044310]

[13]   Yang Z, Yang F, Liu JL, *et al.* Heavy metal transporters: Functional mechanisms, regulation, and application in phytoremediation. Sci Total Environ 2022; 809: 151099.
[http://dx.doi.org/10.1016/j.scitotenv.2021.151099] [PMID: 34688763]

[14]   Ahmad S, Mfarrej MFB, El-Esawi MA, *et al.* Chromium-resistant *Staphylococcus aureus* alleviates chromium toxicity by developing synergistic relationships with zinc oxide nanoparticles in wheat. Ecotoxicol Environ Saf 2022; 230(230): 113142.
[http://dx.doi.org/10.1016/j.ecoenv.2021.113142] [PMID: 34990991]

[15]   Kumar S, Chhabra V, Bishnoi U. Translocation mechanism of heavy metal in plant roots: Concepts & conflicts: A review paper. PharmaInnov J 2022; 11(7): 2320-9.

[16]   Jamla M, Khare T, Joshi S, Patil S, Penna S, Kumar V. Omics approaches for understanding heavy metal responses and tolerance in plants. Curr Plant Biol 2021; 27: 100213.

[http://dx.doi.org/10.1016/j.cpb.2021.100213]

[17] Kim YO, Gwon Y, Kim J. Exogenous cysteine improves mercury uptake and tolerance in Arabidopsis by regulating the expression of heavy metal chelators and antioxidative enzymes. Front Plant Sci 2022; 13: 898247.
[http://dx.doi.org/10.3389/fpls.2022.898247] [PMID: 35755654]

[18] Jogawat A, Yadav B, Chhaya , Narayan OP. Metal transporters in organelles and their roles in heavy metal transportation and sequestration mechanisms in plants. Physiol Plant 2021; 173(1): ppl.13370.
[http://dx.doi.org/10.1111/ppl.13370] [PMID: 33586164]

[19] Sun X, Li P, Zheng G. Cellular and subcellular distribution and factors influencing the accumulation of atmospheric Hg in *Tillandsia usneoides* leaves. J Hazard Mater 2021; 414: 125529.
[http://dx.doi.org/10.1016/j.jhazmat.2021.125529] [PMID: 34030407]

[20] Vardhan KH, Kumar PS, Panda RC. A review on heavy metal pollution, toxicity and remedial measures: Current trends and future perspectives. J Mol Liq 2019; 290: 111197.
[http://dx.doi.org/10.1016/j.molliq.2019.111197]

[21] Kiran , Bharti R, Sharma R. Effect of heavy metals: An overview. Mater Today Proc 2022; 51: 880-5.
[http://dx.doi.org/10.1016/j.matpr.2021.06.278]

[22] Ghori NH, Ghori T, Hayat MQ, *et al.* Heavy metal stress and responses in plants. Int J Environ Sci Technol 2019; 16(3): 1807-28.
[http://dx.doi.org/10.1007/s13762-019-02215-8]

[23] Goyal D, Yadav A, Prasad M, *et al.* Effect of heavy metals on plant growth: an overview.Contaminants in agriculture: sources, impacts and management. Springer Nature Switzerland 2020; pp. 79-101.
[http://dx.doi.org/10.1007/978-3-030-41552-5_4]

[24] Admas T, Kerisew B. Assessment of cytotoxicity and genotoxicity potential of effluents from Bahir Dar Tannery using Allium cepa. Adv Public Health 2022; 2022: 1-10.
[http://dx.doi.org/10.1155/2022/5519304]

[25] Manzoor Z, Hassan Z, Ul-Allah S, *et al.* Transcription factors involved in plant responses to heavy metal stress adaptation.Plant perspectives to global climate changes Academic Press. 2022; pp. 221-31.
[http://dx.doi.org/10.1016/B978-0-323-85665-2.00021-2]

[26] Yüksel EA, Aydin M, Taşpinar MS, Ağar G. Iron toxicity-induced DNA damage, DNA methylation changes, and LTR retrotransposon polymorphisms in Zea mays. Turk J Bot 2022; 46(3): 197-204.
[http://dx.doi.org/10.55730/1300-008X.2682]

[27] Ibrahim M, Nawaz S, Iqbal K, *et al.* Plant-derived smoke solution alleviates cellular oxidative stress caused by arsenic and mercury by modulating the cellular antioxidative defense system in wheat. Plants 2022; 11(10): 1379.
[http://dx.doi.org/10.3390/plants11101379] [PMID: 35631804]

[28] Alam P, Balawi TH, Altalayan FH, Hatamleh AA, Ashraf M, Ahmad P. Silicon attenuates the negative effects of chromium stress in tomato plants by modifying antioxidant enzyme activities, ascorbate–glutathione cycle and glyoxalase system. Acta Physiol Plant 2021; 43(7): 110.
[http://dx.doi.org/10.1007/s11738-021-03276-4]

[29] Giannakoula A, Therios I, Chatzissavvidis C. Effect of lead and copper on photosynthetic apparatus in citrus (*Citrus aurantium* L.) plants. The role of antioxidants in oxidative damage as a response to heavy metal stress. Plants 2021; 10(1): 155.
[http://dx.doi.org/10.3390/plants10010155] [PMID: 33466929]

[30] Khalil R, Haroun S, Bassyoini F, Nagah A, Yusuf M. Salicylic acid in combination with kinetin or calcium ameliorates heavy metal stress in *Phaseolus vulgaris* plant. Journal of Agriculture and Food Research 2021; 5: 100182.

[http://dx.doi.org/10.1016/j.jafr.2021.100182]

[31] Alsahli AA, Bhat JA, Alyemeni MN, Ashraf M, Ahmad P. Hydrogen sulfide (H₂S) mitigates arsenic (As)-induced toxicity in pea (Pisum sativum L.) plants by regulating osmoregulation, antioxidant defense system, ascorbate glutathione cycle and glyoxalase system. J Plant Growth Regul 2021; 40(6): 2515-31.
[http://dx.doi.org/10.1007/s00344-020-10254-6]

[32] Ahmad S, Mfarrej MFB, El-Esawi MA, *et al.* Chromium-resistant *Staphylococcus aureus* alleviates chromium toxicity by developing synergistic relationships with zinc oxide nanoparticles in wheat. Ecotoxicol Environ Saf 2022; 230(230): 113142.
[http://dx.doi.org/10.1016/j.ecoenv.2021.113142] [PMID: 34990991]

[33] Alharby HF, Ali S. Combined role of Fe nanoparticles (Fe NPs) and *Staphylococcus aureus* L. in the alleviation of chromium stress in rice plants. Life (Basel) 2022; 12(3): 338.
[http://dx.doi.org/10.3390/life12030338] [PMID: 35330089]

[34] Al-Rashedy HS, Al-Mtewti WA. Effect of adding Silybum and licorice powder on carbohydrate and protein concentration in beans grown in soil treated with cobalt and copper. Sarhad J Agric 2022; 38(1): 260-5.

[35] Altaf MA, Shahid R, Ren MX, *et al.* Melatonin mitigates cadmium toxicity by promoting root architecture and mineral homeostasis of tomato genotypes. J Soil Sci Plant Nutr 2022; 22(1): 1112-28.
[http://dx.doi.org/10.1007/s42729-021-00720-9]

[36] Aqeel U, Parwez R, Aftab T, Khan MMA, Naeem M. Exogenous calcium repairs damage caused by nickel toxicity in fenugreek (*Trigonella foenum-graecum* L.) by strengthening its antioxidant defense system and other functional attributes. S Afr J Bot 2022; 150: 153-60.
[http://dx.doi.org/10.1016/j.sajb.2022.07.025]

[37] Ashraf MA, Rasheed R, Hussain I, *et al.* Taurine modulates dynamics of oxidative defense, secondary metabolism, and nutrient relation to mitigate boron and chromium toxicity in *Triticum aestivum* L. plants. Environ Sci Pollut Res Int 2022; 29(30): 45527-48.
[http://dx.doi.org/10.1007/s11356-022-19066-5] [PMID: 35147884]

[38] Badawy IH, Hmed AA, Sofy MR, Al-Mokadem AZ. Alleviation of cadmium and nickel toxicity and phyto-stimulation of tomato plant l. by endophytic micrococcus luteus and enterobacter cloacae. Plants 2022; 11(15): 2018.
[http://dx.doi.org/10.3390/plants11152018] [PMID: 35956496]

[39] Barzin G, Safari F, Bishehkolaei R. Beneficial role of methyl jasmonate on morphological, physiological and phytochemical responses of *Calendula officinalis* L. under Chromium toxicity. Physiol Mol Biol Plants 2022; 28(7): 1453-66.
[http://dx.doi.org/10.1007/s12298-022-01213-4] [PMID: 36051237]

[40] Bidast S, Golchin A, Baybordi A, Naidu R. Effects of Fe oxide-based nanoparticles on yield and nutrient content of corn in Cobalt-contaminated soils. Environmental Technology & Innovation 2022; 26: 102314.
[http://dx.doi.org/10.1016/j.eti.2022.102314]

[41] Çavuşoğlu D, Macar O, Kalefetoğlu Macar T, Çavuşoğlu K, Yalçın E. Mitigative effect of green tea extract against mercury(II) chloride toxicity in *Allium cepa* L. model. Environ Sci Pollut Res Int 2022; 29(19): 27862-74.
[http://dx.doi.org/10.1007/s11356-021-17781-z] [PMID: 34981388]

[42] Chen F, Aqeel M, Maqsood MF, *et al.* Mitigation of lead toxicity in *Vigna radiata* genotypes by silver nanoparticles. Environ Pollut 2022; 308: 119606.
[http://dx.doi.org/10.1016/j.envpol.2022.119606] [PMID: 35716894]

[43] Delias DS, Da-Silva CJ, Martins AC, de Oliveira DSC, do Amarante L. Iron toxicity increases oxidative stress and impairs mineral accumulation and leaf gas exchange in soybean plants during hypoxia. Environ Sci Pollut Res Int 2022; 29(15): 22427-38.

[http://dx.doi.org/10.1007/s11356-021-17397-3] [PMID: 34791629]

[44]   Doria-Manzur A, Sharifan H, Tejeda-Benitez L. Application of zinc oxide nanoparticles to promote remediation of nickel by *Sorghum bicolor*: metal ecotoxic potency and plant response. Int J Phytoremediation 2023; 25(1): 98-105.
[http://dx.doi.org/10.1080/15226514.2022.2060934] [PMID: 35452585]

[45]   El-Shehawi AM, Rahman MA, Elseehy MM, Kabir AH. Mercury toxicity causes iron and sulfur deficiencies along with oxidative injuries in alfalfa (*Medicago sativa*). Plant Biosyst 2022; 156(1): 284-91.
[http://dx.doi.org/10.1080/11263504.2021.1985005]

[46]   El-Shehawi AM, Arshi MJB, Elseehy MM, Kabir AH. Sugarcane bagasse acts as a metal absorber in the rhizosphere in mitigating arsenic toxicity in wheat. Rend Lincei Sci Fis Nat 2022; 33(3): 603-12.
[http://dx.doi.org/10.1007/s12210-022-01074-9]

[47]   Escobar-Vargas S, Vargas Aguirre CF, Rivera Páez FA. Arbuscular mycorrhizal fungi prevent mercury toxicity in *Lactuca sativa* (L.) seed germination. Pollution 2022; 8(3): 1014-25.

[48]   Faiz S, Yasin NA, Khan WU, *et al.* Role of magnesium oxide nanoparticles in the mitigation of lead-induced stress in *Daucus carota*: modulation in polyamines and antioxidant enzymes. Int J Phytoremediation 2022; 24(4): 364-72.
[http://dx.doi.org/10.1080/15226514.2021.1949263] [PMID: 34282979]

[49]   Gao H, Huang L, Gong Z, *et al.* Exogenous melatonin application improves resistance to high manganese stress through regulating reactive oxygen species scavenging and ion homeostasis in tobacco. Plant Growth Regul 2022; 98(2): 219-33.
[http://dx.doi.org/10.1007/s10725-022-00857-2]

[50]   González-Moscoso M, Juárez-Maldonado A, Cadenas-Pliego G, Meza-Figueroa D, SenGupta B, Martínez-Villegas N. Silicon nanoparticles decrease arsenic translocation and mitigate phytotoxicity in tomato plants. Environ Sci Pollut Res Int 2022; 29(23): 34147-63.
[http://dx.doi.org/10.1007/s11356-021-17665-2] [PMID: 35034295]

[51]   Hou L, Wang Z, Gong G, *et al.* Hydrogen sulfide alleviates manganese stress in arabidopsis. Int J Mol Sci 2022; 23(9): 5046.
[http://dx.doi.org/10.3390/ijms23095046] [PMID: 35563436]

[52]   Husain T, Suhel M, Prasad SM, Singh VP. Ethylene and hydrogen sulphide are essential for mitigating hexavalent chromium stress in two pulse crops. Plant Biol 2022; 24(4): 652-9.
[http://dx.doi.org/10.1111/plb.13324] [PMID: 34490701]

[53]   İşkil R, Surgun-Acar Y, Çatav ŞS, Zemheri-Navruz F, Erden Y. Mercury toxicity affects oxidative metabolism and induces stress responsive mechanisms in wheat (*Triticum aestivum* L.). Physiol Mol Biol Plants 2022; 28(4): 911-20.
[http://dx.doi.org/10.1007/s12298-022-01171-x] [PMID: 35592475]

[54]   Janeeshma E, Puthur JT. Physiological and metabolic dynamism in mycorrhizal and non-mycorrhizal *Oryza sativa* (var. Varsha) subjected to Zn and Cd toxicity: a comparative study. Environ Sci Pollut Res Int 2023; 30(2): 3668-87.
[http://dx.doi.org/10.1007/s11356-022-22478-y] [PMID: 35953749]

[55]   Kalinhoff C, Calderón NT. Mercury phytotoxicity and tolerance in three wild plants during germination and seedling development. Plants 2022; 11(15): 2046.
[http://dx.doi.org/10.3390/plants11152046] [PMID: 35956524]

[56]   Kang SM, Shahzad R, Khan MA, *et al.* Ameliorative effect of indole-3-acetic acid- and siderophore-producing *Leclercia adecarboxylata* MO1 on cucumber plants under zinc stress. J Plant Interact 2021; 16(1): 30-41.
[http://dx.doi.org/10.1080/17429145.2020.1864039]

[57]   Karmous I, Gammoudi N, Chaoui A. Assessing the potential role of zinc oxide nanoparticles for

mitigating cadmium toxicity in *Capsicum annuum* L. under *in vitro* conditions. J Plant Growth Regul 2023; 42(2): 719-34.
[http://dx.doi.org/10.1007/s00344-022-10579-4]

[58]   Kaur S, Chowhan N, Sharma P, Rathee S, Singh HP, Batish DR. β-Pinene alleviates arsenic (As)-induced oxidative stress by modulating enzymatic antioxidant activities in roots of *Oryza sativa*. Ecotoxicol Environ Saf 2022; 229: 113080.
[http://dx.doi.org/10.1016/j.ecoenv.2021.113080] [PMID: 34929504]

[59]   Kaushik S, Sharma P, Kaur G, *et al.* Seed priming with methyl jasmonate mitigates copper and cadmium toxicity by modifying biochemical attributes and antioxidants in *Cajanus cajan.*. Saudi J Biol Sci 2022; 29(2): 721-9.
[http://dx.doi.org/10.1016/j.sjbs.2021.12.014] [PMID: 35197737]

[60]   Khan T, Bilal S, Asaf S, *et al.* Silicon-Induced Tolerance against Arsenic Toxicity by Activating Physiological, Anatomical and Biochemical Regulation in Phoenix dactylifera (Date Palm). Plants 2022; 11(17): 2263.
[http://dx.doi.org/10.3390/plants11172263] [PMID: 36079645]

[61]   Khan Z, Thounaojam TC, Bhagawati R, Upadhyaya H. Impact of arsenic on the seedlings of Ranjit and Aijung, two most edible rice cultivars of Assam, India. J Stress Physiol Biochem 2022; 18(1): 28-39.

[62]   Li R, Wu L, Shao Y, Hu Q, Zhang H. Melatonin alleviates copper stress to promote rice seed germination and seedling growth *via* crosstalk among various defensive response pathways. Plant Physiol Biochem 2022; 179: 65-77.
[http://dx.doi.org/10.1016/j.plaphy.2022.03.016] [PMID: 35316694]

[63]   Ling GZ, Xiao JL, Yang S, *et al.* The alleviation of manganese toxicity by ammonium in sugarcane is related to pectin content, pectin methyl esterification, and nitric oxide. Glob Change Biol Bioenergy 2022; 14(5): 585-96.
[http://dx.doi.org/10.1111/gcbb.12936]

[64]   Maia CF, da Silva BRS, Batista BL, Bajguz A, Lobato AKS. 24-Epibrassinolide simultaneously stimulates photosynthetic machinery and biomass accumulation in tomato plants under lead stress: Essential contributions connected to the antioxidant system and anatomical structures. Agronomy (Basel) 2022; 12(9): 1985.
[http://dx.doi.org/10.3390/agronomy12091985]

[65]   Mir AR, Alam P, Hayat S. Auxin regulates growth, photosynthetic efficiency and mitigates copper induced toxicity *via* modulation of nutrient status, sugar metabolism and antioxidant potential in Brassica juncea. Plant Physiol Biochem 2022; 185: 244-59.
[http://dx.doi.org/10.1016/j.plaphy.2022.06.006] [PMID: 35717733]

[66]   Mocová KA, Petrová Š, Pohořelý M, Martinec M, Tourinho PS. Biochar reduces the toxicity of silver to barley (*Hordeum vulgare*) and springtails (Folsomia candida) in a natural soil. Environ Sci Pollut Res Int 2022; 29(25): 37435-44.
[http://dx.doi.org/10.1007/s11356-021-18289-2] [PMID: 35066846]

[67]   Najafi-Kakavand S, Karimi N, Ghasempour HR, Raza A, Chaichi M, Modarresi M. Role of jasmonic and salicylic acid on enzymatic changes in the root of two alyssum inflatumnáyr. populations exposed to nickel toxicity. J Plant Growth Regul 2023; 42(3): 1647-64.
[http://dx.doi.org/10.1007/s00344-022-10648-8]

[68]   Nikolaou KE, Chatzistathis T, Theocharis S, Argiriou A, Koundouras S, Zioziou E. Effects of chromium toxicity on physiological performance and nutrient uptake in two grapevine cultivars (*Vitis vinifera* L.) growing on own roots or grafted onto different rootstocks. Horticulturae 2022; 8(6): 493.
[http://dx.doi.org/10.3390/horticulturae8060493]

[69]   Noor I, Sohail H, Hasanuzzaman M, Hussain S, Li G, Liu J. Phosphorus confers tolerance against manganese toxicity in *Prunus persica* by reducing oxidative stress and improving chloroplast

ultrastructure. Chemosphere 2022; 291(Pt 3): 132999.
[http://dx.doi.org/10.1016/j.chemosphere.2021.132999] [PMID: 34808198]

[70]   Noori A, Bharath LP, White JC. Type-specific impacts of silver on the protein profile of tomato (*Lycopersicon esculentum L.*). Int J Phytoremediation 2022; 24(1): 12-24.
[http://dx.doi.org/10.1080/15226514.2021.1919052] [PMID: 34000928]

[71]   Page Z, Tokpah DP, Drame KN, Luther Z, Voor VM, King CF. Mathur J, Chauhan P. Mechanism of toxic metal uptake and transport in plants. Sustain Solutions Elem Defic Excess Crop Plants 2020; 335-49.

[72]   Rather BA, Mir IR, Masood A, Anjum NA, Khan NA. Ethylene-nitrogen synergism induces tolerance to copper stress by modulating antioxidant system and nitrogen metabolism and improves photosynthetic capacity in mustard. Environ Sci Pollut Res Int 2022; 29(32): 49029-49.
[http://dx.doi.org/10.1007/s11356-022-19380-y] [PMID: 35212900]

[73]   Rehman S, Mansoora N, Al-Dhumri SA, *et al.* Associative effects of activated carbon biochar and arbuscular mycorrhizal fungi on wheat for reducing nickel food chain bioavailability. Environmental Technology & Innovation 2022; 26: 102539.
[http://dx.doi.org/10.1016/j.eti.2022.102539]

[74]   Ren QQ, Huang ZR, Huang WL, *et al.* Physiological and molecular adaptations of *Citrus grandis* roots to long-term copper excess revealed by physiology, metabolome and transcriptome. Environ Exp Bot 2022; 203: 105049.
[http://dx.doi.org/10.1016/j.envexpbot.2022.105049]

[75]   Saleem MH, Parveen A, Khan SU, *et al.* Silicon fertigation regimes attenuates cadmium toxicity and phytoremediation potential in two maize (*Zea mays* L.) cultivars by minimizing its uptake and oxidative stress. Sustainability (Basel) 2022; 14(3): 1462.
[http://dx.doi.org/10.3390/su14031462]

[76]   Sardar R, Ahmed S, Yasin NA. Titanium dioxide nanoparticles mitigate cadmium toxicity in *Coriandrum sativum* L. through modulating antioxidant system, stress markers and reducing cadmium uptake. Environ Pollut 2022; 292(Pt A): 118373.
[http://dx.doi.org/10.1016/j.envpol.2021.118373] [PMID: 34662592]

[77]   Šírová K, Vaculík M. Toxic effects of cadmium on growth of Aloe ferox Mill. S Afr J Bot 2022; 147: 1181-7.
[http://dx.doi.org/10.1016/j.sajb.2020.12.026]

[78]   Sperdouli I, Adamakis ID, Dobrikova A, Apostolova E, Hanč A, Moustakas M. Excess zinc supply reduces cadmium uptake and mitigates cadmium toxicity effects on chloroplast structure, oxidative stress, and photosystem II photochemical efficiency in *Salvia sclarea* plants. Toxics 2022; 10(1): 36.
[http://dx.doi.org/10.3390/toxics10010036] [PMID: 35051078]

[79]   Wei C, Jiao Q, Agathokleous E, *et al.* Hormetic effects of zinc on growth and antioxidant defense system of wheat plants. Sci Total Environ 2022; 807(Pt 2): 150992.
[http://dx.doi.org/10.1016/j.scitotenv.2021.150992] [PMID: 34662623]

[80]   Yang S, Ling G, Li Q, *et al.* Manganese toxicity-induced chlorosis in sugarcane seedlings involves inhibition of chlorophyll biosynthesis. Crop J 2022; 10(6): 1674-82.
[http://dx.doi.org/10.1016/j.cj.2022.04.008]

[81]   Zaheer MS, Ali HH, Erinle KO, *et al.* Inoculation of *Azospirillum brasilense* and exogenous application of trans-zeatin riboside alleviates arsenic induced physiological damages in wheat (*Triticum aestivum*). Environ Sci Pollut Res Int 2022; 29(23): 33909-19.
[http://dx.doi.org/10.1007/s11356-021-18106-w] [PMID: 35031990]

[82]   Zhang J, Chen XF, Huang WT, *et al.* Mechanisms for increased pH-mediated amelioration of copper toxicity in *Citrus sinensis* leaves using physiology, transcriptomics and metabolomics. Environ Exp Bot 2022; 196: 104812.
[http://dx.doi.org/10.1016/j.envexpbot.2022.104812]

[83]   Giannakoula A, Therios I, Chatzissavvidis C. Effect of lead and copper on photosynthetic apparatus in citrus (*Citrus aurantium* L.) plants. The role of antioxidants in oxidative damage as a response to heavy metal stress. Plants 2021; 10(1): 155.
[http://dx.doi.org/10.3390/plants10010155] [PMID: 33466929]

[85]   Janeeshma E, Kalaji HM, Puthur JT. Differential responses in the photosynthetic efficiency of *Oryza sativa* and *Zea mays* on exposure to Cd and Zn toxicity. Acta Physiol Plant 2021; 43(1): 12.
[http://dx.doi.org/10.1007/s11738-020-03178-x]

[84]   Guedes FRCM, Maia CF, Silva BRS, *et al.* Exogenous 24-Epibrassinolide stimulates root protection, and leaf antioxidant enzymes in lead stressed rice plants: Central roles to minimize Pb content and oxidative stress. Environ Pollut 2021; 280: 116992.
[http://dx.doi.org/10.1016/j.envpol.2021.116992] [PMID: 33784567]

[86]   Stanislawska-Glubiak E, Korzeniowska J. Effect of salicylic acid foliar application on two wheat cultivars grown under zinc stress. Agronomy (Basel) 2021; 12(1): 60.
[http://dx.doi.org/10.3390/agronomy12010060]

[87]   Yang Y, Han X, Liang Y, Ghosh A, Chen J, Tang M. The combined effects of arbuscular mycorrhizal fungi (AMF) and lead (Pb) stress on Pb accumulation, plant growth parameters, photosynthesis, and antioxidant enzymes in *Robinia pseudoacacia* L. PLoS One 2015; 10(12): e0145726.
[http://dx.doi.org/10.1371/journal.pone.0145726] [PMID: 26698576]

[88]   Wu Z, Wu W, Zhou S, Wu S. Mycorrhizal inoculation affects Pb and Cd accumulation and translocation in Pakchoi (*Brassica chinensis* L.). Pedosphere 2016; 26(1): 13-26.
[http://dx.doi.org/10.1016/S1002-0160(15)60018-2]

[89]   Xie K, Ren Y, Chen A, *et al.* Plant nitrogen nutrition: The roles of arbuscular mycorrhizal fungi. J Plant Physiol 2022; 269: 153591.
[http://dx.doi.org/10.1016/j.jplph.2021.153591] [PMID: 34936969]

[90]   Ben Hassena A, Zouari M, Labrousse P, *et al.* Effect of arbuscular myccorhizal fungi on soil properties, mineral nutrition and antioxidant enzymes of olive plants under treated wastewater irrigation. S Afr J Bot 2022; 148: 710-9.
[http://dx.doi.org/10.1016/j.sajb.2022.05.042]

[91]   Carrillo-Saucedo SM, Gavito ME. Resilience of soil aggregation and exocellular enzymatic functions associated with arbuscular mycorrhizal fungal communities along a successional gradient in a tropical dry forest. Mycorrhiza 2020; 30(1): 109-20.
[http://dx.doi.org/10.1007/s00572-019-00928-9] [PMID: 31836908]

[93]   González-Guerrero M, Oger E, Benabdellah K, Azcón-Aguilar C, Lanfranco L, Ferrol N. Characterization of a CuZn superoxide dismutase gene in the arbuscular mycorrhizal fungus *Glomus intraradices*. Curr Genet 2010; 56(3): 265-74.
[http://dx.doi.org/10.1007/s00294-010-0298-y] [PMID: 20379721]

[94]   Janeeshma E, Puthur JT. Direct and indirect influence of arbuscular mycorrhizae on enhancing metal tolerance of plants. Arch Microbiol 2020; 202(1): 1-16.
[http://dx.doi.org/10.1007/s00203-019-01730-z] [PMID: 31552478]

[95]   Zhang HH, Tang M, Chen H, Zheng CL, Niu ZC. Effect of inoculation with AM fungi on lead uptake, translocation and stress alleviation of *Zea mays* L. seedlings planting in soil with increasing lead concentrations. Eur J Soil Biol 2010; 46(5): 306-11.
[http://dx.doi.org/10.1016/j.ejsobi.2010.05.006]

[96]   Sytar O, Kumar A, Latowski D, Kuczynska P, Strzałka K, Prasad MNV. Heavy metal-induced oxidative damage, defense reactions, and detoxification mechanisms in plants. Acta Physiol Plant 2013; 35(4): 985-99.
[http://dx.doi.org/10.1007/s11738-012-1169-6]

[97]   Rasouli-Sadaghiani MH, Barin M, Khodaverdiloo H, Siavash Moghaddam S, Damalas CA,

Kazemalilou S. Arbuscular mycorrhizal fungi and rhizobacteria promote growth of Russian knapweed (*Acroptilon repens* L.) in a Cd-contaminated soil. J Plant Growth Regul 2019; 38(1): 113-21.
[http://dx.doi.org/10.1007/s00344-018-9815-x]

[98]   Garg N, Kaur H. Response of Antioxidant Enzymes, Phytochelatins and Glutathione Production Towards Cd and Zn Stresses in *Cajanus cajan* (L.) Millsp. Genotypes Colonized by Arbuscular Mycorrhizal Fungi. J Agron Crop Sci 2013; 199(2): 118-33.
[http://dx.doi.org/10.1111/j.1439-037X.2012.00533.x]

[99]   Wang L, Jia X, Zhao Y, Zhang C, Zhao J. Effect of arbuscular mycorrhizal fungi in roots on antioxidant enzyme activity in leaves of *Robinia pseudoacacia* L. seedlings under elevated CO_2 and Cd exposure. Environ Pollut 2022; 294: 118652.
[http://dx.doi.org/10.1016/j.envpol.2021.118652] [PMID: 34890743]

[100]   Yang S, Ling G, Li Q, *et al.* Manganese toxicity-induced chlorosis in sugarcane seedlings involves inhibition of chlorophyll biosynthesis. Crop J 2022; 10(6): 1674-82.
[http://dx.doi.org/10.1016/j.cj.2022.04.008]

[101]   Chen F, Aqeel M, Maqsood MF, *et al.* Mitigation of lead toxicity in *Vigna radiata* genotypes by silver nanoparticles. Environ Pollut 2022; 308: 119606.
[http://dx.doi.org/10.1016/j.envpol.2022.119606] [PMID: 35716894]

[102]   Li J, Sun Y, Jiang X, Chen B, Zhang X. Arbuscular mycorrhizal fungi alleviate arsenic toxicity to *Medicago sativa* by influencing arsenic speciation and partitioning. Ecotoxicol Environ Saf 2018; 157: 235-43.
[http://dx.doi.org/10.1016/j.ecoenv.2018.03.073] [PMID: 29625397]

[103]   Kumutha K, Binodh AK. Accumulation of cadmium in maize roots inoculated with root organ culture of *Rhizophagus irregularis* improving cadmium tolerance through activation of antioxidative defense enzymes. J Appl Biol Biotechnol 2022; 10(5): 84-93.

[104]   Zhan F, Li B, Jiang M, *et al.* Arbuscular mycorrhizal fungi enhance antioxidant defense in the leaves and the retention of heavy metals in the roots of maize. Environ Sci Pollut Res Int 2018; 25(24): 24338-47.
[http://dx.doi.org/10.1007/s11356-018-2487-z] [PMID: 29948717]

[105]   Chaturvedi R, Favas P, Pratas J, Varun M, Paul MS. Assessment of edibility and effect of arbuscular mycorrhizal fungi on *Solanum melongena* L. grown under heavy metal(loid) contaminated soil. Ecotoxicol Environ Saf 2018; 148: 318-26.
[http://dx.doi.org/10.1016/j.ecoenv.2017.10.048] [PMID: 29091834]

[106]   Sharma S, Anand G, Singh N, Kapoor R. Arbuscular mycorrhiza augments arsenic tolerance in wheat (*Triticum aestivum* L.) by strengthening antioxidant defense system and thiol metabolism. Front Plant Sci 2017; 8: 906.
[http://dx.doi.org/10.3389/fpls.2017.00906] [PMID: 28642762]

[107]   Jiang QY, Tan SY, Zhuo F, Yang DJ, Ye ZH, Jing YX. Effect of *Funneliformis mosseae* on the growth, cadmium accumulation and antioxidant activities of *Solanum nigrum*. Appl Soil Ecol 2016; 98: 112-20.
[http://dx.doi.org/10.1016/j.apsoil.2015.10.003]

[108]   Jiang QY, Zhuo F, Long SH, *et al.* Can arbuscular mycorrhizal fungi reduce Cd uptake and alleviate Cd toxicity of *Lonicera japonica* grown in Cd-added soils?. Sci Rep 2016; 6(1): 21805.
[http://dx.doi.org/10.1038/srep21805] [PMID: 26892768]

[109]   Hashem A, Abd Allah EF, Alqarawi AA, Egamberdieva D. Bioremediation of adverse impact of cadmium toxicity on *Cassia italica* Mill by arbuscular mycorrhizal fungi. Saudi J Biol Sci 2016; 23(1): 39-47.
[http://dx.doi.org/10.1016/j.sjbs.2015.11.007] [PMID: 26858537]

[110]   Tan SY, Jiang QY, Zhuo F, *et al.* Effect of inoculation with *Glomus versiforme* on cadmium accumulation, antioxidant activities and phytochelatins of *Solanum photeinocarpum*. PLoS One 2015;

10(7): e0132347.
[http://dx.doi.org/10.1371/journal.pone.0132347] [PMID: 26176959]

[111] Yang Y, Han X, Liang Y, Ghosh A, Chen J, Tang M. The combined effects of arbuscular mycorrhizal fungi (AMF) and lead (Pb) stress on Pb accumulation, plant growth parameters, photosynthesis, and antioxidant enzymes in *Robinia pseudoacacia* L. PLoS One 2015; 10(12): e0145726.
[http://dx.doi.org/10.1371/journal.pone.0145726] [PMID: 26698576]

[112] Goldsbrough P. Metal tolerance in plants: the role of phytochelatins and metallothioneins.Phytoremediation of contaminated soil and water. CRC Press 2020; pp. 221-33.
[http://dx.doi.org/10.1201/9780367803148-12]

[113] Abou Seeda M, Abou El-Nour EZ, Abdallah M, El-Bassiouny H. Impacts of Metal, Metalloid and Their Effects in Plant Physiology: A Review. Middle East J Agric Res 2022; 11(03): 838-931.

[114] Dhalaria R, Kumar D, Kumar H, *et al.* Arbuscular mycorrhizal fungi as potential agents in ameliorating heavy metal stress in plants. Agronomy (Basel) 2020; 10(6): 815.
[http://dx.doi.org/10.3390/agronomy10060815]

[115] Han Y, Zveushe OK, Dong F, *et al.* Unraveling the effects of arbuscular mycorrhizal fungi on cadmium uptake and detoxification mechanisms in perennial ryegrass (*Lolium perenne*). Sci Total Environ 2021; 798: 149222.
[http://dx.doi.org/10.1016/j.scitotenv.2021.149222] [PMID: 34375244]

[116] Ingole NW, Dhawale VR. Development of phytoremediation technology for arsenic removal—a state of art. Devel 2021; 6(1)
[http://dx.doi.org/10.48175/IJARSCT-1342]

[117] Degola F, Fattorini L, Bona E, *et al.* The symbiosis between *Nicotiana tabacum* and the endomycorrhizal fungus *Funneliformis mosseae* increases the plant glutathione level and decreases leaf cadmium and root arsenic contents. Plant Physiol Biochem 2015; 92: 11-8.
[http://dx.doi.org/10.1016/j.plaphy.2015.04.001] [PMID: 25900420]

[118] Kaur H, Garg N. Recent perspectives on cross talk between cadmium, zinc, and arbuscular mycorrhizal fungi in plants. J Plant Growth Regul 2018; 37(2): 680-93.
[http://dx.doi.org/10.1007/s00344-017-9750-2]

[119] Chatterjee S, Kumari S, Rath S, Priyadarshanee M, Das S. Diversity, structure and regulation of microbial metallothionein: metal resistance and possible applications in sequestration of toxic metals. Metallomics 2020; 12(11): 1637-55.
[http://dx.doi.org/10.1039/d0mt00140f] [PMID: 32996528]

[120] Shabani L, Sabzalian MR, Mostafavi pour S. Arbuscular mycorrhiza affects nickel translocation and expression of ABC transporter and metallothionein genes in *Festuca arundinacea*. Mycorrhiza 2016; 26(1): 67-76.
[http://dx.doi.org/10.1007/s00572-015-0647-2] [PMID: 26041568]

[121] Li J, Sun Y, Zhang X, *et al.* A methyltransferase gene from arbuscular mycorrhizal fungi involved in arsenic methylation and volatilization. Chemosphere 2018; 209: 392-400.
[http://dx.doi.org/10.1016/j.chemosphere.2018.06.092] [PMID: 29935468]

[122] Adeyemi NO, Sakariyawo OS, Soremi PA, Atayese MO. Phytoremediation using arbuscular mycorrhizal fungi.Current Developments in Biotechnology and Bioengineering. Elsevier 2022; pp. 73-92.
[http://dx.doi.org/10.1016/B978-0-323-99907-6.00016-5]

[123] Jiang QY, Zhuo F, Long SH, *et al.* Can arbuscular mycorrhizal fungi reduce Cd uptake and alleviate Cd toxicity of Lonicera japonica grown in Cd-added soils?. Sci Rep 2016; 6(1): 21805.
[http://dx.doi.org/10.1038/srep21805] [PMID: 26892768]

[124] Ma Y, Oliveira RS, Freitas H, Zhang C. Biochemical and molecular mechanisms of plant-microb--metal interactions: relevance for phytoremediation. Front Plant Sci 2016; 7: 918.

[http://dx.doi.org/10.3389/fpls.2016.00918] [PMID: 27446148]

[125]   Garcia KG, Gomes VF, Mendes Filho PF, *et al.* Arbuscular mycorrhizal fungi in the phytostabilization of soil degraded by manganese mining. J Agric Sci 2018; 10(12): 192-202.

[126]   Souza LA, López Andrade SA, Ribeiro Souza SC, Schiavinato MA. Evaluation of mycorrhizal influence on the development and phytoremediation potential of *Canavalia gladiata* in Pb-contaminated soils. Int J Phytoremediation 2013; 15(5): 465-76.
[http://dx.doi.org/10.1080/15226514.2012.716099] [PMID: 23488172]

[127]   Shi Z, Zhang J, Lu S, Li Y, Wang F. Arbuscular mycorrhizal fungi improve the performance of sweet sorghum grown in a mo-contaminated soil. J Fungi (Basel) 2020; 6(2): 44.
[http://dx.doi.org/10.3390/jof6020044] [PMID: 32244390]

[128]   Ma Y, Rajkumar M, Oliveira RS, Zhang C, Freitas H. Potential of plant beneficial bacteria and arbuscular mycorrhizal fungi in phytoremediation of metal-contaminated saline soils. J Hazard Mater 2019; 379: 120813.
[http://dx.doi.org/10.1016/j.jhazmat.2019.120813] [PMID: 31254792]

[130]   Zaefarian F, Rezvani M, Ardakani MR, Rejali F, Miransari M. Impact of mycorrhizae formation on the phosphorus and heavy-metal uptake of Alfalfa. Commun Soil Sci Plant Anal 2013; 44(8): 1340-52.
[http://dx.doi.org/10.1080/00103624.2012.756505]

[131]   Mohammad A, Mittra B. Effects of inoculation with stress-adapted arbuscular mycorrhizal fungus *Glomus deserticola* on growth of *Solanum melogena* L. and *Sorghum sudanese* Staph. seedlings under salinity and heavy metal stress conditions. Arch Agron Soil Sci 2013; 59(2): 173-83.
[http://dx.doi.org/10.1080/03650340.2011.610029]

[132]   Souza LA, López Andrade SA, Ribeiro Souza SC, Schiavinato MA. Evaluation of mycorrhizal influence on the development and phytoremediation potential of Canavalia gladiata in Pb-contaminated soils. Int J Phytoremediation 2013; 15(5): 465-76.
[http://dx.doi.org/10.1080/15226514.2012.716099] [PMID: 23488172]

[133]   Wang G, Wang L, Ma F, You Y, Wang Y, Yang D. Integration of earthworms and arbuscular mycorrhizal fungi into phytoremediation of cadmium-contaminated soil by *Solanum nigrum* L. J Hazard Mater 2020; 389: 121873.
[http://dx.doi.org/10.1016/j.jhazmat.2019.121873] [PMID: 31862351]

[134]   Yang Y, Han X, Liang Y, Ghosh A, Chen J, Tang M. The combined effects of arbuscular mycorrhizal fungi (AMF) and lead (Pb) stress on Pb accumulation, plant growth parameters, photosynthesis, and antioxidant enzymes in *Robinia pseudoacacia* L. PLoS One 2015; 10(12): e0145726.
[http://dx.doi.org/10.1371/journal.pone.0145726] [PMID: 26698576]

[135]   Wu QS, Li GH, Zou YN. Roles of arbuscular mycorrhizal fungi on growth and nutrient acquisition of peach (*Prunus persica* L. Batsch) seedlings. J Anim Plant Sci 2011; 21(4): 746-50.

[136]   Souza LA, López Andrade SA, Ribeiro Souza SC, Schiavinato MA. Evaluation of mycorrhizal influence on the development and phytoremediation potential of *Canavalia gladiata* in Pb-contaminated soils. Int J Phytoremediation 2013; 15(5): 465-76.
[http://dx.doi.org/10.1080/15226514.2012.716099] [PMID: 23488172]

[137]   Kapoor R, Bhatnagar AK. Attenuation of cadmium toxicity in mycorrhizal celery (*Apium graveolens* L.). World J Microbiol Biotechnol 2007; 23(8): 1083-9.
[http://dx.doi.org/10.1007/s11274-006-9337-8]

[138]   Chandrasekaran M, Boopathi T, Manivannan P. Comprehensive assessment of ameliorative effects of AMF in alleviating abiotic stress in tomato plants. J Fungi (Basel) 2021; 7(4): 303.
[http://dx.doi.org/10.3390/jof7040303] [PMID: 33921098]

[139]   Li J, Sun Y, Zhang X, *et al.* A methyltransferase gene from arbuscular mycorrhizal fungi involved in arsenic methylation and volatilization. Chemosphere 2018; 209: 392-400.
[http://dx.doi.org/10.1016/j.chemosphere.2018.06.092] [PMID: 29935468]

[140] Watts-Williams SJ, Tyerman SD, Cavagnaro TR. The dual benefit of arbuscular mycorrhizal fungi under soil zinc deficiency and toxicity: linking plant physiology and gene expression. Plant Soil 2017; 420(1-2): 375-88.
[http://dx.doi.org/10.1007/s11104-017-3409-4]

[141] Reddy MS, Prasanna L, Marmeisse R, Fraissinet-Tachet L. Differential expression of metallothioneins in response to heavy metals and their involvement in metal tolerance in the symbiotic basidiomycete *Laccaria bicolor*. Microbiology (Reading) 2014; 160(10): 2235-42.
[http://dx.doi.org/10.1099/mic.0.080218-0] [PMID: 25031424]

[142] Shine AM, Shakya VP, Idnurm A. Phytochelatin synthase is required for tolerating metal toxicity in a basidiomycete yeast and is a conserved factor involved in metal homeostasis in fungi. Fungal Biol Biotechnol 2015; 2(1): 1-3.
[http://dx.doi.org/10.1186/s40694-015-0013-3]

[143] Bahmani-Babanari L, Mirzahosseini Z, Shabani L, Sabzalian MR. Effect of arbuscular mycorrhizal fungus, *Funneliformis fasciculatum*, on detoxification of Nickel and expression of TIP genes in *Lolium perenne* L. Biologia (Bratisl) 2021; 76(6): 1675-83.
[http://dx.doi.org/10.1007/s11756-021-00759-0]

[144] Riaz M, Kamran M, Fang Y, *et al.* Arbuscular mycorrhizal fungi-induced mitigation of heavy metal phytotoxicity in metal contaminated soils: A critical review. J Hazard Mater 2021; 402: 123919.
[http://dx.doi.org/10.1016/j.jhazmat.2020.123919] [PMID: 33254825]

[145] Khanna K, Kohli SK, Bali S, *et al.* Role of micro-organisms in modulating antioxidant defence in plants exposed to metal toxicity Plants Under Metal and Metalloid Stress: Responses, Tolerance and Remediation. Springer 2018; pp. 303-35.

[146] Hildebrandt U, Regvar M, Bothe H. Arbuscular mycorrhiza and heavy metal tolerance. Phytochemistry 2007; 68(1): 139-46.
[http://dx.doi.org/10.1016/j.phytochem.2006.09.023] [PMID: 17078985]

[147] Beneš V, Leonhardt T, Sácký J, Kotrba P. Two P1B-1-ATPases of *Amanita strobiliformis* with distinct properties in Cu/Ag transport. Front Microbiol 2018; 9: 747.
[http://dx.doi.org/10.3389/fmicb.2018.00747] [PMID: 29740406]

[148] Xu FQ, Meng LL, Kuča K, Wu QS. The mechanism of arbuscular mycorrhizal fungi-alleviated manganese toxicity in plants: A review. Plant Physiol Biochem 2024; 213: 108808.
[http://dx.doi.org/10.1016/j.plaphy.2024.108808] [PMID: 38865805]

[149] Shah T, Khan Z, Alahmadi TA, *et al.* Mycorrhizosphere bacteria inhibit chromium uptake and phytotoxicity by regulating proline metabolism, antioxidant defense system, and aquaporin gene expression in tomato. Environ Sci Pollut Res Int 2024; 31(17): 24836-50.
[http://dx.doi.org/10.1007/s11356-024-32755-7] [PMID: 38456983]

CHAPTER 11

Nanoremediation of Heavy-Metal Polluted Soils

Sakshi Verma[1], Shalini Dhiman[2], Renu Bhardwaj[2], Sandeep Kumar[3], Anil Kumar[4], Priyanka Sharma[5,*] and Nitika Kapoor[2,*]

[1] *Department of Zoology, Hans Raj Mahila Maha Vidyalaya, Jalandhar, Punjab, India;*

[2] *Department of Botanical and Environmental Sciences, Guru Nanak Dev University, Amritsar, Punjab, India.*

[3] *Department of Physics, DAV University, Jalandhar, Punjab, India.*

[4] *Serum Institute of India Ltd. Hadapsar, Pune, Maharashtra, India.*

[5] *School of Bioengineering Sciences & Research, MIT Art Design and Technology University, Pune, Maharashtra, India.*

Abstract: Soil contamination with heavy metals poses a significant challenge to environmental sustainability and human health. Conventional remediation techniques, while effective to some extent, are often limited by high costs, lengthy processes, and site-specific challenges. Nanoremediation offers an advanced solution by utilizing nanoparticles' unique properties for efficient remediation. This chapter reviews advancements in nanoremediation, highlighting the diverse range of nanoparticles used for soil remediation, such as metal oxides, carbon-based materials, and biogenic materials. Sustainable approaches, including green and conventional synthesis methods, as well as nanobiotechnology strategies that integrate microbial and plant-based remediation, are emphasized. Further, mechanisms such as adsorption and redox transformations in nanoparticle-heavy metal interactions are explored alongside their potential impacts on soil ecosystems and microbial health. The chapter concludes with future prospects, advocating for scalable and cost-effective nanoremediation technologies and addressing regulatory considerations to ensure environmental restoration.

Keywords: Green Nanotechnology, Heavy metals, Nanoparticles, Remediation, Soil pollution.

* **Corresponding authors Priyanka Sharma and Nitika Kapoor:** School of Bioengineering Sciences & Research, MIT Art Design and Technology University, Pune, Maharashtra, India; E-mails: priya_agn@yahoo.com; Department of of Botanical and Environmental Sciences, Guru Nanak Dev University, Amritsar, Punjab, India; E-mail: nitikaarora8@gmail.com

INTRODUCTION

Soil is an essential part of our ecosystem and fundamental for healthy crop production as it provides anchorage for roots, water and nutrient retention, and a habitat for microorganisms [1]. However, modern industrialization and the widespread application of chemical fertilizers in agricultural practices and other anthropogenic activities are progressively adding to the Heavy Metal (HM) pollution of soil, thereby degrading the soil quality and rendering it unsuitable for the growth of plants and microorganisms [2]. Contamination of soil with HMs has become a global concern due to their potential to accumulate at the various trophic levels of the food chain and, thus, pose a threat to human and environmental health [3]. Therefore, remediation of soil is an emerging challenge in the modern era. Numerous conventional methods have been employed to remove the pollutants from contaminated soil [4]. These include physical methods such as soil washing, vapour extraction, land farming, soil flushing, and chemical remediation by using chemical chelators, immobilization, oxidation, critical fluid extraction, *etc*. The use of conventional procedures is constrained by their high cost, laborious nature, risk of secondary pollutants, and need for excavation of contaminated soil [5]. In this context, bioremediation (the use of plants, microorganisms, and their products to remove pollutants) has been extensively studied and employed to remediate contaminated soils due to its cost-effective, eco-friendly, more efficient, and sustainable approach [6]. The advancements in nanotechnology have added new perspectives to the process of bioremediation of contaminated soil and water [7]. Nanotechnology has been recognised as a promising tool to remediate or restore contaminated soil efficiently by alleviating the toxicity of HM. Nanoremediation involves the use of nanoparticles (NPs), such as engineered and green synthesised NPs, to remove contaminants through various processes such as adsorption, reduction of toxic forms of HM into less toxic forms, photocatalysis precipitation, *etc*. Due to the unique physico-chemical properties, smaller particle size, enormous specific surface area, and high reactivity, NPsact as catalysts and possess the potential to lower the activation energy required for the reduction of compounds [8]. Lately, there have been numerous studies to synthesise and examine the effectiveness of various NPs in the remediation of HM-contaminated soils [9]. Among these, the most common are carbon-based and metal-based NPs such as Fe_3O_4, ZnO, TiO_2, zero-valent iron (nZVI) nanoparticles, and nanocomposites [5]. Diverse chemical methods are used to synthesise NPs, but these are burdened with the limitations of using toxic chemicals in the synthesis process and insufficient knowledge of the release of by-products and their possible toxic effects on human health and the environment [1]. Therefore, the research and industrial focus has been shifted to the green synthesis of NPs by using plants, microorganisms, and their products [10]. The greenly synthesised NPs are cost-effective, environmentally friendly, and more efficient in the remediation of pollutants [7].

Integration of nanotechnology with phytoremediation and microbial remediation has further enhanced the efficiency of bioremediation of pollutants in various environmental matrices [1]. For instance, the phytoextraction capacity of ryegrass (*Lolium perenne L.*) from lead-contaminated soil was enhanced with the use of NHAP (nano-hydroxyapatite) and NCB (Nano-Carbon Black) in a field experiment [11]. Singh and Lee also reported the increased Cd absorption by the *Glycine max* plant in the presence of TiO_2 NPs [12]. The role of NPs in promoting HM accumulation by plants is attributed to their ability to regulate cell wall permeability, heavy metal co-transportation, and transporter gene expression and to prevent the phytotoxicity of HM by scavenging free radicals and enhancing enzymatic antioxidant defense system in plants. However, there are concerns regarding the safe use of NPs in the remediation of contaminated soil and the fate of these NPs in the environment [13]. Thus, the impacts of various NMs on HM uptake and toxicity in various plant species still need to be explored for the large-scale application of nanoremediation. This chapter underscores the potential of nanotechnology in addressing heavy metal contamination in soils, emphasizing its integration with bioremediation techniques and underlying mechanisms. It also discusses environmental implications and challenges in scaling up nanoremediation technologies and sustainable solutions for heavy metal removal in contaminated soils.

Conventional Remediation Techniques and Associated Challenges

Various conventional remediation techniques like excavation, soil washing and stabilization, chemical reduction of reactive forms of HMs electrochemical methods, and bioremediation have been used to remediate the toxic heavy metals from the environment. Physical methods used to remediate the HM-contaminated soils include excavation, soil washing, and soil stabilization. Excavation, which is one of the quick and efficient methods, involves the physical removal of HM-contaminated soil from the site. Further, HM-polluted soils can be restored by washing with some specific wash solutions [14]. HMs can also be stabilized or fixed in soils by using certain immobilising agents, subsequently reducing their mobility and availability to biota [15]. To remediate the areas that are difficult to reach, such as deep soils or under-building soils, chemical methods, including chemisorption, ion exchange, precipitation, flotation, and electrochemical deposition *etc.*, are used. These methods are based on the principle of conversion of reactive forms of HMs to stable forms by using some reducing agents that may be precipitated out at an alkaline pH of soil [16]. Many adsorption techniques are also used to adsorb the reactive forms of HMs on sorbents by non-covalent interactions [17, 3]. In the ion exchange remediation method, HM ions are transferred across the membranes along an ionic concentration gradient [18].

Various electrochemical methods that are employed to remediate HM-contaminated soils are electrosmosis, electrophoresis, and electrolysis. In these methods, a low voltage direct current is passed through the contaminated soil, because of which toxic HMs move towards the electrodes by electromigration. The Asymmetrical Alternating Current Electrochemistry (AACE) method, one of the recent electrochemical methods, is quick and cost-effective and involves the use of Ami-PC electrodes to remove HMs from contaminated soils. Electrochemical methods are more efficient than traditional soil washing methods as they create no secondary contamination and no apparent soil deterioration [19, 20].

Alternate to the physico-chemical methods, green technologies that involve the use of plants and microorganisms are being focused on the remediation of HM-polluted soils. Phytoremediation is one of such eco-friendly techniques used for the reclamation of HMs, radionuclides, and other organic pollutants contaminated soils [21]. Depending upon the types of HMs/contaminants present in the soil, different types of phytoremediation techniques *viz* phytovolatilization, phytostabilization, phytoextraction, rhizodegradation, phytodegradation, and rhizofiltration are used to remediate the polluted soils [22]. Phytostabilization/ phytoimmobilization/ phytotransformation techniques reduce the toxicity of HMs by immobilizing them on the surface of roots, their further absorption and retention by root cells, or by their reduction with the help of microbes present in the rhizosphere. Certain hyperaccumulator plants have also been identified that accumulate an excess of HMs in vacuoles of their cells and thereby further reduce their availability to biota [23, 24]. Bioremediation, which involves the application of microorganisms to contaminated areas, is also one of the novel strategies. Microbes are capable of transforming harmful HMs into non-reactive or less toxic forms and help to clean the contaminated environment [25].

Although all these conventional methods are useful for the remediation of contaminated soils, they are loaded with many limitations and often prove to be inadequate or ineffective. Traditional methods have struggled to address the complexities of soil pollution, which can involve a diverse array of contaminants, varying soil conditions, and the need for long-term, sustained efforts [26, 27]. Furthermore, higher cost, formation of toxic chemical sludge, poor membrane stability, and maintenance of low pH are some major shortcomings of physico-chemical methods [18 - 20, 28]. One of the key limitations of conventional techniques is their reduced efficacy when applied in real-world, complex environmental conditions. Though phytoremediation and microbial remediation are eco-friendly and sustainable methods, both of these are time-consuming and less efficient as complete removal and degradation of HMs to less toxic forms is not possible [3, 29, 30]. Microorganisms used in bioremediation, for example, may perform well in the laboratory but face significant challenges when competing with other soil microbes for resources and adapting to the heterogeneous soil environment [26]. To overcome these limitations, researchers have increasingly turned to innovative technologies, such as nanoremediation, which holds significant promise for more effective and sustainable soil decontamination. Nanomaterials, with their unique physicochemical properties and high surface-to-volume ratios, have shown the ability to adsorb, react with, and transform a wide range of soil contaminants, including persistent organic pollutants, heavy metals, and excess nutrients [22, 31].

Advancements in Nanoremediation of HM-contaminated Soil

Nanotechnology is the branch of science that deals with the manipulation of materials at the nano-scale and has the capability to remediate contaminated soils by using various engineered NPs. It describes the synthesis, structure, and characterization of materials at nanoscale (<100 nm) [32]. Dr. Richard Feynman introduced the concept of nanotechnology in 1959, while Professor Norio Taniguchi first used the term "nanotechnology" in 1974 [33]. NPs, the basic components of nanotechnology, have been defined as particles with a diameter lesser than 100 nm. They have size-dependent physicochemical properties, which make them distinctive from their bulk or sub-micron/micron-sized counterparts [34]. Unique properties of NPs, *viz.* smaller size, greater surface area to volume ratio that provide them high reactivity and physicochemical dynamicity, quantum confinement, higher surface energy, high specific surface area, and selectivity make them potential candidates for nanobioremediation [35, 36]. Further high surface area to volume ratio of NPs helps to adsorb contaminants effectively and degrades them by reducing the toxic valence of metals to a stable metallic state [37]. The detail of various nanomaterials used for the remediation of HM-contaminated soil, the chemical and biogenic synthesis of nanoparticles, as well as the integration of these nanomaterials with bioremediation is emphasized in the section below.

Key Nanomaterials for HM Remediation in Soil

Nanoparticles are materials that have the potential properties to turn down the physical and chemical functionality of toxic contaminants in the soil-plant network owing to their small size and high surface area [1]. These nanomaterials are categorised into different types on the basis of their size and shape, *e.g.*, carbon nanotubes, metallic oxides, polymeric nanomaterials, *etc* [38]. The summary of the most studied nanomaterials used for the removal/detoxification of HM-contaminated environment is illustrated in Fig. (**1**).

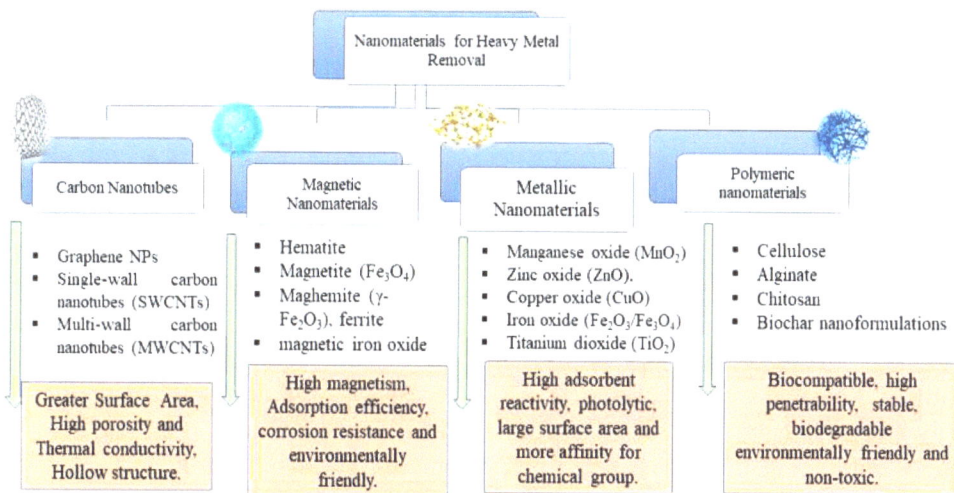

Fig. (1). Pictorial representation of the various nanomaterials used for the remediation of HM-contaminated soil.

Carbon Nanotubes

Carbon nanotubes (CNTs) have been proven as a unique nanomaterial for the removal of pollutants due to their potential to adsorb metal ions, which is attributed to their greater surface area and stable thermal and mechanical strength toward HMs [39]. Some reports indicate the removal of metal ions such as Cd from polluted soil by iron and sulphur nano-sponges of CNTs. Similar studies revealed that a small number of CNTs can immobilize the metal ions, *viz.* Cu^{2+}, Pb^{2+}, Zn^{2+} [40]. The adsorption of metal ions by CNTs is based on the structural component, ion exchange capacity, and electrochemical potential of nanomaterials [41, 42]. However, the removal efficiency of CNTs is enhanced *via* oxidation processes as it modifies the surface of these nanotubes. It was observed that double-oxidized Multiwalled Carbon Nanotubes (MWCNTs) improved the removal ratio of metals like Mn, Cu, and Zn by 78%, 79%, and 48% respectively against oxidized MWCNTs [43]. Additionally, MWCNTs strongly adsorbed numerous HMs on account of the high surface area of the catalyst towards the conversion of Cr (VI) to Cr (III) [44].

Metallic Nanomaterials

Several studies have revealed that metallic and metal oxide NPs, including manganese oxide (MnO_2), zinc oxide (ZnO), copper oxide (CuO), iron oxide

(Fe_2O_3/Fe_3O_4 and titanium dioxide (TiO_2), have the potential to remove toxic metal ions due to their properties, such as reactivity, photolysis, adsorbent capacity, large surface area, and more affinity for chemical groups [45]. Ahmed *et al.* stated that CuNPs caused inhibition of Cd translocation in wheat crops and improved plant growth and nutrient content [46]. Manganese oxide (MnOx) is one of the unique NP that is focussed upon due to its high energy, catalytic activity, and increased adsorption capacity for the removal of pollutants [47, 48]. MnO_2 nanofibers coated with polypyrrole and polyacrylonitrile showed a high removal efficiency of divalent lead ions with 251.90 mg/g adsorption capability [49].

Another important metallic nanomaterial is TiO_2, which possesses notable traits to remove pollutants owing to its self-cleaning ability, mechanical and antibacterial properties, and acts as a catalyst for the degradation of pollutants [50, 51]. Singh and Lee investigated that soil augmented with TiO_2 nanomaterials boosted the removal of Cd 2.6 times more (from 128.5 to 507.6 µg/plant) in *Glycine max* [12]. Likewise, Seisenbaeva *et al.* produced fully crystalline porous TiO_2 and demonstrated that it possesses attractive characteristics as an adsorbent for the removal of dichromate anions [52]. Similar findings also proved that the synthesized TiO_2 nanowire had also shown more capacity for the removal of various metals *viz.*Cd (II), Pb (II), Cu (II) and Zn (II) [53].

Furthermore, bimetallic nanoparticles, such as kaolinite-embedded Fe/Ni NPs have been revealed to eradicate Cu metal ions with an efficiency of more than 90% [54]. Further, it was observed that when CeO_2 nanoparticles coated with 3-mercaptopropyl trimethoxysilane and vinylpyrrolidone helped to eliminate divalent metallic ions like lead and copper. Similarly, the application of ZnO nanofibers coated with polyvinyl alcohol facilitated the removal of toxic ions of U^{6+}, Cu^{2+}, *etc* [55]. Alumina (Al_2O_3) NPs are the second important metal oxides that detoxify contaminants. They show diverse structural forms *viz.*α, β, ɤ, ɵ, and χ with an array of properties such as high thermal conductivity, strength, and interatomic bonding [56]. Among these, the ɤ structural form is a commonly used particle of alumina because of its prospective attributes such as more adsorption affinity, greater surface area, high mechanical strength, and low-temperature requirement [57, 58]. It was noticed that sol-gel-prepared Al_2O_3 exhibited further adsorption and efficiency for the removal of Cd (87%) and Pb (II) (97%) ions. The study conducted on Al_2O_3 NPs validated the removal of Co(II), divalent zinc, Cd, Ni, and other pollutants based on their higher adsorption affinity [59, 60].

Magnetic Nanomaterials

Magnetic NPs are manipulated through magnetic wave fields and are composed of magnetic elements, such as iron, cobalt, and nickel and their alloys, which possess paramagnetic, ferromagnetic, and super-paramagnetic properties [61]. Recently, magnetic NPs have been under intensive research, particularly iron oxide, because of their unique properties, *i.e.*, high magnetism, corrosion resistance, and environmentally friendly [62]. Numerous iron oxide nanomaterials have been used to remediate contaminants in soil like Cr^{6+}, Pb, and As^{5+} [63]. Singh *et al.* studied the effects of magnetic iron oxide NPs under different parameters for the removal of numerous HMs, *viz.* Zn, Cu, Cd, Fe, Cu, Cr, Pb *etc* from polluted soil [64]. They observed that at pH 0.7, the metal removal efficiency of these NPs differed in a range of 69.6 and 99.6%.

However, it is noticed that ferrite NPs showed poor stability and thus, some researchers are working on magnetite composites for the metal segregation from polluted sites [65]. Recent findings have shown that $Fe_3(PO_4)_2$ NPs treatment of contaminated soil helped to immobilize Cd metal ions by up to 70% through the formation of cadmium phosphate [66]. Similarly, Mahanty *et al.* reported the significant application of iron oxide NPs for the removal of different metal ions [Pb (II), Ni (II), Cu (II), and Zn (II)] [67]. Nanomaterials, *e.g.*, zero-valent iron (nZVI), adsorb different types of HMs (Cu, Hg (II), Cr (VI), Ni, Cd) as they provide reactive sites and electrostatic interactions with heavy metals [68, 69]. These nanomaterials showed greater efficiency for Cr detoxification with reduced aggregation and enhanced adsorption potential properties. The efficiency of magnetic NPs for HM removal is directly associated with their size, coating with composites, pH of the solution, and magnetic behaviour of the metal [70, 71].

Polymeric nanomaterials

Polymeric nanomaterials are colloidal particles of size ranging from 1 to 1000 nm and are made up of natural or synthetic polymers that carry active compounds either entrapped within or surface adsorbed onto the polymeric core [72]. Polymeric nanomaterials possess a porous fibre skeleton, high penetrability, and are biodegradable and non-toxic. Polymer-based nanomaterials such as cellulose, alginate, chitosan, and biochar nanoformulations provide excellent adsorption alternatives for the removal of HMs because of their biodegradation behaviour, having specific functional moieties and more tensile nature [73]. Cellulose is the most widely used biopolymer for the adsorption of toxic metal ions due to the presence of hydroxyl functional groups in the glucose ring [74]. Researchers formulated nanocellulose (NC)-Ag beads as a novel, economical process to remove HMs at a rate of 99%. Their results correlated the efficiency rate with the adsorption *via* electrostatic interactions and porosity [75].

Another important eco-friendly biopolymer is chitosan, which has been used extensively because of the amine and hydroxyl group of chitosan structure for

HM immobilization. Although the polymer-nanoformulations have shown widespread potential for the removal of toxic pollutants, more investigations are required to stabilize their synthesis process, recovery techniques, *etc* [76].

Green and Conventional Synthesis of Nanomaterials for Soil Remediation

There are many different approaches for novel NPs synthesis based on particle features such as size, shape, physiochemical and functional properties, and stability. In recent decades, NPs have been formulated *via* chemical, physical, and biological techniques to remediate pollutants from soil [77, 78]. Though numerous nanomaterials were produced by following conventional methods of bottom-up and top-down ways, many of them accumulate in the soil and slowly become toxic in agricultural sectors. Therefore, to minimize the toxicity of chemically synthesised nanomaterials, a green synthesis approach is being utilized. Hence, these methods can be further grouped into different classes: bottom-up, top-down, hybrid, and biogenic synthesis (Fig. **2**).

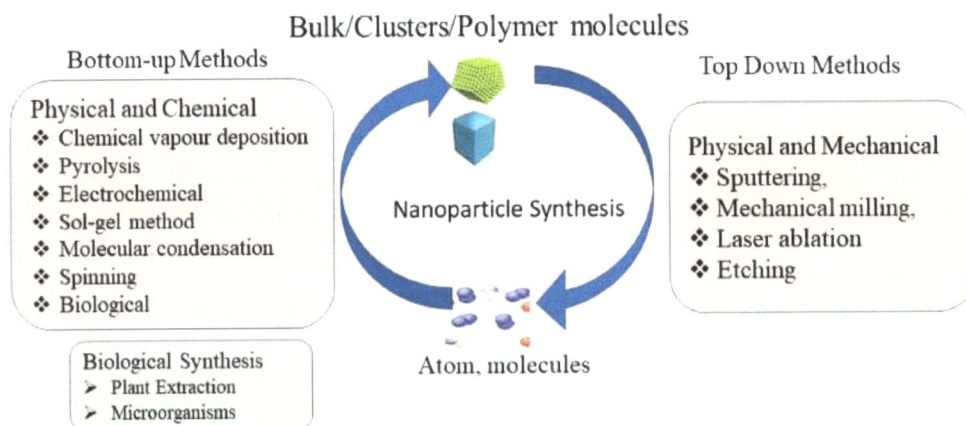

Fig. (2). Various approaches to synthesise nanoparticles

Bottom-up Method

In the bottom-up approach, complex nanomaterials are produced from smaller starting materials such as atoms or molecules and through self-assembly of monomer or polymer molecules *via* noncovalent interactions. This approach follows physical/chemical methods such as electrochemical, laser pyrolysis, flame spraying synthesis, precipitation, spinning, sol-gel processing, Chemical Vapour Deposition (CVD), and biological (plants and microbial assisted) methods for nanomaterial synthesis [79]. Therefore, the bottom-up method can be further categorised into: (1) physical methods, (2) chemical methods, and (3) biological methods (Fig. **2**).

It is noted that the Chemical Vapour Deposition (CVD) process plays a significant role in the production of carbon-based nanomaterials. During the whole process of CVD, a thin film is synthesised at the substrate surface upon the chemical reaction of vapor-phase precursors [80]. In the CVD-based preparation of graphene, Ni and Co catalysts provide multilayer graphene, whereas a Cu catalyst provides monolayer graphene [81]. Overall, CVD is an excellent method for producing high-quality nanomaterials, and it is well-known for the production of two-dimensional nanomaterials [82]. Inert gas condensation is also a bottom-up approach in which metallic or inorganic raw materials are vaporized and condensed on a cold surface in the presence of inert gas to produce NPs. This method can be used to produce a wide variety of NPs with controllable sizes [83]. Another bottom-up process is sputtering, where atoms are ejected from a source by the influence of kinetic energy, and upon ion bombardment, they form a film or nanoparticles [84, 85].

The chemical synthesis of NPs also follows the bottom-up approach. Nucleation, an initial step in chemical methods, principally involves cationic precursors to catalyse reduction to metal monomers. The size of particles controlled by reducing agents and thermally stable particles reach a certain size. Chemically synthesised metal NPs are prepared by using diverse chemical solutions and compositions. It was noticed that the hydrothermal process at high pressure and temperature also produces stable NPs with more yield than the microemulsion method. However, the structural confirmation, *i.e.*, morphology, size, and composition, is uniform [86].

Top-down Method

As compared to the bottom-up method, in the top-down approach, larger/bulk material is reduced to a smaller nonmaterial by mechanical/physical reduction. Therefore, the top-down process comprises mechanical and physical treatment to reduce the complex structure into a smaller nanoparticle *via* different methods, *viz.* sputtering, mechanical milling, laser ablation, *etc*hing, *etc*. This approach is most favourable for the synthesis of 100 nm of NPs and thin fabricating films. A study highlighted that in the process of laser ablation, high energy from a laser beam helped to convert raw material into fine NPs on fast condensation [87].

Mechanical methods are convenient, economical, and effective for producing nanomaterials from bulk sources by employing a shearing mechanism at different phases. These processes are useful to synthesize carbon nanomaterials, alumina, and oxide nanoparticles of copper, magnesium, nickel, aluminium, *etc*. Additionally, another top-down method, *i.e.*, electrospinning is most widely applied for the development of nanostructured materials because of ease. Coaxial electrospinning is commonly used to yield ultrathin nanofibers from polymers *via* liquid-filled capillaries as a shell (viscous) and core (non-viscous) under an electric field [88].

Sputtering is another important approach for the synthesis of nanomaterials through high-energy bombardment (gas or plasma) of bulk material. This approach is suitable for generating thin films of nanofibrils because of the physical ejection of atoms. It can be used to produce magnetrons, radio-frequency diodes, *etc* [89, 90].

Hybrid Method

Hybrid approaches combine both the processes of top-down and bottom-up to formulate nanostructured platforms. These hierarchically organized structures are generally difficult or impossible to fabricate using either top-down or bottom-up techniques. One of the best examples is the photolithograph, which considers layer formation of ions as the bottom-up and *etc*hing as a top-down process [91].

Biogenic or Green Synthesis Methods

In the last few decades, many NPs were synthesised *via* physical, chemical, and conventional approaches of top-down and bottom-up. However, on account of the toxicity produced by chemicals being used and the slow release of toxins from their by-products becomes a matter of serious concern to humankind. Therefore, to minimize the toxicity of chemicals, in the present era, a "biological or green synthesis" approach is highly employed. In bio-synthesis, the biologically originated extracts, for instance, plants and microbes, are used in the formation of NPs [10].

Biological synthesised NPs incorporate bio entities, *i.e.*, microorganisms and plant extract sources, as a material [78, 64]. Many plant biomolecules, including proteins, enzymes, polysaccharides, organic acids, and amino acids utilized for green nanoparticle synthesis, are enriched with metal ions. Some secondary metabolites are also promising candidates for the green synthesis of metallic nanoparticles by causing a reduction of metal cations to NPs (*i.e.*, reducing agents, such as flavonoids and phenols) and by concurrently accomplishing stabilization functions to avoid nanoparticle aggregation [92]. The advantages of using plant extract for NP production over conventional methods are that it is cost-effective, environmentally friendly, and easy to produce at large scale for industrial applications because the biosynthesis process entails low energy, lesser pressure, temperature, *etc* [77].

Currently, scientists are also focusing on the use of microbes for the green synthesis of NPs for multiple purposes in medicine, drug delivery systems, and environmental remediation [93, 94]. Microorganisms such as yeast, blue-green algae, fungi, and bacteria produce diverse metal and metal oxide nanoparticles *via* enzyme reduction, trapping within or outside the microbial cells. Scientific reports have explained that microorganisms under specific conditions are responsible for synthesising specific NPs.

Nanobiotechnology as a Sustainable Approach for Heavy Metal Remediation

Nanobiotechnology offers a sustainable and innovative approach to remediating heavy metal contamination in soils. This interdisciplinary field integrates principles of nanotechnology and biotechnology, leveraging the unique properties of nanoparticles synthesized using biological systems, such as microbes, plants, and enzymes (Fig. **3**) [95]. These biologically derived nanoparticles are eco-friendly and can effectively adsorb, immobilize, or detoxify heavy metals through processes like biosorption, bioprecipitation, and enzymatic transformation [95, 96]. Recent findings have indicated that green synthesised nanomaterials like TiO_2, ZnO, nano-hydroxyapatite, and nano-carbon from *Linum usitatissimum, Glycine max, Lolium temulentum* are more suitable to remove large quantities of Cd, Cu, Pb from soils [7, 11, 12,]. These nanomaterials also mitigate metal toxicity in plants by regulating gene expressions and maintaining homeostasis at physiological and molecular levels [97].

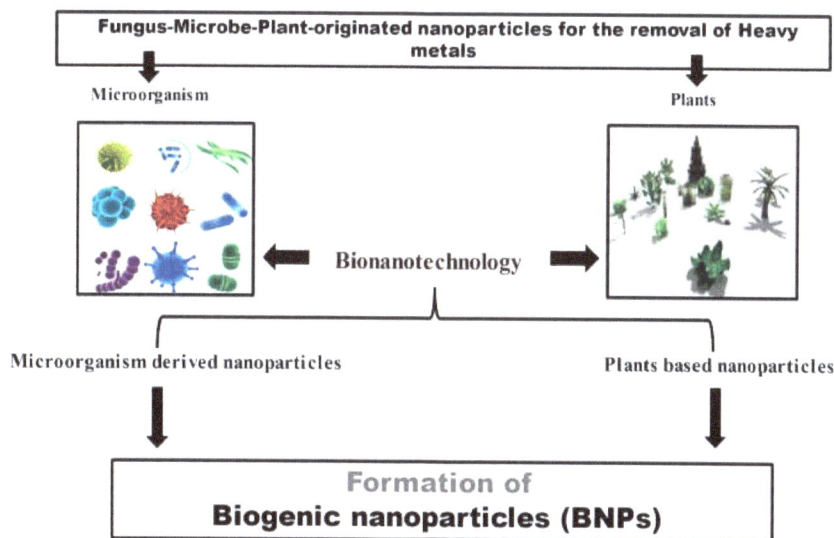

Fig. (3). Origin of Biogenic nanoparticles for the removal of Heavy metals pollutants from the environment.

Compared to conventional remediation methods, nanobiotechnology minimizes environmental impacts by reducing reliance on toxic chemicals and high-energy processes. For instance, biogenic nanoparticles, often synthesized *via* microbial pathways, demonstrate remarkable surface reactivity and specificity for binding heavy metals [98]. Immobilizing microbial cells on magnetic nanoparticles represents a cutting-edge technique that enhances their utility [99]. These biogenic NPs offer several advantages, including chemical stability, nontoxicity, cost-effectiveness, eco-friendliness, and the potential to play a crucial role in the biotransformation of pollutants to less toxic forms [98]. Biogenic magnetic NPs provide intriguing support in environmental pollutant removal because of their vast surface areas, strong surface reactivity, and superparamagnetic characteristics [100]. Their adsorption capacity allows them to act as carriers and detoxify various heavy metals. The adsorption mechanisms involve metal ions binding to surface atoms of nanoparticles, enabling broad-spectrum heavy metal removal. Numerous studies have highlighted the role of microorganism-associated nanoparticles in efficiently removing heavy metals from contaminated environments (Table **1**).

Table 1. Various investigatory reports showing the removal of heavy metal with the help of microorganism-associated nanoparticles.

S.No.	Nanoparticles	Microorganisms involved		Effective Removal of Heavy Metals	References
1	Magnetic Fe_3O_4 nanoparticles (MNPs)		*Lactic* Acid Bacteria (LAB)	Maximum 0.57 mg Cd^{2+}/g and 0.17mg Pb^{2+}/g removed from apple juice.	[101]
2	Palladium nanoparticles (Pd-NPs)		*Shewanella loihica* PV-4	Caused a decrease in Chromium (Cr VI).	[102]
3	Selenium nanoparticles (SeNPs)		*Alcaligenes faecalis* Se03	Reduced 100% selenite (Na_2SeO_3) from the liquid culture medium.	[103]
4	Bacteriogenic nano-selenium (Se^0) nanoparticles	Bacteria	*Citrobacter freundii* Y9	Remediated Hg from Contaminated soil.	[104]
5	CuS nanoparticles		*Shewanella oneidensis* MR-1	Enhanced Cr(VI) adsorption.	[105]
6	Iron oxide nanoparticles		*Agrobacterium fabrum* (SLAJ731)	Removed 197.02 mg g^{-1} Pb(II).	[106]
7	Fe_3O_4 Nanoparticle-Phtalic acid		*Staphylococcus aureus*	Effectively removed 1355 µmol g^{-1}Pb(II), 985 µmol g^{-1} Ni(II), and 795 µmol g^{-1}Cu(II).	[107]

(Table 1) cont.....

S.No.	Nanoparticles	Microorganisms involved		Effective Removal of Heavy Metals	References
8	Magnetic Fe_2O_3 nanoparticle	Fungus	*Coprinus micaceus*	Effectively removed 24.7 mg g^{-1} Co(II) and 26.2 mg g^{-1} Hg(II).	[108]
9	Chitosan- Magnetic γ-Fe_2O_3 nanoparticles		*Aspergillus sydowii*	Enhanced Cu(II) removal capacity up to 119.21 mg g^{-1}.	[36]
10	γ-Fe_2O_3 Magnetic nanoparticles		*Pleurotus eryngii*	Diminished 25.4 mg g^{-1} Co(II) and 30.3 mg g^{-1}Hg(II).	[109]
11	Magnetic chitosan Fe_3O_4 nanoparticle		*Saccharomyces cerevisiae*	Effectively Removed 81.96 mg/g Sr^{2+}.	[110]

Plant-based nanoparticles add another dimension to sustainable remediation by integrating phytoremediation with nanotechnology. These nanoparticles, including carbon-based, zero-valent metals, metal oxides, and nanocomposites, are produced from plant-derived biomaterials like leaves, seeds, fruit peels, and agro-waste (Table **2**). Their diverse sizes and shapes influence their remediation potential, effectively removing heavy metals such as lead (Pb), nickel (Ni), cadmium (Cd), arsenic (As), and chromium (Cr). These nanoparticles are derived from a variety of plant-based biomaterials like leaves, shoots, seeds, flowers, agro-wastes, fruit peels, *etc*. They are produced in different sizes and shapes, which influences their ability to remediate the contaminated sites [111]. Phytonanotechnology plays a major role in plant extracts-assisted conversion of metals into nanoparticles and offers an eco-friendly solution particularly suited for reclaiming agricultural soils. This sustainable approach underscores the significant role nanobiotechnology can play in environmental restoration and pollution management [112].

Table 2. Studies Demonstrating the Efficacy of Plant-Based Nanoparticles in Heavy Metal Remediation

S.No.	Nanoparticle	Plants Species	Plant-based biomaterial used	Effective Heavy Metals Removal	References
1	Copper oxide nanoparticles (CuO NPs)	*Mentha* and *Citrus aurantium*	Mint leaf and orange peel extracts	Enhanced removal of metal ions, maximum up to 88.80 mg g^{-1} Pb, 54.90 mg g^{-1} Ni, and 15.60 mg g^{-1} Cd.	[113]
2	Nano zero valent iron (nZVI)	*Quercus* and *Morus*	Leaf extracts of Oak and mulberry	Caused immobilization of around 73 or 76% Cu and 80 or 81% Ni.	[114]

S.No.	Nanoparticle	Plants Species	Plant-based biomaterial used	Effective Heavy Metals Removal	References
3	Iron oxide nanoparticles (Fe$_3$O$_4$ NPs)	*Ramalina sinensis*	Lichen Extract	Removed 82% Pb and 77% Cd.	[115]
4	Iron nanoparticles (*n*Fe)	*Phoenix dactylifera*	Seed extract	Effective in 90% Cr(IV) removal.	[116]
5	Nano-zero-valent iron composite (nZVI@TP-Mont) nanoparticles	*Camellia sinensis*	Leaf extract of Green Tea	Removed 90.2% of Cr(VI) from contaminated soil.	[117]
6	Iron nanoparticles (*n*Fe)	*Camellia sinensis*	Leaf extract of Green Tea	Removed 97.66% of Cr(VI).	[118]
7	Zero-valent iron (Fe) nanoparticles (ZVI-NPs)	*Verbascum Thapsus*	Leaf extract	Removed 100% of Cr(VI).	[119]
8	Cobalt-zinc ferrites NPs	Floral Honey	Honey originated from Floral extract	Caused 289 mg/g Pb adsorption.	[120]
9	Fe$_2$O$_3$ NPs	*Synechocystis* sp. PCC6803	Microalgae	Removed 69.77 mg/g Cr(VI), 62.63 mg/g Pb(II), 42.12 mg/g Cd(II), and 38.68 mg/g Cu(II) .	[121]
10	Iron nanocomposites (T-Fe$_3$O$_4$)	*Citrus reticulata*	Tangerine peel	95% efficient in removing Pb(II).	[122]
11	Iron nanoparticles (Fe NPs)	*Camellia sinensis*	Leaf extract of Green Tea	100% efficient in removing Pb(II).	[123]
12	α-Fe$_2$O$_3$-Ag nanocomposite	*Psidium guajava*	Leaf extract	71.34 mg/g Cr(VI) removal capacity.	[124]
13	Iron oxide nanoparticles	*Musa* peel extract	Waste banana peel extract	2.715 mg/g As(V) adsorption capacity.	[125]
14	Fe$_3$O$_4$	*Allium cepa* and *Zea mays*	Onion peel and cornsilk extract.	As maximal adsorption capacities of 1.86 and 2.79 mg g^{-1}, respectively.	[126]
15	Iron-based nanoparticles	*Ilex paraguariensis*	yerba mate Leaf extract	80% capacity to remove Cr(VI).	[127]
16	Bimetallic Fe/Ni NPs	*Eucalyptus*	Leaf extract	52.4% Cu removal capacity.	[128]

Mechanistic Insights into Nanoparticle-Heavy Metal Interactions

Numerous studies have illustrated the multiple interaction mechanisms of nanomaterials and HMs in soil. These include immobilization, reduction reactions, ion exchanges, surface complexation, and co-precipitation. The main mechanism underlying the immobilization of HM is adsorption on nanomaterials because of the large surface area, abundant functional groups, and adsorption sites of nanoparticles, which increase their adsorption efficiency [129]. For instance, Lin and his co-workers observed adsorption as the main mechanism of Cd removal by iron oxide NPs when incubated for a long period of time [128]. Similarly, TiO_2, MgO, and ZnO nanoparticles were reported to be efficient adsorbents for the removal of Cr ions from contaminated soil [130]. The iron-based nanoparticles, including Zero-Valent Iron (ZVI), iron sulfide (FeS), and magnetite (Fe_3O_4), were found to effectively immobilize arsenic (As) in As-contaminated soils, thereby reducing the bioavailability and bio-accessibility of As. Surface complexion was concluded to be the key mechanism for As uptake by nanoscale iron oxides and iron sulfide [131]. Furthermore, surface complexation and ion diffusion in the lattices of nano-hydroxyapatite were found to be the mechanisms of Cd interaction with NPs, subsequently reducing the mobility of Cd ions [132].

Transformation of toxic forms of HM into less toxic forms through reduction reactions has been reported to mitigate HM in contaminated water and soils successfully. Reduction reactions involving the use of nZVI have attracted more interest due to their strong reducing ability and are therefore already employed at field scale either in the form of permeable reactive barriers or direct addition through injection wells. nZVI was reported to reduce Cr (VI) to Cr (III) and form complex precipitates such as ferrous chromite ($FeCr_2O_4$) [133]. However, nZVI tends to agglomerate and lose adsorption. The reactive surface consequently loses reducing ability due to its zero valency and magnetic properties. Therefore, many supporting materials have been widely used to prevent aggregation of nZVI and increase its reactivity [51]. Biochar-supported nZVI particles exhibited high removal efficiency for Cr from contaminated soil. Additionally, numerous nanocomposites are extensively researched for their effective role in the remediation of contaminated soils [129]. For instance, the biochar produced from the pol plant, when applied along with nZVI in Cd contaminated paddy field, showed a 93% decrease in Cd content in rice grains [134]. With the application of mineral amendments possessing large surface area and abundant negative charge in HM-contaminated soils, there occurs a strong adsorption, ion exchange, surface inner-sphere complex formation, and co-precipitation of Cd ions, and thus further enhances the immobilization of Cd in contaminated soil [135].

Besides, nanomaterials are used along with phytoremediation in mitigating the HM contaminated soils. Nanomaterials are reported to significantly enhance the phytoremediation efficiency of plants *via* promoting plant growth through increasing the absorption and utilization of nutrients by plants, serving as nano-fertilizers and nano-pesticides for plants, activating the plant reproductive system, enhancing the oxidative defense system of plants as well as alleviating phytotoxicity of HM [7]. Increased Cd uptake by the soybean plant was observed when nano-titanium dioxide was added to Cd-contaminated soil. Furthermore, TiO_2 nanoparticles restricted Cd toxicity in the plants by increasing the photosynthetic rate and growth parameters of the plant, protecting plants from oxidative damage, and capturing free radicals [12]. In a similar study, enhanced phytoextraction potential of ryegrass (*Lolium temulentum* L.) and decreased phytotoxicity of lead (Pb) to the plant was observed with the collective impact of nano-hydroxyapatite (NHAP) and Nano-Carbon Black (NCB) on HM contaminated soil [11]. The regulation of gene expression linked to metal stress, such as transporter gene expression, oxidative stress, modulating cell wall permeability, photosynthetic pathways, and cell division, are some of the proposed mechanisms exerted by nanomaterials to mitigate HM toxicity in plants [136 - 138]. However, various reports in the literature show that NPs have toxic effects on plants as well as soil microbiota [1, 34, 139]. Further, the remediation of soil by nanomaterials is also affected by various factors such as soil type, pH, organic matter and moisture of soil, concentration of contaminants, and type of plants and properties of nanomaterial remediation agents applied. Therefore, in-depth studies focusing on the interaction mechanisms between nanomaterials and various heavy metal contaminants in soil, as well as their impact on the plants and soil microbes, may be useful for the synthesis of nanoparticles and their utilization for sustainable remediation of contaminated soils and sustainable development of agriculture.

Environmental Impacts: Nanoparticles and Soil Ecosystems

NPs interact with soil ecosystems in complex ways, influencing both physical and biological components. NPs possess unique properties, making them effective for applications like remediation. They can significantly influence soil ecosystems, offering both beneficial and harmful impacts [140]. On the positive side, nanoparticles improve soil quality by enhancing nutrient availability and aiding the degradation of organic pollutants [141]. They facilitate the removal of heavy metals, pathogens, and persistent organic compounds, promoting soil fertility and enabling sustainable agricultural practices. It has been observed that a variety of nanomaterials, *viz.* cellulose NMs, carbon nanotubes, metal and metal oxide NMs, *etc.*, improve the growth and yield of various crops by acting as carriers of nutrients and delivering them to plants efficiently through root/leaves [142, 143].

Many reports indicate that metal and metal oxide NPs act as nutrients for plants depending upon their concentrations and promote various physiological and biochemical activities of plants, inducing activation of antioxidative enzymes, enhanced rate of photosynthesis and nitrogen assimilation, *etc* [5]. Exposure of plants to Cu and Zn NPs increases the uptake and bioavailability of Cu and Zn nutrients, thereby promoting the activities of antioxidative enzymes, total protein content, rate of photosynthesis, and improved biomass of plants [144, 145]. Besides, Zn has also been found to reduce N loss to the atmosphere or surface water by inhibiting the transformation of N to N_2O or NO_3^- and further enhancing the uptake and accumulation of N in plants [146]. Foliar spray/soil amendment with CuO NPs also improved the growth and yield of diseased tomatoes and eggplants [147]. However, it is important to mention here that the concentration of NPs to which plants are exposed plays an important role in determining the effects of NPs on plants. For instance, ZnO in a concentration of less than 50 mg kg^{-1} in soil exerted a positive impact on plants, whereas soil containing ZnO in a concentration of more than 500 mg kg^{-1} showed detrimental effects on plants [148]. Similarly, some of the specially tailored inorganic metal NPs in moderate concentrations are found to be effective in increasing soil microbial biomass and their diversity, and activities of soil enzymes (urease, acid phosphatase, catalase, and amylase) that are beneficial for the growth of plants. Crucial microbial phenomena, such as nitrogen fixation, mineralization, and plant growth-promoting activities, are also influenced by NPs depending upon their nature and concentration [149, 150]. Additionally, biogenic nanoparticles can synergize with soil microorganisms to boost their activity and aid in bioremediation [151].

Despite their efficacy in contaminant remediation, nanoparticles pose challenges, such as reduced reactivity during transport or over time and potential adverse effects on soil microorganisms. NPs can alter the soil's physicochemical properties, such as pH, texture, and nutrient availability, which can indirectly affect plant growth and microbial activity [140]. For microbial communities, NPs may exhibit antimicrobial properties, disrupting beneficial soil microbes critical for processes like nutrient cycling and organic matter decomposition [152]. This can lead to shifts in microbial diversity and ecosystem functioning. It was estimated that out of 260,000–309,000 tons of engineered nanomaterials produced in 2010, 9–17% ends up in soils, 2% in water bodies, and 0.1–1.5% in atmosphere [153]. Furthermore, the accumulation and persistence of NPs in soil raise concerns about long-term ecological impacts, including potential toxicity to non-target organisms and bioaccumulation in food chains [34]. In recent studies, it has been reported that deliberate use of NPs may cause their accumulation and surge in their concentration in soil and ultimately alter the physiochemical properties of soil, including pH, nutrient availability, the microbiome of soil, and other various important parameters of soil subsequently affecting the plant growth and development [154]. Numerous studies have reported that modification in the level of nZVI and CuO NPs in soil causes a significant rise in soil pH. Other than these, NPs of Au, Ti, Ag, and Zn have also been reported to alter the soil pH and affect microbial and nematode populations [1, 127]. However, it is imperative to mention here that the implications of NPs vary under different scenarios depending on the complexity and diversity of the environment, physical and chemical properties of NPs, type of soil, contact period, and target species [5, 155]. For instance, the enhanced concentration of NPs in the soil leads to a decrease in the activity of dehydrogenase enzymes that affect nutrient availability and fertility of the soil. Uptake and absorption of NP by microorganisms also influence their normal functioning [34]. Surface complexation, hydrophobic partitioning, and ion exchange are major methods that are involved in the binding and transport of NPs in soil [156].

From the soil, NPs may penetrate the food chain at various trophic levels and cause toxicity to a variety of aquatic and terrestrial plants and animals as well as to humans [157, 1]. Various life stages of plants, including seed germination, growth of root and shoot, rate of photosynthesis, reproduction, and yield, have been affected by the presence of NPs in soil [158 - 160]. A study carried out on carbon NMs *viz.* C70 and MWCNTs indicate that natural organic material aided their uptake, accumulation, and transport along with the uptake of water and nutrients in the xylem [161]. Once these NPs enter the plant cells, they negatively influence various physiological and biochemical activities of plants by disturbing oxidative homeostasis. Cu NPs induced oxidative stress leads to a reduced growth rate of wheat roots by 60% and also induces the formation of lateral roots that may further decrease the yield of thecrop [162]. Particle size, solubility, bio-accessibility, duration of exposure to NPs, and the physico-chemical properties of test plant species are major factors that determine the potent harmful effects of NPs on plants [163]. In addition to these, other abiotic factors, including contact time, light conditions, and soil properties, have been found to markedly affect the behavior of NPs in the soil and their subsequent effects on plants [142, 164]. For instance, in a study carried out by Josko and Oleszczuk, it was observed that the root growth inhibitory effect of ZnO NPs in *Lepidium sativum* was closely related to its concentration in soil and other environmental conditions, including light, soil temperature, and contact time [165].

The impact of NPs on plant growth is also determined by the size of NPs, as indicated by a study [166]. TiO_2, commonly used for photocatalytic treatment of water, is also used for the remediation of contaminated soils. It has been noted that TiO_2 with a diameter of 36 nm was able to accumulate in wheat root parenchyma but not able to reach the stellar system without causing any major detrimental effect on plant growth and development. On the other hand, TiO_2 with a 25 nm diameter accumulated in legume plants after 60 days of exposure, which reduced the growth and biomass of the plants significantly. These variations in plant reactions to TiO_2 may be due to different plant ionomes and the abilities of cells to compartmentalize the NPs in subcellular compartments [166, 167]. Even different parts of the same plant responded differently to NPs. Results of transcriptome profile analyses of ZnO exposed plants of maize indicated that ZnO caused genotoxicity, but gene response patterns in the roots and shoots were different. The number of differentially expressed genes was higher in the roots as compared to the shoots [168]. In addition to terrestrial plants, soil microorganisms and invertebrates are also affected by NPs at the cellular, individual, or community level. Carbon NMs and metal oxide NMs like TiO_2, ZnO, CuO, and nZVI are typical materials that pose a threat to soil biota [169, 170]. Among these, ZnO and CuO have been found to be extremely toxic as they caused almost 100% mortality of *B. subtilis, E. coli, P. fluorescens,* and *P. chlororaphis* [171]. Toxicity and accumulation of NPs are also affected by siderophore secretions of rhizospheric microbes like *Pseudomonas chlororaphis* that decrease the uptake and accumulation of ZnO NPs by *Phaseolus vulgaris* [172]. On the other hand, root exudates of plants also significantly influenced the bioactivities of heavy metal NPs present in contaminated soils, thereby affecting the growth of plants [159]. Hazardous NPs are responsible for genotoxicity in soil microbes, alteration of soil enzyme activities, and ultimately affecting the reproductive capacity of microbes and microbial community structure [173]. Carbon NMs including C70 and reduced graphene oxide, rGO dramatically altered the bacterial community composition and structure [174]. NPs also adversely impact the performance, mortality rate, and reproduction abilities of soil invertebrates like earthworms and collembolans. nZVI at a concentration of more than 500 mg kg^{-1} soil were found to be highly deleterious to earthworms, which may be due to the generation of excess ROS in the body of the earthworm, which in turn causes oxidative stress and affects the physiological activities of earthworms [175, 176].

Nanoparticles pose significant toxic risks to human health also due to their small size and ability to cross biological barriers and interact with cellular structures, potentially leading to toxicity [177]. NPs' exposure to the human body occurs through inhalation, dermal contact, ingestion, or even medical applications and can cause toxicity to different organs of the body, such as reproductive organs, lungs, liver, skin, and brain [178, 179]. NPs have been found to cause permanent damage to cells through oxidative stress [179]. Toxicity is influenced by NP characteristics, including size, composition, surface properties, and aggregation. For instance, carbon-based NPs have been shown to cause size-dependent cytotoxicity and oxidative stress, resulting in potential cellular damage [180]. Additionally, genetic factors in individuals influence susceptibility to NP toxicity. Studies on silica and Fe-containing NPs highlight their toxic potential. Silica NPs have demonstrated various adverse effects depending on the exposure route [181]. Iron-containing NPs have been linked to oxidative stress and bioactivity, evidenced by markers in workers exposed during pigment production [182]. These findings underscore the need for stringent safety protocols to mitigate health risks associated with nanoparticle exposure.

CONCLUSION AND FUTURE PROSPECTS

Research efforts have increasingly focused on developing advanced technologies to address the rising heavy metal contamination in soils caused by extensive anthropogenic activities. Nanoremediation has emerged as a revolutionary approach to addressing the persistent issue of heavy metal contamination in soils. The enhanced remediation efficiency of nano-bioremediation is attributed to the greater surface area provided by the small size of nanoparticles and abundant functional groups on their surface for improved adsorption of pollutants, the ease of regulating the size and shape of nanoparticles, the non-toxic nature of biogenic nanoparticles, reduction in overall cost and remediation time. By leveraging the unique physicochemical properties of nanoparticles, this technique offers improved efficiency in immobilizing, degrading, or transforming toxic heavy metals into less harmful forms. The integration of nanotechnology with sustainable practices like bioremediation further enhances its potential, offering eco-friendly and scalable solutions. Despite its promise, challenges such as the environmental fate of nanoparticles, potential toxicity, and cost-effectiveness remain critical barriers to widespread adoption.

Keeping in view the ecological and health risks associated with the use of nanotechnology, further research needs to be conducted in this field with a special focus on ecological concerns. To facilitate the effective and sustainable use of nanotechnology in environmental remediation, the following key recommendations are proposed:

Green Synthesis Approaches

Emphasis should be placed on sustainable, eco-friendly methods for synthesizing nanoparticles using plant extracts, microorganisms, and waste materials. These approaches minimize the environmental footprint by avoiding toxic chemicals and reducing energy consumption.

Enhanced Remediation Efficiency

Research should aim to optimize nanoparticle longevity and performance by addressing issues like reduced efficiency over time. Combining nanoremediation with complementary techniques, such as bioremediation or electrokinetic remediation could yield more effective, energy-efficient solutions.

Mechanistic Insights and Risk Mitigation

Comprehensive studies on nanoparticle-heavy metal interactions and the long-term impacts of nanoparticles on soil health, microbial diversity, and surrounding ecosystems are needed to validate applications. Advanced tools for ecotoxicological assessments, such as biomarkers and surface-functionalized nanoparticles, can enhance safety and predict long-term impacts [183].

Field-Level Applications and Scaling Up

While laboratory studies have demonstrated the efficacy of nanoparticles, scaling up to field-level applications remains a challenge. Developing industrial-scale processes and pilot projects will be critical for transitioning nanoremediation from experimental to practical use [184].

Collaboration and Technological Integration

Strong collaborations between researchers, industry, and policymakers are crucial for innovation and addressing real-world challenges. The integration of nanotechnology with existing environmental management systems can create comprehensive remediation strategies [183].

Safety and Risk Management

Guidelines for the safe handling, storage, and disposal of nanomaterials must be standardized. Technologies like HEPA filters and spill cleanup procedures should be prioritized to mitigate risks during application [185].

Policy and Regulatory Frameworks

Establishing national and international regulatory bodies to monitor nanoparticle distribution and application is essential. These frameworks should aim to minimize ecological risks while promoting safe deployment practices.

By addressing these challenges and advancing research in the field, nanoremediation has the potential to become a cornerstone technology for restoring polluted soils and achieving long-term environmental sustainability.

REFERENCES

[1]   Mathur S, Singh D, Ranjan R. Remediation of heavy metal(loid) contaminated soil through green nanotechnology. Front Sustain Food Syst 2022; 6: 932424.
[http://dx.doi.org/10.3389/fsufs.2022.932424]

[2]   Rajput VD, Minkina T, Upadhyay SK, *et al.* Nanotechnology in the restoration of polluted soil. Nanomaterials (Basel) 2022; 12(5): 769.
[http://dx.doi.org/10.3390/nano12050769] [PMID: 35269257]

[3]   Azeez NA, Dash SS, Gummadi SN, Deepa VS. Nano-remediation of toxic heavy metal contamination: Hexavalent chromium [Cr(VI)]. Chemosphere 2021; 266: 129204.
[http://dx.doi.org/10.1016/j.chemosphere.2020.129204] [PMID: 33310359]

[4]   Ingle AP, Seabra AB, Duran N, Rai M. Nanoremediation: a new and emerging technology for the removal of toxic contaminant from environment. Microb Biodegrad Bioremediat 2014; 233-50.

[5]   Qian Y, Qin C, Chen M, Lin S. Nanotechnology in soil remediation – applications *vs.* implications. Ecotoxicol Environ Saf 2020; 201: 110815.
[http://dx.doi.org/10.1016/j.ecoenv.2020.110815] [PMID: 32559688]

[6]   Saleem S, Rizvi A, Khan MS. Microbiome-mediated nano-bioremediation of heavy metals: a prospective approach of soil metal detoxification. Int J Environ Sci Technol 2023; 20(11): 12823-46.
[http://dx.doi.org/10.1007/s13762-022-04684-w]

[7]   Usman M, Farooq M, Wakeel A, *et al.* Nanotechnology in agriculture: Current status, challenges and future opportunities. Sci Total Environ 2020; 721: 137778.
[http://dx.doi.org/10.1016/j.scitotenv.2020.137778] [PMID: 32179352]

[8]   Mehndiratta P, Jain A, Srivastava S, Gupta N. Environmental pollution and nanotechnology. Environ Pollut 2013; 2(2): 49.
[http://dx.doi.org/10.5539/ep.v2n2p49]

[9]   Gil-Díaz M, Pinilla P, Alonso J, Lobo MC. Viability of a nanoremediation process in single or multi-metal(loid) contaminated soils. J Hazard Mater 2017; 321: 812-9.
[http://dx.doi.org/10.1016/j.jhazmat.2016.09.071] [PMID: 27720472]

[10]   Andrade-Zavaleta K, Chacon-Laiza Y, Asmat-Campos D, Raquel-Checca N. Green synthesis of superparamagnetic iron oxide nanoparticles with *Eucalyptus globulus* extract and their application in the removal of heavy metals from agricultural soil. Molecules 2022; 27(4): 1367.
[http://dx.doi.org/10.3390/molecules27041367] [PMID: 35209154]

[11]   Liang S, Jin Y, Liu W, Li X, Shen S, Ding L. Feasibility of Pb phytoextraction using nano-materials assisted ryegrass: Results of a one-year field-scale experiment. J Environ Manage 2017; 190: 170-5.
[http://dx.doi.org/10.1016/j.jenvman.2016.12.064] [PMID: 28043023]

[12]   Singh J, Lee BK. Influence of nano-TiO_2 particles on the bioaccumulation of Cd in soybean plants (*Glycine max*): A possible mechanism for the removal of Cd from the contaminated soil. J Environ Manage 2016; 170: 88-96.
[http://dx.doi.org/10.1016/j.jenvman.2016.01.015] [PMID: 26803259]

[13]   Tang Y, Tian J, Li S, *et al.* Combined effects of graphene oxide and Cd on the photosynthetic capacity and survival of *Microcystis aeruginosa.*. Sci Total Environ 2015; 532: 154-61.
[http://dx.doi.org/10.1016/j.scitotenv.2015.05.081] [PMID: 26070025]

[14]   Khan NT, Jameel N, Khan MJ. A brief overview of contaminated soil remediation methods. Biotechnol Ind J 2018; 14(4): 171.

[15]   Kapoor N, Sharma P, Verma S. Microbial remediation of hexavalent chromium from the contaminated soils. Microbes Microb Biotechnol Green Remediat 2022; 527-46.
[http://dx.doi.org/10.1016/B978-0-323-90452-0.00022-0]

[16]   Jiang B, Gong Y, Gao J, *et al.* The reduction of Cr(VI) to Cr(III) mediated by environmentally relevant carboxylic acids: State-of-the-art and perspectives. J Hazard Mater 2019; 365: 205-26.
[http://dx.doi.org/10.1016/j.jhazmat.2018.10.070] [PMID: 30445352]

[17]   Liang H, Song B, Peng P, Jiao G, Yan X, She D. Preparation of three-dimensional honeycomb carbon materials and their adsorption of Cr(VI). Chem Eng J 2019; 367: 9-16.

[http://dx.doi.org/10.1016/j.cej.2019.02.121]

[18]   Rathnayake SI, Martens WN, Xi Y, Frost RL, Ayoko GA. Remediation of Cr (VI) by inorganic-organic clay. J Colloid Interface Sci 2017; 490: 163-73.
[http://dx.doi.org/10.1016/j.jcis.2016.11.070] [PMID: 27912114]

[19]   Reddy KR, Cameselle C. Electrochemical remediation technologies for polluted soils, sediments and groundwater groundwater. John Wiley & Sons, Inc 2009..
[http://dx.doi.org/10.1002/9780470523650]

[20]   Wang Y, Huang L, Wang Z, *et al.* Application of Polypyrrole flexible electrode for electrokinetic remediation of Cr(VI)-contaminated soil in a main-auxiliary electrode system. Chem Eng J 2019; 373: 131-9.
[http://dx.doi.org/10.1016/j.cej.2019.05.016]

[21]   Chirakkara RA, Cameselle C, Reddy KR. Assessing the applicability of phytoremediation of soils with mixed organic and heavy metal contaminants. Rev Environ Sci Biotechnol 2016; 15(2): 299-326.
[http://dx.doi.org/10.1007/s11157-016-9391-0]

[22]   Sarwar N, Imran M, Shaheen MR, *et al.* Phytoremediation strategies for soils contaminated with heavy metals: Modifications and future perspectives. Chemosphere 2017; 171: 710-21.
[http://dx.doi.org/10.1016/j.chemosphere.2016.12.116] [PMID: 28061428]

[23]   Saxena G, Purchase D, Mulla SI, Saratale GD, Bharagava RN. Phytoremediation of heavy metal-contaminated sites: eco-environmental concerns, field studies, sustainability issues, and future prospects. Rev Environ Contam Toxicol 2020; 249: 71-131.
[PMID: 30806802]

[24]   Guo S, Xiao C, Zhou N, Chi R. Speciation, toxicity, microbial remediation and phytoremediation of soil chromium contamination. Environ Chem Lett 2021; 19(2): 1413-31.
[http://dx.doi.org/10.1007/s10311-020-01114-6]

[25]   Okoduwa S, Igiri B, Udeh C, Edenta C, Gauje B. Tannery effluent treatment by yeast species isolates from watermelon. Toxics 2017; 5(1): 6.
[http://dx.doi.org/10.3390/toxics5010006] [PMID: 29051437]

[26]   Febbraio F. Biochemical strategies for the detection and detoxification of toxic chemicals in the environment. World J Biol Chem 2017; 8(1): 13-20.
[http://dx.doi.org/10.4331/wjbc.v8.i1.13] [PMID: 28289515]

[27]   Yang P, Zhang L, Huang Y, Ouyang Z. Research status on soil and water pollution remediation and environmental impact of nanomaterials. IOP Conf Ser Earth Environ Sci. 371(3): 032015.
[http://dx.doi.org/10.1088/1755-1315/371/3/032015]

[28]   Xu X, Huang H, Zhang Y, Xu Z, Cao X. Biochar as both electron donor and electron shuttle for the reduction transformation of Cr(VI) during its sorption. Environ Pollut 2019; 244: 423-30.
[http://dx.doi.org/10.1016/j.envpol.2018.10.068] [PMID: 30352357]

[29]   He Y, Gong Y, Su Y, Zhang Y, Zhou X. Bioremediation of Cr (VI) contaminated groundwater by *Geobacter sulfurreducens*: Environmental factors and electron transfer flow studies. Chemosphere 2019; 221: 793-801.
[http://dx.doi.org/10.1016/j.chemosphere.2019.01.039] [PMID: 30684777]

[30]   Farooqi ZH, Akram MW, Begum R, Wu W, Irfan A. Inorganic nanoparticles for reduction of hexavalent chromium: Physicochemical aspects. J Hazard Mater 2021; 402: 123535.
[http://dx.doi.org/10.1016/j.jhazmat.2020.123535] [PMID: 33254738]

[31]   Klaus-Joerger T, Joerger R, Olsson E, Granqvist CG. Bacteria as workers in the living factory: metal-accumulating bacteria and their potential for materials science. Trends Biotechnol 2001; 19(1): 15-20.
[http://dx.doi.org/10.1016/S0167-7799(00)01514-6] [PMID: 11146098]

[32]   Santos CSC, Gabriel B, Blanchy M, *et al.* Industrial applications of nanoparticles–a prospective overview. Mater Today Proc 2015; 2(1): 456-65.

[http://dx.doi.org/10.1016/j.matpr.2015.04.056]

[33]   Bakker RM, Yuan HK, Liu Z, *et al.* Enhanced localized fluorescence in plasmonic nanoantennae. Appl Phys Lett 2008; 92(4): 043101.
[http://dx.doi.org/10.1063/1.2836271]

[34]   Ameen F, Alsamhary K, Alabdullatif JA, ALNadhari S. A review on metal-based nanoparticles and their toxicity to beneficial soil bacteria and fungi. Ecotoxicol Environ Saf 2021; 213: 112027.
[http://dx.doi.org/10.1016/j.ecoenv.2021.112027] [PMID: 33578100]

[35]   Mauter MS, Zucker I, Perreault F, Werber JR, Kim JH, Elimelech M. The role of nanotechnology in tackling global water challenges. Nat Sustain 2018; 1(4): 166-75.
[http://dx.doi.org/10.1038/s41893-018-0046-8]

[36]   Zhang T, Lowry GV, Capiro NL, *et al. in situ* remediation of subsurface contamination: opportunities and challenges for nanotechnology and advanced materials. Environ Sci Nano 2019; 6(5): 1283-302.
[http://dx.doi.org/10.1039/C9EN00143C]

[37]   Pak T, Archilha NL, de Lima Luz LF. Nanotechnology-based remediation of groundwater. Nanotechnol Charact Tools Environ Health Saf 2019; 145-65.
[http://dx.doi.org/10.1007/978-3-662-59600-5_5]

[38]   Roy A, Sharma A, Yadav S, Jule LT, Krishnaraj R. Nanomaterials for remediation of environmental pollutants. Bioinorg Chem Appl 2021; 2021(1): 1764647.
[PMID: 34992641]

[39]   Alazaiza MYD, Albahnasawi A, Ali GAM, *et al.* Recent advances of nanoremediation technologies for soil and groundwater remediation: A review. Water 2021; 13(16): 2186.
[http://dx.doi.org/10.3390/w13162186]

[40]   Matos MPSR, Correia AAS, Rasteiro MG. Application of carbon nanotubes to immobilize heavy metals in contaminated soils. J Nanopart Res 2017; 19(4): 126.
[http://dx.doi.org/10.1007/s11051-017-3830-x]

[41]   Gupta VK, Agarwal S, Saleh TA. Synthesis and characterization of alumina-coated carbon nanotubes and their application for lead removal. J Hazard Mater 2011; 185(1): 17-23.
[http://dx.doi.org/10.1016/j.jhazmat.2010.08.053] [PMID: 20888691]

[42]   Baysal A, Kuznek C, Ozcan M. Starch coated titanium dioxide nanoparticles as a challenging sorbent to separate and preconcentrate some heavy metals using graphite furnace atomic absorption spectrometry. Int J Environ Anal Chem 2018; 98(1): 45-55.
[http://dx.doi.org/10.1080/03067319.2018.1427741]

[43]   Rodríguez C, Leiva E. Enhanced heavy metal removal from acid mine drainage wastewater using double-oxidized multiwalled carbon nanotubes. Molecules 2019; 25(1): 111.
[http://dx.doi.org/10.3390/molecules25010111] [PMID: 31892164]

[44]   Zhu K, Chen C, Lu S, Zhang X, Alsaedi A, Hayat T. MOFs-induced encapsulation of ultrafine Ni nanoparticles into 3D N-doped graphene-CNT frameworks as a recyclable catalyst for Cr(VI) reduction with formic acid. Carbon 2019; 148: 52-63.
[http://dx.doi.org/10.1016/j.carbon.2019.03.044]

[45]   Aragaw TA, Bogale FM, Aragaw BA. Iron-based nanoparticles in wastewater treatment: A review on synthesis methods, applications, and removal mechanisms. J Saudi Chem Soc 2021; 25(8): 101280.
[http://dx.doi.org/10.1016/j.jscs.2021.101280]

[46]   Ahmed T, Noman M, Ijaz M, *et al.* Current trends and future prospective in nanoremediation of heavy metals contaminated soils: A way forward towards sustainable agriculture. Ecotoxicol Environ Saf 2021; 227: 112888.
[http://dx.doi.org/10.1016/j.ecoenv.2021.112888] [PMID: 34649136]

[47]   Yabuuchi N, Komaba S. Recent research progress on iron- and manganese-based positive electrode materials for rechargeable sodium batteries. Sci Technol Adv Mater 2014; 15(4): 043501.

[http://dx.doi.org/10.1088/1468-6996/15/4/043501] [PMID: 27877694]

[48] Kumar R, Kumar R, Kushwaha N, Mittal J. Ammonia gas sensing using thin film of MnO_2 nanofibers. IEEE Sens J 2016; 16(12): 4691-5.
[http://dx.doi.org/10.1109/JSEN.2016.2550079]

[49] Luo C, Wang J, Jia P, *et al.* Hierarchically structured polyacrylonitrile nanofiber mat as highly efficient lead adsorbent for water treatment. Chem Eng J 2015; 262: 775-84.
[http://dx.doi.org/10.1016/j.cej.2014.09.116]

[50] Bian L, Nie J, Jiang X, *et al.* Selective removal of uranyl from aqueous solutions containing a mix of toxic metal ions using core–shell MFe_2O_4–TiO_2 nanoparticles of montmorillonite edge sites. ACS Sustain Chem& Eng 2018; 6(12): 16267-78.
[http://dx.doi.org/10.1021/acssuschemeng.8b03129]

[51] Wang H, Zhang M, Li H. Synthesis of nanoscale zerovalent iron (nZVI) supported on biochar for chromium remediation from aqueous solution and soil. Int J Environ Res Public Health 2019; 16(22): 4430.
[http://dx.doi.org/10.3390/ijerph16224430] [PMID: 31726717]

[52] Seisenbaeva GA, Daniel G, Nedelec JM, Gun'ko YK, Kessler VG. High surface area ordered mesoporous nano-titania by a rapid surfactant-free approach. J Mater Chem 2012; 22(38): 20374-80.
[http://dx.doi.org/10.1039/c2jm33977c]

[53] Youssef AM, Malhat FM. Selective removal of heavy metals from drinking water using titanium dioxide nanowire. Macromol Symp 2014; 337(1): 96-101.
[http://dx.doi.org/10.1002/masy.201450311]

[54] Cai X, Gao Y, Sun Q, Chen Z, Megharaj M, Naidu R. Removal of co-contaminants Cu (II) and nitrate from aqueous solution using kaolin-Fe/Ni nanoparticles. Chem Eng J 2014; 244: 19-26.
[http://dx.doi.org/10.1016/j.cej.2014.01.040]

[55] Anero MLA, Montallana ADS, Vasquez MR Jr. Fabrication of electrospun poly(vinyl alcohol) nanofibers loaded with zinc oxide particles. Results Phys 2021; 25: 104223.
[http://dx.doi.org/10.1016/j.rinp.2021.104223]

[56] Gudkov SV, Burmistrov DE, Smirnova VV, Semenova AA, Lisitsyn AB. A mini review of antibacterial properties of Al_2O_3 nanoparticles. Nanomaterials (Basel) 2022; 12(15): 2635.
[http://dx.doi.org/10.3390/nano12152635] [PMID: 35957067]

[57] Prabhakar R, Samadder SR. Low cost and easy synthesis of aluminium oxide nanoparticles for arsenite removal from groundwater: A complete batch study. J Mol Liq 2018; 250: 192-201.
[http://dx.doi.org/10.1016/j.molliq.2017.11.173]

[58] Tabesh S, Davar F, Loghman-Estarki MR. Preparation of γ-Al_2O_3 nanoparticles using modified sol-gel method and its use for the adsorption of lead and cadmium ions. J Alloys Compd 2018; 730: 441-9.
[http://dx.doi.org/10.1016/j.jallcom.2017.09.246]

[59] Hojamberdiev M, Daminova SS, Kadirova ZC, Sharipov KT, Mtalo F, Hasegawa M. Ligand-immobilized spent alumina catalyst for effective removal of heavy metal ions from model contaminated water. J Environ Chem Eng 2018; 6(4): 4623-33.
[http://dx.doi.org/10.1016/j.jece.2018.06.070]

[60] Dehghani MH, Yetilmezsoy K, Salari M, Heidarinejad Z, Yousefi M, Sillanpää M. Adsorptive removal of cobalt(II) from aqueous solutions using multi-walled carbon nanotubes and γ-alumina as novel adsorbents: Modelling and optimization based on response surface methodology and artificial neural network. J Mol Liq 2020; 299: 112154.
[http://dx.doi.org/10.1016/j.molliq.2019.112154]

[61] Tadic M, Kralj S, Jagodic M, Hanzel D, Makovec D. Magnetic properties of novel superparamagnetic iron oxide nanoclusters and their peculiarity under annealing treatment. Appl Surf Sci 2014; 322: 255-64.

[http://dx.doi.org/10.1016/j.apsusc.2014.09.181]

[62]   Kargin DB, Konyukhov YV, Biseken AB, Lileev AS, Karpenkov DY. Structure, morphology and magnetic properties of hematite and maghemite nanopowders produced from rolling mill scale. Steel Transl 2020; 50(3): 151-8.
[http://dx.doi.org/10.3103/S0967091220030055]

[63]   Rajput VD, Minkina T, Upadhyay SK, *et al.* Nanotechnology in the restoration of polluted soil. Nanomaterials (Basel) 2022; 12(5): 769.
[http://dx.doi.org/10.3390/nano12050769] [PMID: 35269257]

[64]   Singh A, Chaudhary S, Dehiya BS. Fast removal of heavy metals from water and soil samples using magnetic Fe_3O_4 nanoparticles. Environ Sci Pollut Res Int 2021; 28(4): 3942-52.
[http://dx.doi.org/10.1007/s11356-020-10737-9] [PMID: 32948942]

[65]   El-Gendy NS, Nassar HN. Biosynthesized magnetite nanoparticles as an environmental opulence and sustainable wastewater treatment. Sci Total Environ 2021; 774: 145610.
[http://dx.doi.org/10.1016/j.scitotenv.2021.145610] [PMID: 33609818]

[66]   Manzoor N, Ahmed T, Noman M, *et al.* Iron oxide nanoparticles ameliorated the cadmium and salinity stresses in wheat plants, facilitating photosynthetic pigments and restricting cadmium uptake. Sci Total Environ 2021; 769: 145221.
[http://dx.doi.org/10.1016/j.scitotenv.2021.145221] [PMID: 33736258]

[67]   Mahanty S, Chatterjee S, Ghosh S, *et al.* Synergistic approach towards the sustainable management of heavy metals in wastewater using mycosynthesized iron oxide nanoparticles: Biofabrication, adsorptive dynamics and chemometric modeling study. J Water Process Eng 2020; 37: 101426.
[http://dx.doi.org/10.1016/j.jwpe.2020.101426]

[68]   Liu T, Wang ZL, Sun Y. Manipulating the morphology of nanoscale zero-valent iron on pumice for removal of heavy metals from wastewater. Chem Eng J 2015; 263: 55-61.
[http://dx.doi.org/10.1016/j.cej.2014.11.046]

[69]   Seyedi SM, Rabiee H, Shahabadi SMS, Borghei SM. Synthesis of zero-valent iron nanoparticles *via* electrical wire explosion for efficient removal of heavy metals. Clean (Weinh) 2017; 45(3): 1600139.
[http://dx.doi.org/10.1002/clen.201600139]

[70]   Dehghan Monfared A, Ghazanfari MH, Jamialahmadi M, Helalizadeh A. Adsorption of silica nanoparticles onto calcite: Equilibrium, kinetic, thermodynamic and DLVO analysis. Chem Eng J 2015; 281: 334-44.
[http://dx.doi.org/10.1016/j.cej.2015.06.104]

[71]   Hou L, Liang Q, Wang F. Mechanisms that control the adsorption–desorption behavior of phosphate on magnetite nanoparticles: the role of particle size and surface chemistry characteristics. RSC Advances 2020; 10(4): 2378-88.
[http://dx.doi.org/10.1039/C9RA08517C] [PMID: 35494559]

[72]   Zielińska A, Carreiró F, Oliveira AM, *et al.* Polymeric nanoparticles: production, characterization, toxicology and ecotoxicology. Molecules 2020; 25(16): 3731.
[http://dx.doi.org/10.3390/molecules25163731] [PMID: 32824172]

[73]   Ethaib S, Al-Qutaifia S, Al-Ansari N, Zubaidi SL. Function of nanomaterials in removing heavy metals for water and wastewater remediation: A review. Environments (Basel) 2022; 9(10): 123.
[http://dx.doi.org/10.3390/environments9100123]

[74]   Cai J, Lei M, Zhang Q, *et al.* Electrospun composite nanofiber mats of Cellulose@Organically modified montmorillonite for heavy metal ion removal: Design, characterization, evaluation of absorption performance. Compos, Part A Appl Sci Manuf 2017; 92: 10-6.
[http://dx.doi.org/10.1016/j.compositesa.2016.10.034]

[75]   Suman KA, Kardam A, Gera M, Jain VK. A novel reusable nanocomposite for complete removal of dyes, heavy metals and microbial load from water based on nanocellulose and silver nano-embedded

pebbles. Environ Technol 2015; 36(6): 706-14.
[http://dx.doi.org/10.1080/09593330.2014.959066] [PMID: 25243917]

[76]   Benettayeb A, Seihoub FZ, Pal P, *et al.* Chitosan nanoparticles as potential nano-sorbent for removal of toxic environmental pollutants. Nanomaterials (Basel) 2023; 13(3): 447.
[http://dx.doi.org/10.3390/nano13030447] [PMID: 36770407]

[77]   Jeyaraj M, Gurunathan S, Qasim M, Kang MH, Kim JH. A comprehensive review on the synthesis, characterization, and biomedical application of platinum nanoparticles. Nanomaterials (Basel) 2019; 9(12): 1719.
[http://dx.doi.org/10.3390/nano9121719] [PMID: 31810256]

[78]   Ali MA, Ahmed T, Wu W, *et al.* Advancements in plant and microbe-based synthesis of metallic nanoparticles and their antimicrobial activity against plant pathogens. Nanomaterials (Basel) 2020; 10(6): 1146.
[http://dx.doi.org/10.3390/nano10061146] [PMID: 32545239]

[79]   Kandiah M, Chandrasekaran KN. Green synthesis of silver nanoparticles using *Catharanthus roseus* flower extracts and the determination of their antioxidant, antimicrobial, and photocatalytic activity. J Nanotechnol 2021; 2021(1): 1-18.
[http://dx.doi.org/10.1155/2021/5512786]

[80]   Jones AC, Hitchman ML. Overview of chemical vapour deposition. Chem Vap Depos: Precursor Process Appl 2009; 1: 1-36.

[81]   Ago H. CVD growth of high-quality single-layer graphene. Front Graphene Carbon Nanotubes: Devices Appl 2015; 3-20.
[http://dx.doi.org/10.1007/978-4-431-55372-4_1]

[82]   Machac P, Cichon S, Lapcak L, Fekete L. Graphene prepared by chemical vapour deposition process. Graphene Technol 2020; 5(1-2): 9-17.
[http://dx.doi.org/10.1007/s41127-019-00029-6]

[83]   Zheng K, Branicio PS. Synthesis of metallic glass nanoparticles by inert gas condensation. Phys Rev Mater 2020; 4(7): 076001.
[http://dx.doi.org/10.1103/PhysRevMaterials.4.076001]

[84]   Dhand C, Dwivedi N, Loh XJ, *et al.* Methods and strategies for the synthesis of diverse nanoparticles and their applications: a comprehensive overview. RSC Advances 2015; 5(127): 105003-37.
[http://dx.doi.org/10.1039/C5RA19388E]

[85]   JothiRamalingam R, Periyasami G, Ouladsmane M, *et al.* Ultra-sonication assisted metal chalcogenide modified mesoporous Nickel-cobalt doped manganese oxide nanocomposite fabrication for sono-catalytic dye degradation and mechanism insights. J Alloys Compd 2021; 875: 160072.
[http://dx.doi.org/10.1016/j.jallcom.2021.160072]

[86]   ElFaham MM, Mostafa AM, Mwafy EA. The effect of reaction temperature on structural, optical and electrical properties of tunable ZnO nanoparticles synthesized by hydrothermal method. J Phys Chem Solids 2021; 154: 110089.
[http://dx.doi.org/10.1016/j.jpcs.2021.110089]

[87]   Mostafa AM. Preparation and study of nonlinear response of embedding ZnO nanoparticles in PVA thin film by pulsed laser ablation. J Mol Struct 2021; 1223: 129007.
[http://dx.doi.org/10.1016/j.molstruc.2020.129007]

[88]   Tyagi I, Gupta VK, Sadegh H, Ghoshekandi RS, Makhlouf AS. Nanoparticles as adsorbent; a positive approach for removal of noxious metal ions: a review. Science Technology and Development 2017; 34(3): 195-214.

[89]   Son HH, Seo GH, Jeong U, Shin DY, Kim SJ. Capillary wicking effect of a Cr-sputtered superhydrophilic surface on enhancement of pool boiling critical heat flux. Int J Heat Mass Transf 2017; 113: 115-28.

[http://dx.doi.org/10.1016/j.ijheatmasstransfer.2017.05.055]

[90]   Nam JH, Jang MJ, Jang HY, *et al.* Room-temperature sputtered electrocatalyst WSe$_2$ nanomaterials for hydrogen evolution reaction. J Energy Chem 2020; 47: 107-11.
[http://dx.doi.org/10.1016/j.jechem.2019.11.027]

[91]   Harish V, Ansari MM, Tewari D, *et al.* Nanoparticle and nanostructure synthesis and controlled growth methods. Nanomaterials (Basel) 2022; 12(18): 3226.
[http://dx.doi.org/10.3390/nano12183226] [PMID: 36145012]

[92]   Mittal AK, Chisti Y, Banerjee UC. Synthesis of metallic nanoparticles using plant extracts. Biotechnol Adv 2013; 31(2): 346-56.
[http://dx.doi.org/10.1016/j.biotechadv.2013.01.003] [PMID: 23318667]

[93]   Ovais M, Khalil AT, Ayaz M, Ahmad I, Nethi SK, Mukherjee S. Biosynthesis of metal nanoparticles *via* microbial enzymes: a mechanistic approach. Int J Mol Sci 2018; 19(12): 4100.
[http://dx.doi.org/10.3390/ijms19124100] [PMID: 30567324]

[94]   Salem SS, Fouda A. Green synthesis of metallic nanoparticles and their prospective biotechnological applications: an overview. Biol Trace Elem Res 2021; 199(1): 344-70.
[http://dx.doi.org/10.1007/s12011-020-02138-3] [PMID: 32377944]

[95]   Sunanda , Misra M, Ghosh Sachan S. Nanobioremediation of heavy metals: Perspectives and challenges. J Basic Microbiol 2022; 62(3-4): 428-43.
[http://dx.doi.org/10.1002/jobm.202100384]

[96]   Holmes AB, Gu FX. Emerging nanomaterials for the application of selenium removal for wastewater treatment. Environ Sci Nano 2016; 3(5): 982-96.
[http://dx.doi.org/10.1039/C6EN00144K]

[97]   Gong X, Huang D, Liu Y, *et al.* Remediation of contaminated soils by biotechnology with nanomaterials: bio-behavior, applications, and perspectives. Crit Rev Biotechnol 2018; 38(3): 455-68.
[http://dx.doi.org/10.1080/07388551.2017.1368446] [PMID: 28903604]

[98]   Bhatt P, Pandey SC, Joshi S, *et al.* Nanobioremediation: A sustainable approach for the removal of toxic pollutants from the environment. J Hazard Mater 2022; 427: 128033.
[http://dx.doi.org/10.1016/j.jhazmat.2021.128033] [PMID: 34999406]

[99]   Ebrahiminezhad A, Taghizadeh SM, Ghasemi Y, Berenjian A. Immobilization of cells by magnetic nanoparticles. Immobil Enzymes Cells: Methods Protoc 2020; 427-35.
[http://dx.doi.org/10.1007/978-1-0716-0215-7_29]

[100]   Giese EC, Silva DDV, Costa AFM, Almeida SGC, Dussán KJ. Immobilized microbial nanoparticles for biosorption. Crit Rev Biotechnol 2020; 40(5): 653-66.
[http://dx.doi.org/10.1080/07388551.2020.1751583] [PMID: 32299253]

[101]   Li X, Ming Q, Cai R, *et al.* Biosorption of Cd^{2+} and Pb^{2+} from apple juice by the magnetic nanoparticles functionalized lactic acid bacteria cells. Food Control 2020; 109: 106916.
[http://dx.doi.org/10.1016/j.foodcont.2019.106916]

[102]   Wang W, Zhang B, Liu Q, Du P, Liu W, He Z. Biosynthesis of palladium nanoparticles using *Shewanella loihica* PV-4 for excellent catalytic reduction of chromium(VI). Environ Sci Nano 2018; 5(3): 730-9.
[http://dx.doi.org/10.1039/C7EN01167A]

[103]   Wang Y, Shu X, Zhou Q, *et al.* Selenite reduction and the biogenesis of selenium nanoparticles by *Alcaligenes faecalis* Se03 isolated from the gut of *Monochamus alternatus* (Coleoptera: Cerambycidae). Int J Mol Sci 2018; 19(9): 2799.
[http://dx.doi.org/10.3390/ijms19092799] [PMID: 30227664]

[104]   Wang X, Zhang D, Pan X, *et al.* Aerobic and anaerobic biosynthesis of nano-selenium for remediation of mercury contaminated soil. Chemosphere 2017; 170: 266-73.
[http://dx.doi.org/10.1016/j.chemosphere.2016.12.020] [PMID: 28011305]

[105] Xiao X, Liu QY, Lu XR, *et al.* Self-assembly of complex hollow CuS nano/micro shell by an electrochemically active bacterium *Shewanella oneidensis* MR-1. Int Biodeterior Biodegradation 2017; 116: 10-6.
[http://dx.doi.org/10.1016/j.ibiod.2016.09.021]

[106] Tiwari S, Hasan A, Pandey LM. A novel bio-sorbent comprising encapsulated *Agrobacterium fabrum* (SLAJ731) and iron oxide nanoparticles for removal of crude oil co-contaminant, lead Pb(II). J Environ Chem Eng 2017; 5(1): 442-52.
[http://dx.doi.org/10.1016/j.jece.2016.12.017]

[107] Mahmoud ME, Abdou AEH, Mohamed SMS, Osman MM. Engineered *staphylococcus aureus via* immobilization on magnetic Fe_3O_4-phthalate nanoparticles for biosorption of divalent ions from aqueous solutions. J Environ Chem Eng 2016; 4(4): 3810-24.
[http://dx.doi.org/10.1016/j.jece.2016.08.022]

[108] Özdemir S, Mohamedsaid SA, Kılınç E, Soylak M. Magnetic solid phase extractions of Co(II) and Hg(II) by using magnetized *C. micaceus* from water and food samples. Food Chem 2019; 271: 232-8.
[http://dx.doi.org/10.1016/j.foodchem.2018.07.067] [PMID: 30236672]

[109] Ozdemir S, Mohamedsaid SA, Kilinc E, Yıldırım A, Soylak M. Application of magnetized fungal solid phase extractor with Fe_2O_3 nanoparticle for determination and preconcentration of Co(II) and Hg(II) from natural water samples. Microchem J 2018; 143: 198-204.
[http://dx.doi.org/10.1016/j.microc.2018.07.032]

[110] Yin Y, Wang J, Yang X, Li W. Removal of strontium ions by immobilized *Saccharomyces cerevisiae* in magnetic chitosan microspheres. Nucl Eng Technol 2017; 49(1): 172-7.
[http://dx.doi.org/10.1016/j.net.2016.09.002]

[111] Thilakan D, Patankar J, Khadtare S, *et al.* Plant-Derived Iron Nanoparticles for Removal of Heavy Metals. Int J Chem Eng 2022; 2022(1): 1-12.
[http://dx.doi.org/10.1155/2022/1517849]

[112] Andal V, Kannan K, Selvaraj V, Suba K. Plant-derived nanoparticles for heavy metal remediation. Phytonanotechnol 2022; 59-76.
[http://dx.doi.org/10.1007/978-981-19-4811-4_3]

[113] Mahmoud AED, Al-Qahtani KM, Alflaij SO, Al-Qahtani SF, Alsamhan FA. Green copper oxide nanoparticles for lead, nickel, and cadmium removal from contaminated water. Sci Rep 2021; 11(1): 12547.
[http://dx.doi.org/10.1038/s41598-021-91093-7] [PMID: 34131155]

[114] Slijepčević N, Pilipović DT, Kerkez Đ, *et al.* A cost effective method for immobilization of Cu and Ni polluted river sediment with nZVI synthesized from leaf extract. Chemosphere 2021; 263: 127816.
[http://dx.doi.org/10.1016/j.chemosphere.2020.127816] [PMID: 32835965]

[115] Arjaghi SK, Alasl MK, Sajjadi N, Fataei E, Rajaei GE. Retracted article: green synthesis of Iron oxide nanoparticles by RS Lichen extract and its application in removing heavy metals of lead and cadmium. Biol Trace Elem Res 2021; 199(2): 763-8.
[http://dx.doi.org/10.1007/s12011-020-02170-3] [PMID: 32643097]

[116] Hao R, Li D, Zhang J. Insights into the removal of Cr(VI) from aqueous solution using plant-mediated biosynthesis of iron nanoparticles. Environ Technol Innov 2021; 23: 101566.
[http://dx.doi.org/10.1016/j.eti.2021.101566]

[117] Yang J, Wang S, Xu N, Ye Z, Yang H, Huangfu X. Synthesis of montmorillonite-supported nano-zero-valent iron *via* green tea extract: Enhanced transport and application for hexavalent chromium removal from water and soil. J Hazard Mater 2021; 419: 126461.
[http://dx.doi.org/10.1016/j.jhazmat.2021.126461] [PMID: 34186421]

[118] Hao R, Li D, Zhang J, Jiao T. Green synthesis of iron nanoparticles using green tea and its removal of hexavalent chromium. Nanomaterials (Basel) 2021; 11(3): 650.

[http://dx.doi.org/10.3390/nano11030650] [PMID: 33800123]

[119] Saleh M, Isik Z, Aktas Y, Arslan H, Yalvac M, Dizge N. Green synthesis of zero valent iron nanoparticles using *Verbascum thapsus* and its Cr(VI) reduction activity. Bioresour Technol Rep 2021; 13: 100637.
[http://dx.doi.org/10.1016/j.biteb.2021.100637]

[120] Tatarchuk T, Shyichuk A, Sojka Z, *et al.* Green synthesis, structure, cations distribution and bonding characteristics of superparamagnetic cobalt-zinc ferrites nanoparticles for Pb(II) adsorption and magnetic hyperthermia applications. J Mol Liq 2021; 328: 115375.
[http://dx.doi.org/10.1016/j.molliq.2021.115375]

[121] Shen L, Wang J, Li Z, *et al.* A high-efficiency Fe_2O_3@Microalgae composite for heavy metal removal from aqueous solution. J Water Process Eng 2020; 33: 101026.
[http://dx.doi.org/10.1016/j.jwpe.2019.101026]

[122] Lingamdinne LP, Vemula KR, Chang YY, Yang JK, Karri RR, Koduru JR. Process optimization and modeling of lead removal using iron oxide nanocomposites generated from bio-waste mass. Chemosphere 2020; 243: 125257.
[http://dx.doi.org/10.1016/j.chemosphere.2019.125257] [PMID: 31726263]

[123] Lin Z, Weng X, Owens G, Chen Z. Simultaneous removal of Pb(II) and rifampicin from wastewater by iron nanoparticles synthesized by a tea extract. J Clean Prod 2020; 242: 118476.
[http://dx.doi.org/10.1016/j.jclepro.2019.118476]

[124] Biswal SK, Panigrahi GK, Sahoo SK. Green synthesis of Fe_2O_3-Ag nanocomposite using Psidium guajava leaf extract: An eco-friendly and recyclable adsorbent for remediation of Cr(VI) from aqueous media. Biophys Chem 2020; 263: 106392.
[http://dx.doi.org/10.1016/j.bpc.2020.106392] [PMID: 32417597]

[125] Majumder A, Ramrakhiani L, Mukherjee D, *et al.* Green synthesis of iron oxide nanoparticles for arsenic remediation in water and sludge utilization. Clean Technol Environ Policy 2019; 21(4): 795-813.
[http://dx.doi.org/10.1007/s10098-019-01669-1]

[126] Nikić J, Tubić A, Watson M, *et al.* Arsenic removal from water by green synthesized magnetic nanoparticles. Water 2019; 11(12): 2520.
[http://dx.doi.org/10.3390/w11122520]

[127] García-Gómez C, Fernández MD, García S, Obrador AF, Letón M, Babín M. Soil pH effects on the toxicity of zinc oxide nanoparticles to soil microbial community. Environ Sci Pollut Res Int 2018; 25(28): 28140-52.
[http://dx.doi.org/10.1007/s11356-018-2833-1] [PMID: 30069782]

[128] Lin Y, Jin X, Owens G, Chen Z. Simultaneous removal of mixed contaminants triclosan and copper by green synthesized bimetallic iron/nickel nanoparticles. Sci Total Environ 2019; 695: 133878.
[http://dx.doi.org/10.1016/j.scitotenv.2019.133878] [PMID: 31756849]

[129] Zhang Y, Zhang Y, Akakuru OU, Xu X, Wu A. Research progress and mechanism of nanomaterials-mediated in-situ remediation of cadmium-contaminated soil: A critical review. J Environ Sci (China) 2021; 104: 351-64.
[http://dx.doi.org/10.1016/j.jes.2020.12.021] [PMID: 33985738]

[130] Taghipour M, Jalali M. Effect of clay minerals and nanoparticles on chromium fractionation in soil contaminated with leather factory waste. J Hazard Mater 2015; 297: 127-33.
[http://dx.doi.org/10.1016/j.jhazmat.2015.04.067] [PMID: 25956643]

[131] Zhang M, Wang Y, Zhao D, Pan G. Immobilization of arsenic in soils by stabilized nanoscale zero-valent iron, iron sulfide (FeS), and magnetite (Fe_3O_4) particles. Chin Sci Bull 2010; 55(4-5): 365-72.
[http://dx.doi.org/10.1007/s11434-009-0703-4]

[132] Zhang Z, Li M, Chen W, Zhu S, Liu N, Zhu L. Immobilization of lead and cadmium from aqueous

solution and contaminated sediment using nano-hydroxyapatite. Environ Pollut 2010; 158(2): 514-9.
[http://dx.doi.org/10.1016/j.envpol.2009.08.024] [PMID: 19783084]

[133]   Hou S, Wu B, Peng D, Wang Z, Wang Y, Xu H. Remediation performance and mechanism of hexavalent chromium in alkaline soil using multi-layer loaded nano-zero-valent iron. Environ Pollut 2019; 252(Pt A): 553-61.
[http://dx.doi.org/10.1016/j.envpol.2019.05.083] [PMID: 31181500]

[134]   Qiao J, Liu T, Wang X, *et al.* Simultaneous alleviation of cadmium and arsenic accumulation in rice by applying zero-valent iron and biochar to contaminated paddy soils. Chemosphere 2018; 195: 260-71.
[http://dx.doi.org/10.1016/j.chemosphere.2017.12.081] [PMID: 29272795]

[135]   Zhou H, Zhou X, Zeng M, *et al.* Effects of combined amendments on heavy metal accumulation in rice (*Oryza sativa* L.) planted on contaminated paddy soil. Ecotoxicol Environ Saf 2014; 101: 226-32.
[http://dx.doi.org/10.1016/j.ecoenv.2014.01.001] [PMID: 24507150]

[136]   Gopalakrishnan Nair PM, Chung IM. Cell cycle and mismatch repair genes as potential biomarkers in Arabidopsis thaliana seedlings exposed to silver nanoparticles. Bull Environ Contam Toxicol 2014; 92(6): 719-25.
[http://dx.doi.org/10.1007/s00128-014-1254-1] [PMID: 24652625]

[137]   Hu X, Kang J, Lu K, Zhou R, Mu L, Zhou Q. Graphene oxide amplifies the phytotoxicity of arsenic in wheat. Sci Rep 2014; 4(1): 6122.
[http://dx.doi.org/10.1038/srep06122] [PMID: 25134726]

[138]   Rizwan M, Ali S, Rehman MZ, *et al.* Effects of nanoparticles on trace element uptake and toxicity in plants: A review. Ecotoxicol Environ Saf 2021; 221: 112437.
[http://dx.doi.org/10.1016/j.ecoenv.2021.112437] [PMID: 34153540]

[139]   Lin D, Xing B. Phytotoxicity of nanoparticles: Inhibition of seed germination and root growth. Environ Pollut 2007; 150(2): 243-50.
[http://dx.doi.org/10.1016/j.envpol.2007.01.016] [PMID: 17374428]

[140]   Ogunkunle CO, Oyedeji S, Okoro HK, Adimula V. Interaction of nanoparticles with soil. Nanomater Soil Remediat 2021; 101-32.
[http://dx.doi.org/10.1016/B978-0-12-822891-3.00006-2]

[141]   Zhu Y, Xu F, Liu Q, *et al.* Nanomaterials and plants: Positive effects, toxicity and the remediation of metal and metalloid pollution in soil. Sci Total Environ 2019; 662: 414-21.
[http://dx.doi.org/10.1016/j.scitotenv.2019.01.234] [PMID: 30690375]

[142]   Dimkpa CO. Soil properties influence the response of terrestrial plants to metallic nanoparticles exposure. Curr Opin Environ Sci Health 2018; 6: 1-8.
[http://dx.doi.org/10.1016/j.coesh.2018.06.007]

[143]   Ashfaq M, Verma N, Khan S. Carbon nanofibers as a micronutrient carrier in plants: efficient translocation and controlled release of Cu nanoparticles. Environ Sci Nano 2017; 4(1): 138-48.
[http://dx.doi.org/10.1039/C6EN00385K]

[144]   Ogunkunle CO, Jimoh MA, Asogwa NT, Viswanathan K, Vishwakarma V, Fatoba PO. Effects of manufactured nano-copper on copper uptake, bioaccumulation and enzyme activities in cowpea grown on soil substrate. Ecotoxicol Environ Saf 2018; 155: 86-93.
[http://dx.doi.org/10.1016/j.ecoenv.2018.02.070] [PMID: 29510313]

[145]   Rajput VD, Minkina TM, Behal A, *et al.* Effects of zinc-oxide nanoparticles on soil, plants, animals and soil organisms: A review. Environ Nanotechnol Monit Manag 2018; 9: 76-84.
[http://dx.doi.org/10.1016/j.enmm.2017.12.006]

[146]   Dimkpa CO, White JC, Elmer WH, Gardea-Torresdey J. Nanoparticle and ionic Zn promote nutrient loading of sorghum grain under low NPK fertilization. J Agric Food Chem 2017; 65(39): 8552-9.
[http://dx.doi.org/10.1021/acs.jafc.7b02961] [PMID: 28905629]

[147] Elmer WH, White JC. The use of metallic oxide nanoparticles to enhance growth of tomatoes and eggplants in disease infested soil or soilless medium. Environ Sci Nano 2016; 3(5): 1072-9.
[http://dx.doi.org/10.1039/C6EN00146G]

[148] Reddy Pullagurala VL, Adisa IO, Rawat S, *et al.* Finding the conditions for the beneficial use of ZnO nanoparticles towards plants-A review. Environ Pollut 2018; 241: 1175-81.
[http://dx.doi.org/10.1016/j.envpol.2018.06.036] [PMID: 30029327]

[149] Fang G, Si Y, Tian C, Zhang G, Zhou D. Degradation of 2,4-D in soils by Fe$_3$O$_4$ nanoparticles combined with stimulating indigenous microbes. Environ Sci Pollut Res Int 2012; 19(3): 784-93.
[http://dx.doi.org/10.1007/s11356-011-0597-y] [PMID: 21948126]

[150] Khan ST, Adil SF, Shaik MR, Alkhathlan HZ, Khan M, Khan M. Engineered nanomaterials in soil: Their impact on soil microbiome and plant health. Plants 2021; 11(1): 109.
[http://dx.doi.org/10.3390/plants11010109] [PMID: 35009112]

[151] Verma KK, Joshi A, Song XP, *et al.* Synergistic interactions of nanoparticles and plant growth promoting rhizobacteria enhancing soil-plant systems: a multigenerational perspective. Front Plant Sci 2024; 15: 1376214.
[http://dx.doi.org/10.3389/fpls.2024.1376214] [PMID: 38742215]

[152] Kumar P, Burman U, Kaul RK. Ecological risks of nanoparticles: effect on soil microorganisms. Nanomater Plants, Algae Microorganisms 2018; 429-52.

[153] Jośko I, Oleszczuk P, Futa B. The effect of inorganic nanoparticles (ZnO, Cr$_2$O$_3$, CuO and Ni) and their bulk counterparts on enzyme activities in different soils. Geoderma 2014; 232-234: 528-37.
[http://dx.doi.org/10.1016/j.geoderma.2014.06.012]

[154] Shi J, Ye J, Fang H, Zhang S, Xu C. Effects of copper oxide nanoparticles on paddy soil properties and components. Nanomaterials (Basel) 2018; 8(10): 839.
[http://dx.doi.org/10.3390/nano8100839] [PMID: 30332772]

[155] Li X, Peng T, Mu L, Hu X. Phytotoxicity induced by engineered nanomaterials as explored by metabolomics: Perspectives and challenges. Ecotoxicol Environ Saf 2019; 184: 109602.
[http://dx.doi.org/10.1016/j.ecoenv.2019.109602] [PMID: 31493589]

[156] Gao J, Wang Y, Du Y, *et al.* Construction of biocatalytic colloidosome using lipase-containing dendritic mesoporous silica nanospheres for enhanced enzyme catalysis. Chem Eng J 2017; 317: 175-86.
[http://dx.doi.org/10.1016/j.cej.2017.02.012]

[157] Cedervall T, Hansson LA, Lard M, Frohm B, Linse S. Food chain transport of nanoparticles affects behaviour and fat metabolism in fish. PLoS One 2012; 7(2): e32254.
[http://dx.doi.org/10.1371/journal.pone.0032254] [PMID: 22384193]

[158] Zhang Z, Ke M, Qu Q, *et al.* Impact of copper nanoparticles and ionic copper exposure on wheat (*Triticum aestivum* L.) root morphology and antioxidant response. Environ Pollut 2018; 239: 689-97.
[http://dx.doi.org/10.1016/j.envpol.2018.04.066] [PMID: 29715688]

[159] Huang Y, Zhao L, Keller AA. Interactions, transformations, and bioavailability of nano-copper exposed to root exudates. Environ Sci Technol 2017; 51(17): 9774-83.
[http://dx.doi.org/10.1021/acs.est.7b02523] [PMID: 28771344]

[160] Du W, Tan W, Peralta-Videa JR, *et al.* Interaction of metal oxide nanoparticles with higher terrestrial plants: Physiological and biochemical aspects. Plant Physiol Biochem 2017; 110: 210-25.
[http://dx.doi.org/10.1016/j.plaphy.2016.04.024] [PMID: 27137632]

[161] Lin S, Reppert J, Hu Q, *et al.* Uptake, translocation, and transmission of carbon nanomaterials in rice plants. Small 2009; 5(10): 1128-32.
[http://dx.doi.org/10.1002/smll.200801556] [PMID: 19235197]

[162] Khalaj M, Kamali M, Khodaparast Z, Jahanshahi A. Copper-based nanomaterials for environmental

decontamination – An overview on technical and toxicological aspects. Ecotoxicol Environ Saf 2018; 148: 813-24.
[http://dx.doi.org/10.1016/j.ecoenv.2017.11.060] [PMID: 29197796]

[163] Jośko I, Oleszczuk P, Skwarek E. Toxicity of combined mixtures of nanoparticles to plants. J Hazard Mater 2017; 331: 200-9.
[http://dx.doi.org/10.1016/j.jhazmat.2017.02.028] [PMID: 28273569]

[164] Hou J, Lin D, White JC, Gardea-Torresdey JL, Xing B. Joint nanotoxicology assessment provides a new strategy for developing nanoenabled bioremediation technologies.
[http://dx.doi.org/10.1021/acs.est.9b03593]

[165] Jośko I, Oleszczuk P. Influence of soil type and environmental conditions on ZnO, TiO$_2$ and Ni nanoparticles phytotoxicity. Chemosphere 2013; 92(1): 91-9.
[http://dx.doi.org/10.1016/j.chemosphere.2013.02.048] [PMID: 23541360]

[166] Vijayaraj V, Liné C, Cadarsi S, *et al.* Transfer and ecotoxicity of titanium dioxide nanoparticles in terrestrial and aquatic ecosystems: a microcosm study. Environ Sci Technol 2018; 52(21): 12757-64.
[http://dx.doi.org/10.1021/acs.est.8b02970] [PMID: 30335981]

[167] Giorgetti L, Spanò C, Muccifora S, *et al.* An integrated approach to highlight biological responses of Pisum sativum root to nano-TiO$_2$ exposure in a biosolid-amended agricultural soil. Sci Total Environ 2019; 650(Pt 2): 2705-16.
[http://dx.doi.org/10.1016/j.scitotenv.2018.10.032] [PMID: 30373051]

[168] Xun H, Ma X, Chen J, *et al.* Zinc oxide nanoparticle exposure triggers different gene expression patterns in maize shoots and roots. Environ Pollut 2017; 229: 479-88.
[http://dx.doi.org/10.1016/j.envpol.2017.05.066] [PMID: 28624629]

[169] Schlich K, Beule L, Hund-Rinke K. Single versus repeated applications of CuO and Ag nanomaterials and their effect on soil microflora. Environ Pollut 2016; 215: 322-30.
[http://dx.doi.org/10.1016/j.envpol.2016.05.028] [PMID: 27213573]

[170] Hou J, Wang X, Hayat T, Wang X. Ecotoxicological effects and mechanism of CuO nanoparticles to individual organisms. Environ Pollut 2017; 221: 209-17.
[http://dx.doi.org/10.1016/j.envpol.2016.11.066] [PMID: 27939631]

[171] Jiang W, Mashayekhi H, Xing B. Bacterial toxicity comparison between nano- and micro-scaled oxide particles. Environ Pollut 2009; 157(5): 1619-25.
[http://dx.doi.org/10.1016/j.envpol.2008.12.025] [PMID: 19185963]

[172] Dimkpa CO, Hansen T, Stewart J, McLean JE, Britt DW, Anderson AJ. ZnO nanoparticles and root colonization by a beneficial pseudomonad influence essential metal responses in bean (*Phaseolus vulgaris*). Nanotoxicology 2015; 9(3): 271-8.
[http://dx.doi.org/10.3109/17435390.2014.900583] [PMID: 24713073]

[173] Asadishad B, Chahal S, Akbari A, *et al.* Amendment of agricultural soil with metal nanoparticles: effects on soil enzyme activity and microbial community composition. Environ Sci Technol 2018; 52(4): 1908-18.
[http://dx.doi.org/10.1021/acs.est.7b05389] [PMID: 29356510]

[174] Hao Y, Ma C, Zhang Z, *et al.* Carbon nanomaterials alter plant physiology and soil bacterial community composition in a rice-soil-bacterial ecosystem. Environ Pollut 2018; 232: 123-36.
[http://dx.doi.org/10.1016/j.envpol.2017.09.024] [PMID: 28947315]

[175] Li M, Yang Y, Xie J, Xu G, Yu Y. In-vivo and in-vitro tests to assess toxic mechanisms of nano ZnO to earthworms. Sci Total Environ 2019; 687: 71-6.
[http://dx.doi.org/10.1016/j.scitotenv.2019.05.476] [PMID: 31203009]

[176] Tatsi K, Shaw BJ, Hutchinson TH, Handy RD. Copper accumulation and toxicity in earthworms exposed to CuO nanomaterials: Effects of particle coating and soil ageing. Ecotoxicol Environ Saf 2018; 166: 462-73.

[http://dx.doi.org/10.1016/j.ecoenv.2018.09.054] [PMID: 30296611]

[177] Ravinayagam V, Jermy BR. Nanomaterials and their negative effects on human health. Appl Nanomater Hum Health 2020; 249-73.
[http://dx.doi.org/10.1007/978-981-15-4802-4_13]

[178] Oberdörster G, Oberdörster E, Oberdörster J. Nanotoxicology: an emerging discipline evolving from studies of ultrafine particles. Environ Health Perspect 2005; 113(7): 823-39.
[http://dx.doi.org/10.1289/ehp.7339] [PMID: 16002369]

[179] Kumah EA, Fopa RD, Harati S, Boadu P, Zohoori FV, Pak T. Human and environmental impacts of nanoparticles: a scoping review of the current literature. BMC Public Health 2023; 23(1): 1059.
[http://dx.doi.org/10.1186/s12889-023-15958-4] [PMID: 37268899]

[180] Magrez A, Kasas S, Salicio V, *et al.* Cellular toxicity of carbon-based nanomaterials. Nano Lett 2006; 6(6): 1121-5.
[http://dx.doi.org/10.1021/nl060162e] [PMID: 16771565]

[181] Missaoui WN, Arnold RD, Cummings BS. Toxicological status of nanoparticles: What we know and what we don't know. Chem Biol Interact 2018; 295: 1-12.
[http://dx.doi.org/10.1016/j.cbi.2018.07.015] [PMID: 30048623]

[182] Pelclova D, Zdimal V, Kacer P, *et al.* Oxidative stress markers are elevated in exhaled breath condensate of workers exposed to nanoparticles during iron oxide pigment production. J Breath Res 2016; 10(1): 016004.
[http://dx.doi.org/10.1088/1752-7155/10/1/016004] [PMID: 26828137]

[183] Ganie AS, Bano S, Khan N, *et al.* Nanoremediation technologies for sustainable remediation of contaminated environments: Recent advances and challenges. Chemosphere 2021; 275: 130065.
[http://dx.doi.org/10.1016/j.chemosphere.2021.130065] [PMID: 33652279]

[184] Rather MA, Bhuyan S, Chowdhury R, Sarma R, Roy S, Neog PR. Nanoremediation strategies to address environmental problems. Sci Total Environ 2023; 886: 163998.
[http://dx.doi.org/10.1016/j.scitotenv.2023.163998] [PMID: 37172832]

[185] Ningombam L, Mana T, Apum G, Ningthoujam R, Disco Singh Y. Nano-bioremediation: A prospective approach for environmental decontamination in focus to soil, water and heavy metals. Environ Nanotechnol Monit Manag 2024; 21: 100931.
[http://dx.doi.org/10.1016/j.enmm.2024.100931]

SUBJECT INDEX

A

Abiotic stress 26, 27, 45, 54, 58, 83, 154, 155, 182,183,187, 278
Arbuscular mycorrhizal fungi 251, 265, 274, 275

B

Bioremediation 21, 71, 72, 73, 74, 75, 77, 78, 79, 80, 81, 85, 113, 130, 131, 132, 145, 149, 150, 152, 156, 157, 205, 210, 214, 217, 218, 222, 227, 245, 247, 293, 306, 307, 308, 309, 325, 326, 327
Bio-accumulators 23
Biofilm 115
Biomagnification 2, 51, 205, 210, 260
Biosorption 74, 77, 79, 80, 85, 96, 101, 102, 107, 116, 150, 151, 205, 216, 217, 218, 219, 220, 221, 222, 224, 225, 227, 228, 216
Bioaccumulation 2, 20, 21, 22, 23, 25, 27, 29, 31, 33, 35, 46, 51, 79, 96, 101, 102, 116, 117, 131, 150, 151, 152, 207, 214, 217, 219, 222, 224, 225, 226, 245, 248, 257, 258, 259, 260, 263, 281, 322

C

Cyanoremediation 71, 72, 77
Carcinogenic 49, 111, 205, 206

D

Detoxification 20,21,26, 27, 28, 30, 33, 45, 58,60, 71, 72, 79, 81, 84, 95, 96, 101, 102,106, 113, 115, 116, 130, 131, 132, 139, 145,150, 156, 182, 184, 186, 187, 189, 191, 192, 193,207, 225, 226, 266, 267, 274, 276, 277, 282,285, 289, 291, 312

E

Environmental contamination 1, 212
Environment 1, 2, 4, 8, 10,22, 24, 26, 34, 48, 49,50, 52, 53, 72, 74, 75, 77,79, 81, 86, 95, 101, 103, 105,109, 117, 130,131, 132, 134,136, 137, 148,151, 157, 182, 183, 184, 185, 191,205, 206, 207, 210,212, 213, 214, 221, 225, 249, 274, 306, 307, 308, 309, 323

F

Food chain 2, 21, 51, 72,74, 95, 109,184, 185, 205, 210, 211, 214, 266,306, 323

G

Green Nanotechnology 305

H

Heavy metals (HMs) 1, 20, 21, 22, 23, 24, 25, 26, 28, 29, 30, 33, 35, 45-59, 71-77, 81, 82, 84, 85, 104, 110,131, 132, 138, 149, 151, 152, 153, 154, 156, 205, 206, 207, 210, 211, 212, 214, 220, 221, 222, 225, 275, 283, 285, 286, 290, 291, 306, 308, 309, 312, 320
Hyperaccumulation 11, 21, 24

M

Mycoremediation 71,72, 78, 80, 252, 262
MAPKs 45, 46, 56, 57, 58, 62
Microbes 22, 71, 72,75, 81, 84, 103, 116,130, 132, 140, 151, 152, 153, 154, 220, 247, 255, 256, 258, 260, 263, 265, 308, 309, 315, 316, 324

.

www.ingramcontent.com/pod-product-compliance
Lightning Source LLC
Chambersburg PA
CBHW080019240326
41598CB00075B/119